VLSI Handbook

Handbooks in Science and Technology

Edited by

NORMAN G. EINSPRUCH

College of Engineering
University of Miami
Coral Gables, Florida

NORMAN G. EINSPRUCH (ed.). VLSI Handbook, 1985

In preparation

TZAY Y. YOUNG and KING-SUN FU (eds.). Handbook of Pattern Recognition and Image Processing

VLSI Handbook

Edited by

NORMAN G. EINSPRUCH

College of Engineering
University of Miami
Coral Gables, Florida

1985

ACADEMIC PRESS, INC.
(Harcourt Brace Jovanovich, Publishers)

Orlando   San Diego   New York   London
Toronto   Montreal   Sydney   Tokyo

COPYRIGHT © 1985 BY ACADEMIC PRESS, INC.
ALL RIGHTS RESERVED.
NO PART OF THIS PUBLICATION MAY BE REPRODUCED OR
TRANSMITTED IN ANY FORM OR BY ANY MEANS, ELECTRONIC
OR MECHANICAL, INCLUDING PHOTOCOPY, RECORDING, OR
ANY INFORMATION STORAGE AND RETRIEVAL SYSTEM, WITHOUT
PERMISSION IN WRITING FROM THE PUBLISHER.

ACADEMIC PRESS, INC.
Orlando, Florida 32887

United Kingdom Edition published by
ACADEMIC PRESS INC. (LONDON) LTD.
24–28 Oval Road, London NW1 7DX

Library of Congress Cataloging in Publication Data

Main entry under title:

VLSI handbook.

Includes index.
1. Integrated circuits--Very large scale integration--
Handbooks, manuals, etc. I. Einspruch, Norman G.
TK7874.V563 1985   621.3819'5835   84-20373
ISBN 0-12-234100-7 (alk. paper)

PRINTED IN THE UNITED STATES OF AMERICA

85 86 87 88   9 8 7 6 5 4 3 2 1

Contents

Contributors	xvii
Preface	xxi
Acronyms	xxiii

Chapter 1 Factors Contributing to Increased VLSI Circuit Density
F. W. VOLTMER AND N. W. JONES

I.	Introduction	1
II.	Factors Influencing Circuit Density	2

Chapter 2 Fundamental Principles of Very Large Scale Integrated Circuit Design
AMR MOHSEN, SAI WAI FU, AND CARL SIMONSEN

I.	Introduction	9
II.	VLSI Design Methodology	10
III.	Elements of VLSI Circuit Design	14
IV.	Basics of Layout Design	28
V.	Future Developments	35
	References	35

Chapter 3 Design Automation for Integrated Circuits
AART J. DE GEUS AND SYDNEY B. NEWELL

I.	Introduction	37
II.	A Design through Various Levels of Abstraction	40
III.	A Typical Design Procedure	40
IV.	Semicustom Design Methodologies	44
V.	Building of Cell or Macro Library	46
VI.	Semicustom Layout	47
VII.	Comparison between Semicustom Methodologies	48
VIII.	Trends in Design Automation	48
IX.	Conclusion	50
	Glossary	50
	Bibliography	54

Chapter 4 Computer Tools for Integrated Circuit Design
WILLIAM V. LAWSON

I.	IC Design and Development	55
II.	Applying Computers in the Development Process	57
III.	Availability of CAD Tools	63
	References	65

Chapter 5 VLSI to Go: The Silicon Foundry
ALFRED J. STEIN, GERI ALLISON, AND TONY VALENTINO

I.	Introduction	67
II.	The Silicon Foundry Concept	68
III.	The Foundry Interface	69
IV.	Processing	73
V.	Post Processing	76
	References	78

Chapter 6 Manufacturing Process Technology for MOS VLSI
F. W. VOLTMER AND N. W. JONES

I.	Introduction	79
II.	Directions in Process Technology	81
III.	Process Control	84
	References	86

Chapter 7 Facilities for VLSI Circuit Fabrication
KIRBY G. VOSBURGH

I.	Clean Air	88
II.	Water	89
III.	Provision of Other Supplies	90
IV.	Vacuum	92
V.	Waste Disposal	93
VI.	Physical Considerations	95
VII.	Protection of Personnel, Equipment, and Product	95
VIII.	Personnel Efficiency	96
IX.	Facility Management System	97
	References	97

Chapter 8 MOS VLSI Circuit Technology
CHENG T. WANG

I.	Introduction	99
II.	MOSFET Structures	100
III.	MOS Circuits	103
IV.	Power-Delay Performance of MOS and Bipolar Circuits	107
V.	Conclusion	107
	References	108

Contents

Chapter 9 Bipolar VLSI Circuit Technology
CHENG T. WANG

I.	Introduction	109
II.	Bipolar Transistors	109
III.	Bipolar Digital Gate Circuits	113
	References	120

Chapter 10 CMOS VLSI Technology
DAVID B. SCOTT, KUEING-LONG CHEN, AND RODERICK D. DAVIES

I.	Advantages of Circuit Design with CMOS	122
II.	A State-of-the-Art CMOS Process Flow	124
III.	Problems of Optimization of CMOS Processing	131
IV.	Problems of Interconnects for CMOS	133
V.	Discontinuities in CMOS Technology	135
	Bibliography	138

Chapter 11 New Directions in Microprocessors
JOHN RAITHEL

I.	Introduction	141
II.	Memory Management	142
III.	Cache	145
IV.	Pipelining	146
V.	System Timing	147
VI.	Peripheral Controllers	149
VII.	Current Implementations	150

Chapter 12 VLSI Random Access Memories
JEFF SCHLAGETER, CHING-LIN JIANG, AND ROBERT PROEBSTING

I.	Introduction	151
II.	Static RAM	154
III.	Dynamic RAM	157
IV.	Specialty RAMs	160
	Glossary	165

Chapter 13 VLSI Electrically Erasable Programmable Read Only Memory
S. K. LAI AND V. K. DHAM

I.	Principle of Operation	167
II.	Programming Characteristics	172
III.	Performance and Reliability	174
IV.	Scaling	175
	List of Symbols	176
	References	176

Chapter 14 Electrical Transport Properties of Silicon
W. ROBERT THURBER AND JEREMIAH R. LOWNEY

I.	Introduction	177
II.	Definition of Transport: The Transport Equation	178
III.	Conversion between Resistivity and Dopant Density	178
IV.	Mobility of Charge Carriers	181
V.	Temperature Dependence of Resistivity and Mobility	185
VI.	Dependence of Drift Velocity on Electric Fields	185
VII.	Minority-Carrier Mobility, Lifetime, and Diffusion Length	186
VIII.	Mobility in a MOS Inversion Layer	188
	References	189

Chapter 15 VLSI Silicon Material Criteria
FUMIO SHIMURA AND HOWARD R. HUFF

I.	Introduction	191
II.	Product Characteristics	193
III.	Tabulation of Engineering Properties	194
IV.	VLSI Silicon Product Trends	194
V.	VLSI and ULSI Silicon Product Recommendation	196
	Tables	197
	Figures	230
	References	265

Chapter 16 Characterization of Bulk Silicon Materials
GRAYDON B. LARRABEE AND THOMAS J. SHAFFNER

I.	Polycrystalline Silicon	271
II.	Bulk Single Crystal Silicon	272
III.	Single Crystal Silicon Wafers	272
	References	283

Chapter 17 Growth of Epitaxial Films For VLSI Applications
T. L. CHU

I.	Introduction	285
II.	Epitaxial Growth of Silicon Films	286
III.	Characterization of Epitaxial Silicon Films	296
IV.	Epitaxial Growth of Gallium Arsenide Films	301
	References	303

Chapter 18 Epitaxial Silicon: Material Characterization
S. B. KULKARNI

I.	Introduction	305
II.	Growth Characterization	306
III.	Electrical Characterization	310
IV.	Physical and Optical Characterization	316
V.	Epitaxial Defect Characterization	321
VI.	Epitaxial Defect Measurements	324
	References	325

Chapter 19 Resist Technology in VLSI Device Processing
RONALD C. BRACKEN AND SYED A. RIZVI

I.	Introduction	328
II.	Optical Patterning	328
III.	Multilevel Resists	339
IV.	Electron Beam Patterning	343
V.	X-Ray Resists	347
VI.	Conclusions	349
	References	350

Chapter 20 Electron Beam Lithography
R. KENT WATTS

I.	Introduction	351
II.	Mask Making	352
III.	Direct Writing	354
IV.	Resists	354
V.	Electron Optics	357
VI.	Raster Scan	360
VII.	Vector Scan	362
	References	364

Chapter 21 X-Ray Lithography
R. KENT WATTS

I.	Introduction	366
II.	X-Ray Proximity Printing	366
III.	Sources	367
IV.	Masks	371
V.	Resists	372
VI.	Mask Alignment	376
VII.	Exposure Systems	377
VIII.	Applications	378
	References	380

Chapter 22 Oxides for VLSI
DAVID A. BAGLEE

I.	Introduction	381
II.	Thermal Oxidation	382
III.	Leakage and Breakdown	385
IV.	Oxide Charges	389
V.	Special Considerations	393
	References	399

Chapter 23 Nitrides for VLSI
MICHAEL P. DUANE AND DAVID A. BAGLEE

I.	Introduction	401
II.	Film Formation	402

III.	Electrical Properties	407
IV.	Applications of Nitrides	411
	References	414

Chapter 24 Silicides
MARC-A. NICOLET AND S. S. LAU

	Introduction to Tables	415
	References	433

Chapter 25 Metallization for VLSI
W. H. CLASS AND J. F. SMITH

I.	Introduction	435
II.	Aluminum and Its Alloys	436
III.	Barrier Layers for Metallization	447
	References	452

Chapter 26 Application of Ion Implantation in VLSI
K. L. WANG

I.	Introduction	456
II.	Aspects of Ion Implantation	457
III.	Doping Applications in MOS Technology	466
IV.	Doping Applications in Bipolar Technology	472
V.	Recent Advances and Other Applications	476
	References	484

Chapter 27 Plasma Processing for VLSI
BARBARA A. HEATH AND LEE KAMMERDINER

I.	Introduction	487
II.	Etching	488
III.	Sputtering	494
IV.	Plasma Enhanced Chemical Vapor Deposition	499
	References	502

Chapter 28 Silicon-on-Insulator for VLSI Applications
H. W. LAM, R. F. PINIZZOTTO, AND A. F. TASCH, JR.

I.	Introduction	503
II.	Heteroepitaxy	505
III.	SOI by Thin Film Recrystallization	506
IV.	Formation of Buried Insulating Layers by Ion Implantation	508
V.	Full Isolation by Porous Oxidized Silicon (FIPOS)	509
VI.	Epitaxial Lateral Overgrowth	510
VII.	LPCVD Polysilicon SOI Thin Film Transistors (TFT)	511
VIII.	Grain Boundary Passivation	512
IX.	Three-Dimensional Integrated Circuits	512
X.	Summary	514
	References	514

Contents

Chapter 29 Testing of VLSI Parametrics
CHARLES A. BECKER

I.	Purpose	515
II.	Implications of VLSI	517
III.	Test Types	519
IV.	Test Structures	520
V.	Instrumentation	521
VI.	Data Analysis–Information Retrieval	525
	References	526

Chapter 30 VLSI Testing from Design through Production
CHARLES J. McMINN

I.	Semiconductor Testing	528
II.	Testing with Automatic Test Equipment	531
III.	Test Descriptions	538
IV.	Logic Testing	539
V.	Memory Testing	541
VI.	Testing Throughput	546
VII.	Quality Assurance and Sample Testing	547
	References	550

Chapter 31 VLSI Failure Analysis
LAWRENCE C. WAGNER

I.	Introduction	551
II.	Initial Nondestructive Procedures	552
III.	Input–Output Failures	553
IV.	Single Node Failures	556
V.	Nonfunctional Defects	561
VI.	Conclusion	563

Chapter 32 Radiation Effects and Radiation Hardening of VLSI Circuits
BOBBY L. BUCHANAN

I.	Introduction	565
II.	Radiation-Induced Effects in IC Materials	567
III.	Principal Radiation Effects in Devices and ICs	569
IV.	Radiation Hardening of Semiconductor Devices and ICs	573
V.	Single Event Upset	575
VI.	Radiation Hardness of GaAs ICs	577
VII.	Hardness Trends of Silicon Devices	578
	References	579

Chapter 33 VLSI Imagers
RUDOLPH H. DYCK

I.	Introduction	581
II.	Chip Architectures	582
III.	Fabrication Technologies	589

IV.	Performance Variables versus Design Variables	593
	Bibliography	601
	References	601

Chapter 34 Noise in VLSI
A. VAN DER ZIEL

I.	Introduction	603
II.	Various VLSI Circuits	606
III.	Threshold Voltages	611
IV.	Crosstalk	612
V.	Alpha-Particle- and Cosmic-Ray-Induced Soft Errors in VLSI Circuits	612
	References	613

Chapter 35 Limits to Performance of VLSI Circuits
R. T. BATE

I.	Introduction	615
II.	Fundamental Limits	616
III.	Materials Limits	617
IV.	Device Limits	618
V.	Circuit and System Limits	622
	References	627

Chapter 36 Superconducting Integrated Circuits
THOMAS Y. HSIANG

I.	Model of a Josephson Junction	629
II.	Digital Devices	632
III.	Switching Speed of Josephson Junctions	638
IV.	Other Considerations	640
	References	642

Chapter 37 GaAs Digital Integrated Circuit Technology
T. GHEEWALA

I.	Properties of GaAs	645
II.	GaAs MESFET Devices	647
III.	Planar GaAs Fabrication Process	649
IV.	GaAs MESFET Circuits	651
V.	Performance and Applications	654
VI.	Advanced GaAs Technologies	655
	List of Symbols	656
	References	656

Chapter 38 VLSI in Personal Computers
DAVID J. BRADLEY

I.	Introduction	657
II.	Microprocessor Evolution	658

Contents   xiii

III.	The Evolution of a Personal Computer	662
IV.	The Future of Personal Computers	665
	References	666

Chapter 39   VLSI in the Design of Large Computers
DONALD P. TATE AND DENNIS G. GRINA

I.	Introduction	667
II.	Performance	668
III.	VLSI Design Considerations	671
IV.	Design Tools and Staff	674
V.	Summary	678
	References	678

Chapter 40   Electronic Warfare Applications of VLSI
CHARLES T. BRODNAX

I.	Introduction	681
II.	Electronic Warfare Support Measures	682
III.	Electronic Countermeasures	690
IV.	Electronic Counter-Countermeasures (ECCM)	691
	List of Symbols	693
	References	694

Chapter 41   VLSI in Encryption Applications
CHARLES R. ABBRUSCATO

I.	Introduction	695
II.	Cryptography Overview	696
III.	Data Encryption Standard	696
IV.	Public Keys	702
V.	Commercial Device Implementations	703
VI.	VLSI Impact on Future	704
	References	704

Chapter 42   Application of VLSI to Radar Systems
THOMAS K. LISLE AND JOHN J. ZINGARO

I.	Overview	707
II.	Functional Requirements and Radar Overview	708
III.	Key VLSI Microelectronic Technologies	711
IV.	Summary	719
	References	720

Chapter 43   Medical Applications of VLSI Circuits
JACOB KLINE

I.	Introduction	722
II.	Dual-Chamber Programmable Implantable Pacemaker	722
III.	Digital Hearing Aid	724
IV.	Computerized Tomography	726

V.	Ultrasound Imaging	728
VI.	A Computerized Local Area Network for an Intensive Care Unit	730
VII.	Evoked Potentials	731
VIII.	Neural Stimulators	733
	References	735

Chapter 44 Cardiac Pacer Systems
ROBERT D. GOLD

I.	Introduction	737
II.	Cardiac Cycle and the Pacer Implant	738
III.	System Description	742
IV.	Pacemaker Design Considerations	747
V.	Pacing Modalities	756
VI.	Future Trends in Cardiac Pacing	762
	Bibliography	762
	References	762

Chapter 45 VLSI in a Complex Medical Instrument
JOHN FOSTER

I.	System Architecture	764
II.	Sample Handling and Preparation	765
III.	A/D Conversion and Data Accumulation	770
IV.	Central Data Processor	772

Chapter 46 Impact of VLSI on Speech Processing
RICHARD V. COX

I.	Introduction	775
II.	The Nature of Speech and Speech Processing	777
III.	Spectral Estimation: Algorithms and Hardware	780
IV.	Speech Recognition	781
V.	Speech Coding and Speech Synthesis	782
VI.	The Future	783
	References	784

Chapter 47 Application of VLSI to Pattern Recognition and Image Processing
TZAY Y. YOUNG, PHILIP S. LIU, AND HSI-HO LIU

I.	Introduction	785
II.	Image Processing and Pattern Recognition	787
III.	VLSI Architectures for Pattern Recognition	789
IV.	LSI/VLSI Image Processing: Architectures and Systems	795
	References	799

Contents

Chapter 48 VLSI Approach to FM Detection
SHARBEL E. NOUJAIM

I.	Introduction	801
II.	Frequency Modulation	802
III.	VLSI Approach to Frequency Modulation Detection	803
IV.	System Implementation Example: An Implantable Pulsed Doppler Flowmeter	807
V.	Conclusion	820
	References	820

Chapter 49 VLSI Impact on Modem Design and Performance
LEROY D. YOUNG, JR. AND EDWIN J. HILPERT

I.	Introduction	824
II.	Modem Overview	824
III.	Incentives for VLSI Use in Modems	827
IV.	Custom Implementations	830
V.	Commercial Devices	833
VI.	VLSI and Value-Added Features	834
VII.	Future Trends	835
	References	835

Chapter 50 Impact of VLSI On Distributed Communications
DANIEL HAMPEL

I.	Introduction	837
II.	VLSI Technology and Trends	838
III.	Distributed Communications	841
IV.	Future Trends	849
	References	850

Chapter 51 Applications of VLSI to the Automobile
FRANK S. STEIN

I.	Introduction	851
II.	Driving Forces	852
III.	Automotive Electronics Market	853
IV.	Reliability	854
V.	Engine Controls	855
VI.	Body Computers	859
VII.	Entertainment Systems	861
VIII.	Cellular Telephones	862
IX.	Conclusion	863
	References	864

Chapter 52 How to Protect VLSI Intellectual Property
           HAROLD LEVINE

    I. Patent Protection   866
    II. Copyright Protection   873
    III. Trade Secret Protection   876
    IV. Other Key Aspects in the Protection Program   880
    V. Summary   881
       References   882

Index   883

Contributors

Numbers in parentheses indicate the pages on which the authors' contributions begin.

CHARLES R. ABBRUSCATO (695), Data-Security Department, Racal-Milgo, Inc., Miami, Florida 33166

GERI ALLISON (67), VLSI Technology, Inc., San Jose, California 95131

DAVID A. BAGLEE (381, 401), Advanced Development Activity, Texas Instruments Incorporated, Houston, Texas 77251-1443

R. T. BATE (615), Central Research Laboratories, Texas Instruments Incorporated, Dallas, Texas 75265

CHARLES A. BECKER (515), General Electric Company, Research and Development Center, Schenectady, New York 12301

RONALD C. BRACKEN (327), Solid State Technology Center RCA Corporation, Somerville, New Jersey 08876

DAVID J. BRADLEY (657), IBM, Boca Raton, Florida 33432

CHARLES T. BRODNAX (681), E-Systems, Garland Division, Dallas, Texas 75226

BOBBY L. BUCHANAN (565), Radiation Hardening Technology Branch, Solid State Sciences Division, RADC/ESR, Hanscom Air Force Base, Massachusetts 01731

KUEING-LONG CHEN (121), Texas Instruments, Incorporated, Dallas, Texas 75265

T. L. CHU (285), Southern Methodist University, Dallas, Texas 75275

W. H. CLASS (435), Semi-Alloy Inc., Allied Co., Mount Vernon, New York 10550

RICHARD V. COX (775), AT & T Bell Laboratories (2D-527), Murray Hill, New Jersey 07974

RODERICK D. DAVIES (121), Texas Instruments, Incorporated, Dallas, Texas 75265

V. K. DHAM (167), Intel Corporation, Santa Clara, California 95051

MICHAEL P. DUANE (401), Advanced Development Activity, Texas Instruments Incorporated, Houston, Texas 77251-1443

RUDOLPH H. DYCK (581), Fairchild Camera and Instrument Corporation, CCD Imaging, Palo Alto, California 94304

JOHN FOSTER (763), Coulter Electronics, Hialeah, Florida 33010

SAI WAI FU (9), Intel Corporation, Hillsboro, Oregon 97124

AART J. DE GEUS (37), General Electric Microelectronics Center, Research Triangle Park, North Carolina 27709

T. GHEEWALA (645), GigaBit Logic, Inc., Newbury Park, California 91320

ROBERT D. GOLD (737), Cordis Corporation, Miami, Florida 33102-5700

DENNIS G. GRINA (667), Computer Development Division, Control Data Corporation, St. Paul, Minnesota 55112

DANIEL HAMPEL (837), Technology Development, RCA Government Communications Systems Division, Camden, New Jersey 08102

BARBARA A. HEATH (487), INMOS Corporation, Colorado Springs, Colorado 80935

EDWIN J. HILPERT (823), Racal-Milgo, Miami, Florida 33166

THOMAS Y. HSIANG (629), Department of Electrical Engineering, University of Rochester, Rochester, New York 14627

HOWARD R. HUFF (191), Monsanto Electronic Materials Company, Palo Alto, California 94304

CHING-LIN JIANG (151), Mostek Corporation, Carrollton, Texas 75006

N. W. JONES (1, 79), Intel Corporation, Santa Clara, California 95051

LEE KAMMERDINER (487), INMOS Corporation, Colorado Springs, Colorado 80935

JACOB KLINE (721), Department of Biomedical Engineering, College of Engineering, University of Miami, Coral Gables, Florida 33124

S. B. KULKARNI (305), IBM East Fishkill Facility, Hopewell Junction, New York 12533

S. K. LAI (167), Intel Corporation, Santa Clara, California 95051

H. W. LAM (503), Texas Instruments, Incorporated, Dallas, Texas 75265

GRAYDON B. LARRABEE (270), Texas Instruments Incorporated, Materials Science Laboratory, Dallas, Texas 75265

S. S. LAU (415), Department of Electrical Engineering and Computer Sciences, University of California, San Diego, La Jolla, California 92093

WILLIAM V. LAWSON[1] (55), Applicon, Burlington, Massachusetts 01803

HAROLD LEVINE (865), Sigalos & Levine, Dallas, Texas 75201

THOMAS K. LISLE (707), Systems Development Division, Westinghouse Electric Corporation, Baltimore, Maryland 21203

HSI-HO LIU (785), Department of Electrical and Computer Engineering, College of Engineering, University of Miami, Coral Gables, Florida 33124

PHILIP S. LIU (785), Department of Electrical and Computer Engineering, College of Engineering, University of Miami, Coral Gables, Florida 33124

JEREMIAH R. LOWNEY (177), Semiconductor Materials and Processes Division, National Bureau of Standards, Gaithersburg, Maryland 20899

CHARLES J. McMINN (527), VLSI Systems Division, Megatest Corporation, San Jose, California 95131

AMR MOHSEN (9), Intel Corporation, Aloha, Oregon 97007

Contributors

SYDNEY B. NEWELL (37), General Electric Microelectronics Center, Research Triangle Park, North Carolina 27709

MARC-A. NICOLET (415), California Institute of Technology, Pasadena, California 91125

SHARBEL E. NOUJAIM (801), General Electric Company, Corporate Research and Development Center, Schenectady, New York, 12301

R. F. PINIZZOTTO (503), Ultrastructure Inc., Richardson, Texas 75080

ROBERT PROEBSTING (151), Mostek Corporation, Carrollton, Texas 75006

JOHN RAITHEL (141), Zilog. Inc., Campbell, California 95008

SYED A. RIZVI (327), United Technologies, Mostek Corporation, Carrollton, Texas 75006

JEFF SCHLAGETER (151), Mostek Corporation, Carrollton, Texas 75006

DAVID B. SCOTT (121), Texas Instruments, Incorporated, Dallas, Texas 75265

THOMAS J. SHAFFNER (270), Texas Instruments Incorporated, Materials Science Laboratory, Dallas, Texas 75265

FUMIO SHIMURA (191), Monsanto Electronic Materials Company, St. Louis, Missouri 63167

CARL SIMONSEN (9), Intel Corporation, Aloha, Oregon 97007

J. F. SMITH (435), Materials Research Corporation, Orangeburge, New York 10962

ALFRED J. STEIN (67), VLSI Technology, Inc., San Jose, California 95131

FRANK S. STEIN (851), Delco Electronics Division, General Motors Corporation, Kokomo, Indiana 46901

A. F. TASCH, JR. (503), Motorola Inc., Austin, Texas 78721

DONALD P. TATE (667), Supercomputer Operations, Control Data Corporation, St. Paul, Minnesota 55112

W. ROBERT THURBER (177), Semiconductor Materials and Processes Division, National Bureau of Standards, Gaithersburg, Maryland 20899

TONY VALENTINO (67), VLSI Technology, Inc., San Jose, California 95131

F. W. VOLTMER (1, 79), Intel Corporation, Santa Clara, California 95050

KIRBY G. VOSBURGH (87), General Electric Company, Research and Development Center, Schenectady, New York, 12301

LAWRENCE C. WAGNER (551), Texas Instruments Incorporated, Device Analysis Laboratory, Dallas, Texas 75265

CHENG T. WANG (99, 109), Department of Electrical and Computer Engineering, College of Engineering, University of Miami, Coral Gables, Florida 33124

K. L. WANG (455), Device Research Laboratory, Electrical Engineering Department, University of California, Hilgard Avenue, Los Angeles, California 90024

R. KENT WATTS (351, 365), Bell Laboratories, Murray Hill, New Jersey 07974

TZAY Y. YOUNG (785), Department of Electrical and Computer Engineering, College of Engineering, University of Miami, Coral Gables, Florida 33124

LEROY D. YOUNG, JR. (823), Model Development, Racal-Milgo, Inc., Miami, Florida 33166

JOHN J. ZINGARO (707), Systems Development Division, Westinghouse Electric Corporation, Baltimore, Maryland 21203

A. VAN DER ZIEL (603), Electrical Engineering Department, University of Minnesota, Minneapolis, Minnesota 55455

[1] Present address: Telesis Systems Corporation, Chelmsford, Massachusetts.

Preface

This one-step reference handbook offers engineers and scientists the concise critical facts they need on VLSI (very large scale integration) microelectronics. The entire book is a ready source of information on VLSI circuits, fabrication, and systems applications.

It is now generally recognized and broadly accepted that microelectronics has brought our civilization past the threshold of the second industrial revolution. The first industrial revolution, based on the steam engine, enabled man to multiply his capability to do physical work. In a comparable manner, semiconductor electronics is enabling man to multiply his capacity for performing intellectually based tasks. VLSI is the current embodiment of advanced semiconductor electronics technology. The "VLSI Handbook" is published in an effort to satisfy the need for a systematic compilation of knowledge at the leading edge of this technology to satisfy the needs of those who require readily available answers to rather specifically defined questions.

This handbook is a comprehensive compilation to provide data, performance application information, and guidelines for the entire range of VLSI technology. It will be of value as well to basic and applied researchers interested in the physics and chemistry of materials and processes, to device designers, and to systems designers.

This handbook is organized in a manner wherein chapters are grouped into the fields of design, materials and processes, and examples of specific systems applications. Each of the chapters is prepared by an expert in the field and is written in a way that promotes stand-alone comprehension.

Since it is anticipated that this work will evolve as it goes through future editions, the editor welcomes suggestions for quality enhancement and for other topics for inclusion.

Acronyms

ACU	Array control unit
AES	Auger electron spectroscopy
ARU	Array unit
ASTM	American Society for Testing and Materials
BER	Bit-error rate (modems)
BFL	Buffered FET logic (GaAs)
BIGFON	Broadband integrated glass fiber optic network
C-V	Capacitance-voltage (technique for measuring spreading resistance)
CAD	Computer-aided design
CAE	Computer-aided engineering
CAL	Computer-aided layout
CAM	Content addressable memory
CAM	Computer-aided manufacturing
CAS	Column address strobe
CAT	Computer-aided test
CCD	Charged coupled device (imager)
CD	Dimensional control
CE	Chip enable
CID	Charge injection device (imager)
CIF	Caltech Intermediate Form (database)
CIL	Current injection logic (superconductor junctions)
CML	Current mode logic (bipolar VLSI logic circuit)
CMOS	Complementary metal-oxide semiconductor
CMRT	Cellular mobile radio telephone
CPD	Charge primary device (imager)
CRT	Cathode ray tube
CSMA/CD	Carrier-sense multiple access with collision detection (distributed communications)
CVD	Chemical vapor deposition (epitaxy)
CVE	Chemical vapor (growth) epitaxy
DCL	Direct coupled logic gate
DES	Data encryption standard
DFT	Discrete Fourier transform

DI	Dielectric isolation
DIIC	Dielectric isolated integrated circuit
DLTS	Deep level transient spectroscopy
DMA	Direct memory access
DPSK	Differential phase shift keying
DRAM	Dynamic random access memory
DSP	Digital signal processing
EBIC	Electron beam induced conductivity
ECB	Electronic code book (encryption, DES)
ECL	Emitter coupled logic (bipolar VLSI logic circuit)
EEPROM	Electrically erasable programmable ROM
EPROM	Erasable programmable ROM
FET	Field effect transistor
FFT	Fast Fourier transform
FIIR	Fourier transform infrared spectroscopy(?)
FIPOS	Full isolation by porous oxidized silicon
FLOTOX	Floating gate tunnel oxide
FSK	Frequency shift keying
FTIR	Fourier transform infrared spectrometry (technique for measuring low-level impurities in films)
HBT	Heterojunction bipolar transistor
HEMT	High electron mobility transistor
HMOS	High performance metal oxide semiconductor
ICHD	Intersociety Commission for Heart Disease
IC	Integrated circuit
IGFET	Insulated gate field effect transistor
IMPATT	Impact ionization avalanche transit time
IR	Infrared
ISI	Intersymbol interference (modems)
ITL	Integrated injection logic (bipolar VLSI logic circuit)
JAWS	Josephson Atto-Weber switch
JFET	Junction field effect transistor
JIIC	Junction isolated integrated circuit
LAN	Local area network (patient monitoring, distributed communications)
LDM	Linear delta modulation
LEC	Liquid-encapsulated Czochralski wafers
LEGO	Lateral epitaxial growth over oxide
LLR	Lower level resist
LOCOS	Local oxidation of silicon
LPC	Linear predictive coding
LPC	Linear prediction coefficient (speech processing)
LPCVD	Low pressure chemical vapor deposition
LPE	Liquid phase epitaxy

Acronyms

LSI	Large scale integration
MAC	Multiplier-accumulator (part of a fast Fourier transform)
MBE	Molecular beam epitaxy
MESFET	Metal semiconductor field effect transistor
MF	Metal ion-free developers
MIPS	1 million instructions per second
MMU	Memory management unit
Modem	Modulator–demodulator
MOSFET	Metal-oxide semiconductor field effect transistor
MOS	Metal-oxide semiconductor
MPP	Massively parallel processor
MSI	Medium scale integration
MTBF	Mean time between failures
MTF	Modulation transfer function
MTL	Merged transistor logic (Bipolar VLSI logic circuit)
NAA	Neutron activation (Technique for measuring low-level impurities in films)
NASPE	North American Society for Pacing and Electrophysiology
NES	Noise equivalent signal
NMOS	n-channel metal oxide semiconductor
NVRAM	Nonvolatile random access memory
PCM	Portable conformable masks
PDC	Precipitate dislocation complex
PE	Processing element (pattern recognition)
PECVD	Plasma-enhanced chemical vapor deposition
PG	Pattern generation
PLA	Programmable logic array
PMIPK	(proximity wafer printing)
PMMA	Polymethyl methacrylate
PMOS	p-channel metal oxide semiconductor
PROM	Programmable read-only memory
PSG	Phosphorous doped glass
PSK	Phase shift keying
QAM	Quadrature amplitude modulation
QUITERON	Quasiparticle injection tunneling effect device
RAM	Random access memory
RAS	Row address strobe
RBS	Rutherford backscattering
RC	Resistance-capacitance
rf	Radio frequency (heating)
ROM	Read only memory
RTL	Register transfer level
SAT	Self-aligned transistor

SBT	Silicon bipolar transistor (bipolar VLSI logic circuit)
SDFL	Schottky diode FET logic (GaAs)
SEM	Scanning electron microscope
SEU	Single event upset (radiation damage to ICs)
SIMD	Single instruction-multiple data streams
SIMS	Secondary ion mass spectrometry (technique for measuring low-level impurities in films)
SOI	Silicon-on-insulator
SOS	Silicon-on-sapphire
SPE	Solid phase epitaxy
SPICE	Simulation program for integrated circuit emulation
SQUID	Superconducting quantum interference device
SRP	Spreading resistance probe
SSI	Small scale integration
STL	Schottky transistor logic (bipolar VLSI logic circuit)
SUPREM	A computer simulation program
TDI	Time delay and integrative imagers
TFT	Thin film transistors
TLR	Top level resist
TTL	Transistor-transistor logic (bipolar VLSI logic circuit)
UART	Universal asynchronous receiver-transmitter
ULSI	Ultra large scale integration
V_{FB}	Flatband voltage
V_T	Threshold voltage
VHSIC	Very high speed integrated circuits
VLSI	Very large scale integration
VPE	Vapor phase epitaxy

Chapter 1
Factors Contributing to Increased VLSI Circuit Density

F. W. VOLTMER
N. W. JONES

Intel Corporation
Santa Clara, California

I. Introduction	1
II. Factors Influencing Circuit Density	2
A. Increases in Die Size	2
B. Reductions in Feature Size	3
C. Increase in Device and Circuit Complexity	3
D. Increases in Process Complexity	7

INTRODUCTION I

MOS integrated circuit technology has seen rapid advancement over the last decade resulting in dramatic improvements in both performance and cost per function. Significant evolution in processing and design methodology has led to an increase in chip density and to a corresponding increase in complexity.

In the future VLSI circuit density will continue to increase, undergoing a gradual change from MOS to CMOS technology. Process and design innovations will allow even higher levels of complexity to be achieved. The factors that are expected to continue influencing VLSI circuit density in the 1980s will be reviewed.

II FACTORS INFLUENCING CIRCUIT DENSITY

A Increases in Die Size

A trend of average die size as a function of time and circuit density is shown in Fig. 1. Despite the impact of periodic design rule reduction made possible through process changes, the average die area has increased with time. This increase is a consequence of the general improvement in materials and manufacturing technology. The number of transistors per chip for several typical 8-bit and 16-bit microprocessor families is presented in Fig. 2. As can be seen, the rate of integration has been rapid, which has permitted many new functions to be added to each successive generation of ICs. Memory IC density has increased by a factor of nearly two each year, and there is every reason to believe this trend will continue at least until 1990. Hence, chips containing more than 1,000,000 transistors will probably be available by 1985–1986.

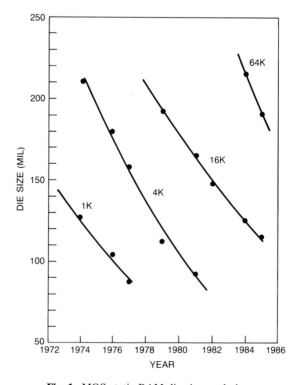

Fig. 1. MOS static RAM die-size evolution.

1. Factors Contributing to Increased Circuit Density

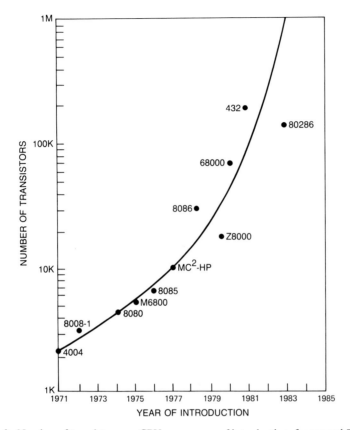

Fig. 2. Number of transistors per CPU versus year of introduction, for several 8- and 16-bit microprocessor families.

Reductions in Feature Size   B

Increased density is critically dependent upon the minimum feature size that can be produced on a chip. Figure 3 shows that roughly an 11% reduction in feature size each year has been achieved during the past two decades. Pursuit of size reduction has had major impact on patterning and vertical scaling technology. VLSI circuit manufacturing will require the introduction of many sophisticated tools such as wafer steppers, multilayer resists, anisotropic dry etch, and low-temperature deposition equipment in order to achieve further reductions in feature size.

Increase in Device and Circuit Complexity   C

NMOS device technology became the workhorse of the 1970s because of its reliability, reasonable manufacturing cost, and scalability. This laid

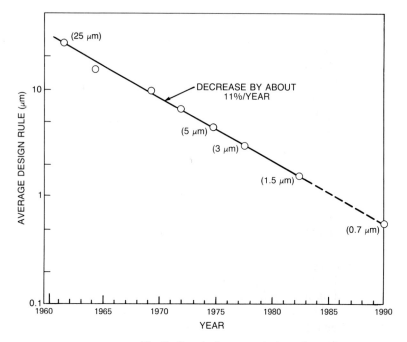

Fig. 3. Trend of average design rule.

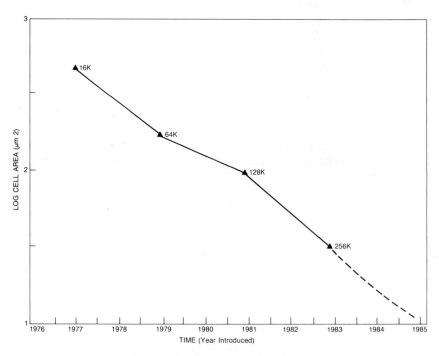

Fig. 4. Impact of scaling an EPROM cell over time.

TABLE I

Practical Application of Scaling

MOS device parameters	Ideal scaling factor	Practical scaling factor
Extrinsic lateral dimensions (Z, L)	$1/S$	$1/S$
Intrinsic lateral dimensions (ℓ_D, X_j)	$1/S$	$\sim 1/S$
Oxide thickness	$1/S$	$\geq 1/S$
Substrate doping	S	$\leq S$
Voltage	$1/S$	~ 1
Current	$1/S$	$\sim 1/S^2$

the foundation for several high-performance generations that were obtained simply by reducing device dimensions, as indicated in Tables I and II and by Fig. 4.

Unfortunately, not all device parameters can be scaled proportionally. The voltage supply, for example, has not been scaled in every generation. This limit on scaling has made device and circuit modeling more important as fundamental limits were approached. Several computer-aided design tools have been developed to better predict specific operational performance. Two-dimensional models are now available that enhance accuracy and, in many cases, make simulation preferable to the more costly experimental laboratory techniques for design verification.

Design complexity has also increased, despite the utilization of fewer devices per function. This increase stems partly from the inability to fully simulate an entire circuit as a unit. Dense memories, for example, can generally be verified only in sections. In addition, most circuits now require several types of transistors on one chip. This feature allows greater design flexibility but complicates manufacture because it necessitates a greater number of mask layers and implant steps and tighter overall parametric control. The emergence of CMOS technology has begun to revolutionize state-of-the-art IC design. Inherent noise immunity and reduced

TABLE II

Typical MOS Scaling Trends

Parameter	1977	1980	1984	1986
Layout density (Gates/mm²)	200	450	700	1000
Speed power product (PJ)	1	0.2	0.05	0.01
Gate delay (ns)	1	0.5	0.25	0.1
Power supply voltage (V)	5	5	5	3
Channel length (μ)	3.3	2.0	1.2	0.7
Gate oxide thickness (Å)	700	400	250	150
Design rules (μ)	3.5	2.0	1.5	1.0

TABLE III
Trends in Process Complexity

Parameter	Al gate PMOS 1969	Si gate PMOS 1970	Si gate NMOS 1972	Depletion NMOS 1976	NMOS 1977	NMOS 1979	CMOS 1983	CMOS 1986	CMOS 1990
Gate length (μm)	20	10	6	6	3.5	2	1.5	1.00	0.75
Junction depth (μm)	2.5	2.5	2.0	2.0	0.8	<0.8	0.45	0.25	0.15
Gate oxide thickness (Å)	1500	1200	1200	1200	700	400	250	150	100
Mask levels	5	5	6	6	7–10	7–10	8–12	10–15	12–18
Diffusions	3	4	5	5	7–9	7–9	12–16	14–18	14–18
Relative complexity	1	1.2	1.3	1.7	2	2.5	3–5	4–5	6

1. Factors Contributing to Increased Circuit Density

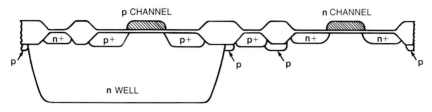

Fig. 5. Cross section of a CMOS device.

power dissipation are two of the dramatic advantages that are expected to make CMOS the mainstream VLSI process of the 1980s.

Increases in Process Complexity   D

Table III illustrates the trend toward increased process complexity and how it is exacerbated by the thrust towards CMOS. This can be better understood by examining the CMOS device structure depicted in Fig. 5. Producing both *n*-channel and *p*-channel devices on one chip has been accomplished through process enhancements. Scaling will continue but is expected to have less effect on development than the addition of new mask levels and other process structural innovations. Multilevel metallization will grow in popularity, along with new metal alloys, as linewidths shrink and electromigration problems intensify. Shallow junctions will probably need new contact metals to overcome the increase in resistance and prevent spiking to the substrate. Major improvements in electrical isolation of devices and in gate interconnect resistance will be required. Additional implant masks and a variety of specialized etch steps also appear inevitable.

Chapter 2

Fundamental Principles of Very Large Scale Integrated Circuit Design

AMR MOHSEN
SAI WAI FU
CARL SIMONSEN

Intel Corporation
Aloha, Oregon

I.	Introduction	9
II.	VLSI Design Methodology	10
	A. Hierarchical Decomposition	11
	B. Computer-Aided Design	13
III.	Elements of VLSI Circuit Design	14
	A. The MOS Transistor	14
	B. The Basic Inverter	16
	C. NAND and NOR Logic Gates	21
	D. Other Unique Circuit Elements in the MOS Technology	21
	E. Distributed Propagation Delay	23
	F. Examples of Design Building Blocks	24
	G. Regular Design Structures	26
	H. Clocking Schemes	28
IV.	Basics of Layout Design	28
	A. Design Rules	29
	B. Physical Layout	30
	C. Automated Layout	33
V.	Future Developments	35
	References	35

I INTRODUCTION

The number of components and transistors that can economically be integrated on a single chip has doubled every one to two years during the last two decades. These two decades have also seen the average dimension of the physical features of integrated circuits scaled down by more than tenfold, as illustrated in Fig. 1 [1,2]. Many of the commercially

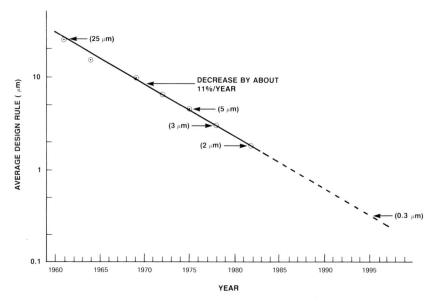

Fig. 1. Past and expected future trend of average design rule.

available very large scale integrated (VLSI) chips, such as 16- and 32-bit microprocessors and 256K-bit memory devices, contain hundreds of thousands of transistors. Many VLSI chips now consist of multiple functions and are therefore made up of complex integrated systems [3]. Physical principles indicate that another order-of-magnitude reduction in size of physical features, to 0.2 μm, is possible [4]. The performance and density of active transistors will then increase by one and two orders of magnitude, respectively [2]. Thus, by the 1990s it will be possible to integrate tens of millions of transistors on monolithic chips.

To effectively manage the development of VLSI chips, hierarchical design methodologies with regular structures and automated design tools are becoming widely used. In this chapter, an overview of VLSI design methodology and the principles of circuit and layout design using the NMOS and CMOS technologies will be presented. NMOS technology, with its higher density and lower cost, has been the dominant VLSI technology in the 1970s and early 1980s. CMOS technology, with its lower power, may become the workhorse technology of the next decade.

II VLSI DESIGN METHODOLOGY

The design of integrated circuits has evolved over the years to become a well-coordinated methodology of the different design stages.

2. Fundamental Principles of Circuit Design

Hierarchical Decomposition   A

Concepts of decomposition and abstraction (breaking a complex problem into manageable pieces and hiding unnecessary details at the interfaces) have been recently borrowed from the structured programming methodologies used in software engineering and applied to VLSI design [5]. Methodologies for hierarchical decomposition of the design along the functional lines into various design levels and regular structures have been developed [3,5]. Levels of VLSI hierarchical design are illustrated in Fig. 2. Each design level, from the top-level functional specification down

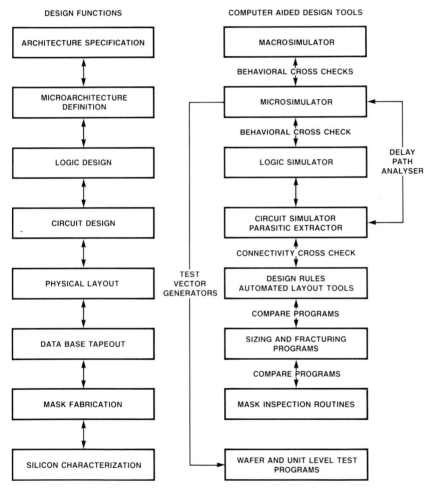

Fig. 2. VLSI design hierarchy and corresponding CAD tools.

to the mask and silicon fabrication, represents a more detailed description of the design.

(a) *Microarchitectural Design.* In the microarchitectural design level, choices of algorithms, degrees of parallelism or pipelining, and organization of functional blocks for speed and density are optimized. During this stage the top-level floor plan of the chip and the coarse block layouts are defined and continuously refined into a more detailed interconnect road map as the design proceeds through the lower level of logic, circuit, and physical layout designs. Figure 3 illustrates an example of the floor plan of the instruction decoder chip of a 32-bit object-based architecture processor [5].

(b) *Logic Definition.* In the logic definition level, logic structures are designed to implement the micro-architectural design.

(c) *Circuit and Layout Design.* In the circuit and layout design phases, the sizes and geometries of the transistors and interconnects are defined using the device models and design rules of the technology to be used for the fabrication of the integrated circuit.

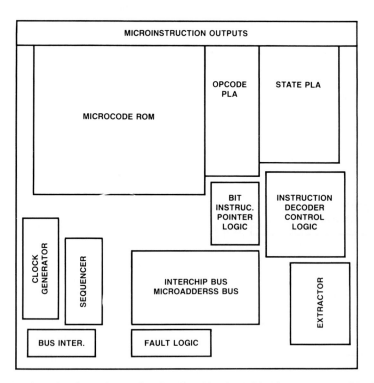

Fig. 3. Floor plan for an instruction decoder chip of a 32-bit microprocessor with object-based architecture.

2. Fundamental Principles of Circuit Design

(d) *Data Base Check.* Once layout is complete, the composite data base is checked for conformity to the layout design rules using design rule check programs. The data base is then processed through sizing and fracturing programs.

(e) *Mask Fabrication.* Polygons on all layers are sized to achieve the required final dimension on the masks. Usually the drawn dimensions are different from the final ones to simplify the design rules and to be less dependent on the process parameter variations. In the fracturing process the composite layout data base, which is drawn with rectangles and polygons, is converted to trapazoids to drive the mask-making equipment.

The hierarchical decomposition of the design into smaller blocks and macros is implemented at the various design levels until the resulting blocks are of manageable sizes. Each block is then implemented independently, and its design can evolve with technological advancement. The various blocks of the hierarchical design are then assembled with careful attention to the interfaces. The hierarchical methodology simplifies not only the implementation but also the verification of complex designs. When the various blocks are reused several times on the chip, the effort needed for simulation, design rule checking, and other validations of correctness are reduced proportionally.

Special emphasis in VLSI design is given to the use of regular array structures built of few basic cells with regular interconnect structures and well-defined communication paths [3,5]. This can save significant effort in the implementation of complex functions. For example, arrays of programmable logic cells and memory cells provide large amounts of function with relatively little design efforts.

Computer-Aided Design   B

Numerous computer-aided design (CAD) tools have been developed to assist the designers in the simulation and verification of the various design functions.

Simulation   1

Macro, micro, logic, and circuit simulators are used to verify the design functionalities at their respective levels. *Macrosimulation* is used to evaluate the architecture design of the VLSI system in performance and functionality. *Microsimulation,* also known as *register transfer language,* models the hardware implementation of the architecture, whereas *logic simulation* validates the logic gate design of the microarchitecture. *Circuit simulation* models the transient circuit response using the transistor and capacitance terminal characteristics and the parasitic nodal capacitances and interconnect resistances extracted from the physical layout. Usually, because of time and computer resource limitations, it is not practical to simulate the circuits of an entire chip. Hierarchical and subcircuit analysis techniques are used to make the circuit simulation manageable. A simplified transient response of the circuit description of the entire chip can be obtained with another simulator, the *delay path analyzer,* which models the gate delay with unit scaled delays and takes significantly less computer resources. Simulation often involves using test cases and worst-case models to stress the design to its logical or physical limits under the projected extreme operating and processing conditions. *Validation* of the equivalence in either behavior or structure of each design level is done with various cross-check programs, as shown in Fig. 3. For example, the output behaviors of the macro, micro, logic, and delay path analyzer simulators for given input conditions are cross checked. The connectivity of the physical layout geometries are verified with the schematics of the circuit simulators. Special programs are used to compare the tapeout data base and the mask features with the composite data base.

2 Testing

Another important consideration in VLSI design, with its present complexity, is to incorporate features that allow effective and economical testing of the finished chip. Means of testing some of the large internal building blocks at the wafer sort level and of testing key nodes and circuits, which are inaccessible at the packaged unit level, are valuable in the debugging and characterization phases. Special programs are used to generate test vectors and patterns compatible with the input and output data obtained from the design microsimulators.

III ELEMENTS OF VLSI CIRCUIT DESIGN

The circuit elements of the MOS VLSI technology are the MOS transistor, the basic inverter, the logic gates, and various other design building blocks.

A The MOS Transistor

The MOS transistor is the basic nonlinear active device of MOS technology; the *n*-channel MOS transistor is illustrated in Fig. 4.

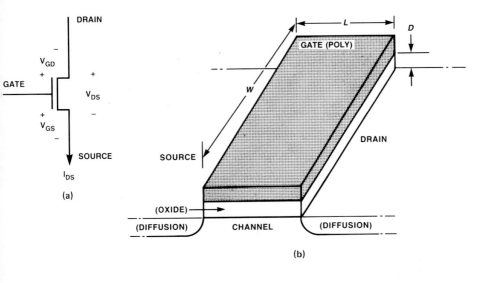

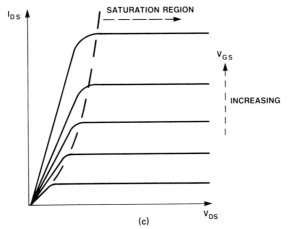

Fig. 4. n-Channel MOS transistor: (a) circuit symbol, (b) cross section, and (c) current–voltage characteristics.

Basic Operation 1

In the absence of charge on the gate, there is no current path from the drain to the source terminals; but when a sufficient positive charge is placed on the gate so that the gate-to-source voltage V_{GS} exceeds a threshold voltage V_T, electrons are attracted to the oxide–silicon interface, forming an inversion layer and a conduction path between drain and source. This use of the charge on the gate to control the inversion layer charge and the conductivity of the path between source and drain is the basic operation of the MOS transistor.

2 Current Characteristics

The relationship of the inversion layer charge Q_I, the average transit time of the electrons under the gate τ, and the current from the drain to source I_{DS} with the transistor terminal voltages is summarized in Table I. The dependence of the drain current I_{DS} on the terminal voltage V_{GS} and V_{DS} is shown in Fig. 4(c). The *p*-channel transistor has identical characteristics but has opposite voltage polarities and directions of current flow; in this transistor, positive holes are the charge carriers in the inversion layer that controls the drain-to-source conduction path. Other physical device phenomena (e.g., the dependence of the carrier mobility on electric field and the threshold voltage on transistor geometries) influence the transistor characteristics [2,6,7].

B The Basic Inverter

The inverter is the basic circuit element in VLSI circuit design. It generates the logical complement of its input signal. The logic threshold voltage V_{INV} is the input voltage that results in an equal output voltage. The logic-1 is represented by a voltage level above V_{INV}, and the logic-0 by a voltage level below V_{INV}. The negative slope of the transfer curve in Fig. 5 is the gain G, which has to be larger than $|1|$ for stable digital operation. Noise margin (NM) is the immunity of a circuit to external disturbances such as variations in power supply voltage and manufacturing processes. For digital circuits, the noise margins for 1 and 0, NM_1 and

TABLE I

MOS Transistor Current Characteristics[a]

Parameter	Definition
Inversion layer charge Q_I	$= C_{OX} \cdot WL \left(V_{GS} - V_T - \dfrac{V_{DS}}{2}\right)$
Transit time τ	$= \dfrac{L}{\text{Velocity}} = \dfrac{L}{\mu E} = \dfrac{L^2}{\mu V_{DS}}$
Drain current I_{DS}	$= \dfrac{Q_I}{\tau} = C_{OX} W L \left(V_{GS} - V_T - \dfrac{V_{DS}}{2}\right) \cdot \dfrac{\mu V_{DS}}{L^2}$
	$= \mu C_{OX} \dfrac{W}{L} \left(V_{GS} - V_T - \dfrac{V_{DS}}{2}\right) V_{DS}$

[a] C_{OX}, the oxide capacitance per units area $= \varepsilon_{OX}/D$; μ, the charge carrier mobility; V_{TH}, the threshold voltage; ε_{OX}, the oxide dielectric constant; D, the oxide thickness.

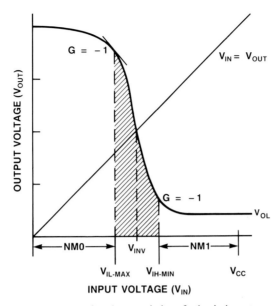

Fig. 5. Transfer characteristics of a basic inverter.

NM_0, are defined as the difference between the input voltages at $G = -1$ and V_{CC} or ground, respectively.

NMOS Technology 1

In the *NMOS technology*, a depletion-type transistor (normally *on* with threshold voltage $V_{TD} < 0$) is used as a pull-up device because of its good current sourcing characteristics. An enhancement transistor (normally *off* with threshold voltage $V_{TE} > 0$) is used as the pull-down device, as shown in Fig. 6. Because the pull-up depletion transistor is always *on*, there is a current drain when the pull-down device is *on*.

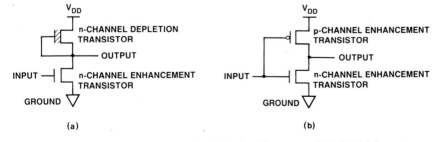

Fig. 6. Circuit configuration of (a) depletion load inverter and (b) CMOS inverter.

2 CMOS Technology

The CMOS technology reduces this static current drain by replacing the depletion transistor with an enhancement *p*-channel transistor (normally *off* with threshold voltage $V_{TP} < 0$). CMOS inverters consume power only during the output transitions.

The dependence of the inverter characteristics on the transistor parameters is summarized in Tables II and III and illustrated in Fig. 7. The dependence of the rise and fall times of an inverter output on the transistor geometries is summarized in Table IV and illustrated in Figs. 8 and 9. The given equations assume linear output capacitance load and step input signals. The output rising and falling transition times of the inverter are proportional to the output capacitance load and to the transit times of the charge carriers under the gates of the inverter transistors. The depletion load inverter is commonly designed with a geometry ratio K equal to 4 for adequate 1 and 0 noise margins and output low level V_{OL}. The fact that the output rising transition is slower than the falling transition of the depletion load inverter, by approximately the geometry ratio K of the transistors, is an inherent characteristic of the ratio-type logic. The

TABLE II
Transfer Characteristic Parameters of Depletion Load Inverter

Parameter	Definition
r	Ratio of electron mobility of depletion and enhancement transistors μ_D/μ_N
W_D, L_D	Width and length of the depletion pull-up transistor
W_N, L_N	Width and length of the enhancement pull-down transistor
$K = \dfrac{L_D W_N}{W_D L_N}$	Inverter geometry ratio
$A = \dfrac{r}{k}$	Normalized geometry ratio
V_{TD}, V_{TN}	Threshold voltages of depletion and enhancement transistors
V_{OL}	Output low voltage of the inverter
V_{IN}	Input voltage
V_{OL}	$\approx \dfrac{\frac{1}{2} A V_{TD}^2}{(V_{IN} - V_{TN})}$
V_{INV}	$\approx V_{TN} - A^{1/2} V_{TD}$

2. Fundamental Principles of Circuit Design

TABLE III
Transfer Characteristic Parameters of CMOS Inverter

Parameter	Definition
r	Ratio of hole and electron mobilities μ_P/μ_N
W_P, L_P	Width and length of the p-channel pull-up transistor
W_N, L_N	Width and length of the n-channel pull-down transistor
V_{TP}, V_{TN}	Absolute threshold voltage of the p- and n-channel transistors
$K = \dfrac{L_P \cdot W_N}{W_P \cdot L_N}$	Inverter geometry ratio
V_{DD}	Supply voltage
$A = \dfrac{r}{k}$	Normalized geometry ratio
$V_{INV} \simeq \dfrac{A^{1/2}(V_{DD} - V_{TP}) + V_{TN}}{1 + A^{1/2}}$	

TABLE IV
Transient Characteristics Parameters of MOS Inverter[a]

Parameter	Definition
Rise time	$\simeq \dfrac{\text{Load capacitance} \times \text{output voltage swing}}{\text{Pull-up current}}$
NMOS rise time	$\simeq \dfrac{F_1}{\mu_D C_{OX}} \cdot \dfrac{L_D}{W_D} \cdot C_L = f_1 \tau_D \dfrac{C_L}{C_D}$
CMOS rise time	$\simeq \dfrac{F_2}{\mu_P C_{OX}} \cdot \dfrac{L_P}{W_P} \cdot C_L = f_2 \tau_P \dfrac{C_L}{C_P}$
Fall time	$= \dfrac{\text{Load capacitance} \times \text{output voltage swing}}{\text{Pull-down current}}$
NMOS and CMOS fall time	$\simeq \dfrac{F_3}{\mu_N C_{OX}} \cdot \dfrac{L_N}{W_N} \cdot C_L = f_3 \tau_N \dfrac{C_L}{C_N}$

[a] C_{OX} is the gate oxide capacitance per unit area; F_s, f_s, the supply voltage and technology dependent constants; C_L, the linear load capacitance; τ_D, τ_P, τ_N, the carrier transient time under the gates of the depletion pull-up, p-channel pull-up and enhancement pull-down transistors, respectively; C_D, C_P, C_N, the gate capacitance of the depletion pull-up, p-channel pull-up and enhancement pull-down transistors, respectively.

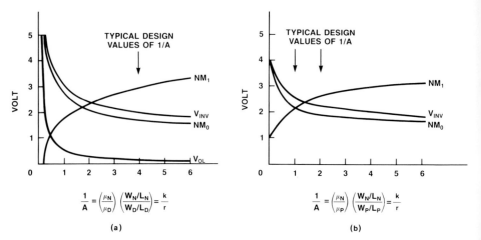

Fig. 7. Transfer characteristics of (a) depletion load inverter and (b) CMOS inverter.

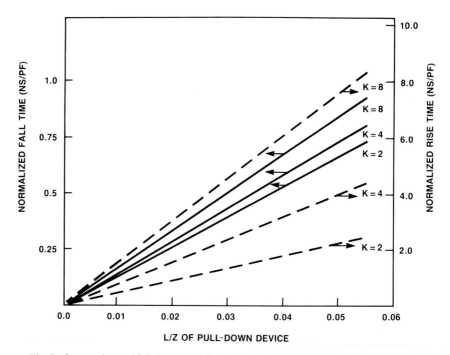

Fig. 8. Output rise and fall time transitions normalized to the load capacitance versus the pull-down transistor L/Z and the inverter geometry ratio K for a depletion load inverter with electrical gate length of 1.2 μm, $(\mu C_{OX})_N = 7.8 \times 10^{-5}$ A/V^2 and $(\mu C_{OX})_D = 6 \times 10^{-5}$ A/V^2.

2. Fundamental Principles of Circuit Design

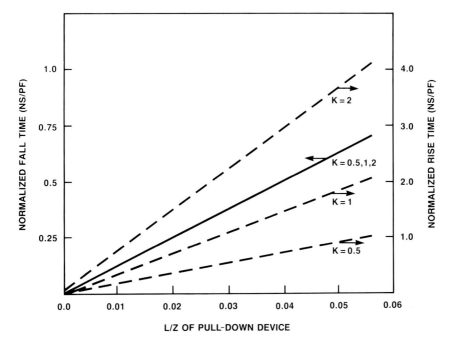

Fig. 9. Output rise and fall time transitions normalized to the load capacitance versus the pull-down transistor L/Z and the inverter geometry ratio K for a CMOS inverter with electrical gate length of 1.2 μm, $(\mu C_{OX})_N = 7.8 \times 10^{-5}$ A/V^2 and $(\mu C_{OX})_P = 2.7 \times 10^{-5}$ A/V^2.

CMOS inverter is usually designed with a geometry ratio K in the range of 1 to $\frac{1}{2}$, which results in less asymmetrical output rise and fall times.

C. NAND and NOR Logic Gates

Combinational logic can be built from NAND gates and inverters or NOR gates and inverters. NAND and NOR gates can be constructed as simple extensions of the inverter. In Fig. 10, examples of two-inputs NAND and NOR gate implementations are shown for both the NMOS and CMOS technologies. The logic threshold voltage V_{INV} of the composite gate can be derived from the basic inverter by substituting the geometry ratio K that takes into account the number of devices in series or in parallel for the pull-up and pull-down legs, as given in Table V.

D. Other Unique Circuit Elements in the MOS Technology

The low-leakage-current characteristics of the gate of the MOS transistors allow their utilization as dynamic storage nodes, delay timers, and

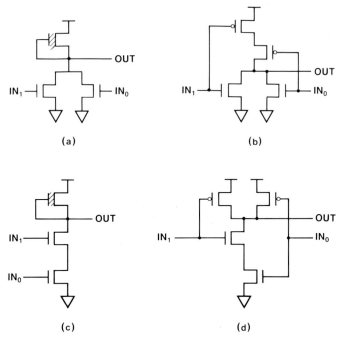

Fig. 10. Circuit configuration of (a) depletion load NOR gate, (b) CMOS NOR gate, (c) depletion load NAND gate, and (d) CMOS NAND gate.

TABLE V

Transfer Characteristics Parameters of MOS Logic Gates[a]

Parameter	Definition
V_{INV} of n-input NMOS NAND	$\simeq V_{TN} - (nA)^{1/2}V_{TD}$
V_{INV} of n-input CMOS NAND	$\simeq \dfrac{(nmA)^{1/2}(V_{DD} - V_{TP}) + V_{TN}}{1 + (nmA)^{1/2}}$
V_{INV} of n-input NMOS NOR	$\simeq V_{TN} - \left(\dfrac{A}{n}\right)^{1/2} V_{TD}$
V_{INV} of n-input CMOS NOR	$\simeq \dfrac{\left(\dfrac{A}{nm}\right)^{1/2}(V_{DD} - V_{TP}) + V_{TN}}{1 + \dfrac{A}{nm}}$

[a] $A = (\mu_D/\mu_N)(W_D L_N / L_D W_N)$ for NMOS, $A = (\mu_P/\mu_N)(W_P L_N / L_P W_N)$ for CMOS, where n is the number of inputs and m the number of inputs changing simultaneously to cause the output to change.

2. Fundamental Principles of Circuit Design

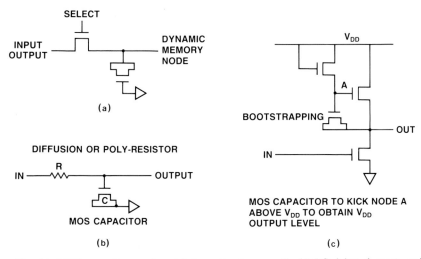

Fig. 11. MOS capacitor used as (a) dynamic storage cell, (b) RC delay element, and (c) bootstrap capacitor of an inverter.

bootstrap kick capacitors, as shown in Fig. 11. The MOS transistor can also be used as a transfer gate to steer-control an input signal. Figure 12 illustrates the use of an enhancement transistor as a transfer gate. In the CMOS technology, an additional parallel *p*-channel gate, driven by the complement of the control signal, allows the output signal to follow the input signal up to the supply voltage V_{DD} with no threshold voltage drop.

Distributed Propagation Delay   E

In VLSI, distributed propagation delay results from interconnect parasitics and chains of passive delay elements. Figure 13 illustrates the distributed delay in the transistors of a *ripple carry chain*. In distributed

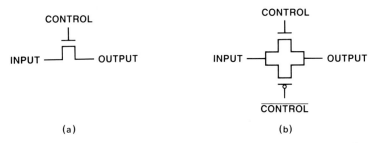

Fig. 12. Transfer gate configuration with (a) NMOS enhancement transistor and (b) CMOS complementary transistor pair.

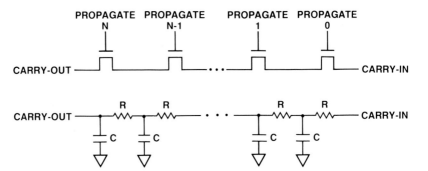

Fig. 13. Distributed delay in the ripple carry chain. The transistors of the chain are represented by an *RC* network.

structures, the delay is proportional to the square of the length of the interconnect line or the square of the number of passive delay elements in the chain. The distributed propagation delay is given approximately by the time constant of the distributed structure (the product of total resistance and total capacitance). One common technique for reducing the distributed propagation delay is to break the passive delay chain into segments and insert active repeaters as shown in Fig. 14. Thus, the propagation delay grows by the linear number of the passive delay segments plus the active repeater delays. The minimum delay is achieved when the distributed propagation delay of each segment is equivalent to the repeater delay.

F Examples of Design Building Blocks

Figure 15 displays a collection of some common building blocks in VLSI design. A *dynamic latch* stores the data by holding the charge on the input capacitance of the inverter. These data should be updated regularly prior to the charge decay by thermal leakage currents. A *static latch* holds the data indefinitely by feeding back the information to its input. A string of dynamic or static latches can be cascaded to form a *shift register*.

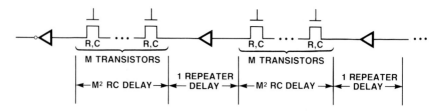

Fig. 14. Repeaters are used to reduce the distributed *RC* delay.

2. Fundamental Principles of Circuit Design

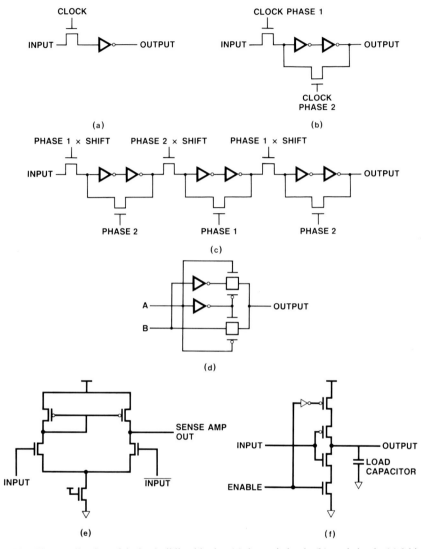

Fig. 15. A collection of design building blocks: (a) dynamic latch, (b) static latch, (c) 3-bit shift register, (d) CMOS exclusive OR gate using transfer gate configuration, (e) CMOS differential sense amplifier, and (f) tristate output driver.

An *exclusive OR gate* uses the steering properties of MOS transistors to save transistor counts relative to inverting logic implementations. An *output driver* is tristated with an enable signal (i.e., disconnected from the supply voltage V_{DD} and ground V_{SS}). A *differential sense amplifier* increases the differences of the input signals and drives a single-ended output.

G Regular Design Structures

Structured designs built from regular arrays of few basic cell structures are commonly used to reduce VLSI design complexity. A large amount of function can be implemented by using a fairly small amount of design effort on the basic regular cells and their interfaces. The types of regular structures depend on the topological properties of the technology and its interconnects. Some of the basic cells of structured designs in the MOS technology are given below.

1 Programmable Logic Array

A programmable logic array (PLA) consists of two plane levels that implement AND and OR functions using the NOR gate configuration [8]. All inputs are brought to the AND plane where they are used selectively as inputs to several gates. The outputs from the gates in the AND plane, called min-terms, run to the OR plane. The outputs from the OR plane are the PLA outputs and may be fedback to the PLA inputs. Figure 16 shows an example of two logic functions implemented by the PLA given by

$$OUT_0 = A \cdot B \cdot C + C \cdot D + A \cdot \bar{B},$$
$$OUT_1 = A \cdot B \cdot C + \bar{A} \cdot C.$$

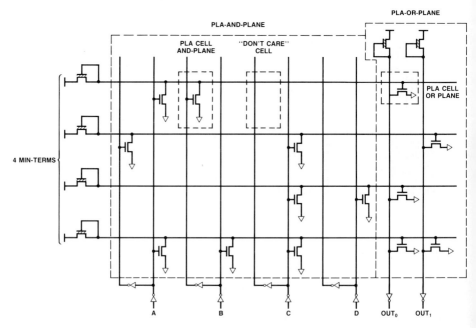

Fig. 16. A PLA implementation of an example of two logic functions.

2. Fundamental Principles of Circuit Design

PLAs are commonly used to implement complex combinational logic, processor controllers, and other finite state machine functions.

Random Access Memory Array (RAM) 2

A RAM consists of an array of memory cells in which data can be stored and read at random [9]. Figure 17 shows a block diagram of a RAM array with static memory cells. RAM arrays are used extensively in VLSI systems for Cache, scratch pad, register, and large data memory storage.

Read Only Memory Array (ROM) 3

A ROM consists of an array of memory cells from which data can only be read [9]. The memory cells are programmed during the design and fabrication of the device. Figure 18 shows a block diagram of a ROM array with one-transistor memory cells. ROM arrays are used for look-up tables, microcode, and program memory.

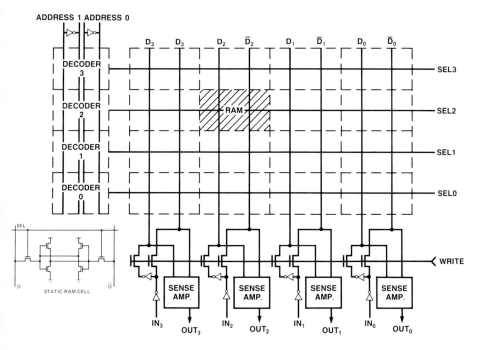

Fig. 17. A 4 × 4 RAM array.

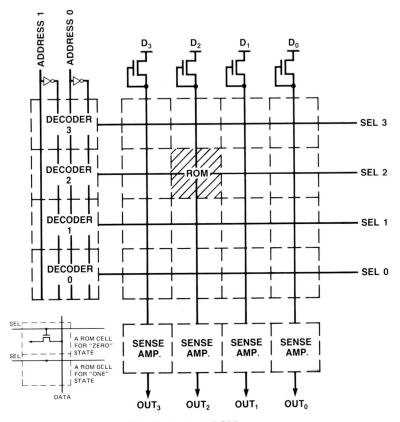

Fig. 18. A 4 × 4 ROM array.

H Clocking Schemes

The flow of data through the transistors and subsystem structures of digital VLSI systems are controlled by clocks (Fig. 19). The clocks govern the times during which the data can flow through the various system stages and the times during which the stages are isolated. In synchronous systems, two-phase nonoverlapping clocks are commonly used (Fig. 20). The complements of the two-phase nonoverlapping clocks in CMOS circuits are generated to drive the p-channel transfer gates.

IV BASICS OF LAYOUT DESIGN

Layout design is the translation of the electrical description of an integrated circuit into layered geometrical patterns from which the photolitho-

2. Fundamental Principles of Circuit Design

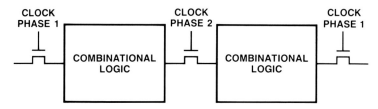

Fig. 19. Synchronous logic design with data flow controlled by a system clock.

graphic masks can be made. The physical layout is implemented in accordance with a set of design rules that define the permissible set of geometries. This ensures the proper function and the manufacturability of the design.

Design Rules A

An important characteristic of VLSI technologies is the separation of wafer fabrication from the design process. This separation requires a set

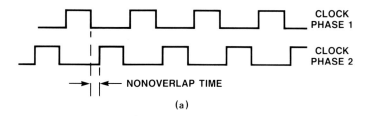

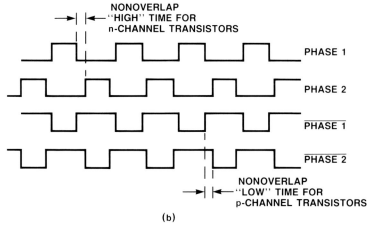

Fig. 20. (a) Two-phase nonoverlapping clock for NMOS circuits and (b) four-phase clocks for CMOS circuits.

of design rules that precisely define the capabilities of the mask and wafer fabrication technologies. The design rules define the limits of the widths, spaces, and overlaps of the geometric patterns on the various layers. As technologies have improved over the last two decades, the average dimension in the design rules has been increasing at the rate of 11% per year as shown in Fig. 1. There is no evidence that this trend is abating; in fact, it is expected that at least another order-of-magnitude shrinkage in linear dimensions is possible before hitting the fundamental physical limitations of device operation and fabrication [4].

Several key limitations define the basic layout design rules:

(a) The minimum width or space limit λ imposed by the photolithographic process.

(b) Control of the critical dimension, which is limited by variations in mask making, photolithography, and material etching processes (critical dimensions typically vary by about 0.2 λ).

(c) The alignment control tolerance, which accounts for the mask runout during its fabrication (skew of the location of the features on the masks) and for the misalignment of printing the masks on the wafers (typically about 0.3 λ).

(d) Process-induced shifts such as lateral diffusions, lateral oxidation growth, and feature widening or narrowing during the patterning process.

(e) Other considerations, including lithography and topological properties of the device structures.

An example of the dependence of the design rule (DR) for the metal to via-contact enclosure on these limitations is shown in Fig. 21. The size of the via-contact is limited by the minimum feature width λ. The metal enclosure of the contact is limited by the alignment tolerance of the metal to the via-contact mask, the control on the critical dimensions of the via-contact opening and metal, and the process-induced shifts (contact hole widening and metal line narrowing during their patterning). The design rule has to ensure proper metal coverage of the contact under the worse case conditions of the above limitations.

B Physical Layout

Figure 22 illustrates the basic layers needed to construct a CMOS integrated circuit and the resulting cross section of the finished silicon.

(a) The first layer defines the well regions whose doping polarity is opposite to that of the substrate. With an n-well process, the starting wafer is p-type with a doping concentration optimized for the performance of the n-channel transistors performance. The n-well regions re-

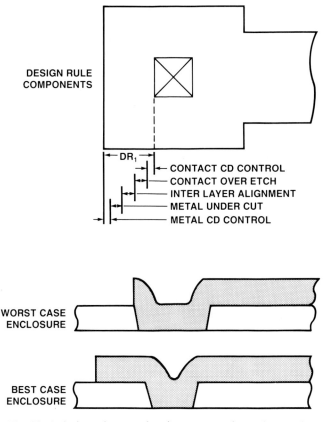

Fig. 21. A design rule example: via-contact enclosure by metal.

ceive an *n*-type implantation during the wafer fabrication process to counter-dope the substrate to allow the fabrication of the *p*-channel transistors.

(b) The second layer defines the active transistor and diffusion areas. The areas not defined by this mask will have thick oxide, which isolates adjacent active areas.

(c) The third mask defines the polysilicon layer, which is used for transistor gates and interconnects. When the polysilicon layer is drawn over an active area, it forms the gate of a transistor.

(d) The fourth and fifth masks are the *n*- and *p*-implant masks, which define the areas that receive the *n*-type (arsenic ions) and *p*-type (boron ions) implants.

(e) The sixth mask defines the contact vias between the metal and the underlying polysilicon and diffusion areas.

(f) The seventh mask defines the metal layer, which is the predomi-

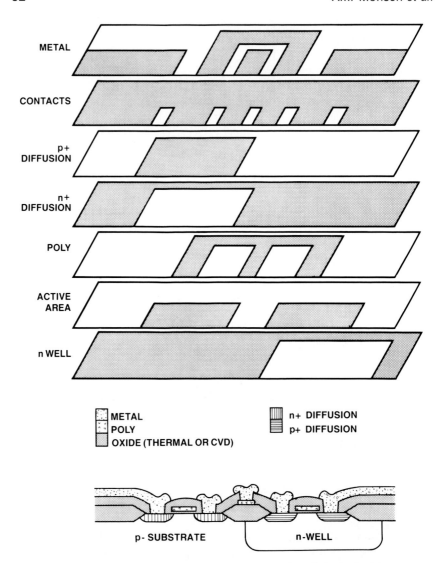

Fig. 22. The photolithographic layers and device cross section of a basic CMOS process.

nant interconnect layer used in most circuits because of its low resistance. At the end of the process, passivation dielectrics are deposited on the wafer to protect it from external contamination.

(g) The eighth mask provides holes in the passivation layer over the bonding pads to allow for probing and bonding.

Figure 23 illustrates the layout of a two-input NOR gate.

2. Fundamental Principles of Circuit Design

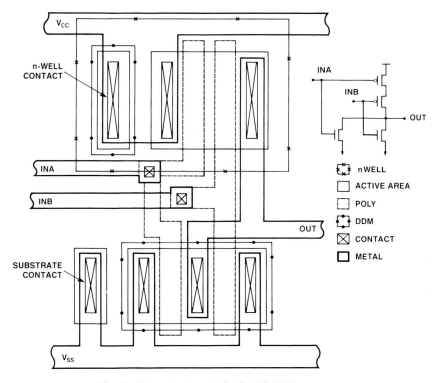

Fig. 23. Example layout of a CMOS NOR gate.

Automated Layout C

A large portion of present VLSI designs are laid out manually with interactive graphics; however, this will be changing with the rapid increase in VLSI design complexity and improvements in automated layout tools. Computer-based layout systems provide designs with faster turnaround and with guaranteed correctness. Available automated layout methods [10] (Fig. 24), that can translate functional specifications into geometric patterns of specialized regular physical structures include those listed below.

(a) *Symbolic layout tools* translate the lines and sticks of the designer into polygons that meet the design rules by construction [11]. The predominant application of symbolic layout is the design of random circuits that do not map efficiently to the other automated layout tools.

(b) In the *standard cell* scheme, the designer selects a set of cells from a library of predesigned cells and defines their interconnection [12]. The

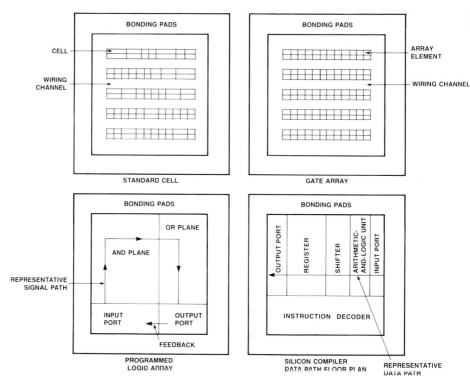

Fig. 24. Examples of automated layout tools: (a) standard cell place and route, (b) gate array, (c) programmable logic array, and (d) silicon compiler with data path floor plan.

standard cell system places the cells and routes the connection as necessary.

(c) Gate array layout systems select cells from arrays of tens of thousands of prefabricated, regularly spaced logic gates on the chip and establish wiring routes among them according to the specified logic functions [13].

(d) Programmable logic array layout systems are used to implement combinatorial logic equations and state machines by selecting the proper signals to be included in the AND and OR array planes and in the feedback path [8].

(e) Silicon compilers provide specification of the mask geometries from a functional description using some parametrized building-block structures and standard floor plans that can be modified to meet particular requirements for speed, power, or bit length. The silicon compiler scheme has been demonstrated on data path systems [14].

University and industrial researchers are working on new layout tools. Circuit synthesis programs are being developed to place and route transistor level structures from scratch [15]. Expert systems that apply artificial intelligence techniques are being developed to assist in the implementation strategies of automated layout tools and top-level architecture design that are now done by the human designer [16].

V FUTURE DEVELOPMENTS

VLSI technology has evolved very rapidly over the last two decades. This has forced VLSI design to deal with levels of complexity whose rate of growth shows no sign of slowing. The projected improvements in the semiconductor processing during the next decade will make it possible to create silicon systems whose complexity was unimaginable a few years ago. For example, in the near future microprocessors will be able to offer the functionality and performance of present main frame computers on a single monolithic chip. VLSI design will continue to refine its methodology to manage this unabated increase in complexity and to better utilize the improvements of computer-aided design tools. Computer-aided design tools with unified data bases will become available to improve the efficiency of the human designer interface. Presently the various simulators often run on different operating and hardware systems with incompatible data bases. This requires a lot of human interface to drive and translate the various design descriptions. Integrated automated layout tools will also merge to provide an environment where existing and new layout tools will work together and have compatible data bases. The implementation of hierarchical and structured approaches will be expanded to extend the capabilities of the design process and simulation tools. New VLSI design methodologies will be developed to manage the design of restructurable logic and wafer scale integration technologies [17] as they become economically feasible.

REFERENCES

1. G. Moore, Are we really ready for VLSI? *ISSCC Dig. Tech. Papers* 54–55 (February 1979).
2. A. Mohsen, VLSI device and circuit design, *Proc. Caltech Conf. Very Large Scale Integration,* 31–54 (January 1979).
3. C. Mead and L. Conway, "Introduction to VLSI Systems." Addison-Wesley, Reading, Massachusetts, 1980.
4. B. Honeison and C. Mead, Fundamental limitations in microelectronics—I. MOS technology, *Solid-State Electron.* **15,** 819–829 (1972).
5. W. A. Lattin, J. A. Bayliss, D. L. Budde, J. R. Rattner, and W. S. Richardson, A methodology for VLSI chip design, *Lambda* (2nd quart.), 34–44 (1981).
6. S. M. Sze, "Physics of Semiconductor Devices," 2nd ed. Wiley (Interscience), New York, 1981.

7. Y. El-Mansy, MOS device and technology constraints in VLSI, *IEEE Trans. Electron Devices* **ED-29** (4), 567–573 (1982).
8. H. Fleischer and L. I. Maissel, An Introduction to Array Logic, *IBM J. Res. Dev.* **19**, 98 (1975).
9. L. Terman, Mosfet memory circuits, *Proc. IEEE* **59**, 1044–1058 (1971).
10. S. Trimberger, Automatic chip layout, *IEEE Spectrum* (June) 38–48 (1982).
11. M. Y. Hsueh, Symbolic layout and compaction for integrated circuits, *Memorandum UCB/REL No. M79/80,* Electronics Res. Lab., Univ. of California, Berkeley, December 1979.
12. G. Persky, O. N. Deutsch, and D. G. Schweiker, LTX, A system for the directed automatic design of LSI circuits, *Proc. 13th Design Automation Conf.* (1976).
13. K. A. Chen, M. Feuer, K. H. Khokhani, N. Nan, and S. Schmidt, The chip layout problem: An automatic wiring procedure, *Proc. 14th Design Automation Conf.* (1977).
14. D. Johanssen, Bristle blocks: A silicon computer, *Proc. 16th Design Automation Conf.*, 310–313, (1979).
15. D. Knapp, J. Greanacki, and A. Parker, An expert synthesis system, *Int. Conf. Computer-Aided Design,* 164–165 (September 1983).
16. M. Stefik, J. Aikins, R. Balzer, J. Benoit, L. Birmbaum, F. Hayes-Roth, and E. Sacerdofi, The organization of expert systems: A prescriptive tutorial, Technical Report VLSI-82-1, Xerox PARC, (1982).
17. J. I. Raffel, On the use of programmable links for restructurable VLSI, *Proc. Caltech Conf. Very Large Scale Integration,* 95–104 (January 1979).

Chapter **3**

Design Automation for Integrated Circuits

AART J. DE GEUS
SYDNEY B. NEWELL

General Electric Microelectronics Center
Research Triangle Park, North Carolina

I. Introduction	37
A. Design	37
B. Fabrication	38
C. Design Automation	39
II. A Design through Various Levels of Abstraction	40
III. A Typical Design Procedure	40
IV. Semicustom Design Methodologies	44
A. The Standard Cell (Polycell) Approach	44
B. The Gate Array Approach	45
V. Building of Cell or Macro Library	46
VI. Semicustom Layout	47
VII. Comparison between Semicustom Methodologies	48
VIII. Trends in Design Automation	48
IX. Conclusion	50
Glossary	50
Bibliography	54

INTRODUCTION I

In the progression from idea to integrated circuit (IC), the work passes through two phases, design and fabrication. Although both phases can be automated, the concern here is with design automation.

Design A

Design encompasses all tasks before manufacturing and results in a set of masks that define the patterns for the layers of the IC. The design process can be subdivided as illustrated in Fig. 1. In the first major design task, *logic specification,* the designer creates a logic diagram that accurately represents the desired electronic function. In the second task, *layout,* the designer creates a physical description of the circuit that shows the locations and dimensions of all circuit elements and their interconnections on the chip.

B Fabrication

Fabrication, the physical creation of the IC, begins with the set of masks generated by the design process and ends with the finished IC, tested and packaged (Fig. 2). In the *processing* task, various multistep

Logic Specification

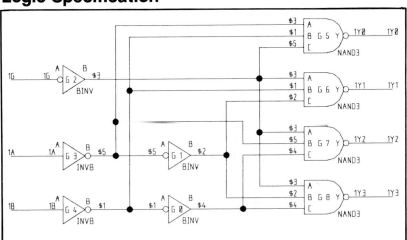

Layout

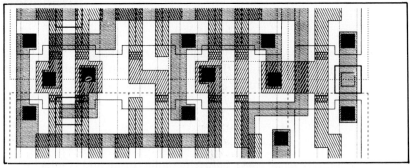

Fig. 1. Overview of IC design process.

3. Design Automation for Integrated Circuits

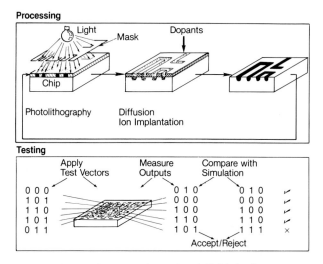

Fig. 2. The two major tasks of IC fabrication.

combinations of oxidation, mask protection, etching, diffusion or ion implantation, and vapor deposition take place to create the IC elements. Next, *functional testing* is done on the finished IC before and after packaging by applying test vectors, which are sets of inputs that thoroughly exercise the circuit. The IC is considered good only if the measured logical output values are identical with the expected output values.

Any error detected during testing will render the IC unusable. Because of the integral nature of ICs, a defective chip cannot be repaired; it must be discarded. Defects have two origins: in manufacture and in design. Manufacturing defects are random and affect a percentage of chips; design errors affect all chips, usually require redesign and refabrication at great expense, and cause delay in getting the chip to the marketplace.

Design Automation C

The main goals of design automation are

(a) a faultless first design that eliminates the need for corrective redesign;

(b) fast turnaround time, so that the chip will reach the marketplace quickly;

(c) small chip area, because a small decrease in area results in a large increase in yield;

(d) high reliability;

(e) performance, a selling point;
(f) low price, also a selling point.

The glossary at the end of this chapter provides definitions of some terms used in IC design.

II A DESIGN THROUGH VARIOUS LEVELS OF ABSTRACTION

Figure 3 shows various levels of abstraction used to describe the IC design process:

(a) description of circuit in large blocks and estimates of chip area;
(b) partitioning of blocks into smaller functional modules;
(c) description of blocks as logic gates and sequential elements;
(d) description of elements as transistors and parasitic elements;
(e) geometric plan showing how transistors and interconnections are to be laid out;
(f) two- or three-dimensional modeling at the device level and computation of electrical characteristics of the transistors;
(g) characterization of impurity profiles (dopant concentration as a function of doping depth).

III A TYPICAL DESIGN PROCEDURE

Figure 4 shows an idealized sequential set of major tasks in the progression from idea to IC. The order of the operations may vary according to the design approach chosen, and some tasks may be performed concomitantly to reduce the turnaround time.

The designer usually begins by providing an objective specification, that is, a detailed description of the intended behavior of the circuit to be implemented as an IC. Then, as in Fig. 4:

(a) The designer plans the global structure of the chip in large behavioral blocks.
(b) A quick estimate of the size of the chip is obtained, and tasks are partitioned between software and hardware.
(c) Behavioral simulation models both hardware and software and checks for design flaws.
(d) Large blocks are partitioned into smaller functions, usually the size of registers, adders, and counters. This level of the description is often referred to as the *register transfer level* (*RTL*).

3. Design Automation for Integrated Circuits   41

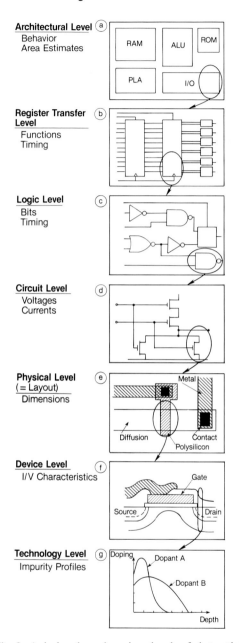

Fig. 3. A design through various levels of abstraction.

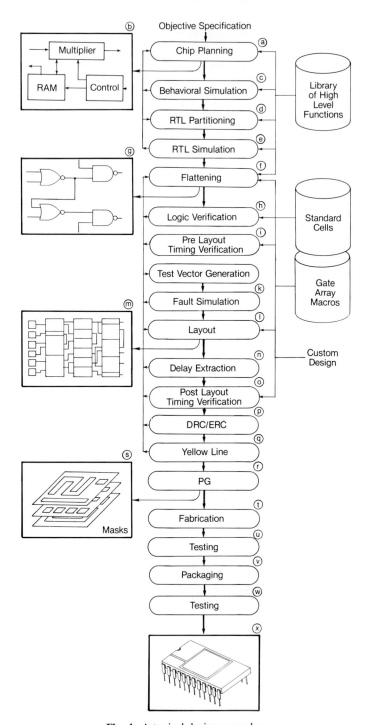

Fig. 4. A typical design procedure.

3. Design Automation for Integrated Circuits

(e)   A functional simulation at the RTL level is performed to verify the correctness of the design. For Tasks (a) through (e), designers can choose from libraries of tested high-level functional blocks.

(f), (g)   In the flattening process (f), functional blocks are expanded into their nonhierarchical gate level components (g). (For small designs, the entry point might be at the RTL, followed immediately by a flattening to logic components.)

(h)   Logic verification checks the flattened logic description for logical integrity and for potential spikes and race conditions. In logic verification, zero or unit propagation delay is assumed for each gate.

(i)   Pre-layout timing verification uses a timing simulator and/or a timing verifier with a set of rise and fall propagation delay times based on gate type and fanout. Zero delay (or a standard nominal delay) is used for interconnections because the dimensions of the interconnections are not known prior to layout.

(j)   Sets of input stimuli used in verification are refined into a set of test vectors that fully exercises the design. (Although automatic test generation (ATG) programs are in full development, most test vector sets are still conceived manually by an experienced designer familiar with the design in question.)

(k)   In fault simulation, stuck-at-one and stuck-at-zero faults are introduced one at a time at each circuit node while applying the test vectors. If the expected and simulated outputs are the same, then the fault is not detected, and the set of test vectors need to be refined. Normally, 100% coverage of all possible faults is too time-consuming to be practical; but often a lower percentage is used, or an attempt is made to change the state of each node at least once (node-wiggling).

(l), (m)   During layout (l), designers strive to optimize overall chip area, turnaround time, and performance constraints in generating the floor plan of the chip (m). Custom layout saves area but is expensive in turnaround time; it is cost-effective only for large volume chips. Semicustom approaches (e.g., standard cells or gate arrays) are more area-intensive but require less turnaround time. In full custom and in semicustom, critical path timing can usually be optimized by placing critical elements close to each other.

(n)   After the interconnect information is known, RC delays caused by resistance and capacitance of interconnect lines can be extracted from the layout.

(o)   These RC delays are then used in post-layout timing simulation and verification.

(p)   Design and electrical rules checking (DRC–ERC) is performed.

(q)   In the yellow line check, designers trace every interconnection with a yellow pencil for a final, manual verification of the design.

(r)   A pattern generation (PG) tape is made.

(s) The PG tape is used by the mask maker to generate the mask set.

(t) The design is fabricated using the mask set. In some semicustom methodologies, only a few masks and processing steps are needed to customize a chip, because most of the layers have been prefabricated.

(u) The chips are tested by applying the set of test vectors generated during the design process and comparing the measured output to the expected values.

(v) The good chips are packaged.

(w) The chips are tested again.

(x) The finished parts are ready.

IV SEMICUSTOM DESIGN METHODOLOGIES

A structured approach to integrated circuit design is a *design methodology*. *Semicustom* means that (a) some part of the design process has been done previously, independent of the particular circuit, and (b) some additional steps are then taken to customize the circuit for a particular application.

Two major semicustom design methodologies are

(a) the standard cell approach, using predesign to shorten the design process; and

(b) the gate array approach, using a combination of predesign and prefabrication.

A The Standard Cell (Polycell) Approach

The term *standard cell* was originally an abbreviation of *standard height cell;* a methodology in which the cells were of variable heights was referred to as a *cell array*. In current usage, *standard cell* usually means *standard functional cell* (where cells are part of a library of standard functions), and *cell array* means a partially preprocessed array of functions that is larger than the macros used in gate arrays.

In the standard cell approach, the designer uses a library of cells representing predefined logic functions, usually logic gates. These cells have been previously laid out using individual transistors; the cell dimensions, performance, and electrical characteristics have been optimized and documented. The designer prepares a logic diagram using the logic functions available in the cell library. The next steps, entering the logic design into the computer system and verifying by simulation, are the same as for any methodology. At the placement step, the designer places the standard cells in rows with variable-width channels between them for interconnections. Placement can be done manually or with varying degrees of computer assistance; routing is almost always done automatically. Every chip must go through the complete mask-making and fabrication process.

Advantages of the standard cell methodology are the following:

(a) *Rapid design turnaround time.* The cells are predesigned and need only to be laid out and connected.

(b) *Flexibility.* High-performance special-purpose functions (e.g., analog) can be implemented by creating a new standard cell.

(c) *Cell optimization.* Because cells are used more than once, time spent on area and performance optimization will be amortized over many designs.

Disadvantages of the standard cell methodology are the following:

(a) *Wasted chip area.* The area occupied by the wiring channels can exceed 50% of the internal chip (that is, exclusive of the I/O area).

(b) *No savings in fabrication time.* The complete fabrication process must be carried out.

The Gate Array Approach **B**

Gate array means an array (matrix) of transistors that can be wired as logic gates. In a gate array, a fixed number (4–20) of uncommitted (unconnected) transistors are pregrouped into a cell. These identical cells are arranged on a chip in rows with a fixed amount of space between them reserved for interconnections. Large numbers of these wafers are prefabricated and stockpiled. To implement a circuit, a designer specifies the metallization layers that will make the connections among the transistors and, therefore, implement the circuit on the array.

Like the standard cell methodology, the gate array approach uses a cell library (usually called a *macro library*) of logic elements predefined by templates that specify interconnections among the fixed transistors. The designer constructs the circuit, enters the design, and verifies it in much the same way as in the standard cell approach.

Advantages of gate arrays are the following:

(a) *Shorter design turnaround time.* The same as for the standard cell methodology.

(b) *Low cost.* Mass production and the need for fewer masks reduce costs.

(c) *Shorter fabrication time.* Only the final metallization and contact layers need to be customized. The initial processing can be done before design has even started.

Disadvantages of gate arrays are

(a) *Wasted chip area.* Gate arrays waste more area than standard cells, because the individual cells cannot be customized; in fact, some transistors may be unused, and entire cells may be unused because the fixed wiring channels may not accommodate total cell occupancy.

(b) *Decreased flexibility.* Fewer circuit functions can be realized, and analog functions are difficult to implement or may be too slow in performance.

(c) *Wiring restrictions.* Because of the fixed wiring-channel width, a circuit may contain too many logic elements for it to be routed.

V BUILDING OF CELL OR MACRO LIBRARY

The basic steps required to design a gate array macro or standard cell are similar to those for designing an integrated circuit. However, designers are often willing to take more time to optimize these logic designs and layouts, because the time spent will be amortized over many designs.

Cell design begins with an *objective specification,* as in Fig. 5, and then proceeds through *logic design, logic and timing simulation, circuit*

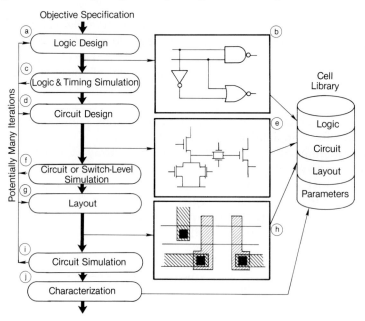

Fig. 5. Design of gate array macro or standard cell.

3. Design Automation for Integrated Circuits

design, and *circuit simulation*. The next steps are circuit *layout*, *extraction of parasitics*, and *post-layout simulation*.

The final step, *characterization*, is by far the most time-intensive activity of cell design. To characterize a cell, designers use simulation to obtain performance data at worst, typical, and best operating conditions of supply voltage, temperature, and processing parameters. Truth tables or state diagrams also need to be verified.

If the cell library is transported to the next lower-scaled technology with smaller device sizes, the logic, circuit, and layout descriptions remain virtually unchanged, but the cells must be completely recharacterized.

VI SEMICUSTOM LAYOUT

Figure 6 diagrams typical semicustom layout procedure. Starting with *verified logic,* modules are physically *placed* on the array and referenced

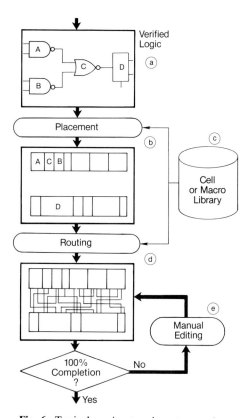

Fig. 6. Typical semicustom layout procedure.

to the cell or macro library. In the next step, *routing*, the physical interconnections among cells or macros are specified; manual *editing* may be required for any connections that cannot be made by the software.

In design automation systems, placement is often done manually by the designer but sometimes can be done partially or totally by software. Macros that are to be interconnected are kept as close to each other as possible by successful placement algorithms, and interconnections are distributed as evenly as possible.

Because routing was the most time-consuming and error-prone stage of integrated circuit design and because relatively simple decision-making algorithms are needed, the first efforts in CAD have focused on this phase.

The goal of an automatic router is 100% routing. Connections not made by the router must be made manually, which is time-intensive and error-prone. Normal completion rates are 99% for gate arrays; for standard cells, 100% routing is common, because the routing channel width can be varied.

VII   COMPARISON BETWEEN SEMICUSTOM METHODOLOGIES

Figure 7 shows a typical standard cell circuit and a typical gate array. Notice that the cells and the wiring channels in the standard cell circuit are of variable width. In the gate array, both cells and wiring channels are of fixed width.

Table I compares the three main design methodologies: full custom, standard cell, and gate array.

VIII   TRENDS IN DESIGN AUTOMATION

Major trends in design automation hardware and software are the following:

(a) The integration of individual tools into a coherent, user-friendly CAD system. The difficulty of this important task lies in maintaining coherence of the redundant data while building an efficient system. An integrated CAD system saves time and prevents errors by automating the data flow between tools. However, designing such a system implicitly requires expertise with all the individual tools and is considered one of the most challenging tasks that face software designers.

3. Design Automation for Integrated Circuits

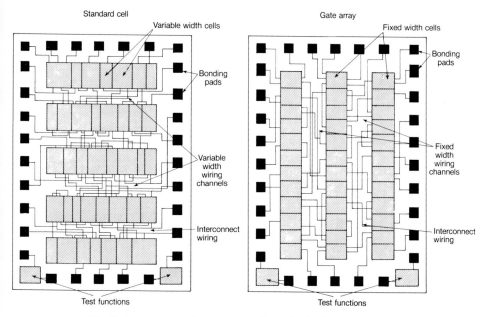

Fig. 7. Comparison of gate array and standard cell technologies.

(b) Workstations—low-priced microcomputers with powerful interactive graphics and complete, integrated design systems. These workstations usually share a central data base and are linked to each other and to a mainframe computer for CPU-intensive tasks. Industries doing semicustom design with little or no experience in the field find these workstations attractive, because they are turnkey systems that minimize the cost and learning curve associated with the development of a CAD system.

(c) More advanced CAD tools. The role of the computer is evolving from analysis into the performance of the synthetic tasks of IC design.

TABLE I

Comparison of Design Methodologies

Characteristic	Gate array	Standard cell	Full custom
Design time	Short	Short	Long
Fabrication time	Short	Long	Long
Turnaround time for minor redesign	Short	Medium	Long
Performance	Lowest	Medium	Highest
Chip area	Large	Medium	Small
Production volume at which cost-effective	Low	Medium	Large

New concepts in artificial intelligence will help computers make decisions based on the automatic interpretation of analytical results.

IX CONCLUSION

The goal of design automation is a coherent, user-friendly IC design system that requires little or no human intervention in the progression from idea to PG tape. Such a system is sometimes referred to as a *silicon compiler,* a term usually restricted to the layout phase of a design.

CAD developers strive to approach the ideal silicon compiler, but designer demands for increased complexity and performance continually move the ideal further from realization. Whether this moving target can ever be struck remains to be seen.

GLOSSARY

Automatic test generation (ATG)   software that automatically generates a set of test vectors for a particular circuit.
Best case analysis   a simulation using estimates for component performance based on the most optimistic temperature, power supply voltage, and process parameters. Best case analysis typically represents the fastest version of a circuit.
CAD   computer-aided design
Cell (standard)   a predefined logic function constructed from individual transistors. The dimensions, performance, electrical characteristics, and geometry of the cell have been optimized and documented.
Cell (gate array)   a group of a fixed number (4–20) of uncommitted (unconnected) transistors that can be efficiently interconnected to make logic gates.
Cell library   a software collection of standard cells or gate array macros.
Channel router   a routing algorithm used primarily for routing between parallel rows of standard cells. Routes are wired by using horizontal paths on one layer and vertical tracks on another layer. A variable channel width guarantees that all connections can be made.
Circuit simulator   software used for electronic circuit simulation and verification at the transistor level. Three basic modes of operation are generally available: direct current (dc), alternating current (ac) and transient. The dc analysis mode yields the steady state voltages and currents in a circuit; the ac analysis mode yields the frequency response of the circuit; and the transient analysis mode computes the response of the circuit in time. For digital designs, dc analysis might be used for gate characterization but transient analysis is the main tool used to verify circuit behavior. The most widely used circuit simulator is the program SPICE from the University of California at Berkeley.
CMOS   complementary metal oxide semiconductor (usually refers to a transistor or a technology using both NMOS and PMOS transistors).
Contact layer   a layer, defined by a mask, that specifies the connections between layers and circuit elements on an integrated circuit.

3. Design Automation for Integrated Circuits

Critical path   the signal path whose total delay limits the performance (speed) of the design.
Custom   a method of IC design in which each step is done individually for a particular circuit.
Custom layout   a method in which individual transistors are manually laid out on a floor plan that represents the integrated circuit.
Design (integrated circuit)   to transform a functional specification into masks that are ready for use in the fabrication of ICs.
Design automation   a design process in which both analytical and synthetic design tasks are performed automatically by computers. Computers carry out a design task by performing a series of iterations and are guided by the analytical results of each iteration to improve the design until the desired specifications are met.
Design methodology   a method of designing an IC.
Design rule checker (DRC)   software for ensuring that all design rules are obeyed by a layout description.
Design rules   a set of geometrical constraints that must be obeyed by the designer to guarantee functionality within the limits of the technology used. Examples of such constraints are minimum metal-to-metal or poly-to-diffusion spacing, minimum metal line or poly line width, or minimum contact areas.
Dies   individual ICs produced by sawing apart multiple ICs that have been fabricated onto a wafer.
Electrical rule checker (ERC)   software that is applied to a layout description to check for simple violations of electrical rules. An ERC also extracts a netlist that can be compared with the original schematic. Some ERC programs also extract transistor parameters (gate width and length) and information on the resistance and capacitance of interconnections.
Electrical rules   a set of constraints based on the electrical requirements of the circuit. Some simple electrical rules include forbidding any shorts between power and ground and requiring that every terminal on every element be connected.
Fabrication   the process of physically creating an IC.
Fault coverage   the percentage of all possible faults that can actually be detected by a set of test vectors. Except for life support applications, fault coverage is usually considerably less than 100% because of the large amount of CPU time required.
Faulted node   a node that is simulated as being stuck at one or stuck at zero to model a potential failure caused by a short circuit to power or ground, respectively.
Fault simulator   software that determines how completely a set of test vectors exercises the design. The design is simulated at the logic level using a set of test vectors while one node (or more in more sophisticated approaches) is faulted. If the output signals obtained through the simulation differ from the output signals that would have been generated by a nonfaulted design, the fault in question can be detected.
Flattening   a procedure by which a hierarchical description is expanded into its lowest-level components.
Floor plan   a diagram showing the location of the main modules on an integrated circuit.
Full custom chip design   a design process in which all elements are individually designed, requiring the full fabrication process for each element.
Gate (logic)   a device with one output and one or more inputs; the output state depends on the states of the inputs.
Gate array   a matrix of identical cells, each containing a number of uncommitted transistors. Gate arrays are prefabricated and stockpiled, and a circuit is constructed by specifying the interconnections among the transistors within and between the cells on one or more final contact and metal layers.
Hardware description language   a formal language used for encoding a network description into a CAD software system. The description usually contains the inputs and outputs of the network, the inputs and outputs of submodules contained in the network, and the interconnections among submodules.

Hierarchical design a design approach in which a large design is decomposed into several simpler, functional modules, each of which can be further decomposed into submodules, and so on until the lowest level submodules are simple enough to implement.

Integrated circuit (IC) a circuit contained on (or in) a continuous piece of solid material (usually silicon) called a die or chip. Components and wiring are fabricated simultaneously onto an IC. (In contrast, a printed circuit requires discrete steps for placing its components and wiring.

Interconnect to make wired connections among transistors or circuit elements on an integrated circuit.

Large scale integration (LSI) a level of circuit complexity in which an integrated circuit contains between 1000 and 10,000 transistors.

Layout a plan for the exact locations of the transistors on a chip and the wiring pattern that will interconnect them.

Logic simulator software that verifies the functional (logical) behavior of a digital circuit at the gate level. Logic simulators cannot be used to detect potential timing problems.

LSI (See large scale integration)

Macro a predefined logic element specified by interconnections among transistors contained in gate array cells.

Macro library a software collection of macros that defines the elements a designer may use in designing a gate array.

Mask a template, usually made of glass and an opaque substance, that determines the pattern for one layer of an integrated circuit. Cut-out portions of the opaque substance permit light to expose the photoresist that coats the underlying wafer.

Mask set the set of masks used for fabricating an entire integrated circuit. One mask is required for each layer.

Master slice (see gate array)

Medium scale integration (MSI) a level of circuit complexity in which an integrated circuit may contain between 100 and 1000 transistors.

Metallization the wiring, usually aluminum, among elements in an integrated circuit.

MSI (see Medium scale integration)

Netlist a description of the interconnections among the elements in a design.

NMOS n-channel metal oxide semiconductor (usually refers to a transistor or a technology).

Objective specification a document that specifies the required behavior and characteristics of a IC before it is designed.

Parasitics electrical elements, not planned in the design, that result from the real-world use of components in an IC; for example, capacitance to ground, and between elements; resistance of interconnects; and back-biased diodes.

Pattern generator a device that is used to make reticles that are used to make a mask. The pattern is generated by means of a file or tape that directs the flashing of a sequence of rectangular exposures onto a photographic plate.

Pattern generator (PG) tape a tape containing all topological information about the layout of a chip, used in a mask-making machine to generate the mask set.

PG tape (see Pattern generator tape)

Photoresist a photosensitive resin applied to a substrate, exposed to light, and developed to selectively mask certain areas from later etching.

Placement the process of laying out blocks of transistors with spaces reserved between the rows for wiring channels.

Polycell (see Standard cell)

Programmable array an array of repeated cells that are independent of any particular circuit implementation. Programmable arrays are partially fabricated chips that can be customized by modifying mask layers.

3. Design Automation for Integrated Circuits 53

Programmable logic array (PLA) a programmable array that contains two rectangular arrays of AND and OR gates. The gates in both arrays can be customized by specifying connections in the final metallization layers.

Real estate a slang expression that means area on the surface of a chip.

Reticle a photographic master copy of the layout of one system layer, usually at a scale ten times the final system chip size.

Routing the process of specifying wiring paths among modules on an IC.

Schematic entry the entering of a design schematic into a CAD system, especially through an interactive graphics terminal.

Semicustom an approach to integrated circuit design that involves some generic steps while still allowing the circuit to be customized for a particular application. Gate arrays and standard cell circuits are two well-established semicustom methodologies.

Silicon compiler software to generate automatically a layout from a functional specification. This concept represents the ultimate goal of design automation and has not been fully realized.

Silicon foundry a facility that fabricates ICs from a set of masks. No testing is performed.

Simulation computer-generated prediction of the behavior of a design or parts of a design. Simulators exist for all levels of the design process and are necessary to verify correct functioning of a design.

Small scale integration (SSI) a level of circuit complexity of up to 100 transistors.

SPICE a circuit simulator developed at the University of California at Berkeley. SPICE performs dc, ac, transient, noise, and sensitivity analysis.

SSI (see Small scale integration)

Standard cell (see Cell)

Stretchable cells standard cells that can be made longer or shorter to accommodate a floor plan.

Stuck-at-one a fault condition in which a node always has a value of logic 1 (high voltage) even though input stimuli are applied that should change the node value to 0. Stuck-at-one implies a short to V_{DD}.

Stuck-at-zero a fault condition in which a node always has a value of logic 0 (low voltage) even though input stimuli are applied that should change the node value to 1. Stuck-at-zero implies a short to V_{SS}.

Submodule a piece of logic smaller than the entire design.

Substrate the supporting material upon or within which an IC is fabricated or to which an IC is attached.

Synthesis the creative act of constructing a part of an IC design.

Test generation (see Automatic test generation)

Test vectors sets of values that, when applied to the inputs of an integrated circuit, generate outputs whose values for a properly working circuit are known.

Timing simulator software used to verify that speed requirements of the circuit will be satisfied and to detect the critical paths of the circuit.

Timing verifier software used without test vectors to verify timing requirements (e.g., setup and hold times, pulse width, and critical path) of a design. Timing verification can be used before layout, using fanout delays associated with each gate, or after layout, using both fanout and interconnect delays. Logic parameters are defined as *stable* or *changing* instead of *high* and *low*.

Top-down design (see Hierarchical design)

Verification the process of determining whether an electronic design will perform according to specifications.

Very large scale integration (VLSI) a level of circuit complexity in which an integrated circuit may contain more than 10,000 transistors.

Via a connection between metal layers of an IC.

VLSI (see Very large scale integration)
Wafer a disc-shaped piece of silicon, sliced from a cylindrical silicon crystal and used to form the basis of ICs. A wafer usually contains several hundred chips.
Worst case analysis a simulation that uses estimates for component performance based on the most pessimistic temperature, power supply voltage, and process parameters. Worst case analysis represents the slowest version of a circuit.
Yellow lining manual verification of a routed design by tracing all interconnections with a yellow pencil.
Yield the number of good dice on a wafer divided by the total number of dice.

BIBLIOGRAPHY

1. D. Eidsmore, Designer's guide to gate arrays, *Digital Design* (May), 60 (1983).
2. A. Rappaport, EDN semicustom IC directory. I: Technology, *EDN* **28** (4), 70 (1983); II: Specifications, *EDN* **28** (4), 99 (1983); III: Customer support, *EDN* **28** (4), 129 (1983).
3. R. Rozeboom, Capable software tools ease semicustom-IC design, *EDN* **4** (10), 185 (1983).
4. All issues of *VLSI Design*.

Chapter 4

Computer Tools for Integrated Circuit Design

WILLIAM V. LAWSON

Applicon
Burlington, Massachusetts

I. IC Design and Development	55
A. The Three Major Design Phases	55
B. Three Subphases	56
II. Applying Computers in the Development Process	57
A. Computer-Aided Engineering	58
B. Computer-Aided Layout	60
C. Computer-Aided Manufacturing	62
D. Computer-Aided Test	63
III. Availability of CAD Tools	63
IV. Conclusion	65
References	65

Modern LSI and VLSI circuits are too complex to design without computer aids, and even simple chips are more efficiently designed with computer-aided design (CAD) tools than by manual methods.

CAD tools have evolved from relatively simple programs that captured and prepared data for mask generation and have evolved into sophisticated computer-based systems that provide assistance at virtually all stages of IC design. Early CAD tools improved designer productivity; later CAD systems completely automate certain stages of the design process.

IC DESIGN AND DEVELOPMENT   I

The Three Major Design Phases   A

The IC development process, consisting of a sequence of activities and a transfer of design data between these activities, can be divided into three major phases: functional design, structural design, and manufacturing (Fig. 1).

B Three Subphases

Within each of the major phases, CAD tools can assist in design capture, design analysis, and design processing (Fig. 2). Design at each phase has one or more representations, which can be descriptive text, procedural descriptions (computer programs), graphics (pictures, block diagrams, schematics, or layout), or some combination of these. The design process involves a series of transitions between the various representations [1,2].

1 Design Capture

Design capture supports the creation and maintenance of the design representations. Because the result of the whole design process is the

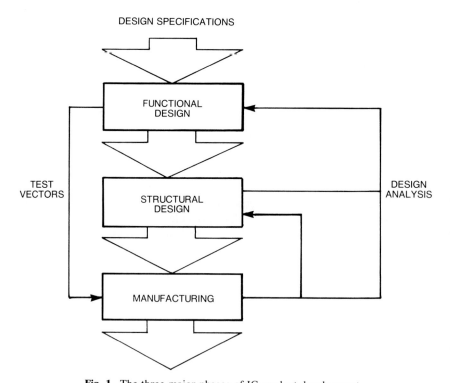

Fig. 1. The three major phases of IC product development.

4. Computer Tools for Integrated Circuit Design

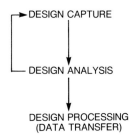

Fig. 2. Subphases of the three major phases of IC development.

geometric artwork, or mask, that is used to control the fabrication of the IC, there has been an historical tendency to equate all design activities with the manipulation of geometries by means of graphics editors. Though many CAD tools are indeed based on interactive graphics editors, many others are based on nongraphic representations.

Design Analysis 2

Design analysis tools improve the likelihood that the fabricated circuit will satisfy the original product specifications. Some design analysis tools evaluate the results of design capture and synthesis in each design phase by comparing them to the specifications for that phase. Other design analysis tools compare the functions and structures that are designed in one phase with the corresponding representations defined in another phase.

Design Processing 3

Certain design processing tools provide for the exchange of design information between the activities in a phase and between phases. Design processing activities include the output of reports and plots and the conversion of data to other design data-base formats.

APPLYING COMPUTERS IN THE DEVELOPMENT PROCESS II

A variety of CAD tools provide assistance throughout the IC development process: computer-aided engineering (CAE), computer-aided layout (CAL), computer-aided manufacturing (CAM), and computer-aided test (CAT).

The CAE and CAL tools are used throughout the functional and structural design phases (See Fig. 3), whereas CAM and CAT tools support manufacturing.

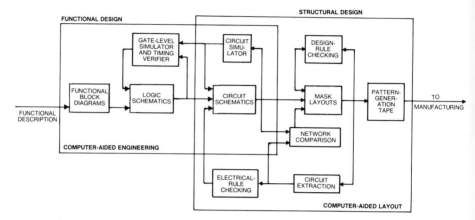

Fig. 3. Relationships among CAD activities.

A Computer-Aided Engineering

Computer-aided engineering (CAE) tools address the needs of engineers in systems, logic, and circuit design. Current CAE software provides tools for hierarchical schematic capture, netlist extraction, logic simulation, circuit simulation, and timing analysis (Fig. 4).

1 Schematic Diagrams

Interactive graphics editing systems provide effective tools for the creation of schematic diagrams, which are a means of graphically defining and

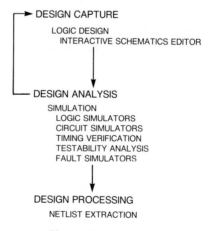

Fig. 4. CAE activities.

4. Computer Tools for Integrated Circuit Design

documenting the gate and switch circuits that implement the product design specifications. Circuits of LSI and VLSI complexity contain the equivalent of thousands of logic gates. Understanding a detailed schematic of even several hundred logic gates without any partitioning of the logic can be difficult; with LSI complexity such schematics become incomprehensible. To cope with complexity, design engineers partition designs into less complex functional blocks [3,4]. Very complex designs usually require further hierarchical partitioning. Engineering-oriented schematic capture tools support this hierarchical representation of the design.

Netlists 2

Early CAD tools for schematic design provided computer-assisted drawing that neglected the electrical nature of the design. Recent tools capture and maintain electrical connectivity information during the design process. Connectivity information, in the form of *netlists,* is the basic input to other CAE tools such as logic and circuit simulators. Netlist information is also used as input to computer-aided layout tools for automated placement and routing and as input to computer-aided test tools for fault-test generation.

Design Analysis Tools 3

Design analysis is essential to the success of later design activities. Tools for CAE-oriented design analysis support early evaluation of the design with respect to logic function and critical timing. Traditional design analysis involves breadboard testing, that is, the evaluation of the design by means of a hardware prototype. Breadboard tests provide the design engineer with direct control of analysis; unfortunately, these techniques are neither cost effective nor accurate for LSI design. Software simulators that provide similar analysis control are available and are gaining popularity. Simulators can be tailored to specific IC technologies, producing more accurate results than breadboard analysis does [5,6].

Logic Simulators 4

Logic simulators model combinatorial operations and simple timing conditions of a circuit. These simulators compare the connectivity information (netlists) extracted from the schematic with models of simple logic elements such as NAND, NOR, and inverter gates. Hierarchical logic simulation uses models of complex functional parts such as registers and flip-flops for comparison. The designer specifies a set of logical inputs, and the simulator calculates the logical outputs. Errors and unexpected results can identify problems in the design. Logic simulators can be used to model designs of VLSI complexity effectively.

5 Circuit Simulators

Circuit simulators model the electrical behavior of the entire design. These programs can accurately generate time waveforms of circuit behavior, but the analysis is costly in computer time and memory. General purpose circuit simulators are typically limited to the analysis of circuits with 300 or fewer gates [3,6]. Although not cost effective for the analysis of entire VLSI circuits, circuit simulation is important for analysis in the hierarchical design approach. Circuit simulations are used to optimize the design of the lowest-level design primitives and the critical functional blocks.

6 Timing Verification

Timing verification tools can provide waveform timing information with an accuracy that is within a few percent of the results of circuit simulation, with greatly increased execution speed. Circuit simulators require accurate solutions to a set of differential equations; timing simulators replace the equation models with delay table models that represent approximate solutions to the circuit equation models. Algorithms based on delay models can be used to analyze the timing of 100,000-gate circuits efficiently [3,6].

B Computer-Aided Layout

Computer-aided layout tools include interactive graphics editors, automated placement and routing software, layout rules checkers, and layout analysis software (Fig. 5).

1 Interactive Layout Editors

Interactive layout editors enable designers to enter the artwork for IC masks directly into computer memory, improving the efficiency of mask design by eliminating the time-consuming process of digitizing hand-drawn artwork.

Even with efficient interactive graphics, custom layout design is not feasible for entire VLSI circuits. As in schematic design, hierarchical

4. Computer Tools for Integrated Circuit Design

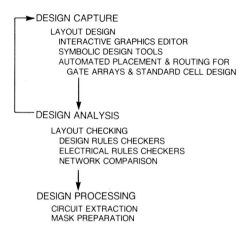

Fig. 5. CAL activities.

partitioning of the layout is used to manage the design complexity. In contrast to the *top-down* partitioning of the functional design phase, structural (layout) design is generally done *bottom-up,* combining basic circuit cells into higher-level functional blocks [1,2,3,4]. This approach to IC layout requires extensive manual design; designers, at workstations, custom design the structures of basic circuit cells and building blocks.

Automated layout, the creation of mask geometry with little or no human intervention, has been developed for certain IC technologies. Gate arrays and standard cell methodologies are the most widely used alternative to the full custom design of every component in a circuit.

Automated Placement and Routing 2

Automated layout methodologies involve the placement and interconnection of predefined structures. Automated placement and routing software for gate array and standard cell designs offer significant time savings over the custom design of equivalent circuits. These methodologies trade design time for the efficient use of silicon. They are best suited to the design of low-density circuits. Silicon compilers and structured design techniques are emerging as two strategies for the complexities of VLSI layout capture.

Layout Verification 3

Design verification is also an important part of the structural design phase. Early techniques of layout analysis required tedious and error-prone visual inspections of hand-drawn layout or computer generated plots; today automated aids are available for several types of layout verification.

Design rules checkers are the most extensively used CAD tools for layout checking. Determined by the requirements of the fabrication process, design rules indicate the size, spacing, and legal geometry configuration for each layer of a chip. Some design rules checking software is used to inspect an entire design and indicate errors for later correction. More recently, dynamic rules checkers have been coupled to interactive graphics editors to prevent errors from occurring in the first place.

Other design aids are used in the structural design phase to check the electrical correctness of the layout. Electrical rules checks find gross circuit errors such as shorts, unconnected components, and illegal connections. Software for electrical parameter extraction makes it possible for designers to detect circuit level defects such as resistance and capacitance problems in complex chips.

4 Circuit Extraction

Circuit extraction tools make another form of layout analysis possible: network comparison with the original logic circuit schematic. Circuit extractors analyze the layout structure and recreate a circuit netlist. Network-comparison software uses the extracted netlist to verify that the layout circuit is equivalent to the logic circuit. Other information such as interconnect length and node capacitances can also be extracted. Timing verification can be rerun with this structural design information to provide an accurate model of final circuit performance.

C Computer-Aided Manufacturing

1 Design Manipulation

A variety of software for design manipulations may be required to prepare layout design data for use in fabrication. The most common manipulations are

(a) sizing operations, which shrink or enlarge mask geometry to compensate for manufacturing processing effects; and

(b) Boolean operations, which derive special fabrication masks from logical combinations of the designed mask geometry.

Other design manipulation software converts the design data into the format required by pattern generators.

Pattern Generation 2

Early pattern generation output programs were among the first CAD applications. Through the 1960s, IC designs were simple enough so that designers could manually draw mask patterns on Rubylith and then photographically reduce them to create fabrication masks. By the early 1970s, due to advances in fabrication technology, feature sizes were too small to be manually prepared for accurate photoreduction. To solve this problem, computer-controlled optical pattern generation equipment was developed.

Manual preparation of data for pattern generation is impractical. Automatic geometry fracturing software converts mask layout geometry into optical and electron beam pattern generator formats.

Computer-Aided Test D

Automatic test equipment is essential for determining whether an IC has been manufactured correctly. Testing involves two activities: test generation, the creation of a set of test input sequences that detect logical faults in the complete chip, and test verification, the evaluation of the test program to determine that it will satisfactorily test the circuit. Although totally automatic test generation for VLSI is impractical, experienced engineers can develop test programs for individual chips that use the results of logic simulations done during the circuit design. For VLSI, test verification by fault simulation must be done during the functional design phase if the final test is to be cost effective. Testability analysis early in the structural design phase is also required if VLSI designs are to be effectively tested. Just as design methodologies have evolved to manage the complexity of VLSI design, other design disciplines are emerging to improve the feasibility of testing fabricated chips [3,5].

AVAILABILITY OF CAD TOOLS III

CAD tools for IC design are available in two forms: CAD software packages or software that is distributed with hardware-based CAD systems.

CAD software can be a single program or a collection of programs to support several design activities. Many IC software tools result from academic or industrial research and are sometimes distributed in the public domain. More often this software is distributed by commercial vendors. Examples of IC CAD software vendors are ECAD, NCA, and

Silvar–Lisco. Vendor-supplied software is frequently designed to supplement IC CAD systems supplied by other vendors.

A   Turnkey CAD Systems

Turnkey CAD systems, such as those from Applicon and Calma, are high-performance computer systems that provide all the hardware and software required to support the design activities of several users. These systems evolved as layout design tools and provide powerful interactive graphics capabilities. More recently, turnkey CAD vendors have extended the capabilities of these systems to include CAE tools.

B   Workstations

Workstations with sufficient processing power to support the design activities of a single user have emerged as popular hardware for the support of CAE software. Dubbed *CAE workstations,* these systems allow individual designers to use IC CAD tools in a distributed computing environment. Network software provides communication among multiple users and access to other computing resources. Some of these systems support mask layout activities for gate array design. Examples of CAE workstation vendors are Daisy Systems, Mentor Graphics, and Valid Logic.

C   Integrated CAD Systems

Neither the turnkey CAD system nor the CAE workstation provides the ideal computing environment for all design activities. The local computing power of a workstation is adequate for many interactive tasks; but other design tasks such as simulations, design rules checks, and test generation require host processors with greater computer power.

Until recently, these hardware-based distinctions were dictated by the available systems. Fully integrated systems, combining the aspects of turnkey CAD systems and CAE workstations, were developed only for in-house use by companies such as Bell Labs, IBM, and RCA. Integrated CAD systems that support a division of tasks between host processor and distributed computing resources are now being introduced. Future CAD systems will provide tools that operate on workstations and host processors, permitting more efficient and productive division of design tasks.

CONCLUSION IV

The design activities described in this chapter are essential elements of the IC design process. A large collection of effective CAD tools is currently available to support these activities, with new developments continually extending the capabilities and performance of these tools. Simultaneously, fully integrated CAD systems are emerging that offer improved performance and result in increased productivity; this marks the beginning of a new era in CAD for IC design.

REFERENCES

1. R. W. Hon and C. H. Sequin, A guide to LSI implementation (2nd ed.) *SSL-79-7, Xerox PARC* (January 1980).
2. W. W. Lattin, J. A. Bayliss, D. L. Budde, J. A. Rattner, and W. S. Richardson, *Utel Corporation*, pp. 34–44.
3. O. G. Folberth (ed.), Special issue on VLSI design: Problems and tools, *Proc. IEEE* **71**, No. 1, 1–166 (January) (1983).
4. C. Mead and L. Conway, "Introduction to VLSI Systems." Addison-Wesley, Reading, Massachusetts, 1980.
5. E. H. Frank and R. F. Sproull, Testing and debugging custom integrated circuits, *Computing Surveys* **13** (4), 425–451 (1981).
6. A. R. Newton, The simulation of large-scale integrated circuits, *Electronics Res. Lab. Report No. ERL-M78/52, University of California, Berkeley* (July 1980).

Chapter 5

VLSI to Go: The Silicon Foundry

ALFRED J. STEIN
GERI ALLISON
TONY VALENTINO

VLSI Technology, Inc.
San Jose, California

I.	Introduction	67
II.	The Silicon Foundry Concept	68
III.	The Foundry Interface	69
	A. Data Base Transfer	70
	B. The Design Interface	71
	C. Circuit Simulation	71
IV.	Processing	73
	A. Processing Mode Selection	73
	B. Technologies Available	75
V.	Post Processing	76
VI.	Conclusion	78
	References	78

INTRODUCTION I

Increasingly large numbers of system engineers would prefer to integrate their electronic designs on custom VLSI chips rather than depend on off-the-shelf components implemented on a printed circuit board. The benefits of custom VLSI are summarized nicely by Smith [1]: "Custom circuits offer several well-known advantages over equivalent designs implemented with circuits: lower power consumption, greater reliability, space savings, lower product assembly and test costs, and design security."

In the past, custom VLSI has been impractical for most system designers because of its extremely long development times, involving large design teams, and because of the high volume silicon commitment required by the silicon foundry manufacturers. Additionally, most manufacturers that would accept customer-owned tooling for processing custom integrated circuits were only purported silicon foundries. In fact, these facilities were usually an offshoot business for a standard parts house burdened by inactive processing lines. This meant that the enormous benefits of custom VLSI were not available to most small- and medium-sized customers. Recently, however, several silicon processing suppliers that can offer complete services have entered the market and are willing to address the needs of this growing marketplace.

II THE SILICON FOUNDRY CONCEPT

The term silicon foundry is defined succinctly by Jansen and Fairbairn [2]: "A silicon foundry is a facility which fabricates an integrated circuit from a design supplied by an independent party."

Because of the proliferation of advanced VLSI CAE–CAD systems, it is now possible for independent parties and custom houses to rapidly design custom circuits. The recent advent of several silicon foundries makes the production of such designs a practical reality.

Engineers are now taking advantage of these design tools and foundry services. It is estimated that the custom MOS marketplace is currently

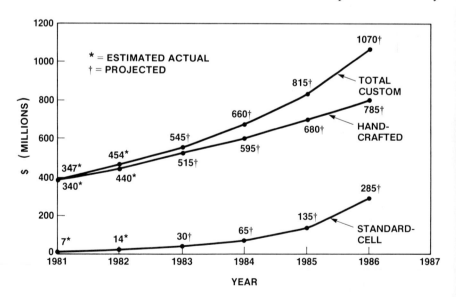

Fig. 1. U. S. market for custom MOS ICs (Source: HTE Management).

growing at a rate of 32% a year. At this rate, custom MOS will be a $1 billion business by the end of 1985 (Fig. 1).

A true silicon foundry is designed to profit from small as well as large orders and to support customers throughout the design and production of their circuits. The most advanced foundries provide a unique combination of semiconductor design and fabrication services for systems designers in all electronic industries; these services may include training in the use of and access to CAE–CAD tools. Modern silicon foundries also offer rapid turnaround of silicon processing, with typical cycle times of four to six weeks. This rapid processing is due in part to automated computer-controlled fabrication equipment and standardized interfaces.

THE FOUNDRY INTERFACE   III

A silicon foundry should support custom VLSI development far more than by merely being willing to accept customer-owned tooling. Carver Mead, professor of computer science and electrical engineering at the California Institute of Technology (Caltech), envisions silicon foundries that quickly and transparently process prototype and other small orders and that have standardized design rules, standard formats for data interchange that can be interpreted in only one way, and standard process control monitors for process evaluation [3].

Mead's view that foundries should be flexible and standardized promotes the possibility that customers can secure alternative sources for their custom circuits. This implies that foundries have developed formal agreements that cover the cross-licensing of technology, thus ensuring their customers that one set of design rules can be run in more than one foundry. Unfortunately for customers and foundries alike, such alternate sourcing agreements have not yet been formed, primarily because of inertia. Until they are formulated, it will be impossible for customers to be confident that their designs can be produced in more than one foundry. Typically, a customer requires at least two alternates to be sure that at least one back-up source of silicon is available.

A flow chart of a typical foundry interface is shown in Fig. 2. A foundry should present customers with good documentation that describes input and processing requirements. Basic process descriptions should include the technologies offered, wafer sizes, production minimums, prototyping information, and testing and packaging capabilities. The key criteria for customers is that the foundry be flexible in terms of the interface it presents, the processes it can run, the volumes it will accept, and the support services it provides.

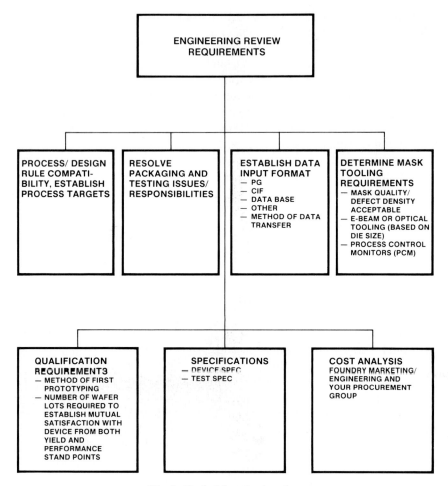

Fig. 2. Typical foundry interface.

A Data Base Transfer

A modern foundry accepts custom IC layout descriptions in a variety of formats such as hard tooling (mask sets or working plates), pattern generation (PG) tapes, and advanced data bases. These input methods are characterized by unique benefits and burdens.

E-beam 1X masks are often used for multiproduct wafer runs to simplify the placing of several different designs on a single wafer. Also, if a foundry has working plates, it can use them to verify performance and production compatibility with a small prototype run.

5. VLSI to Go: The Silicon Foundry

Pattern generation tapes are usually supplied to foundries in one of three formats: Mebes, David Mann, or Electromask [4]. Foundries prefer to accept design formats on PG tapes rather than data base tapes to avoid the need to convert the data base to an acceptable PG format.

Common *data base formats* include: Calma, Applicon, and Caltech Intermediate Form (CIF). Several vendors specialize in converting a data base from one format to another, including VLSI Technology, Inc., NCA, OCTAL, and ECAD.

VLSI Technology, Inc. provides an additional service by offering an electronic interface that allows circuit designs to be transmitted directly to the factory via a high-speed packet switching network. *VTInet*™, as the network is called, can be used to send design descriptions in the form of pattern generators, data base files, or Caltech Intermediate Format. The network is linked to the company's foundry production control and scheduling system so that customers can easily track the progress of their circuit.

B The Design Interface

There are a host of CAE–CAD systems dedicated to the various phases of VLSI design. The state-of-the-art is the integrated approach with compatible tools supporting the entire design cycle from systems functional description through implementation to design verification. These tools simplify design so that engineers can concentrate on the development of products. Today's designers are provided with tools for chip planning, logic design, physical layout, verification, and simulation. Ideally, such an integrated system should support custom design in several ways, including

(a) a library of flexible, standard cells,
(b) symbolic layout software for designing custom cells,
(c) a composition system for the layout and autorouting of the cells, and
(d) a layout editor for optimizing a design so that the highest performance and silicon utilization is obtained.

The product of this integrated design system should be a design data base that can be used to create the production mask.

C Circuit Simulation

The automated design functions described above need to be supported by simulation and verification routines to ensure that the fabricated circuit will function as intended. Several simulation packages are currently in use. Most are derivatives of SPICE, which was originally developed at the University of California at Berkeley. Along with SPICE, simulation packages include ISPICE (used mainly for bipolar models), HSPICE, ASPEC (a traditional MOS simulator), and R-CAP.

1 Process Specifications

A foundry's process specifications are the input to the simulation models. These specifications include layout rules for geometric design (state-of-the-art basic design rules are shown in Table I), processing limits, and basic electrical rules. This information is usually sufficient input for modeling; however, as geometries and design rules shrink, models must be more closely linked to processing.

2 Simulation Models

Hailey [5] recommends that designers ask foundries the following questions about their simulation models:

(a) Are models provided for minimum, typical, and maximum transistor sizes? MOSFETs are especially sensitive to the effects of narrow widths and short channels.

(b) Are models provided for high and low temperatures as well as for room temperature? Many simulators cannot predict performance over a wide range without special models for high-temperature effects.

(c) Are the models guaranteed over the full voltage range expected for the circuit design?

(d) Are the models characterized for all expected process variations?

TABLE I

Basic Design Rules

Rule	NMOS	HMOS I	HMOS II	CMOS
Transistor length (μm)	5	3	2	3
Gate oxide (Å)	850	650	400	500
Selective enhancement implant	NO	YES	YES	—
CMOS wells	—	—	—	n-well twin-well
Contact size (μm)	←——————— 3 × 3 ———————→			
Metal pitch (μm)	←——————— 8 ———————→			

5. VLSI to Go: The Silicon Foundry

Model Characterization 3

VLSI Technology, Inc. is in the process of characterizing the three models we currently use (SPICE, HSPICE, and ASPEC) by comparing predicted results with the actual performance of fabricated circuits. This method is expensive (especially if several iterations are required), but the results are precise. Many foundries provide customers with access to their simulation models. VLSI Technology, Inc. has integrated modeling with its CAE–CAD design system. For example, a SPICE file can be created directly from a circuit residing in our automated design system.

PROCESSING IV

Foundries fabricate silicon in two basic modes: dedicated and multiproduct wafers. *Dedicated-wafer processing* is traditional full-ownership semiconductor manufacturing. A single part type is repeated across the wafer (Fig. 3).

In *multiproduct-wafer processing,* individual customer designs are distributed across the wafer (Fig. 4). Tooling and wafer costs are significantly reduced.

Processing Mode Selection A

The multiproduct approach is attractive when producing prototype parts, and when making small production runs. The customer must decide when it makes economic sense to go from multiproduct- to dedicated-wafer processing. Production costs are not the only consideration. If volumes are low enough, the customer can test each part by inserting it in a trial target system to evaluate its performance. As volumes increase to the extent that dedicated-wafer production is warranted, a test program must be written and each chip tested at the foundry. A typical flowchart for multiproduct-wafer processing is shown in Fig. 5.

Process control monitors are stepped into dedicated and multiproduct wafers. A process control monitor (PCM) is a chip containing special process specific circuitry such as contacts, resistors, and transistors. The PCMs are tested for voltage thresholds, oxide breakdown, poly doping levels, and so on. They are used to verify that the processing is correct.

VLSI Technology, Inc. also places *canaries* or yield measurement monitors on a customer's wafers. A yield measurement monitor is a stan-

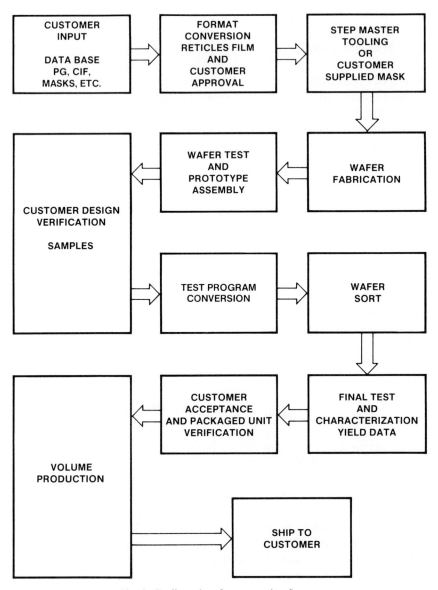

Fig. 3. Dedicated-wafer processing flow.

dard circuit, such as a 64K ROM, with a yield history. The yield history of the yield measurement monitor is compared with the actual yield obtained from a given production run to verify the soundness of the production process.

5. VLSI to Go: The Silicon Foundry   75

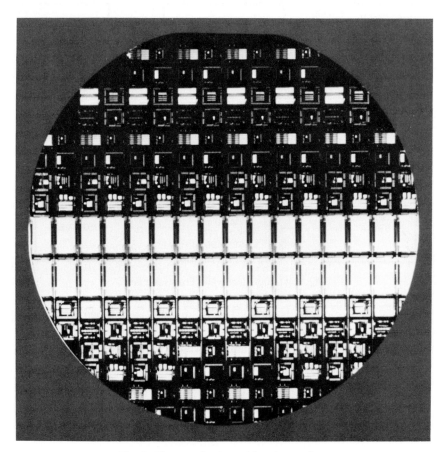

Fig. 4. Photograph of a multiproduct wafer.

Technologies Available B

Foundries are known for offering conservative processing technologies; however, a few, such as VLSI Technology, Inc., are now offering state-of-the-art technologies. The range of technologies currently available is shown in Table II. Whatever the geometries offered, foundries recommend that the designer not push the technology, in an effort to optimize performance, to such an extent that yield problems result. On the other hand, too conservative an approach can result in an inordinately large die with low yields (Fig. 6). The ideal is to offer up to state-of-the-art levels but not attempt the leading edge. For example, VLSI Technology,

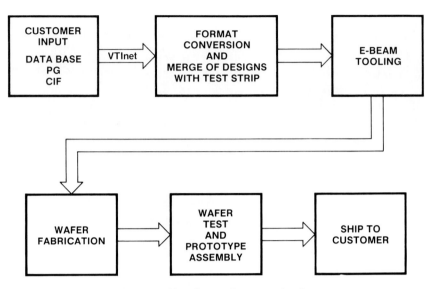

Fig. 5. Multiproduct-wafer processing flow.

Inc. is currently able to produce circuits with over 100,000 transistors. The company offers NMOS, HMOS, and HCMOS (n well, p well, and twin well) technologies. HMOS and HCMOS circuits are now processed with both 3- and 2-μm geometries. In late 1985, the company plans to offer both technologies at the 1.5-μm level.

V POST PROCESSING

Post processing of custom VLSI circuits includes testing and packaging. Customers can supply foundries with test programs or have the foundry generate a program based on device specifications and data pat-

TABLE II

Processing Technologies Available in Today's Foundries

5–7 μm metal-gate NMOS
3–5 μm silicon-gate NMOS
2–3 μm silicon-gate HMOS I and II
3–8 μm metal-gate PMOS
3–5 μm silicon-gate PMOS
3–7 μm metal-gate CMOS
2–5 μm silicon-gate CMOS
Bipolar technologies including I^{2}L, ECL, and Schottky

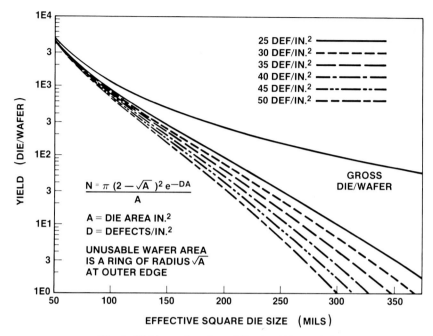

Fig. 6. Yield as a function of die size (4-in. wafer).

tern information. Table III contains a checklist for evaluating test programs before submission to a foundry.

Packaging alternatives range from high-volume, low-cost plastic for the consumer industry to specialized ceramic packaging for high-reliability applications. In addition to standard dual in-line plastic and ceramic packages, foundries also offer chip carrier and pin-grid arrays. The recent trend in custom VLSI packaging is toward higher pin counts. This trend is a reflection of the increased complexity of custom circuits. A reversal of this trend can be seen as chips become more highly integrated with greater on-chip intelligence (Fig. 7).

TABLE III

Test Program Evaluation Checklist

1. Check the software and assembly language program compatibility with foundry's tester.
2. Check the program test flow and functional correlation to standard devices and parts.
3. Check program and electrical specification for consistency; report any discrepancy.
4. Review for standard format (VTI) test time, binning, instructions, and software.
5. Review product engineering aids or yield analysis, gross functional tests, gross dc tests, wafer sort, and final test.
6. Check the program structure for margin analysis capability.

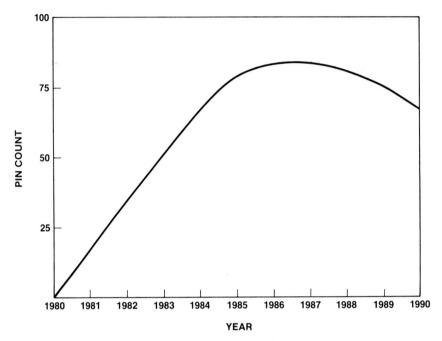

Fig. 7. Projected pin complexity.

VI CONCLUSION

Silicon foundries provide engineers with a practical route for implementing their custom VLSI designs. Independent parties are well served by the design tools, simulation models, processing technology, and post processing technology at foundries. The continued proliferation of these tools and services will make the custom VLSI approach even more attractive in the future.

REFERENCES

1. R. P. Smith, Mead–Conway: A method for systems in silicon, *Computer Design*, 221–225 (May 1982).
2. W. Jansen and D. Fairbairn, The silicon foundry: Concepts and reality, *Lambda* (1st quart.), 16–26 (1981).
3. C. Mead and L. Conway, "Introduction to VLSI Systems," 2nd ed. Addison-Wesley, Reading, Massachusetts, 1980.
4. S. McMinn, Semiconductor manufacturing considerations for VLSI designers, *VLSI Design*, 16–18 (July/August 1982).
5. S. Hailey, Interfacing with silicon foundries: A circuit-design perspective, *VLSI Design*, 52–54 (July/August 1983).

Chapter 6

Manufacturing Process Technology for MOS VLSI

F. W. VOLTMER
N. W. JONES

Intel Corporation
Santa Clara, California

I. Introduction	79
II. Directions in Process Technology	81
A. Horizontal Scaling	81
B. Vertical Scaling	83
III. Process Control	84
References	85

INTRODUCTION I

The current development of VLSI is being accompanied by significant changes in integrated circuit technology. Though these changes are, in general, evolutionary, they have a profound impact on manufacturing. In this chapter, the changes that are contributing to increased circuit density will be reviewed, directions in manufacturing process technology will be summarized, and the manufacturing consequences will be discussed.

The overall integrated circuit manufacturing flow is illustrated in Fig. 1. The flow can be separated into four parts:

(a) substrate preparation,
(b) circuit fabrication,
(c) wafer sorting, and
(d) assembly and test

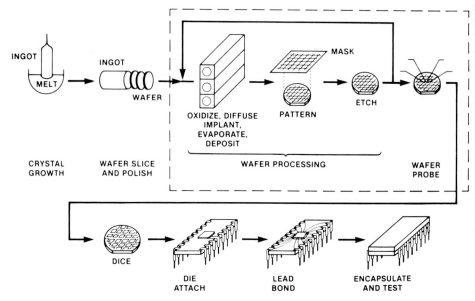

Fig. 1. Integrated circuit manufacturing process flow (reproduced from reference [1]).

It is advances in the wafer-processing part of the manufacturing flow that will allow continued increase in circuit density.

Figure 2 presents the historical trends and future projections for circuit density. The increase in circuit density is broken down into major elements:

(a) design cleverness,
(b) reduced feature size, and
(c) increased die size

The process changes that are associated with each of these elements lead to increased difficulties in manufacturing. This stimulates changes in manufacturing philosophy and techniques, especially in the areas of lithography, process control, and automation.

The wafer-processing part of the manufacturing flow can be considered as the repetitive sequential operations of film deposition, pattern definition, and patterning (see Fig. 1). Increased density through feature size reduction is causing changes in pattern definition and patterning technology (horizontal scaling), and as a necessary consequence vertical dimensions are also being reduced (vertical scaling). Improved facilities, equipment, and processes are reducing defect densities and are allowing the average die size to increase.

6. Manufacturing Process Technology for MOS VLSI

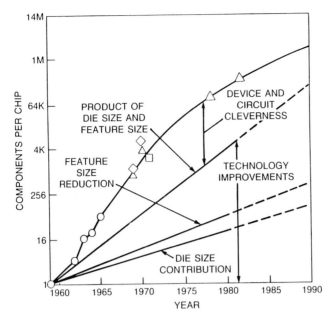

Fig. 2. Trends in integrated circuit density (after G. Moore).

II. DIRECTIONS IN PROCESS TECHNOLOGY

A. Horizontal Scaling

The use of horizontal scaling to achieve reduced feature size depends on the lithographic tools available and on resist technology. Figure 3 depicts the trends in minimum feature size, the current capability in production being about 1.5 μm. The trends in pattern definition techniques and the corresponding feature sizes are listed in Table I. The conversion from full wafer exposure to local field exposure through the availability and use of wafer steppers, along with positive resist and anisotropic plasma etching, has allowed the achievement of features of less than 1.5 μm in the resist and better than 0.5-μm registration. Extension of these techniques and the application of topology planarization will allow further reduction of feature size, both in the resist and on the wafer, to the submicron range.

Realizing <0.5-μm features and 0.25-μm registration will probably involve another change in the lithographic tools, with the current contenders being x-ray and e-beam lithography. These approaches have seri-

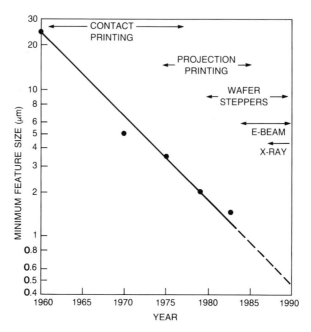

Fig. 3. Trend in minimum feature size (reproduced from reference [1]).

ous technical obstacles, and the transition to their use will not occur as quickly as the conversion from contact printers to projection printers or from projection printers to steppers. E-beam lithography is slow, costly, and limited by beam spreading in the resist. Suitable masks and resists are yet to be developed for x-ray lithography, and alignment is still done by optical techniques, limiting the improvements in registration.

TABLE I

Pattern Definition Trends[a]

Period	Feature size (μm)	Printing	Resist	Etching
1960s	24.0–6.0	Contact	Negative	Wet chemical
1970s	6.0–3.0	Contact	Negative/positive	Wet chemical
		Hybrid contact/projection		Plasma
		Projection		
Early 1980s	3.0–1.5	Contact	Negative/positive	Wet chemical
		Projection		Plasma
		Hybrid projection/stepper		
Late 1980s	1.5–0.5	Projection	Positive	Wet chemical
		Hybrid projection/stepper		Dry
		E-beam/x-ray		

[a] Reproduced from reference [1].

6. Manufacturing Process Technology for MOS VLSI

The etching requirements of patterned films for VLSI were realized through the conversion from wet chemical to dry plasma etching which, when properly carried out, results in anisotropic etching. Though the anisotropy of the etching aids in achieving smaller features, plasma etching is not as selective as wet chemical etching nor is it as insensitive to particulate defects. Thus, significantly more process control is required. Dry plasma etching, in fact, refers to a number of techniques including plasma, sputter, and reactive ion etching. Processes can be tailored by using the increased number of process variables available with dry etching.

Vertical Scaling B

Vertical scaling, the deposition of ever thinner films onto wafers is required by the continued device scaling and the associated smaller horizontal feature sizes. This requirement for thinner films results in the continued reduction of temperatures to maintain adequate control of thickness, especially for oxide. The deposition of high-quality, thin oxides at ever lower temperatures will require considerable development. The reduced distortion that is required for the smaller feature sizes is also driving the hot processing toward lower temperatures. As a result, high-temperature diffusion and thin-film operations are being replaced by low-temperature operations such as ion implant or CVD.

The use of ion implantation is continually expanding and has found greater use in forming source/drains, buried layers, and plugs, and for gettering. These applications are in addition to the technique's traditional use for resistors, isolation, and threshold adjustments. For high-dose ap-

TABLE II

Properties of Some Refractory Metals and Their Disilicides[a]

Material	Melting point (°C)	Electrical resistivity ($\mu\Omega$-cm)	Primary oxide
Si	1420		SiO_2
Al	660	2.8	Al_2O_3
Mo	2620	5.3	—
Ta	2996	13.1, 15.5	—
Ti	1690	43, 47	—
W	3382	5.3	—
$MoSi_2$	1870	21.5	MoO_3
$TaSi_2$	2400	8.5	Ta_2O_5
$TiSi_2$	1540	16.1	TiO_2
WSi_2	2050	12.5	WO_3

[a] Reproduced from reference [1].

plications, the cost of implantation is no longer so prohibitive because of the development of 10–15-mA "pre-dep" high-current ion implanters. The major technical difficulties that must be overcome still lie in maintainance of resist integrity under high beam current and problems in annealing the substrate damage caused by the implant process.

A major consequence of thinner films and smaller feature size is the increased impedance of interconnects, which results in longer RC time constants. Numerous refractory metal silicides are being investigated as replacements for polysilicon interconnects because they offer a tenfold reduction in sheet resistivity. Properties of some of the silicides is given in Table II.

III PROCESS CONTROL

Each of the contributors to increased circuit density (i.e., increased complexity, reduced feature size, and increased die size) necessitates more stringent process control. Sensitivity to defects increases as the die size and number of layers where defects are introduced increase and as

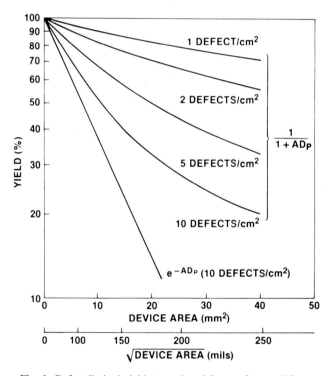

Fig. 4. Defect-limited yield (reproduced from reference [1]).

6. Manufacturing Process Technology for MOS VLSI

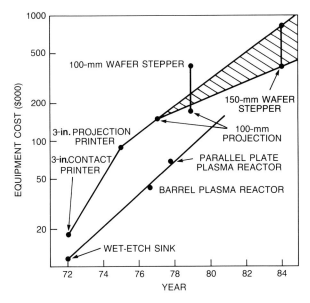

Fig. 5. Manufacturing equipment cost.

feature size decreases. Figure 4 depicts the results of calculations for two models for detect-limited yield:

$$Y = 1/(1 + AD) \quad \text{or} \quad Y = \exp(-AD),$$

where Y = yield, A = device area, and D = defect density.

For VLSI circuits on new technologies in which the die size is on the order of 40 mm^2 and the defect density is 5–10 d/cm^2, the initial defect-limited yield is 20–30%. The yield is further reduced by parametric yield losses. New techniques that reduce defect-limited yield and improve parametric control will be applied to the manufacturing area. Design improvements, such as the use of redundancy, will reduce the sensitivity of VLSI circuits to defects. Inspection tools will also necessarily change because current techniques, which use inspectors and microscopes, are inadequate to inspect large die areas for critical defects that have linear dimensions approaching 0.5 μm. Continued equipment automation and mechanization and reduced defect levels are necessary to overcome the low initial yield. As the tools for increasing circuit density become more complex, their cost increases substantially. Each new generation of manufacturing tools has been considerably more expensive than its predecessor, as shown in Fig. 5. Higher initial yields are necessary to support this increased cost and still allow continued reduction in the cost per function.

REFERENCES

1. N. G. Einspruich (ed.), "VLSI Electronics, Microstructure Science," Vol. 1. Academic Press, New York, 1981.

Chapter 7

Facilities for VLSI Circuit Fabrication

KIRBY G. VOSBURGH

General Electric Company
Research and Development Center
Schenectady, New York

I. Clean Air	88
A. Makeup Air	88
B. Air Specifications	89
C. Personnel Discipline	89
II. Water	89
A. Deionized Water	89
B. Cooling Water	90
III. Provision of Other Supplies	90
A. Gases	90
B. Electrical Power	91
C. Equipment	91
D. Materials	92
E. Personnel	92
IV. Vacuum	92
A. House Vacuum	92
B. Housecleaning Vacuum	92
C. Process Equipment Vacuum Pumps	93
V. Waste Disposal	93
A. Chemical Drainage	93
B. Fume Exhaust	93
C. Solvent Exhaust	95
D. Floor Drainage	95
VI. Physical Considerations	95
A. Room Materials	95
B. Vibration	95
VII. Protection of Personnel, Equipment, and Product	95
A. Personnel Safety	95
B. Equipment Fire Protection	96
C. Zones of Cleanliness	96
VIII. Personnel Efficiency	96
A. Work Flow	96
B. Personnel Amenities	97
IX. Facility Management System	97
References	97

The major factors in the design of a facility for VLSI circuit fabrication are the support of process equipment, process activities, personnel, and safety. We shall discuss these in terms of the various services provided.

I CLEAN AIR

Various approaches to clean air supply are shown in Table I. Air quality is measured by its freedom from particulate contamination and chemical carryover and its controlled temperature and humidity.

The choice of a clean air system depends on the activity in the fabrication area. For a factory with a fixed process, flexibility is less important than installation and operating cost, whereas the reverse may be true for a research or multipurpose facility. The sensitivity of the product to random particulate defects is the driving factor. Evolution of the system for capacity expansion or quality upgrade must be considered.

A Makeup Air

To maintain positive room pressure, the flow of makeup air must exceed the sum of the outward flow of fume exhaust and leaks from the

TABLE I

Approaches to a Clean Air Supply for VLSI Fabrication Plants

Type of Air System	Description	Characteristics
Global built-in	Full HEPA ceiling, full grille floor with returns	Costly to install and run Permits flexible placement of equipment Service penetration difficult
Partial built-in	HEPA filters over work areas, wall or floor returns	Low cost Inflexible work areas Service penetration easy
Modular stations	Stand-alone laminar flow units with plenum connections for makeup air and fume exhaust	Flexibility Complex controls High cost
Wafer channel	Wafers (or cassettes of wafers) that travel among equipment in isolated track environment	Requires full automation of wafer handling
Wafer transfer box	Wafers moved among equipment in high-purity box	Requires novel interfaces Robot automation possible

7. Facilities for VLSI Circuit Fabrication

room. Particular care must be taken in designing the ventilation for areas where toxic or process-contaminating materials are used; otherwise, contaminants will be recycled uncountable times past the wafers and personnel.

Intakes for room makeup air should be located upwind and as far as feasible from the scrubber fume exhaust. The main air-handling system must be more robustly designed than a conventional air conditioning system in two respects: (a) it must maintain specifications in temperature and humidity under essentially all external climate conditions, and (b) it must have a very low probability of unplanned downtime. These conditions imply a design with substantial reserve capacity and redundancy.

Air Specifications B

Typical air specifications for LSI/VLSI clean rooms are Class 100 (or better) for photolithography rooms and Class 1000 (or better) in other work areas [1]. Clean room class numbers refer to the number of particles greater than 0.3 μ in a cubic foot of air. Temperature is generally controlled to $\pm 0.5°C$ in lithography, $\pm 1°C$ elsewhere. Relative humidity control of $\pm 2.5\%$ will suffice for most applications; but as with all general facility specifications, the product, desired yield, and tooling should determine specific targets.

Personnel Discipline C

The amount of particulates in clean-room air is a function of the discipline of the process operators and of garb and cleaning procedures. All personnel who enter a clean room should be thoroughly trained in clean-room behavior; strict supervision and enforcement of rules are required to maintain cleanliness.

WATER II

Deionized Water A

Table II shows target specifications for pure water, as derived by Balazs and Poirier [2]. Deionized water systems are designed to remove dissolved minerals (by chemical treatment), particulates (by filtration),

TABLE II

Pure Water Specifications

Test	Concentration			
	Attainable	Acceptable	Alert	Critical
Residue (ppm)	0.1	<0.3	>0.5	>1.0
TOC[a] (ppb)	<50	<200	>200	>400
Particulates[b] (number/L)	<500	<1000	>2500	>5000
Bacteria (number/100 mL)	0	<5	>10	>50
Dissolved SiO_2 (ppb)	<3[c]	<5	>20	>50
Resistivity[d] (MΩ-cm)	18			

[a] Total oxidizable carbon

[b] In-line samples are counted at 100× on a Leitz Metallopian microscope. Batch sample counts are generally 4 times higher. All counts are subjective and can be used as relative values. Different counting methods may be relatively comparable.

[c] Detection limit

[d] In-line resistivity measurement should be 18 MΩ-cm 90% of the time, with 17 MΩ-cm minimum.

and organic growth (by UV sterilization and filtration). Recirculation loops can be used, though in-line monitoring and buffer storage are required to avoid contamination of the main supply. Point-of-use filtration is often employed, though it entails additional maintenance (weekly filter sanitizing). Deionized water systems must be kept operational, at all times to avoid contamination.

B Cooling Water

Several types of semiconductor process equipment require cooling water. Though requirements vary, this water is usually treated to avoid corrosion of equipment. The treatment chemicals, however, can make the water unsuitable for equipment, such as an rf sputter deposition system, that needs very low-conductivity water.

III PROVISION OF OTHER SUPPLIES

A Gases

Depending on the scale and intended use of the facility, high-pressure *nitrogen, hydrogen, oxygen,* and other gases may be needed throughout

TABLE III
House Gas Specifications

Gas	Purity	Dew point	Comments
Nitrogen	99.998%	−100°F	<5 ppm oxygen
Oxygen	99.96%	−90°F	
Hydrogen	99.999%	−100°F	<1 ppm oxygen <2 ppm nitrogen
Air	—	−100°F	100 psig at use point

the facility. Table III lists typical gas specifications. Specially cleaned pipe is required. Point-of-use filters can be used to lessen the chance of contamination.

Dewar flasks of liquid nitrogen can be located beside the equipment using the gas, or gas can be supplied through insulated or vacuum jacket pipe from a central source adjacent to the process areas. The Dewar approach is more flexible and avoids costly insulated pipe runs, but at the expense of increased operating costs, the inconvenience of changing Dewars, and the risk of contaminating the clean room.

B Electrical Power

The primary electrical power requirements are dictated by the choice of process equipment. It is not uncommon for equipment to have power requirements that differ from specification, because process tools are under continual developmental engineering. The specification and acceptance of such equipment require particular attention.

Many tools are controlled by microprocessors, and many pieces of equipment have power systems that use solid state switches. Switching transients of hundreds or even thousands of volts are thus encountered, even on nominally isolated lines, and the insertion of isolating transformers at each equipment is desirable. Effective rf shielding of equipment is important to protect microprocessors and other electronics from stray signals.

C Equipment

Since integrated circuit process technology evolves rapidly, the occasional replacement or addition of equipment should be anticipated. Ideally, a double-door air-locked room should be constructed. A separate clean air supply will be required so that the room can be run either clean or unclean. When new equipment is delivered, it can be immediately moved into this area, cleaned and prepared for installation, and then moved into the clean room with no perturbation of operations.

D Materials

Process materials can be transported through pass-through chambers or rooms. When possible, liquid or gaseous bulk materials should be supplied from an associated semiclean room through piping. Minimum storage, particularly of chemicals in glass bottles, should be maintained inside the clean room.

E Personnel

Access into the clean process areas must be strictly controlled. Except for emergency entry and egress, personnel should enter through a gowning area. Gowning and cleaning procedures should be implemented to meet the needs of the product. For VLSI fabrication, the body is fully covered, gloves are often worn, and face guards may be worn. The gowning area is designed to serve as an air lock so that positive pressure is maintained in the clean room. Air showers are generally used, as are shoe cleaners and tacky floor mats, to remove particulates before entering the clean area. Lockers and hanging racks should be provided for the storage of garments and personal effects. The gowning area should be designed so that it is difficult to enter the clean room unless the gowning sequence has been followed.

IV VACUUM

A House Vacuum

Thirty inches of Hg is usually sufficient for house vacuum.

B Process Vacuum

Many pieces of equipment (e.g., gas phase etchers) require roughing and backing vacuum. To lessen hazards to personnel and potential contamination, these should be fed to the fume exhaust. The use of mechanical pumps in the confines of the clean room should be avoided to minimize particulate contamination by rotating external parts such as belts.

C. Housecleaning Vacuum

An alternative to the traditional wet mopping and wiping of clean-room floors and walls is the use of built-in clean vacuum systems that have dirt receptacles and that exhaust externally. In the system design, care should be taken to avoid stirring up contamination. Hose runs should be short and air leaks avoided.

D. Process Equipment Vacuum Pumps

The vacuum pumps associated with process equipment (e.g., gas phase etchers and sputtering and evaporation apparatus) can be located away from the equipment in an associate semiclean space or basement to avoid contamination and simplify maintenance. The pumps should be exhausted to the fume exhaust, unless hydrogen may be present, for which external exhaust is preferable.

V. WASTE DISPOSAL

A. Chemical Drainage

Liquid waste must be collected, stored, and treated in a fashion consistent with environmental laws and standards, which vary with location and time. The waste material contains metals, salts, and organic solvents and is similar to general chemical laboratory waste except that it is often highly dilute, because of the large number of water rinses in an integrated circuit fabrication process. Waste water treatment involves a dilute incoming stream that may make treatment control difficult. Hydrofluoric acid is used as an etchant, and appropriate piping (e.g., PVC or PVDF) and treatment procedures must be used.

B. Fume Exhaust

Gaseous chemical exhaust loading has become proportionately larger in comparison with liquid chemical waste as the process technology has shifted from MSI to VLSI, as a result of the use of gas phase etchers. These etchers (plasma, reactive ion, reactive ion beam, etc.) generate highly reactive discharge streams; they depend on their ability to produce gaseous products to remove the materials being etched. Typical etch gases and etch products that are encountered in the exhaust streams are listed in Table IV. In addition, organic materials, primarily from degradation of the photoresists used in masking operations, are seen in the fume exhaust from the etchers and from resist stripping equipment (plasma ashers).

The fume exhaust is used to isolate and protect personnel by establishing an airflow pattern that prevents accidental chemical release. For example, the housings for the mercury arc lamps in photolithographic equipment are exhausted in this fashion to prevent mercury contamination from (inevitable) lamp explosions. Likewise, when extremely toxic gases (arsine, phosphine, etc.) are used as dopant or implantation sources, the fume exhaust is employed. This use of the exhaust system, though practical for the primary considerations, does shift the safety problem to the centralized treatment of the exhaust. Here, at least, significant dilution of the contaminant does occur, but attention must be paid to the consequences of a major accident in this system. Fume exhausts are usually treated with water scrubbers. Several dangerous gases require special scrubber systems [3]. Fume exhausts and scrubbers must be kept operational at all times.

Fume exhaust ducting should be coated with an impervious material such as epoxy or constructed of a material such as fiberglass-reinforced plastic, and procedures employed for connection to the ducts should leave no metal surfaces vulnerable to chemical attack.

TABLE IV
Typical Etch Gases and Etch Products Present in Fume Exhausts

Etch gases	Etch products
Boron trichloride, BCl_3	Silicon tetrachloride, $SiCl_4$
Chlorine, Cl_2	Silicon tetrafluoride, SiF_4
Carbon tetrachloride, CCl_4	Aluminum chloride, $AlCl_3$
Tetrafluoromethane, CF_4	Molybdenum hexafluoride, MoF_6
Nitrogen trifluoride, NF_3	Perchloroethylene, C_2Cl_4
Hexafluoroethane, C_2F_6	Trichloroethylene, C_2Cl_3
Sulfur hexafluoride, SF_2	Methyl chloride, CH_3Cl
Fluoroform, CHF_3	Trichloroethane, $C_2H_3Cl_3$
Chloroform, $CHCl_3$	Hexachloroethane, C_2Cl_6
Hydrogen chloride, HCl	Carbon monoxide, CO
Dichlorodifluoromethane, CCl_2F_2	Carbon dioxide, CO_2
Oxygen, O_2	Hydrogen chloride, Cl_2
Ammonia, NH_3	Phosgene, $CoCl_2$

C. Solvent Exhaust

Depending on the process technology, a separate solvent exhaust may be required.

D. Floor Drainage

All apertures in the floors should have lips to contain spills. Floor drains should be piped into the liquid chemical treatment system.

VI. PHYSICAL CONSIDERATIONS

A. Room Materials

Floors, walls, and ceilings should be of smooth nonshedding material such as vinyl plastic. Special materials have been developed for cleanroom floors. Ceilings should be coated with a nonporous material. If a suspended ceiling system is used, the tiles should be covered on all surfaces (particularly edges). Secure and easy-to-install clips should be used to prevent tile "blow up" due to the positive room pressure. Light fixtures should be designed to avoid perturbation of air flow and to avoid dust accumulation.

B. Vibration

Successful control of vibration depends on overall attention to design. Among the factors to be considered are sources of vibration (such as the air conditioning fan motors), the degree of isolation of floor structures, the isolation of key equipment from the floor, and the sensitivities of the process steps to various vibration amplitudes and frequencies. Submicron fabrication with beam tools may require that low-frequency vertical vibration amplitudes be less than $0.3~\mu$.

VII. PROTECTION OF PERSONNEL, EQUIPMENT, AND PRODUCT

A. Personnel Safety

Primary safety hazards in an integrated circuit processing area include toxic gases, toxic and corrosive chemicals, and fire. Gases should be stored and used in separately ventilated and monitored cabinets designed for this purpose. Arsine, phosphine, and silane should be routed through double-walled, positive-pressure jacket welded piping with monitors in the jacket flow. Chemical safety is dependent on well-designed facilities and well-trained and disciplined personnel.

Safety showers and eye washers should be placed near wet chemical processing areas. This equipment should be tested periodically to make sure it is effective. Lest the acids and gases be thought to be the greatest potential problems, one should remember that many of the organic solvents used with photoresists are highly toxic or flammable. Solvents, acids, and other chemicals should be transported and stored in appropriate containers, and the possibility of explosive or toxic interactions should be decreased by proper segregation of containers. Fire safety is dependent on access to exits from the clean room (normally mandated by fire codes). Procedures for handling false alarms (garb policy, for example) should be developed.

B Equipment Fire Protection

For electronic or sensitive optical tools, consider chemical fire extinguishers (such as Halon) rather than sprinklers.

C Zones of Cleanliness

Whenever possible, the areas of high cleanliness should be restricted to active process workstations. This reduces cost and alleviates the chance of airstream contamination. This can be accomplished by installing walls just behind the operating face of processing equipment so that the bulk of the equipment is located literally in a separate room. In addition to the benefits in the clean room, such an arrangement permits access to the back of the equipment for maintenance without having to enter the clean area or risk contamination by the maintenance activities. The design tradeoff, however, is that such arrangements are far less flexible in layout than open areas.

VIII PERSONNEL EFFICIENCY

A Work Flow

Most integrated circuit process facilities involve a circular flow of material around a facility; rarely is a production line truly linear. Thus the facility layout should be planned to optimize this flow; a common layout would place a photolithography suite in the center of a roughly circular facility. This concept has the additional advantage that the cleanest areas are isolated from the facility exits.

Personnel Amenities B

Clean-room operation makes severe demands on personnel. In addition to the stringency of execution of the process procedures, behavior is circumscribed in other ways. Forbidden activities include smoking, eating, and wearing cosmetics. Thus, attention should be paid to the psychological support of the operators. Areas for breaks should be located in proximity to the facility entrance. "Goldfish bowl" windows should be avoided. The fabrication area should be made as visually open as possible; the use of glass for air return ducts is helpful.

Since access to the facility is strictly controlled, ample means of communication should be provided by a conveniently placed intercom system. Intercom design should take into account the possibility of a higher than normal level of background noise due to high room air flow and to operating equipment.

FACILITY MANAGEMENT SYSTEM IX

A computer-based facilities management system should have the following capabilities.

(a) Fire protection
(b) Facility security/access
(c) Life safety: toxic gas alert, evacuation, and lighting
(d) Clean air system: temperature in clean areas, humidity in clean areas, sensors for control, and operational transducers
(e) Interface with process management system: process status, facilities information data logging, and process alerts
(f) Miscellaneous monitoring: gas bottle weight

REFERENCES

1. P. R. Austin, "Design and Operation of Clean Rooms." Business News Publ. Co., Troy, Michigan, 1970.
2. M. Balazs and S. Poirier, Those confusing pure water 'specifications': Setting the record straight, *Microelectronic Manufacturing and Testing*, 22 (February 1984).
3. G. K. Herb, R. E. Caffrey, E. T. Eckroth, Q. T. Jarrett, C. L. Fraust, and J. A. Fulton, Plasma processing: Some safety, health and engineering considerations, *Solid State Tech.*, 185–193 (August 1983).

Chapter 8
MOS VLSI Circuit Technology

CHENG T. WANG

Department of Electrical and Computer Engineering
College of Engineering
University of Miami
Coral Gables, Florida

I. Introduction	99
II. MOSFET Structures	100
A. NMOSFET	100
B. PMOSFET	102
C. CMOS Transistors	102
III. MOS Circuits	103
A. The NMOS Inverter	103
B. The CMOS Inverter	104
C. The NMOS Gate Circuit	104
D. The CMOS Gate Circuit	105
IV. Power-Delay Performance of MOS and Bipolar Circuits	107
V. Conclusion	107
References	108

INTRODUCTION   I

MOS integrated circuits are perhaps the most important circuit technologies in VLSI. The application of the scaling law not only reduces the size of the MOSFET (metal-oxide semiconductor field-effect transistor), allowing a higher packing density, it also sharply increases the performance of the transistor. But the performance of its bipolar counterpart increases only gradually when the device dimensions are similarly reduced. This allows the speed of the MOS circuits to gradually catch up to that of the bipolar circuits.

The most important type of MOS circuit is CMOS (complementary MOS), which dissipates little power in quiescent states (logic 1 or 0) and dissipates appreciable power only when it is switching between logic 0 and 1. This low power consumption is a very attractive feature in VLSI. Because hundreds of thousands of transistors are packed into a tiny chip, the removal of the heat generated by the circuits has always been a big concern for engineers.

II MOSFET STRUCTURES[†]

A NMOSFET

Figure 1 shows a typical processing sequence of a self-aligned silicon gate NMOSFET. [When the gate oxide is thinner than 700 Å, and the channel length is less than 3 μm, designations such as HMOS (high-performance) and SMOS (scaled MOS) are used by various manufacturers.] The steps in a typical process sequence are the following.

(a) A patterned silicon nitride (Si_3N_4) layer, placed on top of the p substrate, defines the boron (p^+) field implant region which serves the purpose of isolation.

(b) The field oxide is grown over the p^+ region, resulting in a recessed, nonplanar oxide. This is known as the local oxidation of silicon (LOCOS) process, which can result in ICs with higher packing densities and better performance. Planar oxidation techniques such as SWAMI [2], which does not have the bird-beak structure of the LOCOS process, have also been reported. These techniques may lead to an even higher packing density.

(c) Si_3N_4 is removed, and the gate oxide is grown. The polycrystalline silicon gate electrode is deposited and patterned. An n-type dopant is implanted to create the drain and source regions and to increase the conductivity of the poly-gate. Because implantation of n-type dopant uses the poly-gate as the mask, it is known as a *self-aligned process*.

(d) SiO_2 is deposited, and windows are opened for the contacts. This final step includes metal deposition, interconnection patterning, and passivation.

1 Enhancement Mode Devices

Without the gate bias, the path between the drain and source has two n^+p diodes, back-to-back in series. The only current that can flow when

---

[†] See Hodges and Jackson [1].

8. MOS VLSI Circuit Technology

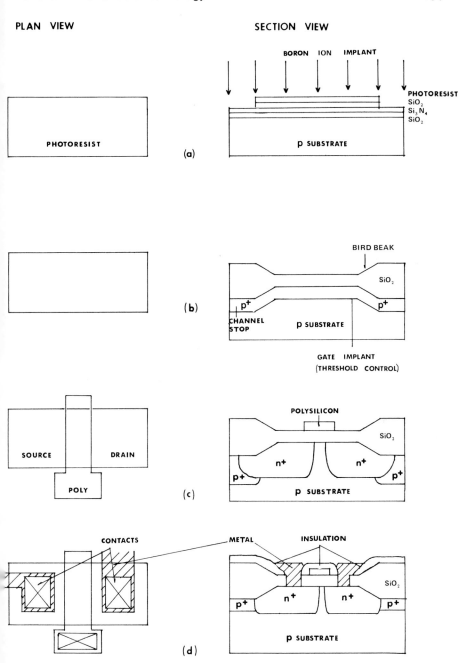

Fig. 1. Process sequence for NMOSFET (see text).

the drain is biased is the very small diode leakage current; so the transistor is considered to be *off*. As the gate bias increases to a certain positive voltage, it induces electrons at the oxide–silicon interface and creates an inverted (*p*- to *n*-type) conduction channel. Appreciable current flows through this channel when the drain is biased; so the transistor is considered to be *on*. The gate voltage that turns on the transistor is known as the *threshold voltage*. This transistor is a normally-off device; it is called an *enhancement-mode device* because a positive gate voltage will enhance the conductivity of the channel.

2 Depletion-Mode Devices

A device that is normally on is called a *depletion-mode device*. It has an *n*-type surface channel. Negative gate voltage is used to deplete the conducting channel and turn off the transistor.

B PMOSFET

Figure 2 shows a PMOSFET. The preceding discussion of NMOS also applies to PMOS, if the threshold voltage polarities and the designations *n*-type and *p*-type are interchanged. PMOS is seldom used alone as the active element in VLSI because of its low hole mobility.

C CMOS Transistors

Figure 3 is the cross-sectional view and circuit symbol for the CMOS transistor, which combines a *p*-channel and an *n*-channel device inside the *p*-well. If the gates of the two devices are connected, only one FET can be conducting under dc conditions because of the different threshold voltage

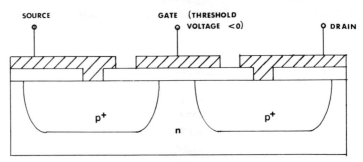

Fig. 2. Cross-sectional view of a PMOS transistor.

8. MOS VLSI Circuit Technology

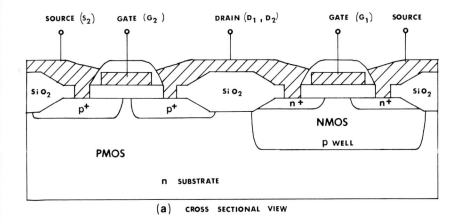

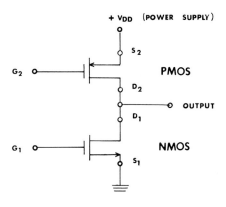

Fig. 3. CMOS transistors.

polarities of these two devices. Other structures such as *twin-tub* (*p*-well and *n*-well) [3] and *stacked CMOS* [4,5] (PMOS is on top of NMOS, also known as a *three-dimensional device*) have also been reported.

MOS CIRCUITS III

The NMOS Inverter A

Figure 4a shows the NMOS inverter. Q_2 in the NMOS circuit is a depletion-mode transistor that serves as the active load. With the high

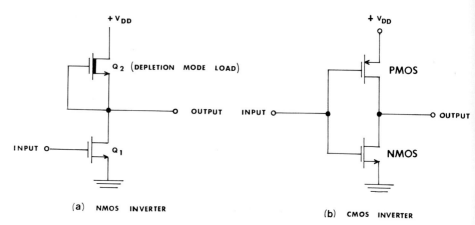

Fig. 4. NMOS and CMOS inverters.

input voltage (logic 1), Q_1 is on. A large current can flow through Q_1 and Q_2 and cause a large voltage drop across Q_2. Thus, the output voltage, which is the voltage difference between the power supply voltage V_{DD} and the voltage across Q_2, is low (logic 0). Conversely, if the input is low, Q_1 is off. The small voltage drop across Q_2 ensures that the output voltage is almost clamped to V_{DD} (logic 1). Because of the current flow, power is dissipated in quiescent states, as well as during the switching.

B The CMOS Inverter

In the CMOS inverter circuit (Fig. 4b), the gates of PMOS and NMOS are connected. Due to the polarity difference in the threshold voltages, one of the transistors is off in quiescent states, and no current can flow through the circuit. The power consumption of a CMOS gate circuit thus approaches zero under dc conditions. However, current is drawn from the power supply to charge the output capacitive load as it switches from low to high. The power consumption of CMOS circuits thus increases linearly with the switching rate of the data.

C The NMOS Gate Circuit

Figure 5 shows typical NMOS gate circuits. In Figure 5(a), the output is high only when both inputs are low (Q_1, Q_2 off), which is a *NOR function*. Additional input can be added to this circuit by connecting an NMOS in parallel with Q_1 and Q_2. In Fig. 5(b), the output is low only when Q_1 and

8. MOS VLSI Circuit Technology

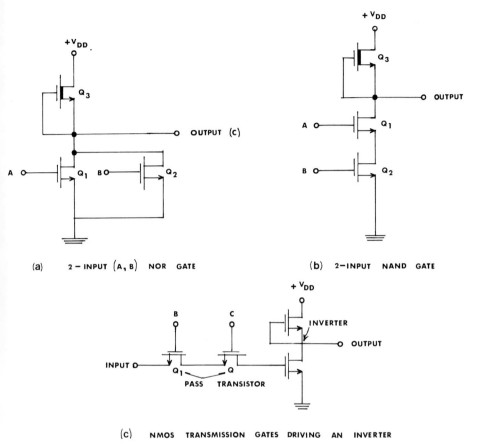

Fig. 5. NMOS gate circuits.

Q_2 are both on, which is a *NAND function*. Additional input can be incorporated if we connect a NMOS in series with Q_1 and Q_2. Figure 5(c) is an *NMOS transmission gate* circuit. The input signal travels through the pass transistors Q_1 and Q_2 if the control signals at B and C are both high. This allows a timing control for the logic circuit. Transmission gates (pass transistors) are frequently used in the memory circuits and sequential circuits such as the latches to control the signal flow.

The CMOS Gate Circuit   D

Figure 6 shows typical CMOS gate circuits. In Fig. 6(a), additional input can be added to the NOR gate circuit by connecting NMOS in parallel with Q_1 and PMOS in series with Q_3. For a NAND gate, addi-

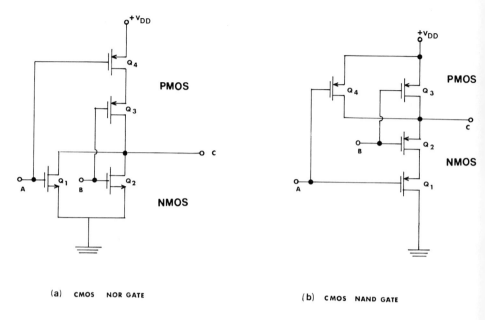

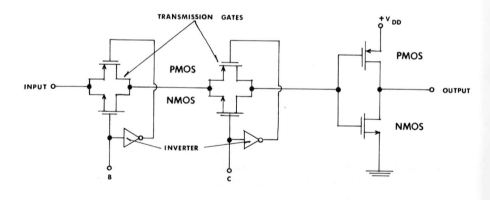

Fig. 6. CMOS gate circuits.

tional input can be incorporated by connecting NMOS in series with Q_1 and PMOS in parallel with Q_3. As mentioned before, CMOS has become the most important technology in VLSI because of its low power consumption. With scaled device dimensions, CMOS circuits can also achieve a very high switching speed.

8. MOS VLSI Circuit Technology

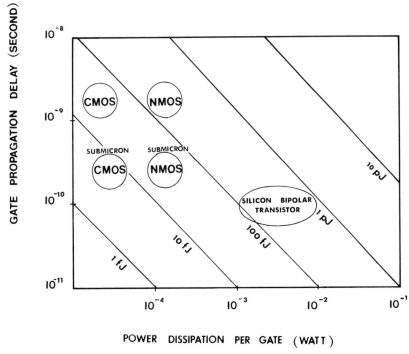

Fig. 7. Power-delay continuums for the state-of-the-art VLSI circuit technologies.

IV POWER-DELAY PERFORMANCE OF MOS AND BIPOLAR CIRCUITS

Figure 7 compares the power-delay performances of MOS and bipolar circuits. As can be seen, submicron CMOS (channel length is smaller than 1 μm) is an excellent choice for VLSI and has a subnanosecond gate delay time and a power consumption on the order of only 10^{-5} W per gate. One serious drawback for MOS circuits, however, is the smaller off-chip current-driving capability, which limits the off-chip data rate to below 20 Mb/s. For this reason, bipolar silicon circuits are still dominating in high-speed mainframe computers and signal processing systems; however, technologies such as gallium arsenide (GaAs) FET, which has higher electron mobility, are being evaluated.

V CONCLUSION

The development of MOS circuit technologies has made the logic and memory cheap, which ushers us into the VLSI era. As the processing techniques continue to improve, we can expect a million or more gates on the chip. With the technologies such as the submicron CMOS, we will not only have a densely packed but also a high-performance circuit. However, as the circuit density increases, the problems associated with packaging and on-chip interconnections may become the limiting factors towards the higher integration. Also, semi-custom or custom design methodologies will have to be used to handle circuits of this complexity.

REFERENCES

1. D. A. Hodges and H. G. Jackson, "Analysis and Design of Digital Integrated Circuits." McGraw-Hill, New York, 1983.
2. K. Y. Chiu *et al.*, The sloped-wall SWAMI—A defect-free zero bird's-beak local oxidation process for scaled VLSI technology, *IEEE Trans. Electron Devices* **ED-30**, 1506–1511 (1983).
3. L. C. Parrillo *et al.*, Twin-tub CMOS II—An advanced VLSI technology, *IEDM, Tech. Dig.* 706 (1982).
4. T. T. Gibbons and K. F. Lee, One-gate-wide MOS inverter on laser-recrystallized polysilicon, *IEEE Electron Devices Lett.* **EDL-1** 117–118 (1980).
5. J. Colinge *et al.*, Stacked transistor CMOS (ST-MOS), an NMOS technology modified to CMOS, *IEEE Trans. Electron Devices* **ED-29**, 585–589 (1982).

Chapter 9

Bipolar VLSI Circuit Technology

CHENG T. WANG

Department of Electrical and Computer Engineering
College of Engineering
University of Miami
Coral Gables, Florida

I. Introduction	109
II. Bipolar Transistors	109
A. *npn* and *pnp* Transistors	109
B. The Schottky Barrier Transistor	112
III. Bipolar Digital Gate Circuits	113
A. TTL Circuits	113
B. ECL Gate Circuits	115
C. CML Gate Circuits	116
D. IIL (MTL) Gate Circuits	116
E. STL Gate Circuits	119
F. Comparison Logic Circuits	119
References	120

INTRODUCTION   I

The silicon bipolar junction transistor (BJT) has been one of the most widely used devices in IC technology. Its high bulk electron mobility made high-speed applications its exclusive province until the scaled-MOS technology recently emerged.

BIPOLAR TRANSISTORS[†]   II

npn and *pnp* Transistors   A

Figure 1 shows a typical *npn* transistor. The typical process sequence is as follows:

[†] See [1,2,3].

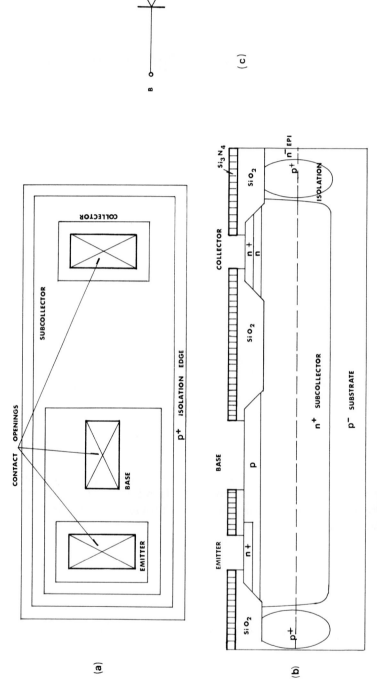

Fig. 1. The *npn* bipolar transistor: (a) top view, (b) cross section, (c) circuit symbol.

9. Bipolar VLSI Circuit Technology

(a) The device is fabricated on a *p*-substrate. Windows are opened for arsenic diffusion, which forms the subcollector n^+ regions. It reduces the series resistance of the collector, measured from the collector contact on the surface to the epitaxial collector of the *npn* transistor.

(b) Boron (p^+) is diffused in a selected area that serves as part of the isolation region. A n^- epitaxial layer is deposited, and the recessed oxide is formed by the LOCOS process. The p^+ region under the recessed oxide is called the *channel stop;* it prevents shorting between neighboring n^+ subcollector islands. This is known as *oxide isolation.*

(c) The base is either diffused or implanted; a layer of Si_3N_4 is deposited, and contacts are defined.

(d) The emitter window is opened, and arsenic is either diffused or implanted. Then the other contacts are opened.

Figure 2 shows a lateral *pnp* transistor. It has a similar process sequence, so both types of transistors can be fabricated on the same chip, which is required for certain circuits.

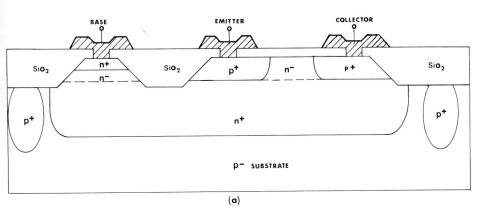

(a)

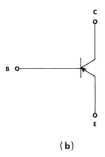

(b)

Fig. 2. The *pnp* transistor: (a) cross section, (b) circuit symbol.

B The Schottky Barrier Transistor

A Schottky barrier (Schottky-clamped) transistor, often used in the digital circuit, is shown in Fig. 3. The Schottky barrier diode is placed over the region where the base metal contacts the *n* epitaxial layer. It serves as a clamping diode across the collector junction and prevents the transistor from deep saturation. Since the Schottky diode is a majority carrier device, it does not suffer the minority storage delay time found in a *pn* junction diode. This helps improve the switching speed of logic circuits using the Schottky transistors.

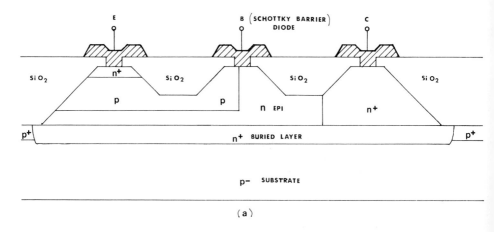

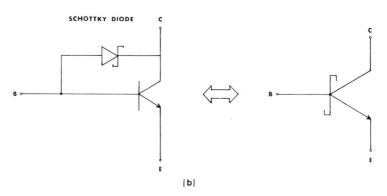

Fig. 3. Schottky-clamped *npn* transistor: (a) cross section, (b) circuit symbol.

9. Bipolar VLSI Circuit Technology

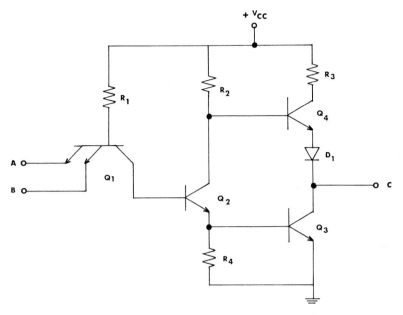

Fig. 4. TTL NAND gate circuit.

BIPOLAR DIGITAL GATE CIRCUITS†   III

The performance of a particular logic family is usually described by its gate propagation delay versus gate power dissipation curve. We shall describe four types of logic circuits (TTL, transistor–transistor logic; ECL, emitter coupled logic; IIL, integrated interjection logic; and STL, Schottky transistor logic) and compare their power-delay curves.

TTL Circuits   A

Figure 4 shows the most popular *NAND gate circuit*. The input is a multi-emitter transistor. If inputs A and B are both at logic one, the current flows through R_1 and the collector junction of Q_1. This drives transistors Q_2 and Q_3 into saturation. The output voltage is V_{CE} (sat.), which is about 0.1 V (logic 0). On the other hand, if any of the inputs is low, the current is diverted to the emitter of Q_1. Due to the lack of base drive current, Q_2, Q_3 are switched off. The high voltage at the collector of

† See [4].

Q_2 then turns on Q_4 and D_1, and this circuit sources the current to the following gate stage. The output voltage at the terminal C is about 3.5–3.6 V, which is the power supply voltage (+5 V) minus the voltage drops across the emitter junction of Q_4 and diode D_1. The purpose of the phase-splitting circuit Q_2 is to make sure that the output totem-pole transistors Q_4, Q_3 are not turned on at the same time in quiescent states (0 to 1). This reduces the power consumption. The Q_4 is the so-called active load, which, though it occupies less area on the IC chip than a passive load such as a resistor, can source more current to the following stage. The higher the sourcing current, the lower is the charging time for a capacitive load. This reduces the propagation delay time of the gate circuit.

To reduce the delay time further, Schottky transistors are often employed. Figure 5 shows an *advanced Schottky (AS TTL) NAND gate* circuit. The input transistor has been replaced by the Schottky diodes to save space. The inclusion of the Q_1 improves the switching characteristic of the gate, because Q_1 and Q_3 will be turned on at the same time instead of at separate instants as in a standard TTL NAND gate (Fig. 4).

Another type of logic circuit is the *advanced low-power Schottky*

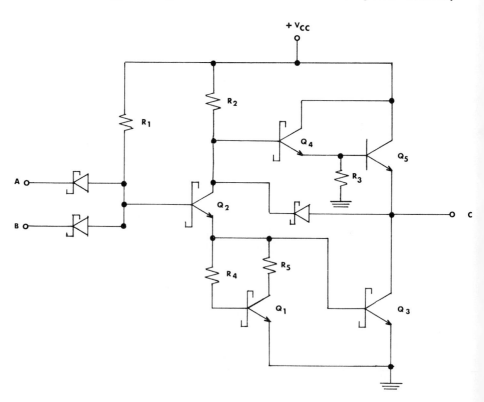

Fig. 5. Advanced-Schottky TTL NAND gate circuit.

NAND gate *(ALS TTL)*, which uses larger resistance values for low power consumption.

ECL Gate Circuits B

The logic swing (output voltage difference between logic 1 and 0) for a TTL circuit is about 3.5 V. ECL circuits, however, reduce this swing to about 0.8 V, which takes less time. In addition, ECL circuit design keeps transistors from saturation. This is known as *nonsaturated logic*. It has a higher switching speed because of the absence of the delay time that is associated with the removal of the overdrive charge in the base region when the transistor comes out of saturation.

Figure 6 is a typical ECL NOR/OR gate circuit. The input stage is a differential amplifier with coupled emitters and a constant current source Q_4. If both inputs are at logic 0 (voltage is smaller than the reference voltage V_R), Q_3 is *on*. Since Q_5, Q_6 are always *on* and serve as the voltage level shifter, the high current that flows through Q_3 and R_2 reduces V_{C3} and hence the output voltage at terminal C (logic 0). On the other hand, if any input goes high (logic 1), Q_3 will be *off*, and the output is at logic 1. This is an OR gate function. If the output is taken from the D terminal, it will be a NOR gate function because of the complementary nature of the collector voltages of Q_2 and Q_3. Since a voltage difference of only a few

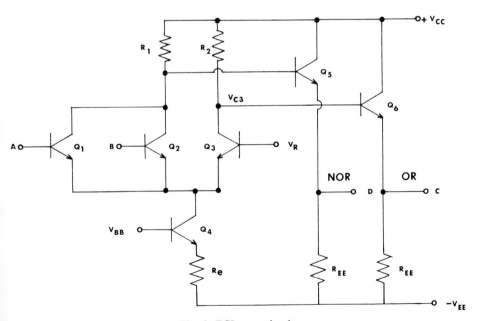

Fig. 6. ECL gate circuit.

tenths of a volt at the input can switch Q_3 from *on* to *off*, the switching speed of ECL is very fast.

ECL has 10K and 100K series, which differ (Fig. 6) in their temperature compensation circuits, which stabilize the switching characteristics against the variation in device temperature. Since Q_5 and Q_6 are always *on* and in the active region, the ECL circuit consumes a large amount of power and generates excessive heat.

C CML Gate Circuits

To overcome the preceding drawback, the CML (current mode logic) circuit is often used. CML gates are ECL gates without the output emitter followers Q_5, Q_6 (Fig. 7). The outputs are taken directly from collectors Q_2 and Q_3. Thus the CML gate dissipates less power. The state-of-the-art switching speed of ECL/CML gates is 84 ps per gate delay and is considered to be the fastest bipolar LSI or VLSI circuit [5].

D IIL (MTL) Gate Circuits

Because of its low power consumption and high packing density, IIL (I^2L) has become a major bipolar candidate for answering the challenge by MOSFETs in VLSI.

1 Typical I^2L Gate

The typical I^2L circuit is shown in Fig. 8. The lateral *pnp* transistor supplies the current (injector current) to the following switching *npn* transistor. Since the base of the *pnp* shares the same region with the emitter of the *npn* transistor, and the collector of the *pnp* shares the same region with the base of the *npn*, this logic is also known as *merged transistor logic (MTL)*.

If the input node A is at logic 1 (high voltage), the injector current turns on the *npn* transistor Q_2. The output is at logic 0. When the input voltage is low (logic 0), the injector current is diverted and sunk by the *npn* transistor of the previous stage. Q_2 will be *off* due to the lack of base drive current, and the output is at logic 1. It is basically an inverter.

2 Simple I^2L NOR Gate

A simple NOR gate CKT is shown in Fig. 9. The output is high only when both inputs are low. Q_1 and Q_2 will both be *off* under this input

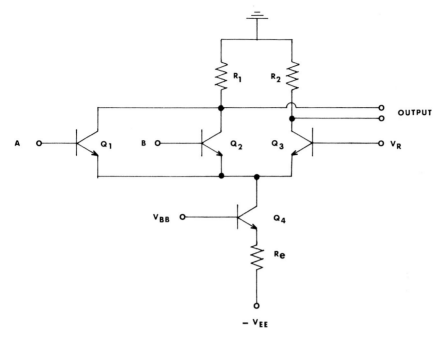

Fig. 7. CML gate circuit.

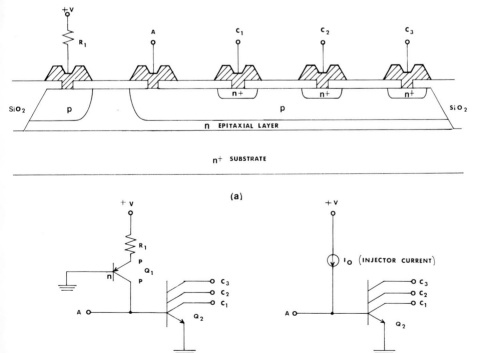

Fig. 8. I²L gate: (a) cross section, (b) circuit symbols.

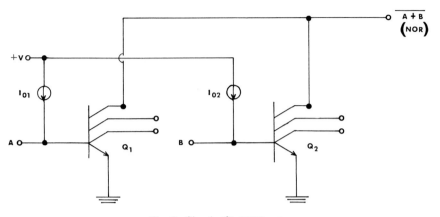

Fig. 9. Simple I²L NOR gate.

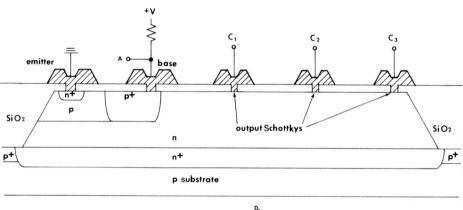

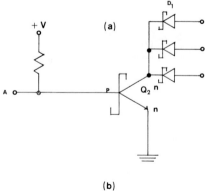

Fig. 10. STL gate circuit: (a) cross section, (b) circuit symbol.

condition. If any of the inputs goes high, the *npn* transistor will be *on,* and the output will be low. Since the emitter of the *npn* transistor is connected to the substrate instead of being on the surface, it reduces the surface area for the emitter interconnection. Together with the merged nature of the structure, I^{2}L can achieve denser packing of the gate circuits than TTL and ECL.

Power Dissipation

The power dissipation of an I^{2}L gate is simply the product of the injector current and power supply voltage (0.7–1 V). The delay time is generally inversely proportional to the injector current. Hence, the power-delay product in I^{2}L remains essentially constant over a large range of injector currents. I^{2}L/MTL has been shown to be capable of providing subnanosecond (0.3–0.8 ns) delay times at high injector current [5,6,7] and power delay products on the order of 10 fJ at low speed.

STL Gate Circuits

To improve the switching speed of I^{2}L further, STL can be used (Fig. 10). It eliminates the *pnp* transistor and uses only the resistor to limit the current supply. The collector junction of the *npn* transistor is clamped by the Schottky barrier diode to keep it from saturation. The output is taken from the Schottky diode which is connected to the collector of *npn* transistor. This arrangement increases the output voltage at logic 0 from 0.1 V (V_{CE} sat.) to 0.5 V (Q_2, D_1 conducting). It reduces the logic swing from 0.7 V to 0.3 V, thus increasing the switching speed. An STL circuit with 0.4 ns minimum delay time and 30 fJ power-delay product at low speed has been reported [6].

Comparison Logic Circuits

The power dissipation and the propagation delay times of different logic circuits are compared in Fig. 11. The ECL/CML provides the highest switching speed, but it also dissipates the most power. I^{2}L or STL can provide adequate speed with much less power dissipation per gate. Their high packing density is the most desirable feature in VLSI. TTL, because of its large logic swing, has the advantage of a large noise margin. Its popularity and circuit maturity are far beyond any other circuit technologies. To take advantage of the high speed of ECL and the high packing

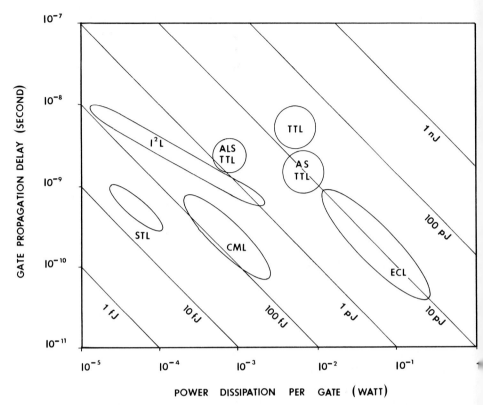

Fig. 11. Power-delay continuums for the state-of-the-art bipolar VLSI circuit technologies.

density of I²L, an integrated system can be built. Those parts of the VLSI circuit that require high speed can be made using ECL, whereas the parts that need less speed can be made of I²L for high packing density [5].

REFERENCES

1. I. Magdo, Vertical *p–n–p* for complementary bipolar technology, *IEEE Trans. Electron Devices* **ED-27,** 1394–1396 (1980).
2. J. Guerena *et al.*, Oxil, a versatile bipolar VLSI technology, *IEEE Trans. Electron Devices* **ED-27,** 1397–1401 (1980).
3. A. Glaser and G. Subak-Sharpe, "Integrated Circuit Engineering." Addison-Wesley, Reading, Massachusetts, 1977.
4. D. A. Hodges and H. G. Jackson, "Analysis and Design of Digital Integrated Circuits." McGraw-Hill, New York, 1983.
5. T. Nakamura *et al.*, Integrated 84 ps ECL with I²L, *ISSCC Tech. Dig.* (1984).
6. S. A. Evans *et al.*, A 1-μm bipolar VLSI technology, *IEEE Trans. Electron Devices* **ED-27,** 1373–1379 (1980).
7. D. D. Tang *et al.*, Subnanosecond self-aligned I²L/MTL circuits, *IEEE Trans. Electron Devices* **ED-27,** 1379–1384 (1980).

Chapter 10
CMOS VLSI Technology

DAVID B. SCOTT
KUEING-LONG CHEN
RODERICK D. DAVIES

Texas Instruments
Dallas, Texas

I. Advantages of Circuit Design with CMOS	122
A. Reduction of Active Power	122
B. Ease of Design	122
C. Applications	123
II. A State-of-the-Art Process Flow	124
III. Problems of Optimization of CMOS Processing	131
A. Drain Current versus Channel Length	131
B. Threshold Voltage versus Gate Length	132
C. Performance versus Scalability	133
IV. Problems of Interconnects for CMOS	133
A. Dual-Layer Metallization	133
B. Chip and Technology Costs	134
C. Reduction of Gate Electrode Resistance	134
D. Reduction of Source/Drain Diffusion Resistance	134
V. Discontinuities in CMOS Technology	135
Bibliography	138

CMOS was proposed in 1962 by C. T. Sah and Frank Wanlass in a paper delivered to the International Solid-State Circuits Conference. Except for a relatively few specialized micropower applications, well over a decade of neglect of the concept followed while metal-gate PMOS and then silicon-gate NMOS were developed. The recent attention to CMOS technology and circuit techniques for mainstream integrated-circuit products has occurred for two reasons: First, diverse circuit design requirements are being met by various aspects of CMOS capabilities; second, the higher cost of processing CMOS is coming into line with the rising costs of producing circuits with competing scaled NMOS and bipolar technologies.

I ADVANTAGES OF CIRCUIT DESIGN WITH CMOS

A Reduction of Active Power

The reduction of active power is the single most important reason for the recent strong interest in CMOS. Commonly available packaging technology limits total chip dissipation to approximately one watt to avoid excessively high temperatures, which would lead to metallization electromigration and other reliability problems, and which would cause circuit-design difficulties due to high leakage currents. This one-watt limit is becoming hard to meet with NMOS and bipolar technologies. A study reported by NEC (1978 ISSCC) demonstrated that selective replacement of the depletion load transistors used in NMOS driver circuits with *p*-channel transistors can significantly reduce the overall power-delay product of a chip. In their 16-bit enhancement/depletion NMOS microprocessor design, replacement of only 180 out of a total 3239 depletion load transistors with *p*-channel transistors enabled a reduction of overall chip power by 40% even though the clock frequency was increased from 6 to 10 MHz. This improvement in power-delay product is typical of that achievable by implementing a logic design in CMOS rather than NMOS.

B Ease of Design

The key additional advantages in realizing circuits with CMOS basically all relate to ease of design. Sah and Wanlass' original concept of CMOS was to use the flexible "static ratioless" circuit technique with matching pairs of *n*- and *p*-channel transistors in logic gates (Table I).

1 Circuit Timing

Static means that no clocking is required to store charge dynamically on transistor input nodes as a means of synchronizing circuit timing. This, along with only the pullup or pulldown transistors being biased *on* at any instant, results in the well-known ability to realize microwatt standby-power circuits and enables class of applications in portable equipment and in pseudo-nonvolatile memory boards with static RAM and small self-contained batteries.

2 Noise Immunity

Ratioless means that the choice of transistor widths during circuit design is not first-order critical for successful operation; this yields an inher-

TABLE I
Application Areas that Draw on CMOS Design Advantages[a]

Design advantages	Static RAM	Gate array	Logic/ microprocessor	Analog/telecom	DRAM
Reduced operating power	×	×	⊗	×	
Circuit design flexibility		⊗		⊗	
Single-stage bitline/bussline pullups without node booting	×		×		×
Microwatt standby power	⊗	×	×		
High noise immunity		×	×		
Simpler dynamic circuits			×		⊗
Potential barrier to alpha particles	×				×

[a] ⊗ denotes key feature; × denotes important (but not key) feature.

ently high noise-immune circuit. It contrasts with commonly used enhancement/depletion NMOS circuitry in which careful selection of transistor layout-geometry (W/L) is needed. This high noise immunity of CMOS can be directly used to advantage in harsh environments or, alternatively, it can be used for "uncommitted circuit" design, which is the basis for gate arrays and standard-cell libraries. The use of NMOS or PMOS is very difficult in such situations.

High Layout-Density

CMOS also makes possible the design of simple circuits of high layout-density (dynamic, ratioless logic) that are less cumbersome than competing NMOS and PMOS dynamic multiphase-clocking schemes. This is achieved by using NMOS-intensive multiple-input gates sharing a single pair of clocked *n*- and *p*-channel pullup and pulldown transistors.

Applications

Design simplicity and low power are also the keys to *analog, data-conversion* (D/A, A/D), and *telecommunications* circuit (telecom) realization with CMOS. Operational amplifiers are easier to design because of wider margins for parameter variation, they operate at lower power with higher gain, and they can swing their outputs nearer the voltage supply potentials than is possible with competing NMOS approaches. It is possible to design associated control logic and clocks more conveniently in CMOS than in the bipolar analog-circuit approach. Individual-telephone subscriber-line interface circuits continue to be realized in bipolar technology rather than MOS because of the difficult high-voltage requirements. CMOS, however, in the form of *charge redistribution* or *switched capacitor* type circuits, is evolving toward being the technology of choice for applications such as coder–decoder (CODEC) and precision standalone A/D and D/A circuits for use in digitally switched telephone systems and modems, and for many kinds of instrumentation interfaces.

A recent area of CMOS application is in *dynamic memories*. The availability of a p-channel transistor can eliminate the need to boost the voltage of selected driver transistors above the standard 5-V power supply voltage, as is common in NMOS design. Boosting is undesirable in advanced MOS technologies because transistor reliability may be jeopardized by operating at the resulting 7–12 V levels. It is widely felt that even the highly standard 5-V supply value will eventually have to be decreased to 3 V to avoid these problems. Column decoders designed with static rather than dynamic circuits, convenient "bitline shorting" circuits to balance matched lines running to the memory array, and the chance to put the entire memory core in a diffused "well" to repel the stray electrons and holes that are generated in the substrate by strikes from alpha-particles and cosmic rays, are other strong advantages of dynamic RAM design with CMOS.

The various advantages of CMOS to the circuit designer as described above are summarized in Table II.

II A STATE-OF-THE-ART CMOS PROCESS FLOW

A CMOS process sequence is designed to effect a tradeoff between the circuit designer's requirements and minimum process complexity. A variety of process sequences are currently in use in the industry. As an example, we describe the implementation of a Twin Well (or Tank) process fabricated on a p^- epitaxial layer that was grown on a heavily doped p^+ substrate (see Fig. 1).

(a) A silicon nitride layer (Fig. 1a) is deposited on a thermally grown pad oxide layer, and photoresist is then applied and patterned to define the n-Tank (or n-Well) regions. After nitride etch, phosphorus is implanted for the n-Well.

(b) The exposed silicon is oxidized, (Fig. 1b) and the nitride is stripped to leave the thermally grown oxide in place to be a self-aligned mask for the p-Tank implant. The depression left after the n-Tank oxide

10. CMOS VLSI Technology

TABLE II

Circuit Design Advantages of CMOS

CMOS	NMOS
Static	
(+) Microwatt standby power	(−) Require 2-stage push-pull (and often bootstrapping) for acceptable drivers
(+) Flexible design	
(+) High noise immunity	(−) design not flexible: gate arrays not practical
(−) Multiple input gates require many devices, consume chip area	
Dynamic	
CMOS with 1-phase "domino" logic for large gates, alu blocks, etc.	
NMOS with 2- or 4-phase clocking	NMOS with 2- or 4-phase clocking

strip is used to align subsequent levels. Both tanks are now simultaneously diffused.

(c) Nitride is deposited on the thermally grown pad oxide and patterned and etched, as in (Fig. 1c) with the moat (transistor active region) mask. An unmasked boron implant forms a channel stop in the p-Tank. After photoresist strip, the exposed silicon is thermally oxidized to form the field regions.

(d) The nitride and pad oxide layers are removed, (Fig. 1d) and the gate oxide is grown. A boron implant sets the n-channel threshold voltage.

(e) A photoresist mask defining the p-Tank regions is used for additional implanted boron to set the p-channel threshold voltage, as in Fig. 1e.

(f) An optional contact from the gate to the n-moat regions may be made by selectively etching holes in the gate oxide (Fig. 1f).

(g) After *buried contacts* have been defined (Fig. 1g) polysilicon is deposited and doped with phosphorus. In advanced processes, a refractory metal silicide is also deposited on the cleaned polysilicon surface to provide a low-resistivity gate interconnect. The gate is next patterned and etched. Note that the silicon in the region of the buried contact has also been etched because no oxide was present to stop the silicon etch. The silicide is annealed after gate-electrode definition.

(h) The n-channel source and drain regions are formed (Fig. 1h). Aluminum is often used to mask this heavy arsenic or phosphorus implant.

(i) The p-channel regions are implanted (Fig. 1i). Often a photoresist mask is not necessary for this process if the n-channel source and drain regions can tolerate being boron counterdoped. The boron implant dose must be at least a factor of three lower than the n-type implant. An oxide

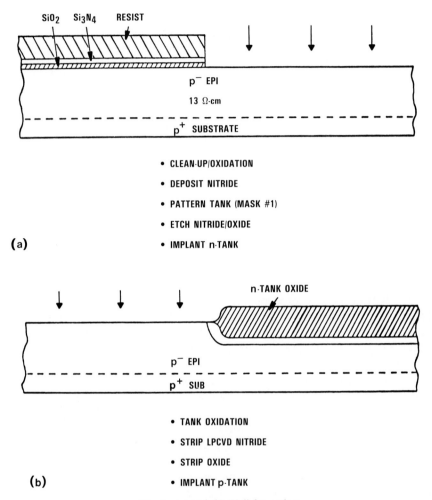

Fig. 1. (a) and (b) Well formation.

is now deposited over the source and drain regions and these p^+ junctions are now annealed.

(j) Phosphorus-doped glass (PSG) is deposited on the slice and densified (Fig 1j), and contact holes are patterned and etched; PSG is reflowed to slope the contact walls. This aids step coverage of aluminum into the contact holes.

(k) The aluminum is patterned, as shown in Fig. 1k. A protective overcoat is deposited on the slice and is patterned to open up the bond pad areas.

The process just described is typical of a modern CMOS process in that it has a polycide gate for low resistivity and one interconnect enhance-

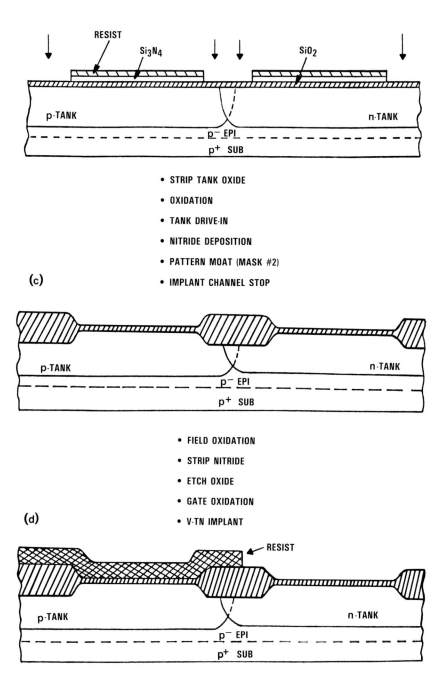

Fig. 1. (c) and (d) Field oxide formation; (e) threshold adjust.

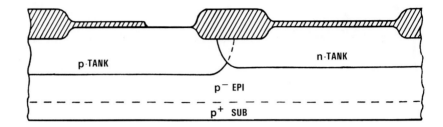

(f)
- PATTERN BURIED CONTACT (MASK #4)
- ETCH GATE OXIDE

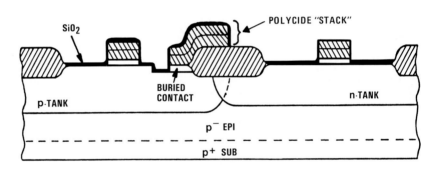

(g)
- DEPOSIT POLY
- DOPE POLY
- "DEGLAZE" DOPED OXIDE
- DEPOSIT SILICIDE
- PATTERN GATE, (MASK #5)
- ETCH POLYCIDE
- ANNEAL POLYCIDE

Fig. 1. (f) Buried contact patterning; (g)–(i) transistor formation.

10. CMOS VLSI Technology

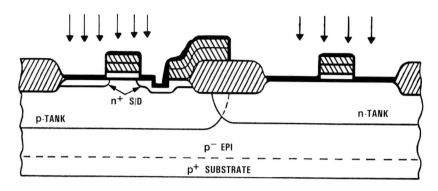

(h)
- IMPLANT p+ S/D
- DEPOSIT OXIDE
- ANNEAL p+ S/D

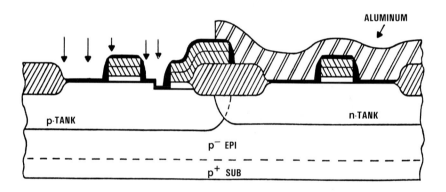

(i)
- DEPOSIT ALUMINUM
- PATTERN n+ S/D (MASK #6)
- ETCH ALUMINUM
- IMPLANT n+
- ANNEAL n+
- STRIP ALUMINUM

Fig. 1. (*continued*)

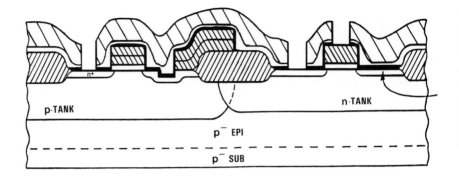

(j)
- DEPOSIT PSG
- REFLOW PSG
- PATTERN CONTACTS (MASK #7)
- ETCH CONTACTS
- REFLOW CONTACT WALLS

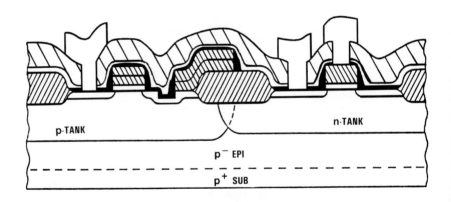

(k)
- DEPOSIT ALUMINUM
- PATTERN ALUMINUM (MASK #8)
- ETCH ALUMINUM
- SINTER ALUMINUM
- DEPOSIT PROTECTIVE NITRIDE
- PATTERN (MASK #9)
- ETCH NITRIDE
- BACKSIDE GRIND

Fig. 1. (j) and (k) Metallization and overcoat formation.

10. CMOS VLSI Technology

ment requiring an additional mask (buried contact mask). Often, however, additional interconnect enhancements are required, depending on the circuit. Further, the choice of doping concentrations and oxide thicknesses is often determined by performance and density considerations.

PROBLEMS OF OPTIMIZATION OF CMOS PROCESSING III

In defining a process, one must consider a multitude of *tradeoffs*. These include process complexity, device performance, and overall scalability. Figure 2 shows a comparison of the complexity of various processes. Modern NMOS processes have become comparable in complexity to those of CMOS. The performance and scalability of *p*-channel devices will be used as an example of necessary tradeoffs.

Drain Current versus Channel Length A

To compare the performance of the devices fabricated using these various technologies, the drain current was measured with a 5-V drain-to-source voltage and with the gate-to-source voltage being 4 V greater than the threshold voltage, as illustrated in Fig. 3. Plotting drive current inversely as a function of channel length allows comparison of the current-carrying capability of different *p*-channel devices. All *p*-channel devices studied had an n^+ poly gate electrode.

Figure 3 shows that the drive current increases as the *effective* channel

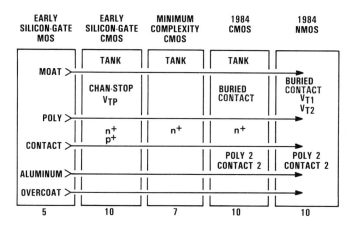

Fig. 2. Challenges in implementing CMOS: cost.

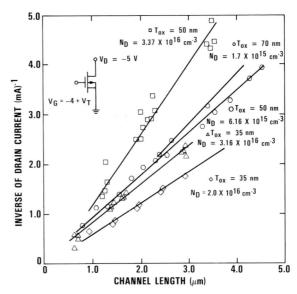

Fig. 3. Comparison of transistor drive current in various CMOS process designs.

length is decreased. However, for a given oxide thickness, the drive current degrades as the bulk-doping concentration increases. In fact, it is clear for both the 50-nm gate oxide devices and the 35-nm gate oxide devices that an increase in bulk-doping concentration caused a degradation in the current-carrying capability of the device. In one case a device with 70-nm gate oxide had a higher current-carrying capability than a device with 50-nm gate oxide due simply to the current degradation caused by high bulk doping. Thus, in terms of raw device performance, it is desirable to have bulk doping as low as possible for p-channel devices.

B   Threshold Voltage versus Gate Length

Though it appears that high bulk doping causes degradation of the p-channel current-carrying capability, Fig. 4 shows that it has a positive impact on the ability of the device to turn off when short channel lengths are involved. In Fig. 4, simulated threshold voltages (using the MINIMOS program) are shown as a function of gate length for different bulk concentrations. At low bulk-doping concentrations and short gate lengths, a drastic reduction in threshold voltage magnitude is noted. Thus, not only the device performance but also the ability of the device to turn off must be comprehended for successful operation.

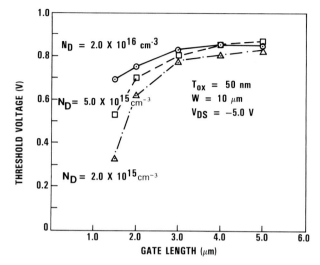

Fig. 4. Short-channel effects in p-channel transistors bulk doping concentrations.

Performance versus Scalability C

Figures 3 and 4 indicate that there is a tradeoff between device performance and scalability. If a process is designed to have devices functioning with 1.5 μm gate lengths, then the doping concentration needs to be high enough to withstand the short channel effects. However, a device that is overdesigned in this regard will suffer performance disadvantages when compared with devices fabricated with better optimization.

PROBLEMS OF INTERCONNECTS FOR CMOS IV

Though the topic of interconnections is vital in virtually all IC technologies, it is particularly so in CMOS, because there are more transistors to hook up for a given function than in NMOS and because the common applications of CMOS, such as gate arrays and standard-cell logic, demand more flexible wiring.

Dual-Layer Metallization A

Most logic (or gate) arrays require two levels of metallization to enable the efficient routing of circuit customization. This is particularly true of circuits for which the routing is computer-aided. Development of high-density dual-layer aluminum metallization is more difficult for MOS than for bipolar technology because of the more difficult topology presented by the gate electrodes of MOS. One decision is whether or not to use copper doping of aluminum on the first level to suppress stress-relieving "hillock" spike formation, a classic yield killer caused by puncturing the overlying interlevel oxide. Inclusion of copper has presented problems in dry etching to achieve fine lines, whereas exclusion of the copper requires subtle and difficult-to-control processing tricks to suppress or withstand hillock formation.

B Chip and Technology Costs

The key issue in determining the extent of popularity of gate arrays in the spectrum of logic forms will likely rest on chip cost and the technology level needed to realize high gate-counts without overly tight metallization layers and grossly large chips. Many workers believe that the standard-cell library, custom-design approach may eventually emerge as being more cost effective than the gate arrays.

C Reduction of Gate Electrode Resistance

The reduction of gate electrode resistance and, hence, memory bitline or microprocessor-controlling RC-time delay is a well-known goal for all advanced MOS technologies. Use of tantalum disilicide, titanium disilicide, molybdenum disilicide, tungsten disilicide, and others has been reported in MOS circuits, though few production circuits employing silicides are yet available. The key challenges are deposition uniformity, adhesion, and etchability.

D Reduction of Source/Drain Diffusion Resistance

A silicide application that is somewhat unique to CMOS, however, is that of reducing the very high p^+ source/drain (S/D) diffusion resistance that occurs when shallow boron junctions for device scaling are used. Greater than 100 ohm/Sq often results for junctions reduced to 0.4 μm or below as required for minimization of transistor shortchannel effects. This virtually eliminates the p^+ S/D as a useful conductor in general circuit layout. Application of metal silicides to transistor sheet resistance reduction has been experimentally achieved with titanium, platinum, and tungsten silicides as shown in Fig. 5.

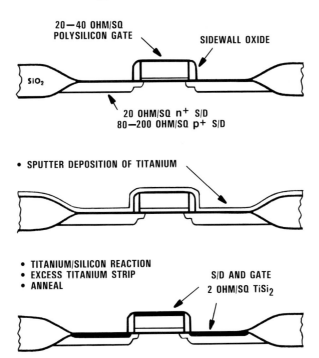

Fig. 5. Direct reaction of silicides to achieve low-resistance gate and source/drain regions simultaneously.

A *gate sidewall oxide* is formed prior to siliciding to prevent shorting of S/D to gate. This is done with a deposited or thermally grown layer that is anisotropically etched; the vertical projection of the oxide layer on the gate sidewall presents a thicker layer than does the oxide on the gate and S/D surface. In a single operation both S/D regions and gates are simultaneously clad with a metal silicide or refractory metal. This results in both the gate and the S/D regions having sheet resistances below 5 ohm/Sq as compared with 15–25 ohm/Sq unsilicided gates and >50–100 ohm p^+ S/D and >20 ohm/Sq n^+ S/D regions; such reduction in parasitic resistance is crucial to achieving maximum speed in VLSI applications such as memories and large microprocessors.

DISCONTINUITIES IN CMOS TECHNOLOGY V

The optimization of "conventional" CMOS technology will require four or five more years. The factors that are expected to propel the technology beyond the 1 Mbit DRAM and 256K SRAM level of circuits are true discontinuities in that they represent radical changes in process design. Three such discontinuities are discussed below.

(1) Trench Isolation. The formation of deep grooves by plasma or reactive-ion etching, with subsequent polysilicon or oxide refilling to achieve substrate planarity, holds great promise for reducing n^+ S/D to p^+ S/D spacing across a well boundary from the present 5–9 μm values obtained with standard technology, down to the 1–3 μm range commensurate with the other layout rules achieved in scaled MOS technology. This layout rule at present severely limits applications such as the "full CMOS" SRAM in which both transistor types are required in each six transistor cell, and it is a limitation in the layout of tight-pitch memory periphery circuits such as decoders and sense amplifiers. Only recently has the etching and lithographic capability necessary to define the required 1–2 μm openings become generally available. A typical processing sequence is shown in Fig. 6. In which the etched trench is refilled with polysilicon and then etched back to planarize the wafer surface.

The ultimate form of trench isolation would allow butting of n^+ and p^+ source/drains directly against the trench for maximum density, eliminating the need for a traditional field oxide. However, the lightly doped p-tube surface on the n-channel side tends to invert due to the leaching of boron doping into the oxide along the trench. With extra n^+ S/D-to-trench layout spacing, allocated in conjunction with the retention of the usual

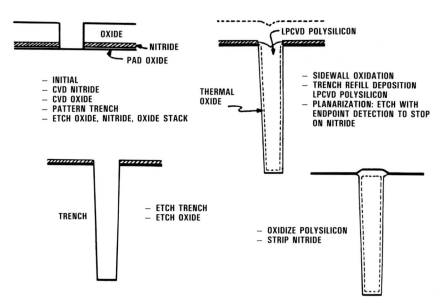

Fig. 6. (a) The introduction of a deep, refilled groove between n- and p-channel transistors to improve layout density.

10. CMOS VLSI Technology

"locos" field oxide to eliminate this transistor-to-transistor leakage source, a substantial density increase has still been demonstrated. Besides a density enhancement, a substantial increase in latchup resistance has also been achieved.

(2) *Stacking Transistors to Achieve Multilevel Layout.* It has been proposed that entire planes of transistors and independent interconnects might be stacked, but this will probably not be practical for some time. A more immediate possibility is to stack a *p*-channel device topside, which would need to do little more than shut off to a low leakage state. This would allow extremely dense RAM cells of static and other types by totally eliminating the traditional *n*-to-*p* device-isolation problem. Figure 7 shows a cross section of this concept in which the gate electrode for a CMOS inverter is shared by both transistors of the pair, with the common output connection of transistor drains seen at right and the supply connections to the transistor sources seen at left. This provides a memory cell that in many ways resembles a standard four-transistor NMOS polysilicon-load cell. Because only leakage makeup currents need be supplied in this type of circuit, even the extremely low carrier mobilities, associated with polysilicon-like silicon quality, will do—so long as the transistor can be shut off to minimize power dissipation in a large, 256K or perhaps 1 Mbit, SRAM. Such an SRAM would be superior by achieving the density associated with a conventional (stacked) polysilicon load-resistor type and offering the microwatt (as opposed to milliwat) standby power associated with the much larger (nonstacked) conventional "full CMOS" six transistor cell.

(3) *The Emulation and Extension of SOS with a More Tractable, Affordable Substrate.* Silicon dioxide would be used as an insulator with the familiar silicon crystal substrate used in bulk MOS production (Fig. 8). Such a form (SOI) would not suffer from the SOS problems of very high substrate cost, high incidence of breakage during processing due to thermal stress, and low thermal conductivity that cause packaging problems in achieving adequate thermal conductivity. This capability would offer complete immunity from latchup and would achieve high density (though lower than that of stacked layouts).

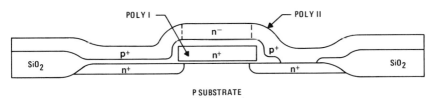

Fig. 7. Stacking a load transistor on top of conventional topology would provide another source of density increase. Some *p*-channel transistors have been reported with a gate electrode common to both the *n*- and *p*-channel devices of an inverter.

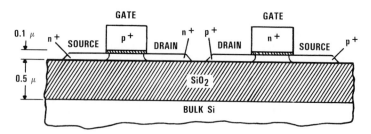

Fig. 8. Although silicon-on-sapphire has almost universally been rejected as a mainstream VLSI contender, silicon-on-oxide (a) may yet become a cost-effective CMOS approach that is latchup free. The n^+/p^+ spacing problem would be eliminated without having to put in grooves. (b) As scaling of layout features proceeds toward 1 μm, however, the saturation of interconnection capacitance reduction due to fringing fields may preclude performance that is dramatically better than bulk technology.

Two main SOI options exist: the recrystallization of deposited silicon and the direct formation of a buried insulator layer by way of ion implantation of the substrate. Recrystallization using lasers, electron beams, graphite-strip heaters, and various UV and visible lamp irradiation has been reported.

Implantation of oxygen or nitrogen to form a silicon dioxide or silicon nitride "in situ" buried insulator is attractive to avoid the problems of depositing a pure polysilicon layer and then recrystallizing it. However, there is a serious issue of practicality because very large doses of the chosen specie are required, as much as 100 times that of the already high S/D implant doses of MOS processing. Although high-quality insulated silicon layers have been formed with this technique, the future of buried insulators of this type will surely rest on the availability of a new type of high-beam-current implanters to achieve cost-effectiveness.

ACKNOWLEDGMENT

Stephen Saller is gratefully acknowledged for his careful reading of the chapter and for the valuable technical assistance that he has given.

BIBLIOGRAPHY

CMOS Well Formation

R. S. Payne et al., Twin-tub CMOS—A technology for VLSI circuits, *Int. Electron Devices Meet. Tech. Dig.* (1980).

R. D. Rung et al., A retrograde p-Well for higher density CMOS, *IEEE Trans. Electron Devices* **ED-28** (10), (1981).

K. Hashimoto et al., Counterdoped well structure for scaled CMOS, *Int. Electron Devices Meet. Tech. Dig.* (1983).

10. CMOS VLSI Technology

Latchup

Several papers in *Int. Electron Devices Meet. Tech. Dig.* (1982).
D. Estreich, Ph. D. dissertation, Stanford University, Palo Alto, California, 1980.
Int. Solid-State Circ. Tech. Dig. (1982); (1983).
IEEE J. Solid-State Circ. (October 1982); (December 1983).

Stacked CMOS Technology

J. F. Gibbons and K. F. Lee, One-gate-wide CMOS inverter on laser-recrystallized films, *Solid-State Electron.* **15**, 789–799 (1972).
J. P. Colinge and E. Demoulin, A high-density CMOS inverter with stacked transistors, *IEEE Electron Device Lett.* **EDL-2** (10), (October 1981).

Trench (Groove) Isolation

R. D. Rung *et al.*, Deep trench isolated CMOS devices, *Int. Electron Devices Meet. Tech. Dig.* (1982).

SOI

K. Izumi *et al.*, High-speed CMOS using buried SiO_2 layer formed by ion implantation, *11th Conf. Solid-State Devices (Tokyo) Dig. Tech. Papers*, (1979).

Chapter 11

New Directions in Microprocessors

JOHN RAITHEL

Zilog, Inc.
Campbell, California

   I. Introduction   141
  II. Memory Management   142
      A. Memory Allocation   143
      B. Memory Protection   143
      C. Virtual Memory   144
 III. Cache   145
      A. Types of Cache   145
      B. The Burst Mode   146
 IV. Pipelining   146
  V. System Timing   147
      A. Bus Scaling   147
      B. Wait States   148
      C. Clock Chips   148
      D. Advantages of Clock Chips over Wait States   148
 VI. Peripheral Controllers   149
VII. Current Implementations   150

INTRODUCTION   I

    Two major influences have directed the development of microprocessors: the correction of problems or limitations arising from their use and the incorporation of features of existing larger computers. For instance, an early microprocessor that manipulated data four bits at a time was quickly found to be too limited in speed and memory-addressing capabilities for many applications. The move to an 8-bit processor was one obvious result, and often the addressing capability was simultaneously increased to 16 bits to support a much greater memory space. To support this new CPU, the entire microcomputer system had to be redesigned. It became necessary to create micro-support circuits, or peripheral controllers on a chip, that matched the size and speed of the CPU. These support chips began to replace much of the cumbersome external circuitry, and entire computers were reduced to a single board.

The progression from a 4- to an 8-bit processor is one example of the move toward larger computer concepts, and that progression continues today. New generation microcomputers feature 16- and 32-bit processors, something that was formerly available only with minicomputers. Increased data manipulation size (word size) means faster processing and fewer transactions with memory and peripherals.

Integrating more features and more performance into an integrated circuit leads to reduced system size. This is what allows minicomputer performance in a microcomputer. If the micros simply relied on multichip implementation to promote higher performance, they would only become more like minicomputers. But the microcomputer concept of miniaturization continues. Several thousand transistors could fit on an 8-bit chip, but a 16-bit chip contains tens of thousands, and a 32-bit chip contains over a hundred thousand transistors. Now, with peripheral controllers right on the CPU chip, the number of external components and their associated bus connections is greatly reduced. The discussion that follows illustrates several advances in computer concepts in the microcomputer environment.

II MEMORY MANAGEMENT

Memory management becomes increasingly important as microcomputer applications become more sophisticated. The concept of memory management was borrowed from the larger computers to handle the demands of more sophisticated memory transactions. Memory management, on a separate chip or within the CPU, permits the efficient allocation and protection of a very large memory. Memory allocation is accomplished by taking the logical address contained in the program and mapping it to an area of physical memory. Memory protection is accomplished by instructing the memory management unit to associate attributes (e.g., read only) with particular addresses.

With memory management, the logical addresses assigned to a program are converted to physical addresses that describe an actual memory location. Address attribute information, such as read only, program, or data space, can be maintained by the memory management unit. For example, the 2-byte logical address can be concatenated with an 8-bit descriptor register from the memory management unit to form a 3-byte address with associated attribute information. Memory management is illustrated in Fig. 1.

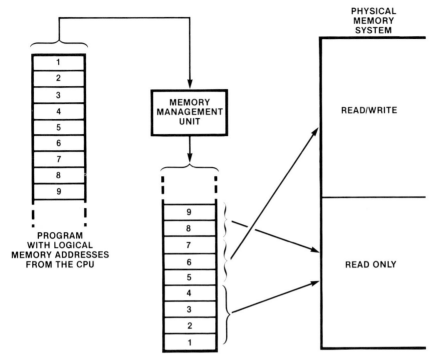

Fig. 1. Memory management. The use of memory management to translate logical addresses to physical address spaces is illustrated.

Memory Allocation   A

Memory allocation has become increasingly important as microcomputers have moved into multiuser, multitasking environments. With memory management performing the allocation, many tasks can run concurrently without competing for specific physical addresses. Two programs that contain the same logical addresses can be mapped to separate areas of memory without programmer intervention. Also, different applications can share resources. A compiler, for example, can be used by two programmers concurrently, or a table can be read by two programs running simultaneously. This is practical only when the shared resource is protected.

Memory Protection   B

Memory protection promotes system organization and integrity. Particular applications or functions can be assigned specific attributes, and read or write permission can be selectively granted. This protection ensures that a malfunctioning application or inappropriate action does not access system resources or privileged information.

C Virtual Memory

Memory management has the ability to handle virtual memory schemes. A program can contain logical addresses involving a range as large as the CPU's addressing capabilities, even though the main memory is considerably smaller. With virtual memory, the operating system is responsible for swapping pages (a fixed-size block) of memory between the fast and expensive main memory and the slower and relatively inexpensive secondary memory. This is transparent to the running program. When memory management encounters an address outside of the main memory during program execution, it aborts the instruction, and the operating system swaps a page of the main memory with a page of the secondary memory containing the requested address. The CPU then resumes program execution with the previously aborted instruction (Fig. 2). Like other components in the microcomputer system, memory management is moving from the board to the CPU chip.

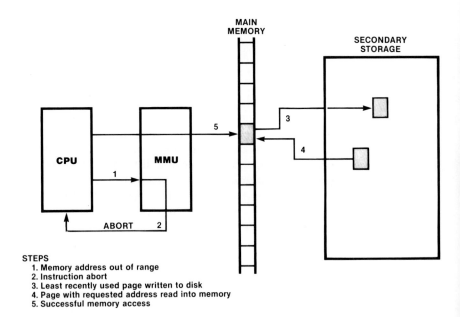

Fig. 2. Virtual memory. Illustrated is the mechanics of virtual memory: An out-of-range address is aborted and tried again.

CACHE   III

The *cache* is a portion of random access memory (RAM) on the CPU chip. On-chip RAM access is faster than an external memory transaction because no external bus transaction is involved. The cache differs from standard on-chip RAM in that it duplicates certain external memory addresses, rather than providing a source of new memory. When the CPU accesses a memory location, a copy is retained in the cache, and subsequent references to this location are then supplied by the cache. Because the cache can supply data faster than memory, system speed is improved.

A memory read or write results in a cache *hit* or *miss*. A cache hit occurs when the addressed location is duplicated in the cache; no external memory transaction is required on cache hits unless the cache is being written to, in which case the corresponding location in memory must also be updated. A cache miss means that the memory location accessed is not duplicated in the cache, and a bus transaction to external memory is required.

The likelihood of a cache hit is dependent on the algorithm used to load the cache. One of the most efficient cache-loading algorithms keeps the most recently referenced addresses in the cache. When a cache miss occurs, the memory location addressed is automatically duplicated in the cache, overwriting the least recently used data. The probability of the most recently used address being accessed again is high, due to program looping, frequently used variables, and so on. The larger the cache, the greater the possibility of a cache hit, but very large caches become impractical because of the high cost of producing such complex chips.

Types of Cache   A

A cache can hold instructions, data, or instructions and data. This can be fixed in a system, or it can be programmable, depending on the needs of an application. The cache can also be combined with a "burst" mode of operation to produce a highly efficient means of updating the cache.

A cache that contains only instructions avoids the necessity of updating memory because instructions are read and not written to. Except for a limited number of applications, however, it is impossible to contain all the instructions in the cache, so cache misses will occur on some instruction references. An instruction-only type of cache will also produce cache misses because instructions typically operate on data, and when data are required a cache miss occurs.

A cache containing only data is practical only when it contains fixed addresses, because otherwise every write to the cache requires a memory update, defeating the streamlining provided by the cache.

The most efficient use of a cache is for instructions and data, because it can operate for relatively extended periods without a cache miss. For example, the instructions in a program loop and the data it manipulates can be contained entirely in the cache, and changes made to data can be updated in memory while processing with the cache continues uninterrupted.

B The Burst Mode

A new feature available with some of the new-generation microprocessors is the burst mode of operation. A system operating in burst mode can generate a single memory address and access several consecutive memory locations, rather than generate an address for each memory location. This reduces the number of external bus transactions.

Combining the burst mode of operation with a cache offers further advantages. Updating the cache with a burst transaction every time a memory reference is made, automatically duplicates consecutive addresses in the cache. This promotes system efficiency because adjacent addresses in memory are often the next required in processing; they may be the next instructions in a stored program or the next data in a table.

IV PIPELINING

Processing a typical instruction can be divided into three steps: *instruction fetch, instruction decode,* and *instruction execution*. The number of processor clock cycles required to complete an instruction is dependent on the size and nature of the instruction. Older microprocessors processed one instruction at a time: all three steps had to be performed on one instruction before the instruction fetch for the next could begin. New generation microprocessors allow multiple instructions to be queued.

Pipelining allows subsequent instruction fetches to occur while the CPU is processing an instruction by handling interface processing separately from data processing. This frees data processing from bus-speed constraints and allows instruction fetches to proceed without waiting for a previous instruction execution to be completed (Fig. 3).

The instruction fetch in the simplest pipeline is separate from the instruction execution. The hardware that fetches the instruction decodes it and then performs another fetch without waiting for processing to occur. The fetched and decoded instructions are queued to a depth of two or more. When the processor has completed one instruction execution, it

11. New Directions in Microprocessors

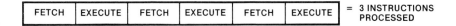

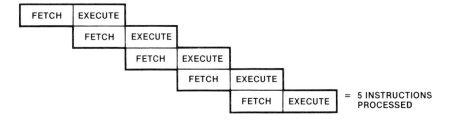

Fig. 3. Pipelining. A simple pipeline is illustrated in which the next instruction is fetched while the previous instruction is executing.

begins execution of the next instruction in the queue. More sophisticated CPUs can divide tasks even further—for example, instruction fetch, instruction decode, address calculation, operand fetch, and execution.

Though this arrangement is more efficient than traditional sequential methods, pipelining combined with cache operations is even more effective. With simple pipelining between memory and the CPU, short instructions (e.g., register to register load) result in lost time, because the processor finishes before the next instruction is fetched and decoded. When on-chip cache is used, the cache can quickly provide what is otherwise accessible only with a slower external memory transaction.

V. SYSTEM TIMING

Today, a system need not require a massive overhaul when a fast CPU has been added. The historical problem of speed differences between system components is being solved with bus scaling, wait states, and programmable clock chips.

A. Bus Scaling

Bus scaling is a property of some of the new CPUs. Though the CPUs themselves are very fast, they can interface with the external bus at a fraction of their own speed. For example, a 24-MHz CPU might be programmed to interface at 12 or 6 MHz. Thus, bus scaling allows slower peripherals or memory to exist in the same system as a much faster CPU, still permitting the CPU to operate at its maximum speed internally. If faster memory and peripherals are acquired, only the bus scaling fraction needs to be reprogrammed.

B Wait States

Programmable wait states are another means of slowing a system to meet timing requirements. The CPU automatically inserts wait states directly into the bus transaction, giving slower peripherals or memory the time necessary to complete their functions. A wait state can be as short as one clock cycle or as long as needed, eliminating the need for external wait state generation circuitry.

C Clock Chips

The programmable clock chip promises to reduce board space and enhance system speed and efficiency. The more advanced clock chips solve most of the problems of interfacing fast CPUs and slow peripherals, of shortcomings of wait-state logic, and of the requirements of variable clock cycles.

Clock chips operate at high internal speeds and have fast (e.g., 10 ns) rise and fall periods, creating high resolution and great flexibility for system clock needs. They also provide a *clock-stretching* capability that enables the insertion of clock half-cycles (single oscillator cycles) anywhere in a processor's machine cycle. This is a double advantage over wait-state logic.

D Advantages of Clock Chips over Wait States

Half-cycle resolution provides significant increases in system efficiency compared with the relatively long full-cycle delays of wait states. The capability of stretching any clock cycle of a processor's machine cycle is not available with wait-state logic, in which the wait input is not sampled each cycle. This feature can be used to allow a peripheral interrupt daisy-

11. New Directions in Microprocessors

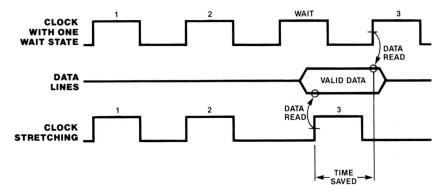

Fig. 4. Clock stretching and wait states. A timing diagram shows how clock stretching can save time compared with the use of traditional wait states.

chain to settle, for example, between machine cycles. And clock chips can provide the output necessary for CPUs with variable duty cycles, that is, machine cycles in which a specific clock period always has a duration different from that of another clock period. Finally, clock chips can provide reset control to ensure sufficient durations of the reset state for proper system initialization.

Figure 4 compares a typical machine cycle with a wait state with the same machine cycle delayed by a clock chip input.

PERIPHERAL CONTROLLERS VI

In the continuing progress toward system miniaturization, certain peripheral devices such as Direct Memory Access (DMA) and communications controllers are following the trend of integrating formerly discrete system elements into the CPU chip.

A DMA controller moves data with minimal CPU intervention, freeing the CPU to spend more time on processing. When the controller is on the chip, a system is faster and needs fewer chips, which means lower cost, less board space, and greater ease of system design and debug. Sophisticated on-chip DMA controllers provide multiple channels and powerful addressing capabilities. Similarly, on-chip communications controllers reduce board space. An on-chip Universal Asynchronous Receiver–Transmitter (UART), for example, can receive and transmit serial communications in a variety of common protocols with a minimum of CPU intervention.

VII CURRENT IMPLEMENTATIONS

These new features in microprocessors are being implemented to varying degrees by chip manufacturers. Intel Corporation has introduced its 80286, which features an on-chip MMU and a pipelined mode of operation, and its 80186, which features timing control and on-chip peripherals. Motorola Semiconductor Division has announced its 68020, offering an instruction cache on-chip as well as a pipelining scheme and burst mode. Zilog has unveiled its Z80,000, offering on-chip cache, memory management, pipelining, timing control, and burst mode, and its Z800 with peripherals, memory management, timing control, burst mode, and programmable cache, all on-chip.

Chapter 12

VLSI Random Access Memories

JEFF SCHLAGETER
CHING-LIN JIANG
ROBERT PROEBSTING

Mostek Corporation
Carrollton, Texas

I. Introduction	151
A. Random Access Memory Bit-Growth	151
B. Definitions and Comparisons	152
C. Basic RAM Organization	153
D. Voltage Requirements	153
II. Static RAM	154
A. Memory Cell	154
B. Decoders	156
III. Dynamic RAM	157
A. Background	157
B. Refresh	158
C. Address Multiplexing	159
D. Soft Errors	160
E. Available Devices	160
F. Outlook for Dynamic RAMs	160
IV. Specialty RAMs	160
A. Dual-Port RAM	161
B. Video RAM	162
C. Content Addressable Memory	163
D. Self-Contained Battery Backup Nonvolatile RAM	164
E. Nonvolatile Shadow RAM	164
Glossary	165

INTRODUCTION   I

Random Access Memory Bit-Growth   A

Since 1970, semiconductor memories have become pervasive and bit-growth has increased in geometric proportions, as shown in Fig. 1. As

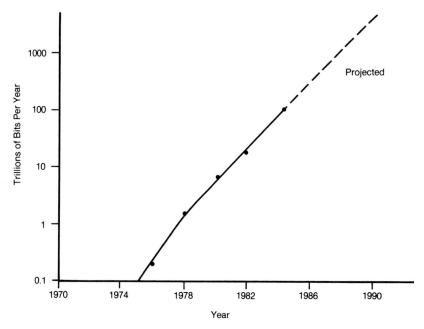

Fig. 1. Actual and projected memory bit growth versus year.

memory cost per bit declines, new applications become economically feasible, which continues to fuel bit-growth. The goal has been to create the highest density of bits in the smallest package at the lowest cost. The 64K dynamic RAM market in 1984 is estimated to be about 750 million units, with a total revenue of about $2.7 billion.

B Definitions and Comparisons

1 Random Access Memory

Random access memory (RAM) is also referred to as Read–Write (R/W) memory. Here data are stored (written) or retrieved (read), in comparable intervals of time. *Random access* means reading or writing data in any location, in any sequence, upon the application of appropriate address inputs.

2 MOS and Bipolar Technologies

Though MOS and bipolar technologies are referred to in this chapter, the focus is on MOS. Definitions of these terms and others are included in the glossary.

Comparison of RAM Technologies   3

A comparison of various RAM technologies is difficult because of differences in applications; however, Table I compares ranges of access time, density, quiescent power, and operating power available in RAMs in production in 1984.

The density of RAM has increased approximately by a factor of four every three years during the last 15 years. In 1984, 1.5-μm lithography has made possible a 256K dynamic RAM cell that occupies about the same area as a human blood cell.

Basic RAM Organization   C

An example of a simple RAM organization is shown in Fig. 2. This is a *by one* ($\times$1) memory because there is one input and one output. An input and output can share one pin (common I/O).

For an array of 8 $\times$ 8 or 64 bits, three row (X) addresses are required to select one of 8 words, and three column (Y) addresses are needed to select the desired bit in the word. Read/write (R/W) and chip enable (CE) control signals are also required.

Voltage Requirements   D

The voltage required for RAM operation has declined from 15 V to 5 V over the last decade as transistor channel length, junction depth, and gate oxide thickness have been reduced by approximately five times. Continued increases in memory density will require further voltage reductions.

Most MOS RAMs now operate off of V_{cc} = 5 V and V_{ss} = 0 V, which is down from four power pins a decade ago.

TABLE I

1984 RAM Density and Performance

RAM family	Access time (ns)	Highest density	Quiescent power (mW)	Operating power (mW)
Static NMOS	25–150	64K	500	500
Static CMOS	55–200	64K	0.05	500
Dynamic NMOS	80–200	256K	20	500
TTL bipolar	15–80	16K	750	750
ECL bipolar	5–25	16K	750	750

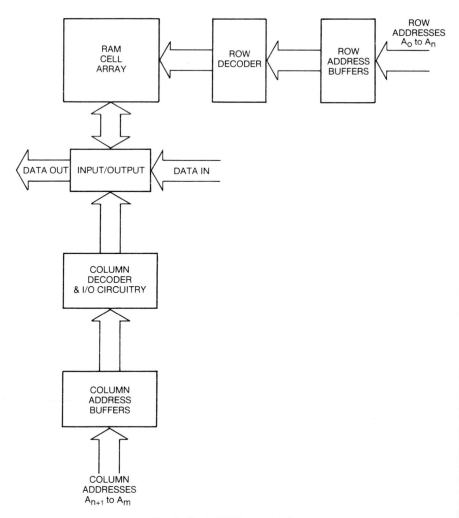

Fig. 2. Basic RAM organization.

II STATIC RAM

A modern static RAM consists of an array of static memory cells; control signals for enable, write, and read functions; address buffers and decoders; sensing amplifiers; and data input and output circuits.

A Memory Cell

The key element of a static RAM is the *static memory cell,* which can store binary digit data as long as power is supplied to the chip. The most

12. VLSI Random Access Memories

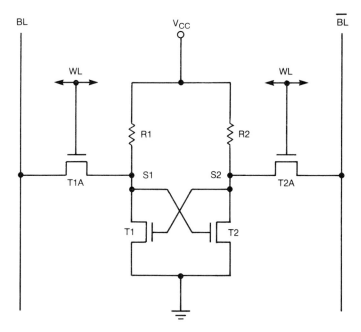

Fig. 3. The *n*-channel static RAM cell with polysilicon load resistors.

commonly used NMOS static RAM cell is shown in Fig. 3. It consists of two cross-coupled *n*-channel transistors T1 and T2, two intrinsic polysilicon load resistors R1 and R2, and two *n*-channel access transistors T1A and T2A. The resistance of R1 and R2 is usually much higher than $10^6 \, \Omega$, resulting in low cell power dissipation. The access of the cell is controlled by the wordline (WL). Data input and output are provided through a pair of bit lines (BL and $\overline{BL}$).

Writing Data 1

To write into the cell, WL has to be pulled to a high potential, turning T1A and T2A *on* and connecting the storage nodes S1 and S2 to the bit lines. Data are written into S1 from BL and complementary data into S2 from $\overline{BL}$.

The cell enters its store mode when WL goes to a low potential, isolating the cell from the bit lines. If a logic 1 is stored, S1 is at a high potential and S2 is low. In the cross-coupled transistor arrangement, S2 is held low by T2 while S1 is sustained to V_{cc} through the load resistor R1, and T1 is turned *off* by the low potential at S2. If a logic 0 is stored, the reverse is true. This data storage approach does not need refreshing.

2 Reading Data

To read data that is stored in the cell, WL has to go high. In the meantime, the bit-line pair should be pre-charged to a high potential. If a logic 1 is being read, BL remains high while $\overline{BL}$ is being pulled to lower potential. Thus, a differential signal is developed between BL and $\overline{BL}$. The column decoder connects the differential signal from the selected bit-line pair into the output circuit. A sense amplifier converts the differential signal into digital data.

3 Types of Cells

For a smaller die size, and therefore lower cost, polysilicon load resistors are commonly used in static RAM cells. For applications that require very low power dissipation, the full-CMOS static RAM cell shown in Fig. 4 is used. The full-CMOS cell usually contains twice the area of a comparable polysilicon resistor cell described above.

B Decoders

The binary coded addresses to the RAM are buffered and decoded to provide the row and column selections needed for the read and write

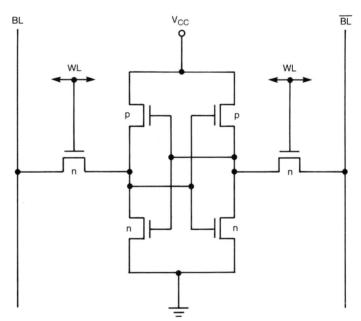

Fig. 4. Full CMOS static RAM cell.

12. VLSI Random Access Memories

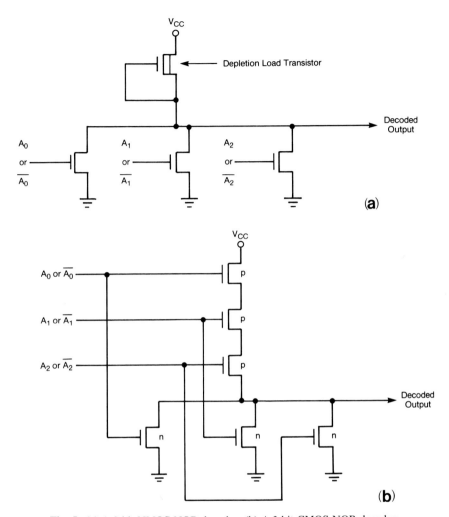

Fig. 5. (a) A 3-bit NMOS NOR decoder. (b) A 3-bit CMOS NOR decoder.

operations. A commonly used decoder is the NOR decoder, NMOS and CMOS versions of which are shown in Fig. 5(a) and (b), respectively. The CMOS NOR requires no dc power.

DYNAMIC RAM   III

Background   A

Dynamic MOS RAM, like static RAM, provides memory in which any arbitrary bit or word can be written or read in any cycle. However, the dynamic RAM cell is much simpler and, therefore, occupies much less area on an integrated circuit than is required by a static RAM cell. Typically, for a given set of process layout rules (feature sizes), a dynamic RAM cell occupies about one-quarter of the area of a comparable NMOS static RAM cell. Therefore, at a given time, dynamic RAMs typically have four times as many bits as static RAMs, yet cost about the same. This lower cost per bit dictates the use of dynamic RAM for most large memory applications. However, dynamic RAMs require more support circuitry to retain data than do static RAMs.

B Refresh

The extra support circuitry required by dynamic RAMs is obvious from an examination of the dynamic RAM cell (Fig. 6). The data are stored as charge on a capacitor. To write data, the selected word line (WL) is enabled, and the selected bit line (BL) is driven to the appropriate voltage. A logic 1 might be stored as +5 V and a logic 0 as 0 V. After writing, the word line is disabled, and the capacitor charge reflects the data stored in the cell.

Any capacitor, including the memory storage capacitor, discharges over a period of time. The self-discharge time of memory cell capacitors is relatively short, on the order of milliseconds to seconds depending on temperature. To retain data indefinitely, it is necessary to read the data periodically (before it has had time to discharge significantly) and to re-

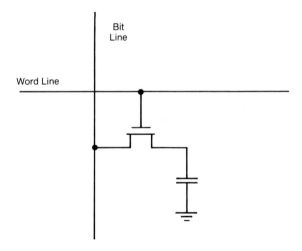

Fig. 6. One transistor (and one capacitor) RAM cell.

store the stored voltage to the previously written level. If a logic 1 is written and then read before it has decayed below 2 V, it is interpreted as 1. It is then automatically restored to its original 5-V level, only to decay once again. This restoration process, called *refresh,* must be done to every memory cell, typically once every 2–4 ms. Fortunately, an entire row is refreshed simultaneously so that individual cells need not be separately refreshed. The 256K Dynamic RAMs typically require 64 to 256 refresh cycles every 2–4 ms.

Some dynamic RAMs, especially ×8 (8 input–outputs) have built-in *refresh counters* to simplify the overhead circuitry required for refreshing. Such circuits need only be interrupted from normal read or write cycles and a refresh command given to the circuit. This command refreshes all bits in the row selected by the refresh counter and increments the counter to the next row. Every dynamic RAM in the memory system requires 64, 128, or 256 such interruptions and refresh commands every 2–4 ms to retain data. Typically, this will take 1–2% of real time, leaving 98 or 99% of the time available for useful memory operations.

Address Multiplexing C

The dynamic memory cell establishes a very small signal on the bit line that is detected by a sensitive sense amplifier. The bit line must not be disturbed until sensing is done, therefore the column selection must be delayed relative to row selection. This necessary delay means that column addresses are not needed by the circuit as early in a cycle as are the row addresses. This fact permits dynamic RAMs to use address multiplexing without access time penalty. With this concept, first the row addresses are presented to the circuit and locked in or strobed by a signal called *row address strobe* (RAS). This signal initiates the cycle enabling sensing and refreshing of all cells in the addressed row. Next, the column addresses are strobed by a signal called *column address strobe* (CAS), which transfers data from the selected column to the data output buffer. Address multiplexing permits both the row and the column addresses to be input to the circuit on the same pins. A 256K RAM (2^{18} bits) requires nine row addresses and nine column addresses. Using the same nine pins for both row and column addresses permits this (as well as earlier ×1 dynamic RAMs) to use a 16-pin dual-in-line package that allows greater RAM packing density on a board. Address multiplexing also permits page mode operation—multiple column cycles in the same row during a single row (RAS) cycle to give faster access and cycle times. Another high-bandwidth mode, called *nibble mode,* allows sequential accessing of multiple columns of the selected row.

D Soft Errors

Because memory cells are so small, alpha particles can cause the loss of data in any MOS memory. Alpha articles, emitted by the decay of trace amounts of radioactive materials in integrated circuit packages, enter the semiconductor and generate electron–hole pairs. These electrons cause a very temporary, very high leakage current on the nearest circuit node, possibly a memory cell or data line. The total charge lost can be more than the charge stored on the affected memory cell, in which case data are lost. A single alpha particles causes the loss of only one bit. The average time between such data losses is typically more than 1 million hours per error for a dynamic RAM containing 64K bits. Many memory systems incorporate *error correction*, in which extra bits check and correct errors when they occur.

E Available Devices

Most dynamic RAMs to date use address multiplexing, are packaged in a 16-pin dual-in-line package, and are organized with a single input and output. As RAM density has increased to 256K bits, the 256K ×1 organization has too many words for many microprocessor systems. Therefore, at the 256K level, and to a lesser extent at the 64K level, ×4 and ×8 organizations are available.

F Outlook for Dynamic RAMs

The capacity of dynamic RAMs has increased by a factor of four every three years since the 1K dynamic RAM was introduced in 1970. This trend is expected to continue through at least the 16 Mbit RAM level of integration, although the time between generations may be longer than before. As device geometries shrink, access times and cycle times will also decrease. Figure 7 shows the decrease of dynamic RAM cell size versus time.

IV SPECIALTY RAMS

A recent trend in MOS memories is to add features other than those usually found in a random access memory.

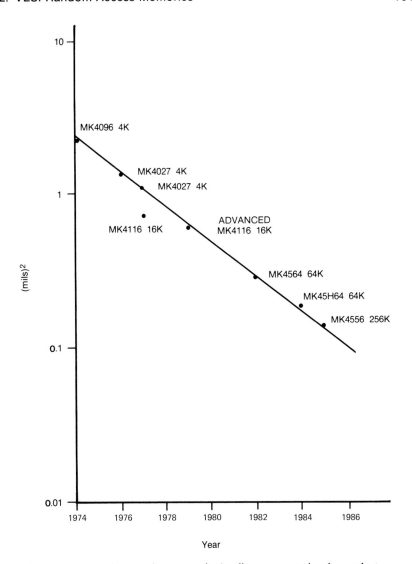

Fig. 7. One transistor (and one capacitor) cell area versus time by product.

Dual-Port RAM A

One such feature is the addition of a second input/output port so that two addresses can be independently read or written simultaneously. Such a device is called a *dual-port RAM*. A general purpose dual-port RAM has two sets of control signals, addresses, and data inputs/outputs for inter-

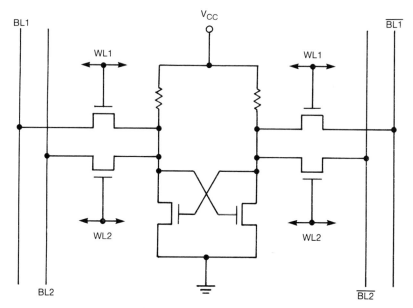

Fig. 8. Dual-port RAM cell.

facing two asynchronous systems that share RAM. This type of RAM can use a true dual-port memory cell, as shown in Figure 8, or it can use a single-port cell with contention logic circuitry. To provide synchronization and communication, it usually supports interrupt operations. Because a dual-port RAM connects two asynchronous systems, some form of system discipline is required to maintain meaningful communication between the two ports. Dual-port RAMs of density from 512 to 1024 bytes were available in 1984.

B Video RAM

A video RAM is a specialized dual-port RAM in which the first port is a standard input/output port. The second port is usually a read-only serial-format port, in which all bits in a row of memory cells are read in sequence for display on a CRT. In a single memory cycle this device transfers the serial data from a row of memory cells into a shift register during the horizontal retrace time of the CRT. Then, as the serial bits are output during the horizontal trace, the normal input/output port is available for reading or writing. Thus, the memory is available for reading and writing almost 100% of the time and yet is also available for video display.

Content Addressable Memory

Another modification to RAMs is the addition of logic to the memory cells creating forms of *smart memory*. One form of smart memory is *content addressable,* or *associative,* memory. A content addressable memory can be addressed either by standard physical location address or by content (stored data). The application of a content addressable memory can be seen by the following problem. Suppose 100 million people have their fingerprints on file in a computer data base. To find who belongs to a given fingerprint, the file would be searched, one set of prints at a time, until the match was found—after an average of 50 million attempts. A content addressable memory, on the other hand, would be addressed by the relevant content: the desired set of fingerprints. In one machine cycle, the matching data word would be flagged. In the next cycle, the entire word would be read: name, address, and so on. Content addressable memories can be very useful for some types of data processing. Other forms of smart memory, in which the memory itself can manipulate data, are also possible.

Fig. 9. Battery backup nonvolatile RAM (Courtesy of Mostek Corporation).

D Self-Contained Battery Backup Nonvolatile RAM

The advances in CMOS circuit design and technology allow the integration of high-density static RAM and analog voltage-sensing circuitry on one chip. Furthermore, the progress in battery technology has led to reliable, long-life, coin-sized lithium batteries. This combination of silicon IC and new battery technology produces a true nonvolatile RAM in a single package (Fig. 9). The 1983 introduction of a 2K×8 ZERO-POWER™ nonvolatile RAM, the MK48Z02, was a milestone in this area.

E Nonvolatile Shadow RAM

Another approach to nonvolatile RAM combines an electrically programmable cell and an ordinary static RAM cell. The device operates as a typical static RAM when the power is on. When power failure occurs, the chip immediately saves the data stored in each static RAM cell into its nonvolatile backup, or shadow, cell. When the power is turned back on, the chip recalls the data saved in each shadow cell and transfers them to its static cell (Fig. 10). The trend in this field is toward single power supply

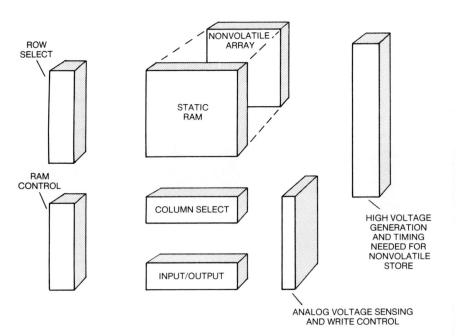

Fig. 10. Nonvolatile shadow RAM architecture.

12. VLSI Random Access Memories

and user friendly features. The density of shadow RAM is typically a factor of four less than that of a CMOS static RAM.

GLOSSARY

Bipolar technology utilizing electron and hole charge carriers with current flow in one direction.
CMOS complementary MOS technology utilizing both n-channel and p-channel MOS transistors.
Dynamic memories requiring refreshing to keep cell capacitances charged.
ECL emitter coupled logic (fast bipolar).
K refers to 1024 (binary 2^{10}).
Lithography refers to the average line width and spacing created on integrated circuit masking layers.
NMOS metal oxide semiconductor (MOS) technology utilizing n-channel transistors.
Refresh the periodic rewriting of stored information to prevent its loss.
Static memories requiring no refreshing.
TTL transistor–transistor logic (bipolar).
Volatile stored information is lost if power is removed.

Chapter 13

VLSI Electrically Erasable Programmable Read Only Memory

S. K. LAI
V. K. DHAM

INTEL Corporation
Santa Clara, California

I. Principle of Operation	167
A. Memory Cell Structure	167
B. Fowler–Nordheim Tunneling	169
C. Coupling Equations for Program and Erase	170
II. Programming Characteristics	172
A. Threshold Voltage Window as a Function of Program Voltage and Time	172
B. Cycling and Electron Trapping	173
III. Performance and Reliability	174
A. Retention	174
B. Read Disturb	174
C. Reliability	175
IV. Scaling	175
List of Symbols	176
References	176

PRINCIPLE OF OPERATION   I

Memory Cell Structure   A

An electrically erasable programmable read only memory (EEPROM) is a programmable read only memory (ROM) that can be erased and reprogrammed electrically in circuit. There are various technologies for EEPROMs, but the most widely used is based on floating gate tunnel oxide (FLOTOX) [1]. It is a technology that was developed from the floating gate EPROM technology, which has proven manufacturability and reliability. The technology is also compatible with standard NMOS and CMOS technologies, allowing fabrication of standard high-performance circuits in association with the memory.

1 Operation of the FLOTEX EEPROM

The basic structure of the FLOTOX EEPROM is shown in Fig. 1. The floating gate is surrounded by high-quality insulators (silicon dioxide) and is not connected electrically to any other electrodes. The significant difference between a FLOTOX EEPROM and an EPROM is the tunnel oxide, in the range of 100 Å thick, which is the key element in the EEPROM. In programming, a high voltage is applied to the top gate formed by Poly 2, with the source and drain grounded. A fraction of the high voltage is coupled capacitively across the tunnel oxide. Electrons flow from the drain to the floating gate through Fowler–Nordheim tunneling [2]. As electrons build up on the floating gate, they lower the electric field across the tunnel oxide, reducing the electron flow. Thus, the electron flow is a self-limiting process. In the reverse process of erase, the top gate is grounded, the source is floating, and high voltage is applied to the drain. Again, a fraction of the high voltage is coupled across the tunnel oxide, but in the opposite direction. Electrons flow from the floating gate to the drain, giving net positive charge on the floating gate.

2 Floating Gate in a MOSFET

The floating gate also forms the gate of a MOSFET, which is used to sense the charge state. In the programmed state, there is negative charge on the floating gate, resulting in the shift of the threshold voltage in the positive voltage direction. After erase, there is net positive charge on the floating gate. The threshold voltage is then shifted to a negative value.

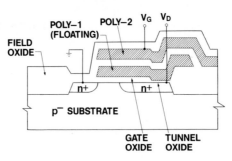

Fig. 1. Cross-section structure of a FLOTOX EEPROM showing the different poly and oxide layers, together with the different terminals.

EEPROM in an Array

The floating gate transistor is the basic memory element for the EEPROM. However, when used in an array, the single element is not sufficient for the proper function of the memory, and a second transistor called the *select transistor* is required. There are two important reasons for the select transistor. First, there is the problem of *disturb*. Without the select transistor, high voltages applied to one drain also appear on the drains of other cells in the same column and would give *erase* for the nonselected cells; the select transistor disconnects nonselected drains. Second, the floating gate transistor can be erased to a negative threshold. Without the select transistor, cross connection occurs in an array. The select transistor is a necessary part of the EEPROM memory and limits the scaling of the memory cell for high density.

Fowler–Nordheim Tunneling

The principle mechanism for the transfer of electrons through silicon dioxide is by the Fowler–Nordheim tunneling process [2]. At very high electric field, the barrier for electron tunneling is reduced to the point that significant numbers of electrons can flow across an otherwise perfect insulator. The *tunneling probability* can be calculated from quantum mechanics; the *tunneling current* in the Fowler–Nordheim regime is given by the following equation:

$$J = \alpha E^2 \exp(-\beta/E), \tag{1}$$

where $\alpha = 9.625 \times 10^{-7}$ and $\beta = 2.765 \times 10^8$.

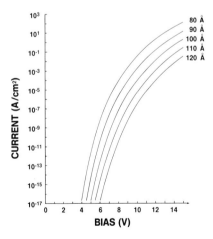

Fig. 2. Calculated Fowler–Nordheim tunneling current for different tunnel oxide thicknesses. The curves are very close to experimental ones under positive biases. There is a voltage drop in silicon for negative biases, and the IV curves are shifted to higher voltages.

The equation has been well proven experimentally. Calculated *IV curves for tunnel oxides* from 80 to 120 Å are shown in Fig. 2. It can be seen that the current changes by many orders of magnitude over the voltage range. The very steep IV slope is the basis of operation for the FLOTOX EEPROM. Programming an EEPROM in 10 ms takes an average current density of 0.001 A/cm^2, even though the peak current may be much larger. The voltage that gives such a current density is a good measure of the voltage that has to be coupled across the tunnel oxide for programming. The IV slope becomes very steep at low voltages. To achieve the desired ten-year retention goal, there has to be approximately an 11-decade difference in tunnel current from program and erase to store. The voltages at the current level of 10^{-14} A/cm^2 are approximately the maximum voltage on the tunnel oxide that can exist during storage or read conditions for the different oxide thicknesses. The above represents the design points for the operation of the tunnel oxide.

C Coupling Equations for Program and Erase

1 The Read and the Program and Erase Conditions

The operation of the EEPROM sense transistor can be understood by analyzing it in two regimes. In the *read condition,* the transistor is a floating gate MOSFET with the threshold voltage determined by the charge on the floating gate. The only significant difference in the performance of the transistor is that the tunnel oxide gives high drain coupling, with the result that the threshold voltage of the transistor is sensitive to the drain voltage. In fact, the change in threshold voltage of the transistor as a function of the drain voltage is a measure of the drain coupling.

In the high-field region, during the *program and erase condition,* the cell operates in a different regime. For all practical purposes, the transistor action is no longer important, and the cell is essentially a number of capacitors tied to the floating gate (Fig. 3). The symbols are defined in the List of Symbols.

The performance of the cell can be understood by analyzing the dynamics of the charge transfer to the floating gate through capacitive volt-

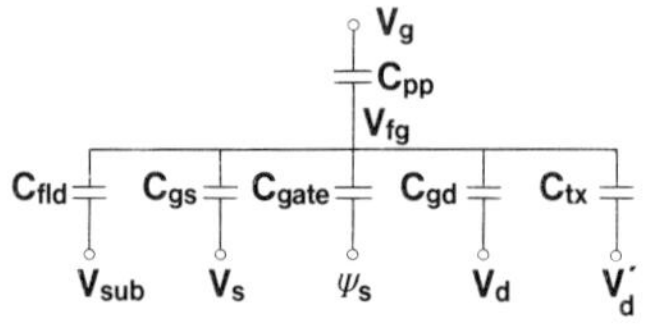

Fig. 3. Capacitor network representing the basic operation of the cell during high-voltage program or erase. The symbols are defined in the List of Symbols.

13. VLSI Electrically Erasable Programmable Read Only Memory

age coupling. The coupling equations are

$$V_{fg} = \frac{Q_{fg}}{C_t} + V_g \frac{C_{pp}}{C_t} + V_{sub} \frac{C_{fld}}{C_t} + V_s \frac{C_{gs}}{C_t}$$

$$+ V_d \frac{C_{gd}}{C_t} + V'_d \frac{C_{tx}}{C_t} + (V_{fb} + \psi_s) \frac{C_{gate}}{C_t}, \qquad (2)$$

$$V_{fg} - V_{fb} = \frac{Q_{channel}}{C_{gate}} + \psi_s. \qquad (3)$$

These equations were used in a computer simulation that gave an accurate prediction of the cell performance. The only complication occurs during erase, when the transistor surface potential changes from accumulation to inversion. The semiconductor capacitance under the gate oxide decreases as a function of time, resulting in a decrease in the coupling ratio of voltages to the floating gate as a function of time.

Program Coupling Ratio and Tunnel Oxide Thickness 2

Figure 4 shows the simulated effect of program coupling ratio and tunnel oxide thickness on the change in threshold voltage during programming: for any required program window (as determined by sensing and

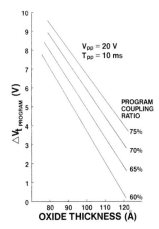

Fig. 4. The change in the sense transistor threshold voltage from a neutral state after programming, as a function of oxide thickness, and for different program coupling ratios (from stimulation). The program voltage is 20 V, and the program pulse width is 10 ms. A higher program coupling ratio usually means a larger cell area. In principle, small changes in the threshold voltage can be sensed. In real design, one has to consider changes with processing, window distribution, temperature variation, window closing after many cycles, and other reliability requirements. Typically, a few volts are required to satisfy all the conditions.

margin requirement), the thinner the oxide, the lower the coupling ratio required. Lower coupling ratio has the advantage of smaller cell area. On the other hand, for yield and reliability, it is desirable to use a thicker tunnel oxide, and larger coupling ratio is required, giving a larger cell. The design of an EEPROM is a tradeoff between oxide thickness and cell area.

II PROGRAMMING CHARACTERISTICS

A Threshold Voltage Window as a Function of Program Voltage and Time

The change in threshold voltage after program and erase is a sensitive function of the *program and erase voltage* used (Fig. 5). For a given coupling ratio, there exists a simple relationship: For every volt increase in program voltage, the threshold voltage change will also increase by one volt. The relationship for erase is similar: The change in threshold voltage is slightly more than one volt per volt increase in drain voltage. The result is that for every volt increase in operating voltage, the cell window is increased by *two* volts. Figure 5 also shows the dependence of threshold voltage on tunnel oxide thickness. For every 10 Å change in tunnel oxide thickness and a given operating voltage, the total program and erase window is changed by approximately three volts.

The threshold voltages are also sensitive functions of the *programming*

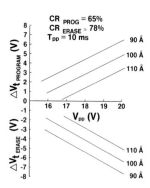

Fig. 5. Changes in program and erase threshold voltages as a function of applied high voltages for different oxide thickness (from simulation). The program coupling ratio (CR) is 65%, whereas the erase coupling ratio is approximately 78%. The pulse width is 10 ms. The one volt for one volt relationship is shown clearly. Also, it can be seen that because of the higher erase coupling ratio, the change in erase threshold voltage is larger for a given applied voltage.

13. VLSI Electrically Erasable Programmable Read Only Memory

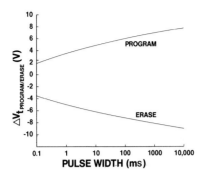

Fig. 6. Changes in program and erase threshold voltages as a function of pulse width. The conditions were the same as in Fig. 5, with the oxide being 100 Å thick and the applied voltage 20 V.

time. The dependence of the program and erase window as a function of the logarithmic of pulse width is shown in Fig. 6. There is an initial period of faster change, and then slower changes occur after 10 to 100 ms. The programming time was chosen to be 10 ms in the early generations of EEPROMs, but the trend now is to reduce that to shorter times. Again, there is a tradeoff between window and programming time.

Cycling and Electron Trapping B

An EEPROM is designed to be programmed and erased more than once. A typical specification calls for up to 10,000 program and erase cycles. Ideally, there should not be any degradation of the performance and reliability of the cell with cycling; but in reality, cycling tends to degrade the quality of the tunnel oxide. One of the degradation processes is due to the trapping of electrons in the tunnel oxide [3]. The effect of electron trapping is the buildup of a permanent negative charge in the

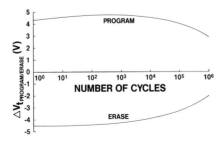

Fig. 7. Changes in program and erase window as a function of the number of program and erase cycles (from experiments). The window closing after 10,000 cycles was due to electron trapping in the tunnel oxide.

oxide, which reduces the electric field for the same applied voltage. The current will be reduced, giving a smaller window. The effect shows up as the window-closing phenomenon, as shown in Fig. 7. Significant closing occurs after about 10,000 cycles. It is important in designing EEPROM to have a large enough initial window to allow for the required number of operation cycles.

III PERFORMANCE AND RELIABILITY

A Retention

The ten-year retention goal for EEPROM, mentioned earlier, is not a problem if the devices are properly designed and processed. In fact, retention is used as a tool for evaluating the quality of the memory and forms the basis for screening the product. Any cell that shows a retention problem over some accelerated test conditions will probably fail the retention goal and should be discarded. Typical retention problems arise from the failure of the tunnel oxide in some defect-related mechanism.

B Read Disturb

There is a condition called *read disturb* that produces unwanted changes in the threshold voltage. Read disturb occurs when voltages are applied to the cell for reading. Although the applied voltage is very low, there may still be some finite tunnel currents that, over a very long time, change the threshold voltage. Read disturb can be evaluated by accelerated testing using the one volt for one volt relation (Fig. 8). In the test, a

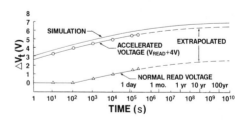

Fig. 8. Changes in the threshold voltage of a partially erased cell as a function of time under normal read conditions and also under accelerated test conditions, together with the result of simulation. The 4-V higher test voltage gives a 4-V higher threshold voltage change. Together with the time dependence, one can use the accelerated test conditions to evaluate read disturb in the cell.

voltage four volts higher than the normal read voltage is applied to the gate, and the change in threshold voltage as a function of time is recorded. The expected change from normal threshold voltage can be obtained by subtracting 4 V from the accelerated threshold voltage. It can be seen that even though there are changes in threshold voltages due to read disturb, the change is so small over ten years that it does not affect data retention. It is important in the design to evaluate the read disturb and have sufficient margin for the disturb condition.

C. Reliability

The most important quality for a successful product is that it be reliable over the lifetime of the product. Problems in the reliability of EEPROM are usually related to insulators [4]. First, because of the high voltages used in the circuits, there is a high probability for failure due to time-dependent breakdown of the periphery gate oxides. Second, there is a finite probability that the tunnel oxide will fail because of the high electric field across it. Typical failures are catastrophic, with the tunnel oxide becoming very leaky. At this time, the failure rate for FLOTOX EEPROM is typically in the range of 0.4% or less per thousand cycles. It is expected that improvement in the quality and cleanliness of the process will decrease the failure rate and improve the endurance of the memory.

IV. SCALING

The big push in VLSI is to scale devices to smaller dimensions so that more devices can be packed into an area to give higher memory density or more logic functions. The scaling of the EEPROM cell is complicated. The most fundamental requirement for EEPROM scaling is that the electric field across the tunnel oxide has to be constant to maintain the reliability and performance of the cell [5]. Cell areas and dielectric thicknesses are then adjusted to satisfy the constant field condition. Given the very thin tunnel oxide that is being used, there is not very much room for scaling of its thickness. Furthermore, for oxides down to about 60 Å, direct tunneling may limit the retention of the cell, not to mention the more basic question of yield and manufacturability of the oxide. This is the factor that most limits the scaling of EEPROM. When tunnel oxide thickness is not reduced, the operating voltage cannot be reduced. The minimum channel length of the transistors are not reduced, nor are the isolation distance between devices. So far, most of the improvement in cell size has come from improvement in *lithography* and innovations in *cell layout*. In lithography, the tunnel oxide area provides the most leverage. A smaller tunnel oxide area gives a smaller cell area.

LIST OF SYMBOLS

C_{fld}	Floating gate over field oxide capacitance
C_{gate}	Floating gate over gate oxide capacitance
C_{gd}	Floating gate to drain capacitance (excluding tunnel area)
C_{gs}	Floating gate to source capacitance
C_{pp}	Poly to poly oxide capacitance
C_t	$C_{fld} + C_{gate} + C_{gd} + C_{gs} + C_{pp} + C_{tx}$
C_{tx}	Tunnel oxide capacitance
E	Electric field
J	Current density
$Q_{channel}$	Charge in the channel region
Q_{fg}	Charge on the floating gate
T_{pp}	Program and erase pulse width
V_{fb}	Flat band voltage
V_d	Drain voltage
V_d'	Effective drain voltage in tunnel oxide region
V_{fg}	Floating gate voltage
V_g	Gate voltage to poly 2
V_{pp}	Program and erase voltage
V_s	Source voltage
V_t	Transistor threshold voltage
ψ_s	Surface potential for channel

REFERENCES

1. W. S. Johnson, G. Perlegos, A. Renninger, G. Kuhn, and T. R. Ranganath, *ISSCC Dig. Tech. Papers,* 152–153 (1980).
2. M. Lenslinger and E. Snow, *J. Appl. Phys.* **40,** 278–283 (1969).
3. C. S. Jenq, T. R. Ranganath, C. H. Huang, H. S. Jones, and T. T. L. Chang, *IEDM Tech. Dig.,* 388–391 (1981).
4. B. Euzent, C. Jenq, J. Lee, and N. Boruta, *Proc. 19th Annu. Reliability Phys. Symp.,* 11–16 (1981).
5. J. Lee and V. K. Dham, *IEDM Tech. Dig.,* 589–591 (1983).

Chapter 14
Electrical Transport Properties of Silicon*

W. ROBERT THURBER
JEREMIAH R. LOWNEY

Semiconductor Materials and Processes Division
National Bureau of Standards
Washington, D.C.

I.	Introduction	177
II.	Definition of Transport: The Transport Equation	178
III.	Conversion between Resistivity and Dopant Density	178
IV.	Mobility of Charge Carriers	181
	A. Hall Mobility	181
	B. Conductivity Mobility	181
V.	Temperature Dependence of Resistivity and Mobility	185
VI.	Dependence of Drift Velocity on Electric Fields	185
VII.	Minority-Carrier Mobility, Lifetime, and Diffusion Length	186
VIII.	Mobility in an MOS Inversion Layer	188
	References	189

INTRODUCTION I

This chapter discusses

(a) the transport of electrons and holes in silicon,
(b) the dopant density and temperature dependencies of majority mobilities,
(c) the effects of high electric fields,
(d) the differences between minority and majority mobilities,
(e) transport in an MOS inversion layer.

---

* This contribution was prepared as part of the Semiconductor Technology Program at the National Bureau of Standards.

II DEFINITION OF TRANSPORT: THE TRANSPORT EQUATION

For a cubic crystal such as silicon, one can obtain from Boltzmann transport theory the charge–current density equations for electrons and holes:

$$\mathbf{J}_n = q(\mu_n n \mathbf{E} + D_n \nabla \mathbf{n}), \tag{1}$$

$$\mathbf{J}_p = q(\mu_p p \mathbf{E} - D_p \nabla \mathbf{p}), \tag{2}$$

where $\mathbf{J}_n$ and $\mathbf{J}_p$ are the electron and hole charge–current densities, respectively; μ_n and μ_p are the electron and hole mobilities, respectively; n and p are the electron and hole densities, respectively; D_n and D_p are the electron and hole diffusion coefficients, respectively; $\mathbf{E}$ is the electric field, and q is the electronic charge. A relation exists between the diffusion coefficients and mobilities [1]:

$$\frac{D_i}{\mu_i} = \frac{2kT}{q} \frac{F_{1/2}(E_F/kT)}{F_{-1/2}(E_F/kT)}, \qquad i = n,p, \tag{3}$$

where $F_{1/2}$ and $F_{-1/2}$ are Fermi–Dirac integrals, E_F is the Fermi energy with respect to the appropriate carrier-band edge (the sign of the energy axis is reversed for holes), k is Boltzmann's constant, and T is absolute temperature. In the limit of nondegenerate statistics $D_i/\mu_i = kT/q$. Therefore, without loss of generality, we can omit a separate discussion of diffusion coefficients.

Two continuity equations exist:

$$dn/dt = G - (n - n_0)/\tau_n(n,p) + (1/q) \nabla \cdot \mathbf{J}_n, \tag{4}$$

$$dp/dt = G - (p - p_0)/\tau_p(n,p) - (1/q) \nabla \cdot \mathbf{J}_p, \tag{5}$$

where G is any nonequilibrium generation rate of electron–hole pairs (e.g., by light), $\tau_n(n,p)$ and $\tau_p(n,p)$ are the recombination–generation lifetimes of electrons and holes, respectively (which can depend on the values of both n and p), and n_0 and p_0 are the thermal equilibrium values of electron and hole density. Together with Poisson's equation for the electric field, Eqs. (1)–(5) constitute the basis for the solution of the carrier densities of any semiconductor device from which one may obtain a current–voltage relationship.

III CONVERSION BETWEEN RESISTIVITY AND DOPANT DENSITY

The ASTM Standard Practice F723 [2] gives conversions between resistivity and dopant density based on recent data for boron- and phosphorus-doped silicon. These conversions are shown in graphical form in Fig. 1

14. Electrical Transport Properties of Silicon

and in computational form in Table I. ASTM F723 shows the conversions in tabular form. For phosphorus-doped material, the ASTM conversion differs by no more than 15% from the curve determined for n-type silicon by Irvin [3], the most used relationship prior to 1982. However, for boron-doped material the ASTM conversion differs from Irvin's curve for p-type silicon by up to 50%.

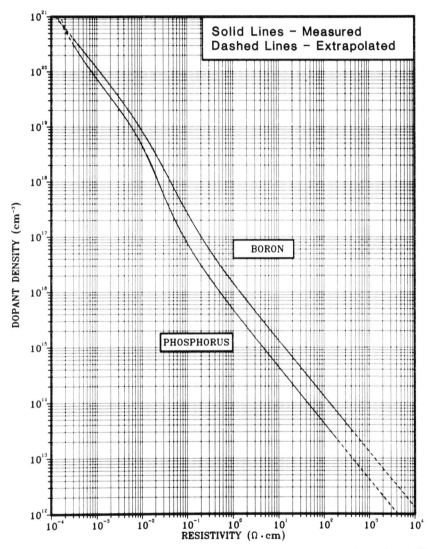

Fig. 1. Dopant density versus resistivity at 23°C for boron- and phosphorus-doped silicon. The solid lines show the resistivity to dopant density conversion for the regions of observed data. Dashed lines are shown in regions of extrapolation from the data. (Reproduced from Standard Practice F723 [2] with permission from the American Society for Testing and Materials, 1916 Race Street, Philadelphia, PA 19103.)

TABLE I
Conversions between Resistivity and Dopant Density for Silicon

	Resistivity ρ (Ω cm) to dopant density N (cm^{-3})	Dopant density N (cm^{-3}) to resistivity ρ (Ω cm)
Boron-doped	$N = \dfrac{1.330 \times 10^{16}}{\rho} + \dfrac{1.082 \times 10^{17}}{\rho[1 + (54.56\rho)^{1.105}]}$	$\rho = \dfrac{1.305 \times 10^{16}}{N} + \dfrac{1.133 \times 10^{17}}{N[1 + (2.58 \times 10^{-19} N)^{-0.737}]}$
Phosphorus-doped	$N = \dfrac{6.242 \times 10^{18}}{\rho} \times 10^Z$	$\rho = \dfrac{6.242 \times 10^{18}}{N} \times 10^Z$
	$Z = \dfrac{A_0 + A_1 x + A_2 x^2 + A_3 x^3}{1 + B_1 x + B_2 x^2 + B_3 x^3}$	$Z = \dfrac{A_0 + A_1 y + A_2 y^2 + A_3 y^3}{1 + B_1 y + B_2 y^2 + B_3 y^3}$
	$x = \log_{10} \rho \qquad A_0 = -3.1083$	$y = (\log_{10} N) - 16 \qquad A_0 = -3.0769$
	$B_1 = 1.0265 \qquad A_1 = -3.2626$	$B_1 = -0.68157 \qquad A_1 = 2.2108$
	$B_2 = 0.38755 \qquad A_2 = -1.2196$	$B_2 = 0.19833 \qquad A_2 = -0.62272$
	$B_3 = 0.041833 \qquad A_3 = -0.13923$	$B_3 = -0.018376 \qquad A_3 = 0.057501$

Differences between the two conversions are discussed in considerable detail in the ASTM document. In the experimental work [4,5] supporting these conversions, the capacitance–voltage (C–V) technique was used for dopant densities of less than 10^{18} cm^{-3}, because it measures the fully ionized net dopant density. For dopant densities from 10^{18} to 10^{19} cm^{-3}, neutron-activation analysis, spectrophotometric analysis, and the nuclear-track technique were used; and for those above 10^{19} cm^{-3}, Hall-effect measurements were used to obtain the carrier density, which is equal to the dopant density because the bound dopant state has disappeared at these densities. However, for specimens heavily doped with phosphorus, the fraction of electrically active atoms decreases rapidly for densities greater than 10^{20} cm^{-3} because of the formation of complexes [6]. In the heavily doped range in Fig. 1, the phosphorus density is equated to the carrier density because the latter determines the electrical properties.

IV MOBILITY OF CHARGE CARRIERS

A Hall Mobility

When the dopant is fully ionized, the mobility determines the temperature dependence of the electrical conductivity σ of an extrinsic semiconductor, that is, $\sigma = q\mu_n n$ for n-type material. The Hall mobility μ_H is defined by $\mu_H = |R_H|\sigma$ where R_H is the Hall coefficient given by $R_H = -r/qn$ in n-type material. The Hall proportionality factor r is a function of the band structure and scattering mechanisms. Thus, the relationship between the two mobilities is $\mu_H = r\mu_n$. Values of r are given in Table II. Typical curves of Hall mobility as a function of temperature are given in Fig. 2 for both n- and p-type silicon. For dopant densities of less than 10^{15} cm^{-3}, the mobility at room temperature is determined mainly by lattice scattering. With increased dopant density the mobility is reduced because of ionized impurity and carrier–carrier scattering. At lower temperatures these additional scattering mechanisms greatly reduce the mobility even for lightly doped material.

B Conductivity Mobility

The conductivity mobility, rather than the Hall mobility, is usually needed for electron transport expressions. It is important to note that the carrier density is not equal to the dopant density in the 10^{17}–10^{19} cm^{-3}

TABLE II

Hall Proportionality Factor r at 300 K

Dopant density	n-type		p-type	
	Value	References	Value	References
$<10^{19}$ cm^{-3}	1.0–1.3	[7–10]	0.7–0.9	[7,11,12]
$>10^{19}$ cm^{-3}	1.0–1.1	[9,10]	0.7–0.8	[13]

range for dopants with activation energies of several tens of millivolts because of partial deionization of the dopant atoms. Since this deionization depends on the activation energy of the dopant level, the effect is more pronounced for a deeper level, such as indium, than for a shallower level, such as boron or phosphorus.

Taking deionization into account, Thurber et al. [4,5] calculated conductivity mobility values from their resistivity and dopant density measurements on phosphorus- and boron-doped silicon for dopant densities from 10^{13} to 10^{20} cm^{-3}. Measurements of mobility in the heavily doped range using incremental sheet resistance and Hall effect have been re-

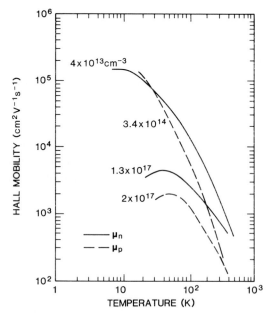

Fig. 2. Hall mobility of electrons and holes as a function of temperature for several values of dopant density. (From references [8,14,15].) For lightly doped material the mobility near room temperature varies as $T^{-2.44}$ for electrons and $T^{-2.20}$ for holes.

ported by Masetti et al. [16] for B-doped silicon, by Masetti and Solmi [17] and Finetti et al. [18] for P-doped silicon, and by Masetti et al. [16] and Solmi et al. [19] for As-doped silicon. In the overlapping regions, these results for diffused and implanted-annealed silicon are in good agreement with those of Thurber et al. for as-grown material. Masetti et al. [16] derived empirical models that gave satisfactory fits to the combined data for carrier densities from 10^{13} to 5×10^{21} cm^{-3}. For B-doped silicon their expression is

$$\mu_p = \mu_0 e^{-p_c/p} + \frac{\mu_{\max}}{1 + (p/C_r)^\alpha} - \frac{\mu_1}{1 + (C_s/p)^\beta}. \tag{6}$$

For As- and P-doped silicon, their expression is

$$\mu_n = \mu_0 + \frac{\mu_{\max} - \mu_0}{1 + (n/C_r)^\alpha} - \frac{\mu_1}{1 + (C_s/n)^\beta}. \tag{7}$$

The parameters in these equations are given in Table III. Figure 3 is a plot of these equations for the three dopants. The most heavily doped region ($>5 \times 10^{20}$ cm^{-3}), which is above the solubility limit of the dopants, is based on measurements of metastable layers formed by laser annealing, which in general are not stable during device fabrication. Note that the mobility in As-doped silicon decreases faster with carrier density than that in P-doped silicon. Most previous models did not have different parameters for these two n-type dopants. Sb-doped silicon in the 10^{19} cm^{-3} range [20] has a higher electron mobility than does P-doped silicon of the same carrier density. The variation between dopants is a consequence of their departure from ideal point charges. The decrease in mobil-

TABLE III

Parameters for the Fit of the Carrier Mobility versus Carrier Density at 300 K, Using Eqs. (6) and (7)[a]

	Arsenic	Phosphorus	Boron	Units
μ_0	52.2	68.5	44.9	cm^2/V·s
$\mu_{\max}$	1417	1414	470.5	cm^2/V·s
μ_1	43.4	56.1	29.0	cm^2/V·s
C_r	9.68×10^{16}	9.20×10^{16}	2.23×10^{17}	cm^{-3}
C_s	3.43×10^{20}	3.41×10^{20}	6.10×10^{20}	cm^{-3}
α	0.680	0.711	0.719	
β	2.00	1.98	2.00	
p_c	—	—	9.23×10^{16}	cm^{-3}

[a] Reprinted with permission from [16] G. Masetti, M. Severi, and S. Solmi, *IEEE Trans. Electron Devices* **ED-30**, 764 (1983), © 1983, IEEE.

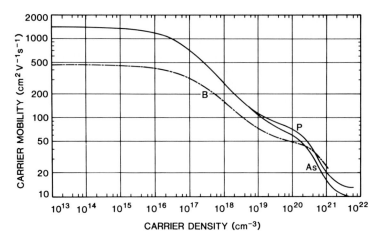

Fig. 3. Carrier mobility versus carrier density for silicon doped with boron, phosphorus, or arsenic. These curves were calculated from Eqs. (6) and (7) using the parameters given in Table III.

ity correlates with the strength of the scattering by the ions and, therefore, with the depth of the bound-state energy of the ions at low density. Figure 4 is a graph of carrier density versus resistivity for the dopants boron, phosphorus, and arsenic. These curves were generated by using Eqs. (6) and (7) for the mobility.

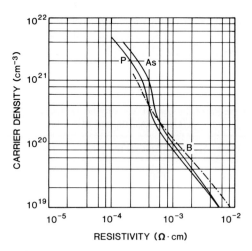

Fig. 4. Carrier density versus resistivity for silicon doped with boron, phosphorus, or arsenic. The curves were calculated using Eqs. (6) and (7) for the carrier mobility. (After Masetti et al. [16] © 1983 IEEE.)

V TEMPERATURE DEPENDENCE OF RESISTIVITY AND MOBILITY

The temperature dependence of the *resistivity* near room temperature in lightly doped silicon primarily reflects changes in the mobility with temperature. However, in the 10^{17}–10^{19} cm^{-3} range, the carrier density changes because of deionization of the dopant atoms, which varies with temperature. Bullis et al. [21] measured the temperature coefficient of resistivity near room temperature, and the ASTM Standard Method F84 [22] has a table of coefficients based on that work for the range 18–28°C as a function of resistivity for *n*- and *p*-type silicon. Calculations and experimental results covering a much wider temperature range are also available [23–25].

Arora et al. [26] have arrived at empirical relations for the *mobility* as a function of both dopant density and temperature. These relations give reasonable fits except in the heavily doped range where the expressions do not model the decrease seen in Fig. 3. For hole mobility in (cm^2/V · s) in n-doped silicon:

$$\mu_h = 54.3 T_n^{-0.57} + \frac{1.36 \times 10^8 T^{-2.23}}{1 + [N/(2.35 \times 10^{17} T_n^{2.4})]0.88 T_n^{-0.146}}, \quad (8)$$

where $T_n = T/300$, T is in kelvins, and N is the dopant density in (cm^{-3}). Similarly, for the electron mobility in P-doped silicon:

$$\mu_e = 88.0 T_n^{-0.57} + \frac{7.4 \times 10^8 T^{-2.33}}{1 + [N/(1.26 \times 10^{17} T_n^{2.4})]0.88 T_n^{-0.146}}. \quad (9)$$

VI DEPENDENCE OF DRIFT VELOCITY ON ELECTRIC FIELDS

At high electric fields, when the drift velocity becomes comparable to the thermal velocity, the linear relationship of velocity with field becomes sublinear with saturation occurring near 10^5 V/cm. Jacoboni et al. [27], in their review of some charge-transport properties of silicon, consider impurity and temperature effects on the drift mobility in connection with the electric field dependence. The drift velocity in high-purity bulk silicon as a function of field at four different temperatures is shown in Fig. 5 for electrons and holes. These curves were generated using the equation [28]

$$v_d = v_m \frac{E/E_c}{[1 + (E/E_c)^\beta]^{1/\beta}}, \quad (10)$$

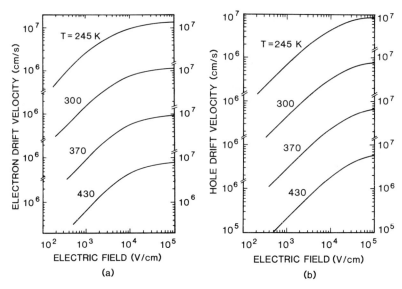

Fig. 5. Electron (a) and hole (b) drift velocity in high-purity silicon as a function of electric field at four different temperatures for carriers traveling in the [111] direction. The curves were obtained from Eq. (10) using the parameters in Table IV. (After Jacoboni *et al.* [27] © 1977, Pergamon Press, Ltd.)

where values of the parameters that gave the best fits to experimental data for the fields and temperatures shown in Fig. 5 are given in Table IV.

VII MINORITY-CARRIER MOBILITY, LIFETIME, AND DIFFUSION LENGTH

It has been customary to assume that the minority-carrier *mobility* is equal to the mobility that the carrier would have if it were a majority

TABLE IV

Parameters for the Electric Field and Temperature Dependence of Electron and Hole Drift Velocities in High-Purity Silicon Using Eq. (10)[a]

	Electrons[b]	Holes[b]	Units
v_m	$1.53 \times 10^9 \times T^{-0.87}$	$1.62 \times 10^8 \times T^{-0.52}$	cm/s
E_c	$1.01 \times T^{1.55}$	$1.24 \times T^{1.68}$	V/cm
β	$2.57 \times 10^{-2} \times T^{0.66}$	$0.46 \times T^{0.17}$	—

[a] Reprinted with permission from [27] C. Jacoboni, C. Canali, G. Ottaviani, A. Alberigi-Quaranta, *Solid-State Electron.* **20**, 77 (1977), © 1977, Pergamon Press, Ltd.

[b] T is in kelvins.

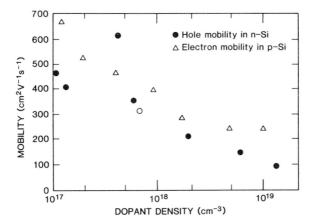

Fig. 6. Minority-carrier mobilities versus dopant density at 300 K for n- and p-type silicon as measured by Dziewior and Silber [29]. The triangles and solid circles were determined by measuring the complex diffusion length of minority carriers generated by optical excitation. The open circle for hole mobility was obtained from static diffusion length and photoluminescence–decay lifetime measurements.

carrier at the same dopant density. This is a good approximation when the mobilities are dominated by phonon scattering but breaks down at doping densities at which ion and carrier scattering predominate. The only known measurements of minority mobility in heavily doped material are those of Dziewior and Silber [29] shown in Fig. 6. Even though there is considerable scatter in the data, it is clear that these mobilities, especially the hole mobility, are larger than the majority-carrier mobilities seen in Fig. 3. These results agree fairly well with a theoretical study by Bennett [30] which concludes that the main reason for the larger value of minority compared with majority mobility is that the extrinsic dopant-ion potential is repulsive to minority carriers, whereas it is attractive to majority carriers.

In bipolar devices, which are controlled by minority carriers, the excess minority-carrier density relaxes to the thermal equilibrium value according to the equation $r = r_0 + r'e^{-x/L}$, where r, r_0, and r' are the total, equilibrium, and excess minority-carrier densities, respectively, x is the distance into the quasi-neutral material from the junction, and L is the *diffusion length* equal to $\sqrt{D_r \tau_r}$, where τ_r is the minority-carrier recombination lifetime, and D_r is the minority-carrier diffusion coefficient. For minority carriers, D_r/μ_r usually equals kT/q because their density is normally nondegenerate.

The *lifetime* depends on the injection ratio (ratio of excess minority carriers to majority carriers) for recombination through traps as discussed in Shockley–Read–Hall theory [31]. If the injection ratio is much less than unity, one can use the zero-injection limiting case to obtain the lifetime. However, it is important to recognize that this lifetime can be very different from that which occurs at higher injection ratios or in a space-charge region. Above a majority dopant density of about 5×10^{18} cm^{-3}, Auger recombination becomes important in silicon devices and causes the lifetime to decrease at a rate nearly equal to n^{-2}, where n is the majority-carrier density. Several references exist for the Auger coefficients [32,33], but there is still disagreement by as much as an order of magnitude regarding their values.

VIII MOBILITY IN AN MOS INVERSION LAYER

The mobility in an inversion or accumulation layer is a function of many variables including crystal orientation, substrate doping, normal and tangential electric fields, fixed oxide charge, surface roughness, and temperature. Cooper and Nelson used a time-of-flight technique to measure the drift velocity of electrons [34] and holes [35] at the interface between lightly doped (100) silicon and silicon dioxide as a function of both tangential and normal electric fields. They found that the saturation velocities in a uniform applied tangential field were independent of the normal component of electric field and had values close to those observed in bulk silicon. They obtained a fit of their measured drift velocities using the empirical expressions

$$\mu = \frac{\mu_0}{[1 + (E_n/E_c)^c]}, \qquad (11)$$

$$v = \frac{\mu E_t}{[1 + (\mu E_t/v_s)^\alpha]^{1/\alpha}}, \qquad (12)$$

where E_n is the normal field, E_t is the tangential field, v_s is the saturation drift velocity, and μ_0, E_c, c, and α are parameters determined by the fit

TABLE V

Parameters for the Fits to the Inversion Layer Mobility, Eq. (11), and Drift Velocity, Eq. (12), at the (100) Si–SiO$_2$ Interface

	Electrons	Holes	Units
μ_0	1105	342	cm^2/V·s
E_c	30.5×10^4	15.4×10^4	V/cm
c	0.657	0.617	—
v_s	9.23×10^6	1.0×10^7	cm/s
α	1.92	0.968	—

14. Electrical Transport Properties of Silicon

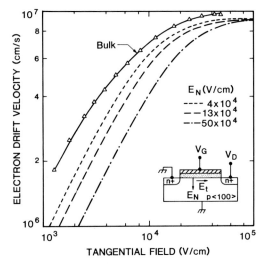

Fig. 7. Electron drift velocity in an inversion layer versus tangential component of electric field for several normal fields. The curves were calculated from Eqs. (11) and (12) using the parameters given in Table V. Also shown are the data of Canali *et al.* for electrons traveling in the [011] direction in the bulk. (After Cooper and Nelson [34].)

and are given in Table V. Figure 7 shows the velocity for electrons given by Eq. (12) plotted as a function of tangential field with normal field as a parameter. The data of Canali *et al.* [36] for electrons traveling in the [011] direction in the bulk are also plotted in Fig. 7.

The above expressions assume that the carriers are traveling parallel to the surface. However, since the current path is curved near the source and drain, short-channel MOS devices require more detailed modeling.

REFERENCES

1. S. M. Sze, "Physics of Semiconductor Devices," p. 29. Wiley (Interscience), New York, 1981.
2. Standard practice for conversion between resistivity and dopant density for boron-doped and phosphorus-doped silicon, ASTM designation F723, *in* "Annual Book for ASTM Standards," vol. 10.05 (November 1984).
3. J. C. Irvin, *Bell System Tech. J.* **41**, 387 (1962).
4. W. R. Thurber, R. L. Mattis, Y. M. Liu, and J. J. Filliben, *J. Electrochem. Soc.* **127**, 1807 (1980).
5. W. R. Thurber, R. L. Mattis, Y. M. Liu, and J. J. Filliben, *J. Electrochem. Soc.* **127**, 2291 (1980).
6. R. B. Fair, *J. Electrochem. Soc.* **125**, 323 (1978).
7. J. Messier and J. M. Flores, *J. Phys. Chem. Solids* **24**, 1539 (1963).
8. P. Norton, T. Braggins, and H. Levinstein, *Phys. Rev. B* **8**, 5632 (1973).
9. W. R. Thurber, *J. Electron. Mater.* **9**, 551 (1980).
10. F. Mousty, P. Ostoja, and L. Passari, *J. Appl. Phys.* **45**, 4576 (1974).
11. J. F. Lin, S. S. Li, L. C. Linares, and K. W. Teng, *Solid-State Electron.* **24**, 827 (1981).

12. F. Szmulowicz, *Appl. Phys. Lett.* **43,** 485 (1983).
13. W. R. Thurber, R. L. Mattis, Y. M. Liu, and J. J. Filliben, *Nat. Bur. Std. Spec. Publ. 400-64* (May 1981).
14. F. J. Morin and J. P. Maita, *Phys. Rev.* **96,** 28 (1954).
15. R. A. Logan and A. J. Peters, *J. Appl. Phys.* **31,** 122 (1961).
16. G. Masetti, M. Severi, and S. Solmi, *IEEE Trans. Electron Devices* **ED-30,** 764 (1983).
17. G. Masetti and S. Solmi, *Solid-State Electron Devices* **3,** 65 (1979).
18. M. Finetti, P. Negrini, S. Solmi, and D. Nobili, *J. Electrochem. Soc.* **128,** 1313 (1981).
19. S. Solmi, M. Severi, R. Angelucci, L. Baldi, and R. Bilenchi, *J. Electrochem. Soc.* **129,** 1811 (1982).
20. R. A. Logan, J. F. Gilbert, and F. A. Trumbore, *J. Appl. Phys.* **32,** 131 (1961).
21. W. M. Bullis, F. H. Brewer, C. D. Kolstad, and L. J. Swartzendruber, *Solid-State Electron.* **11,** 639 (1968).
22. Standard method for measuring resistivity of silicon slices with a collinear four-probe array, ASTM designation F84, *in* "Annual Book of ASTM Standards," vol. 10.05 (November 1984).
23. P. Norton and J. Brandt, *Solid-State Electron.* **21,** 969 (1978).
24. S. S. Li and W. R. Thurber, *Solid-State Electron.* **20,** 609 (1977).
25. L. C. Linares and S. S. Li, *J. Electrochem. Soc.* **128,** 601 (1981).
26. N. D. Arora, J. R. Hauser, and D. J. Roulston, *IEEE Trans. Electron Devices* **ED-29,** 292 (1982).
27. C. Jacoboni, C. Canali, G. Ottaviani, A. Alberigi-Quaranta, *Solid-State Electron.* **20,** 77 (1977).
28. C. Canali, G. Ottaviani, and A. Alberigi-Quaranta, *J. Phys. Chem. Solids* **32,** 1707 (1971).
29. J. Dziewior and D. Silber, *Appl. Phys. Lett.* **35,** 170 (1979).
30. H. S. Bennett, *Solid-State Electron.* **26,** 1157 (1983).
31. W. Shockley and W. T. Read, *Phys. Rev.* **87,** 835 (1952); R. N. Hall, *Phys. Rev.* **87,** 387 (1952).
32. L. Passari and E. Susi, *J. Appl. Phys.* **54,** 3935 (1983).
33. J. Dziewior and W. Schmid, *Appl. Phys. Lett.* **31,** 346 (1977).
34. J. A. Cooper, Jr., and D. F. Nelson, *J. Appl. Phys.* **54,** 1445 (1983).
35. D. F. Nelson, J. A. Cooper, Jr., and A. R. Tretola, *Appl. Phys. Lett.* **41,** 857 (1982).
36. C. Canali, C. Jacoboni, F. Nava, G. Ottaviani, and A. Alberigi-Quaranta, *Phys. Rev.* **12,** 2265 (1975).

Chapter 15
VLSI Silicon Material Criteria

FUMIO SHIMURA

Monsanto Electronic Materials Company
St. Louis, Missouri

HOWARD R. HUFF

Monsanto Electronic Materials Company
Palo Alto, California

I. Introduction	191
II. Product Characteristics	193
III. Tabulation of Engineering Properties	194
IV. VLSI Silicon Product Trends	194
V. VLSI and ULSI Silicon Product Recommendation	196
Tables	197
Figures	230
References	265

INTRODUCTION I

The silicon material modifications that take place during IC fabrication are extremely complex. This is especially the case for VLSI circuit fabrication processes.* Reduced feature size, larger chip area, increased number of dielectric and conductor layers, reduced power-delay product, and, especially, reduced leakage current requirements are driving today's silicon wafers to ever more stringent specifications. This brief handbook article summarizes several important considerations for the development of silicon wafer products for VLSI applications.

Section II presents a representative summary of currently specified material characteristics, ASTM diagnostic procedures, instrumentation,

---

* Gases, chemicals, and masking disorders will not be discussed, although their importance for VLSI certainly cannot be minimized.

and brief comments. The characteristics are classified in terms of the electrical, chemical, structural, and mechanical properties of silicon. These characteristics are strongly interdependent, however, and their separation is a simplification to facilitate analysis. A brief representative summary of physicochemical diagnostic techniques that are often used to characterize incoming and/or in-process material, as well as their limits of applicability, are then presented.

Section III tabulates several *engineering properties of silicon* useful for calculating material modifications during circuit fabrication. Although scientific aspects are referred to and occasionally noted for currently active areas of research, such as the mixed vacancy/self-interstitial diffusion of common dopants in silicon, the emphasis is mainly an engineering approach. For example, the *solid solubility of impurities in silicon* determined by diffusion from an oxidizing ambient source is presented. These data are considered to be more appropriate for VLSI processing than are solubilities determined from grown-crystal bulk doping followed by sufficient time–temperature annealing to induce the bulk equilibrium solubility. The data in this section are also classified in terms of the electrical, chemical, structural, and mechanical properties to facilitate their comparison with Section II. We have also included a brief comparison of the physicochemical properties of silicon with germanium and gallium arsenide. Finally, several engineering formulae are also included to facilitate calculation of several process-induced modifications occurring in silicon.

Section IV briefly summarizes the evolving *target values for VLSI silicon products* and their characteristics. Specification of the appropriate electrical parameters measured on test-device structures fabricated in silicon is anticipated to ensure a more useful match with circuit requirements. For example, an oxide-coated wafer might complement the conventional emphasis on incoming surface visual properties, often difficult to correlate with IC performance, by emphasizing electrical parameters. Flat-band voltage shift, minority carrier generation lifetime, and oxide breakdown voltage distribution, developed for a subsequent metallization or polysilicon pattern applied to the oxide-coated wafer by the customer, might facilitate the development of more appropriate starting material specifications and correlation with circuit performance. Relevant engineering references, which may guide the development of future industrial standards, are included in both this and additional sections of this chapter.

Section V extends this discussion by considering the *silicon product as a multizone system* uniquely configured and engineered to the specific requirements of each customer. Several benefits of this product system view, perhaps including the integration of several IC process steps into the wafer supplier's facility, will be summarized. Indeed, the multizone system integration approach is useful for the development of silicon products for ULSI, the Ultra Large Scale Integration era. Closer cooperation between the silicon wafer and IC manufacturers, however, is essential to ensure the availability of superior wafers for consistent circuit performance.

PRODUCT CHARACTERISTICS II

Crystal growth and wafer shaping are, of course, of critical importance and have recently been reviewed [1,2]. It is anticipated that knowledge of these techniques, including the interdependence of impurities, microdefects, and thermal history as well as the methodology of polysilicon preparation,† will become even more important in the VLSI era.

The specification of the silicon material characteristics depends on the specific VLSI product being fabricated. There are, however, several general approaches. For example, one can write the basic equations for IC processes, devices, and circuits and deduce the required material parameters. That is, one can correlate:

(a) Specific interfacial charge with surface orientation
(b) Body-effect resistivity with bulk resistivity
(c) Refresh performance with minority carrier generation lifetime
(d) Active device dimensions with local site flatness

The device and circuit parameters at the present time, however, are as often dependent on the detailed IC processes and their sequence as on the starting silicon parameters. The modification of the material characteristics and the resulting structure sensitivity of the IC electronic characteristics form the basis of an extensive literature [3–6] and are not further discussed here. Our approach presents an alternative method based on a general materials science point of view.

Table I‡ summarizes several currently specified silicon material characteristics and their ASTM designations. The characteristics are classified in terms of the electrical, chemical, structural, and mechanical properties of silicon. Tables II–V present further details for each of these topics, including common instrumentation and relevant comments. Of course, tolerances are required for these parameters. In these sections, however, we shall mainly emphasize the concept. Tables VI and VII and Figs. 1–4§ illustrate a representative selection of physicochemical diagnostic techniques often used to characterize incoming and/or in-process material, as well as their limits of applicability.

---

† The purity requirement, compensation level, and related handling procedures for polysilicon are extremely important but beyond the scope of this handbook article.
‡ All tables (I–XXXIX) appear on pp. 197–229.
§ All figures (1–63) appear on pp. 230–264.

A complete tabulation of the physicochemical properties should include additional topics such as the optical, elastic, and thermodynamic properties of silicon [7–9]. However, it is our purpose in this brief chapter to emphasize only those characteristics required for useful specification of silicon for subsequent VLSI fabrication. Therefore, a number of characteristics, although scientifically important, are nevertheless second-order considerations for most of today's evolving silicon material requirements. More complete material information and specification data requiring low-temperature diagnostics such as the degree of compensation, electron and hole capture cross sections, and detailed assessment of various dopant–impurity–structural complexes should be necessary, however, especially for potential future low-temperature circuit applications.

III TABULATION OF ENGINEERING PROPERTIES

The conversion of a silicon wafer into a functional IC requires a knowledge of the behavior and modification of the material during circuit fabrication [14,15]. Several engineering formulae in addition to detailed silicon data are, therefore, included to facilitate calculation of several process-induced modifications occurring in the material. These data include *solubility, oxidation, diffusion, epitaxy, ion implantation,* and several related *thermophysical properties*. The data are generally also presented in the format of *electrical, chemical, structural,* and *mechanical properties* to facilitate their comparison with Section II.

For convenience, we initially summarize several useful physical constants in Table VIII and the basic physicochemical properties of silicon in comparison with germanium and gallium arsenide in Table IX. Anisotropic properties of silicon are summarized in Table X. Table XI presents the selected engineering formulae, and Tables XII–XXXVIII summarize the engineering data noted above. A selected number of figures (Figs. 5–61) accompany Tables XII–XXXVIII. Detailed thermomechanical data related to plastic deformation, however, are extremely structure-sensitive, and great care, therefore, must be exercised in their use.

IV VLSI SILICON PRODUCT TRENDS

Emphasis on the visual, cosmetic character of the polished, front surface and etchant artifacts and stains on the back surface has been extensive. Although these characteristics are certainly indicative of the quality

15. VLSI Silicon Material Criteria

of the material supplier, their specific correlation with circuit performance is often difficult to establish. Improvements continue, as they should; but the expeditious growth of a thermal oxide film after polishing is expected to retain the surface cleanliness and facilitate the specification of electrical, rather than visual, parameters. This approach might indeed lead to more effective correlation of starting-material characteristics with circuit performance and to the development of more meaningful starting-material specifications. For example, the flat-band voltage shift, minority carrier generation lifetime, and oxide breakdown voltage distribution might eventually be specified for a subsequent metallization or polysilicon pattern applied to the oxide-coated wafer by the customer. This approach, to be further discussed in Section V, would also ensure retention of a hydrophilic, rather than hydrophobic, surface and would significantly reduce inadvertent transfer of metallics to the silicon wafer surface.

Another area expected to receive significant emphasis is crystal identity and the reproducibility of wafer characteristics [130], such as carbon, oxygen, oxygen radial gradient, and the role of noninfrared active oxygen. Specification of the change in the infrared active oxygen spectra for a defined thermal process appears certain to come. This change denotes the propensity for the generation of bulk precipitate–dislocation-complexes (PDC), stacking faults, and associated structural irregularities for internal gettering purposes. Detailed analysis will be required of oxygen control and the exploitation of its temperature-dependent solubility concentration, ± 2 ppma, resistivity variation ($\leq 10\%$), metal concentration, wafer cleanliness, and particle density, as well as local site flatness (1.0 μm/20 $\times$ 20-mm^2 field, 99% usable wafer area). These target values are summarized in Table XXXIX.

Finally, utilization of larger diameter wafers (150–200 mm) will further emphasize the importance of thermomechanical properties, including the diameter, thickness, and orientation flat tolerances. A diameter tolerance of less than ± 0.2 mm is anticipated whereas use of the second-orientation flat should be avoided to facilitate more accurate lithographic pattern alignment. Laser marking will be extensively utilized. Proper selection of the desired wafer thickness (see Fig. 62) and control of wafer bow and warpage, as well as wafer edge contour, and the utilization of gettering procedures will also become more important in the case of larger diameter wafers.

It may not be so straightforward, however, to achieve the target values for VLSI silicon products tabulated in Table XXXIX. The accomplishment strongly depends on the development of advanced diagnostic techniques [131] as well as on the individual fabrication process technologies. State-of-the-art diagnostics, such as the detection of surface particles, evaluation of the native oxide, and the related analysis of trace impurities, may not be sufficient for VLSI requirements. The development of improved diagnostic techniques is indeed a major driving force for establishing VLSI quality silicon.

V VLSI AND ULSI SILICON PRODUCT RECOMMENDATION

The development of both back-surface damage and polysilicon films facilitates external gettering. These techniques can be regarded as controlled zones of silicon with unique characteristics on the back surface of the wafer, certainly different from that of bulk silicon. Internal oxygen precipitation gettering can likewise be regarded as distinctly different from unprocessed material or material with less than a critical value of oxygen. The development of a surface-denuded zone and localization of the IC within the upper several microns or so of the silicon surface can also be similarly viewed as distinct zones. Therefore, conversion of the silicon wafer into a multizone system, uniquely configured and engineered for the customer's desired VLSI function requirements, is recommended (see Fig. 63) [5].

A logical extension of the above would be the inclusion of an oxide film as discussed in Section IV. Utilization of an oxide–nitride film sandwich, appropriately adjusted for local stress considerations, might then follow. The inclusion of a silicon epitaxial layer before the dielectric films are fabricated offers another possibility. Finally, a structure with a buried blanket n^+ implant prior to epi and dielectric film fabrication could form the basis of an advanced bipolar trench isolation product [135].

Clearly, the permutation of this concept permits opportunities for the development of a multizone silicon system appropriate to the needs of the individual customer. The multizone system integration approach is indeed the path toward the development of silicon products for ULSI, the Ultra Large Scale Integration era [136]. Closer cooperation between the silicon wafer and IC manufacturers, however, it is essential to ensure the availability of superior wafers for consistent circuit performance.

TABLE I
Currently Specified Silicon Material Characteristics and ASTM Designations[a]

Material characteristics	Property	ASTM procedure
Electrical properties	Conductivity type	F42
	Resistivity and resistivity gradient	F43, 81, 84, 397 533, 672, 673
	Minority carrier recombination lifetime	F28
Chemical properties	Dopant density	F723
	Interstitial oxygen	F120, 121
	Substitutional carbon	F120, 123
	Surface cleanliness	F154, 523
	Hydrophilic surface	F21
Structural properties	Wafer orientation	F26
	Flat orientation	F26
	Dislocation density	F47
	Oxidation induced defects	F416
Mechanical properties	Diameter	F613
	Flatness	F691
	Thickness and total thickness variation (TTV)	F657
	Bow and warp	F534, 657
	Edge contour	In progress
	Surface feature	F24, 154, 515, 523

[a] From reference [10] and updates.

TABLE II
Electrical Properties

Property	ASTM procedure	Instrumentation	Comment
Conductivity type	F42	Hot probe thermal EMF	The test is reliable on n- or p-type silicon with room temperature resistivities up to 1000 ohm·cm Silicon in various forms can be measured: crystal, sawn, ground, lapped, etched, or polished wafers
Resistivity and resistivity gradient	F43, 81, 84, 397, 533, 672 673	Four-point probe, non-contact resistivity gauge	Resistivity is a measure of electrical carrier concentration, which is a basic property of semiconductor electronics; it determines a material's resistance to the flow of electricity Radial resistivity gradient is the percentage variation in resistivity between the center and selected outer regions of a wafer (normally half-radius $R/2$ and 6 mm from the wafer edge)
Minority carrier recombination lifetime	F28	Photoconductive decay measurement system	The minority carrier recombination lifetime is defined as the time required for the number of minority carriers, generated by a flashlight, to decay to $1/e$ of its value at the beginning of the observation for a properly sized and shaped specimen

15. VLSI Silicon Material Criteria

TABLE III

Chemical Properties

Property	ASTM procedure	Instrumentation	Comment
Dopant density	F723	Graphic method Tabular method Computational method	The conversion from resistivity to electrical carrier concentration taken as equal to the dopant density in crystals is important for a variety of applications from material inspection to process and device modeling These methods are established for the boron- and phosphorus-doped crystals at room temperature using appropriate values for the hole and electron mobility
Interstitial oxygen and substitutional carbon	F120, 121, 123	Dispersive or Fourier transform infrared spectrometer (FTIR)	The amounts of oxygen and carbon in silicon wafers are important material parameters and must be specified for all device technologies. The ASTM procedures were specifically developed for 2-mm slugs measured in dispersive spectrometers. Application to wafers and FTIR spectrometers requires the utmost caution Use of the FTIR procedure is fraught with assumptions and requires careful correlation studies about the software used to establish the baseline of the IR spectrum and correction for multiple internal and back-surface reflections These techniques will only measure the infra-

(*Continued*)

TABLE III (Continued)

Property	ASTM procedure	Instrumentation	Comment
			red active modes of the contaminants; noninfrared active oxygen may also play an especially important role as one of several impurities possibly acting as nuclei for subsequent precipitation
Surface cleanliness	F154, 523	Microscope with TV monitor	Surface stains, contamination, and haze are generally detrimental to IC performance (see also *Surface feature* in Table V)
Hydrophilic surface	F21	Spray gun atomizer	A hydrophobic surface is very active for picking up contaminants and particulates; a hydrophilic surface reduces these phenomena

15. VLSI Silicon Material Criteria

TABLE IV

Structural Properties

Property	ASTM procedure	Instrumentation	Comment
Wafer orientation	F26	X-ray unit	The wafer orientation is a most significant parameter as regards a host of physicochemical properties
			Bipolar applications generally specify $4 \pm 0.5°$ off the $\langle 111 \rangle$ orientation towards the nearest $\langle 110 \rangle$ direction to minimize both lithographic pattern distortion and the growth of epitaxial defects; MOS applications are generally specified within $\pm 1°$ of the desired orientation
Flat orientation	F26	Laser reflection screen	The flat orientation is essential to ensure optimal scribe separation of die by saw or laser techniques after electrical test
Dislocation density	F47	Wright etchant	Preferential etch techniques to assess crystallographic perfection are specific to orientation, doping density, etc.
			Today's unprocessed wafers are essentially dislocation-free on a macroscopic level after etching, as observed through an optical microscope
Oxidation induced defects	F416	Oxidation furnace Wright etchant	The detection of haze and related surface and/or bulk defects is essential to anticipate subsequent wafer responsivity to process influences during VLSI fabrication

TABLE V

Mechanical Properties

Property	ASTM procedure	Instrumentation	Comment
Diameter	F613	Two-point micrometer	Crystals are grown to "constant diameter" using either analog or digital automatic diameter control methodologies; Centerless grinding procedures are subsequently used to achieve exact diameter control; Eccentricity is measured by using a 3-point contact procedure
Flatness	F691	Flatness gauge	Wafer flatness is an important parameter for VLSI fabrication. The peak-to-valley measurement has routinely been quoted. The local site flatness criterion for wafer stepper use is becoming more important. Data are usually quoted over the field of view.
Thickness and total thickness variation (TTV)	F657	Noncontact thickness gauge	The diameter-to-thickness ratio is important to minimize thermal slip. TTV is the difference between the maximum and minimum values.
Bow and warp	F534, 657	Noncontact wafer gauge	The concavity or convexity of a polished silicon wafer is averaged between the front- and back-surface measurements to determine bow;

TABLE V (Continued)

Property	ASTM procedure	Instrumentation	Comment
			Half the difference between the largest and smallest of the differences of the paired displacements is taken as a measure of the initial warp.
Edge contour	In progress	Projector with TV monitor	The contour of the wafer edge is important as regards the onset of thermal slip, as well as *crown* in both epitaxial growth and resist coverage
Surface feature	F24, 154, 515, 523	Fluorescent lamp	The use of a high-intensity lamp for more accurate assessment of residual surface defects is especially advantageous for scratches, particulates, stain, and contamination The back surface condition is also important as regards unintentional contamination of IC Fab lines

TABLE VI

Comparison of Imaging Resolution as a Function of Technique and Detected Species[a]

Best resolutions	Technique	Detected species
1–5 μm	Secondary ion mass spectroscopy (SIMS)	Secondary ions
1–2 μm	X-ray topography (XRT)	Diffracted X rays
0.2–2 μm	Electron microprobe (EM)	X rays
0.2 μm	Optical microscopy (OM)	Light
300 Å	Scanning Auger microscopy (SAM)	Auger electrons
50 Å	Electron energy loss spectroscopy (EELS)	Transmitted electrons
10–30 Å	Secondary electron microscopy (SEM)	Secondary electrons
2–5 Å	Transmission electron microscopy (TEM)	Diffracted electrons

[a] From reference [11].

TABLE VII
Physicochemical Diagnostic Techniques for Semiconductors[a]

Appellation	Acronym	Input probe	Detected species	Application	Beam resolution[b]
Auger electron spectroscopy	AES	Electron	Auger electron	Elemental analysis	ϕ: 0.1–1 mm Z: ~10–20 Å
Analytical electron microscopy	AEM	Electron	Transmitted diffracted electrons	Micro area: EDX EELS DLTS Channeling pattern	ϕ: a few Å
Cathode luminescence	CL	Electron	Photon	Electrically active crystal defects Precipitates Impurities Free carrier diffusion length	ϕ: >0.5 μm Z: ~1 μm
Deep level transient spectroscopy	DLTS	Forward bias pulse voltage	Capacitance current	Impurities Electrically active crystal defects	Trap Conc. $\geq 10^{10} \sim 10^{11}$ cm^{-3}
Electron beam induced current	EBIC	Electron	Induced current	Electrically active crystal defects Diffusion length	ϕ: a few μm Z: a few μm
Electron energy loss spectroscopy	EELS	Electron	Reflected, transmitted electrons	Microprecipitates[c] Surface states	ϕ: >100 Å Z: Monoatomic layer
Electron probe microanalysis (energy dispersive X-ray spectroscopy, wavelength dispersive X-ray spectroscopy)	EPMA (EDX, WDX)	Electron	Characteristic X rays	Elemental chemical analysis Contaminants	ϕ: >0.5 μm z: >0.3 μm
Electron spectroscopy for chemical analysis. (X-ray photoemission spectroscopy)	ESCA (XPS)	X rays	Photoelectron	Elemental chemical analysis Chemical shift Band structure	Z: >10 Å

Extended x-ray absorption fine structure	EXAFS	White X ray	Scattered X ray	Structural analysis of amorphous Si	Z: a few μm
Field emission microscopy	FEM	Electric field	Neutral atom	Crystal structure	ϕ: ~10 Å Z: Monoatomic layer
Field ion microscopy	FIM	Electric field	Ionized atom	Crystal structure	ϕ: ~10 Å Z: Monoatomic layer
High voltage electron microscopy	HVEM	Electron	Transmitted, diffracted electrons	Microdefects	ϕ: >10 Å
Ion microprobe (mass) analysis	IM(M)A	Ion	Secondary ions	Elemental chemical analysis	ϕ: >0.1 μm Z: >10 Å
Infrared absorption spectroscopy	IR	Photon	Photon	Interstitial oxygen, substitutional carbon concentration	>0.1 cm^{-1}
Low energy electron diffraction	LEED	Electron	Diffracted electrons	Thin film crystal structure Surface adsorbed layer	Z: a few atomic layers
Nuclear magnetic resonance	NMR	Electromagnetic wave	Electromagnetic wave	H, F in amorphous Si	
Photoluminescence	PL	Light	Light	Impurities Point defects	ϕ: a few μm
Rutherford backscattering	RBS	H$^+$, He$^+$ ions	Scattered ions	Structural analysis Impurity diffusion Elemental chemical analysis	Z: 50 ~ 200 Å
Reflective high energy electron diffraction	RHEED	Electron	Scattered electrons	Surface structural analysis	Z: a few atomic layers
Scanning auger microscopy	SAM	Electron	Auger electron	Surface elemental chemical analysis	ϕ: ≥500 Å Z: >5 Å
Scanning electron microscopy	SEM	Electron	Secondary electrons	Surface morphology	ϕ: >30 Å
Spreading resistance	SR	Current	Voltage	Resistivity profile	ϕ: a few μm

(*Continued*)

TABLE VII (Continued)

Appellation	Acronym	Input probe	Detected species	Application	Beam resolution[b]
Scanning transmission electron microscopy	STEM	Electron	Transmitted, diffracted, secondary electrons	(see AEM)	ϕ: ~10 Å Z: >20 Å
Transmission electron microscopy	TEM	Electron	Transmitted, diffracted electrons	Microdefects Crystal structure	ϕ: a few Å
X-ray diffractometry	XD	X ray	Diffracted X ray	Bulk crystallinity Defects	Z: >0.1 μm
X-ray diffuse scattering	XDS	X ray	Scattered X ray	Point defects	Z: >0.1 μm
X-ray rocking curve	XRC	X ray	Diffracted X ray	Diffused layer perfection Epitaxial mismatch Bulk crystallinity	Z: >0.1 μm
X-ray topography	XRT	X ray	Diffracted X ray	Defects Whole wafer picture	Z: ≤1 mm

[a] From reference [12].
[b] Representative judgment. ϕ: diameter; Z: penetration depth.
[c] Microdefects include geometrical and chemical irregularities; microprecipitates refer to chemical irregularities.

TABLE VIII

Basic Physical Constants

Quantity	Symbol	Number	MKSA unit	CGS unit
Speed of light	c	2.998	$\times 10^8$ m·s^{-1}	$\times 10^{10}$ cm·s^{-1}
Boltzmann's constant	k	1.381	$\times 10^{-23}$ J·K^{-1}	$\times 10^{-16}$ erg·K^{-1}
Electron rest mass	m_0	9.109	$\times 10^{-31}$ kg	$\times 10^{-28}$ g
Electronic charge[a]	e	1.602	$\times 10^{-19}$ C	$\times 10^{-20}$ emu
Planck's constant	h	6.626	$\times 10^{-34}$ J·s	$\times 10^{-27}$ erg·s
Avogadro's number	N_A	6.023	$\times 10^{23}$ mol^{-1}	$\times 10^{23}$ mol^{-1}
Gas constant	R	8.314	$\times$ J·mol^{-1}·K^{-1}	$\times 10^7$ erg·mol^{-1}·K^{-1}
Permittivity of vacuum	ε_0	8.854	$\times 10^{-12}$ F·m^{-1}	$\times 10^{-14}$ F·cm^{-1}

[a] Absolute value.

TABLE IX

Physicochemical Properties[a] of Some Semiconductor Materials at 300 K[a,b]

Properties	Ge	Si	GaAs
Atoms or molecules/cm^3	4.42×10^{22}	5.0×10^{22}	2.21×10^{22}
Atomic or molecular weight	72.60	28.09	144.63
Breakdown field (V/cm)	~10^5	~3×10^5	~4×10^5
Crystal structure	Diamond	Diamond	Zincblende
Density (g/cm^3)	5.3267	2.328	5.32
Dielectric constant	16.0	11.9	13.1
Effective density of states in conduction band N_c (cm^{-3})	1.04×10^{19}	2.8×10^{19}	4.7×10^{17}
Effective density of states in valence band N_v (cm^{-3})	6.0×10^{18}	1.04×10^{19}	7.0×10^{18}
Effective mass m^*/m_0			
Electrons	$m_l^* = 1.64$, $m_t^* = 0.082$	$m_l^* = 0.98$, $m_t^* = 0.19$	0.067
Holes	$m_{lh}^* = 0.044$, $m_{hh}^* = 0.28$	$m_{lh}^* = 0.16$, $m_{hh}^* = 0.49$	$m_{lh}^* = 0.082$, $m_{hh}^* = 0.45$
Electron affinity χ (eV)	4.0	4.05	4.07
Energy gap (eV)	0.66	1.12	1.42
Intrinsic carrier concentration (cm^{-3})	2.4×10^{13}	1.45×10^{10}	1.79×10^6
Intrinsic debye length (μm)	0.68	24	2250

(Continued)

TABLE IX (Continued)

Properties	Ge	Si	GaAs
Intrinsic resistivity ($\Omega \cdot$cm)	47	2.3×10^5	10^8
Lattice constant (Å)	5.64613	5.43095	5.6533
Linear coefficient of thermal expansion $\Delta L/L \, \Delta T$ (°C^{-1})	5.8×10^{-6}	2.6×10^{-6}	6.86×10^{-6}
Melting point (°C)	937	1420	1238
Minority carrier lifetime (s)	10^{-3}	2.5×10^{-3}	$\sim 10^{-8}$
Mobility (drift, lightly doped) (cm^2/V·S)			
Electrons	3900	1350	8500
Holes	1900	475	400
Optical phonon energy (eV)	0.037	0.063	0.035
Phonon mean free path λ_0 (Å)	105	76 (electron) 55 (hole)	58
Specific heat (J/g·°C)	0.31	0.7	0.35
Thermal conductivity (W/cm·°C)	0.6	1.5	0.46
Thermal diffusivity (cm^2/s)[c]	0.36	0.92	0.25
Vapor pressure (mmHg)	7.5×10^{-3} at 1330°C; 7.5×10^{-4} at 760°C	7.5×10^{-3} at 1650°C; 7.5×10^{-19} at 900°C	7.5×10^{-1} at 1050°C; 7.5×10^{-3} at 900°C

[a] From reference [16].
[b] Several temperature-dependent variations of these properties are presented in subsequent Figs. 48–57.
[c] Thermal diffusivity = (thermal conductivity)/(density × specific heat)

TABLE X
Anisotropic Properties of Crystalline Silicon[a]

Tensor rank	Property	Symbol	Relates	Form of relation	Isotropic in silicon
0	Density	δ	Mass to volume	A scalar to a scalar	Yes
	Heat capacity	C	Heat transferred to temperature changes	A scalar to a scalar	Yes
1	Pyroelectric coefficient	P_i	Electrical polarization to temperature change	A vector to a scalar	No
2	Electrical conductivity	σ_{ik}	Current density to applied field	A vector to a vector	Yes
	Electrical mobility	μ_{ik}	Current density to applied field and number of carriers		
	Thermal	k_{ij}	Heat transferred to temperature gradient	A vector to a vector	Yes
	Diffusion coefficient	D_{ij}	Current density to concentration gradient	A vector to a vector	Yes
	Thermal expansion	α_{ij}	Elongation to temperature change	A scalar and second-rank tensor	Yes
3	Piezoelectric coefficient	d_{ijk}	Polarization to applied stress	A vector to a second-rank tensor	No
4	Elastic constants	C_{ijkl}	Stress to elongation	Two second-rank tensors	No
	Piezoresistance	π_{ijkl}	Change of resistivity to applied stress	Two second-rank tensors	No

[a] From references [17,18].

TABLE XI

Selected Engineering Formulas

Fluid Mechanics
  Reynolds number: $R = Lv/\nu$,
    where L is a typical dimension of the system, v a measure of the velocities that prevail in stationary flow, and ν the kinematic viscosity
  Taylor number: $T = \Omega^2 \cdot L^4/\nu^2$,
    where Ω is angular velocity
  Prandtl number: $P = \nu/K$,
    where K is thermal conductivity
Crystal Growth
  Normal freezing [19]: $C/C_0 = k_e(1 - g)^{k_e-1}$,
    where C is the impurity concentration in the crystal after the fraction g of the initial melt volume has solidified (constant number of impurities), C_0 the impurity concentration in the melt before freezing, k_e the effective distribution coefficient ($= C_s/C_1$, where C_s is the concentration in solid silicon, and C_1 is the concentration in liquid silicon), and g the solidified fraction of the original melt volume
  Effective distribution coefficient [20]: $k_e = k_0/\{k_0 + (1 - k_0) \exp(-Vd/D)\}$,
    where k_0 is the equilibrium distribution coefficient, V the freezing velocity (subsequent research has shown that this is really the microscopic growth rate) [21-23], d the thickness of the diffusion layer in the melt at the melt-solid interface ($= 1.6 \times D^{1/3} \times \nu^{1/6} \times \omega^{-1/2}$), ω the angular velocity of the crystal, and D the diffusion constant of the impurity in the melt
Crystallographic
  Interplanar spacing: $d_{hkl} = a/(h^2 + k^2 + l^2)^{1/2}$,
    where d is interplanar spacing, h, k, l are plane indices, and a is the crystal constant
  Interplanar angle: $\cos \phi = (h_1h_2 + k_1k_2 + l_1l_2)/\{(h_1^2 + k_1^2 + l_1^2)(h_2^2 + k_2^2 + l_2^2)\}^{1/2}$
  Bragg condition: $2d \sin \theta_B = n\lambda$,
    where θ_B is the Bragg angle, n an integer, and λ the wavelength
Diffusion
  Diffusion constant: $D = D_0 \exp(-Q/kT)$,
    where D_0 is the pre-exponential factor, Q the activation energy of diffusion, k Boltzmann's constant, and T the absolute temperature
  Distribution of an impurity in silicon from an infinite source: $C(x,t) = C_s \operatorname{erfc}\{x/2(Dt)^{1/2}\}$,
    where x is distance or depth diffused, t time, and C_s the surface concentration
  Distribution of an impurity in silicon from a finite source: $C(x,t) = \{Q_0/(\pi Dt)^{1/2}\} \exp\{x/2(Dt)^{1/2}\}^2$,
    where Q_0 is the total amount of impurity per unit area
Thermomechanical Properties
  Thermomechanical stresses [24]: $\bar{\sigma}_r = \sigma_r/\alpha E(T_0 - T)$, $\sigma_\theta = \sigma_\theta/\alpha E(T_0 - T)$,
    where σ_r is the radial tensile stress, σ_θ the angular tensile stress, α the linear thermal expansion coefficient (2.6×10^{-6} C^{-1} for silicon), E: Young's modulus (1.9×10^{12} dynes/cm^2 for silicon), T the temperature, and T_0 the initial temperature
  Longitudinal stress introduced into the film [25]: $\sigma = \{E/6(1 - \nu)\}(h_s^2/h_f)(1/R)$,
    where ν is Poisson's ratio, h_s the thickness of the substrate, h_f the thickness of the film, and R the radius of curvature of the bent structure
  Dislocation velocity [26]: $V = (D_0/kT)F = (D_0/\sigma b^2)/kT \exp(-U/kT)$,
    where σ is the applied stress, F the force on the dislocation, D_0 the diffusion constant of the crack, b Burger's vector of the dislocation (absolute value), and U the activation energy for dislocation motion

TABLE XI (*Continued*)

Avrami equation [27]: $P = 1 - \exp(-a \cdot t^n)$,
  where P is the fraction of transformed material from state A to state B, a the rate constant, t time, and n a parameter characterizing the nature of solid state transformation

Electrical Properties

Intrinsic concentration: $n_i = (N_c N_v)^{1/2} \exp(-E_g/2kT)$,
  where N_c is the effective density of states in conduction band [$\equiv 2(2\pi m_0 kT/\hbar^2)^{3/2}(m_e^*/m_0)^{3/2}$], N_v the effective density of states in valence band [$\equiv 2(2\pi m_0 kT/\hbar^2)^{3/2}(m_e^*/m_0)^{3/2}$], and E_g the energy gap

Mass action relationship (nondegenerate statistics): $np = n_i^2$,
  where n_i is the intrinsic concentration, n the electron concentration, and p the hole concentration

Normalized Fermi energy: $E_F = (kT/|e|) \ln(N_a/n_i)$,
  where N_a is the acceptor dopant concentration and $|e|$ the absolute value of the electronic charge

Resistivity (linear four-point probe): $\rho = 2(\pi V/I)/[(1/S_2) + (1/S_3) - (1/(S_1 + S_2)) - (1/(S_2 + S_3))]$,
  where V is the applied voltage, I the applied current, and S_1, S_2, and S_3 are probe spacings

Spreading resistance: $R = \rho/4a$,
  where ρ is the bulk resistivity and a the radius of the circular contact spot between the metal probe and the semiconductor surface

Van der Pauw method [28]: $R_s = (\pi/2 \ln 2)(R' + R'')f(R'/R'')$,
  where R_s is the sheet resistance, R' the potential difference between the contacts C and D per unit current through the contacts A and B, R'' the potential difference between the contacts A and D per unit current through the contacts B and C, and $f(R'/R'')$ is Van der Pauw's function

Minority carrier recombination lifetime, from surface photovoltaic (SPV) technique [29]: $\tau_r = L^2/D$,
  where L is the diffusion length of minority carrier and D the diffusion constant of minority carrier [diffusion length L obtained from the x intercept (linear extension) of the graph of relative intensity for a constant SPV (for varying incident wavelength) versus the corresponding reciprocal absorption coefficient α^{-1}; D is obtained from the Einstein relation $D/\mu = kT/|e|$, where μ is minority carrier mobility]

Minority carrier generation lifetime

  Heiman technique [30,31]: $1/\tau_g = (2N_a/\Delta t n_i)[(C_{ox}/C_0) - (C_{ox}/C_f) + \ln\{(C_{ox}/C_0) - 1\}/\{(C_{ox}/C_f) - 1\}]$,
    where C_{ox} is the system capacitance under accumulation (equal to oxide capacitance), C_0 the system capacitance $t = 0^+$ (after applying inversion voltage), C_f the system capacitance when inversion is achieved, and Δt the storage time; time for system capacitance to recover from C_0 to C_f

  Constant capacitance technique [32,33]: $\tau_g = |e|An_i(W_1 - W_2)/(I_1 - I_2)$,
    where A is the surface area of device, W_1 and W_2 are the depletion region widths of the two scans, and I_1 and I_2 are the generation currents of the two scans

  Zerbst technique [34]: $\tau_g = (2n_i/mN_a)(C_{ox}/C_f)$,
    where m is the Zerbst slope in graph of $-d(C_{ox}/C)^2/dt$ versus $\{(C_f/C) - 1\}$ taken over a region delimited by experimental judgement and C the system capacitance

  Surface recombination velocity [34]: $s = (N_A \varepsilon \varepsilon_0/2n_i C_{ox}) \cdot b$,
    where b is the y intercept, ε the silicon permittivity, and ε_0 the permittivity in vacuum

Intrinsic Debye length [35]: $L_D = (\varepsilon \varepsilon_0 kT/2e^2 n_i)^{1/2}$

TABLE XII

Properties of Various Impurities in Silicon[a]

Element	Group	Crystal equilibrium distribution coefficient[b]	Site position	Electric state	Tetrahedral covalent radius Å	Ionic radius Å
Lithium (Li)	IA	1×10^{-2}	—	D	1.23	0.60(+1)
Copper (Cu)	IB	4×10^{-4}	I,S	A	1.35	0.96(+1), 0.69(+2)
Silver (Ag)	IB	$\sim 1 \times 10^{-6}$	—	A,D	1.53	1.26(+1)
Gold (Au)	IB	2.5×10^{-5}	—	A,D	1.50	1.37(+1)
Zinc (Zn)	IIB	$\sim 1 \times 10^{-5}$	—	A	1.31	0.74(+2)
Cadmium (Cd)	IIB	$\sim 1 \times 10^{-6}$	—	A	1.48	0.97(+2)
Boron (B)	IIIA	8×10^{-1}	S	A	0.88	0.20(+3)
Aluminum (Al)	IIIA	2×10^{-3}	S	A	1.26	0.50(+3)
Gallium (Ga)	IIIA	8×10^{-3}	S	A	1.26	1.13(+1), 0.62(+3)
Indium (In)	IIIA	4×10^{-4}	S	A,D	1.44	1.32(+1), 0.81(+3)
Thallium (Tl)	IIIA	1.7×10^{-4}	S	A	1.47	1.40(+1), 0.95(+3)
Carbon (C)	IVA	$7 \pm 1 \times 10^{-2}$	S	N	0.77	2.60(−4), 0.15(+4)
Germanium (Ge)	IVA	3.3×10^{-1}	S	N	1.22	0.93(+2), 0.53(+4)
Tin (Sn)	IVA	1.6×10^{-2}	S	N	1.40	2.45(−3), 0.62(+5)
Nitrogen (N)	VA	7×10^{-4}	I	N,D	0.70	1.71(−3), 0.11(+5)
Phosphorus (P)	VA	3.5×10^{-1}	S	D	1.10	2.12(−3), 0.34(+5)
Arsenic (As)	VA	3×10^{-1}	S	D	1.18	2.22(−3), 0.47(+5)
Antimony (Sb)	VA	2.3×10^{-2}	S	D	1.36	2.45(−3), 0.62(+5)
Bismuth (Bi)	VA	7×10^{-4}	S	D	1.46	1.20(+3), 0.74(+5)
Oxygen (O)	VIA	1.4 ± 0.3	I	D	0.66	1.40(−2), 0.09(+6)
Sulfur (S)	VIA	$\sim 1 \times 10^{-5}$	—	D	1.04	1.84(−2), 0.29(+6)
Chromium (Cr)	VIB	1.1×10^{-5}	—	A,D	—	0.69(+3), 0.52(+6)
Titanium (Ti)	IVB	2×10^{-6}	—	D	—	0.90(+2), 0.68(+4)
Vanadium (V)	VB	4×10^{-6}	—	A,D	—	0.74(+3), 0.59(+5)
Manganese (Mn)	VIIB	$\sim 1 \times 10^{-5}$	—	D	—	0.80(+2), 0.46(+7)
Iron (Fe)	VIII	8×10^{-6}	—	A,D	—	0.76(+2), 0.64(+3)
Cobalt (Co)	VIII	8×10^{-6}	—	A,D	—	0.74(+2), 0.63(+3)
Nickel (Ni)	VIII	3×10^{-5}	—	A	—	0.72(+2), 0.62(+3)
Molybdenum (Mo)	VIB	4.5×10^{-8}	—	D	—	0.68(+4), 0.62(+6)
Tantalum (Ta)	VB	1×10^{-7}	—	D	—	0.73(+5)
Platinum (Pt)	VIII	—	—	A,D	—	0.96(+2)

[a] From reference [36–39].
[b] Listing based on a volume basis rather than on a molar basis.

TABLE XIII

Diffusion Coefficient and Activation Energy of Impurities in Silicon[a]

Group	Element	D_0 (cm²/s)	Q (eV)	Reference
IA	H[b]	9.4×10^{-3}	0.48	40
IA	Li	2.5×10^{-3}	0.66	41
IB	Cu	4.7×10^{-3}	0.43	42
	Ag	2.0×10^{-3}	1.60	43
	Au (i)	2.4×10^{-4}	0.39	44
	Au (s)	2.75×10^{-3}	2.04	44
IIB	Zn	1×10^{-1}	1.40	45
IIIA	B	9.1×10^{-2}	3.36	46
	Al	1.385	3.39	46
	Ga	3.74×10^{-1}	3.41	46
	In	7.85×10^{-1}	3.63	46
	Tl	1.37	3.70	46
IVA	C	0.33	2.92	47
	Ge	1.535×10^{3}	4.65	48
	Sn	3.2×10	4.25	49
IVB	Ti	2.0×10^{-5}	1.50	50
VA	N	8.7×10^{-1}	3.29	51
	P	3.85	3.66	46
	As	3.8×10^{-1}	3.58	46
	Sb	2.14×10^{-1}	3.65	46
	Bi	1.08	3.85	46
VIA	O	2.3×10^{-1}	2.56	52
	S	9.2×10^{-1}	2.20	53
VIB	Cr	1×10^{-2}	1.00	54
VIIB	Mn	1.42×10^{-1}	1.30	55
VIII	Fe	1.3×10^{-3}	0.68	56
	Co	9.2×10^{4}	2.80	57
	Ni	2×10^{-3}	0.47	58

[a] Additional useful surveys of diffusion coefficients in silicon have been published in a number of references [38,59–63] and L. C. Kimerling and J. R. Patel, in "VLSI Electronics Microstructure Science," Vol 12 (N. G. Einspruch and H. R. Huff, eds.). Academic Press, New York (1985).

[b] For molecular hydrogen (H_2), the data of $D_0 = 9.4 \times 10^{-3}$ cm²/s and $Q = 2.7$ eV are discussed by J. W. Corbett in *Bull. Am. Phys. Soc.* **28**, 1326 (1983).

TABLE XIV

One Hour Diffusion Length[a] of Impurities in Silicon at Several Temperatures

Element	Diffusion length (μm)					
	700°C	800°C	900°C	1000°C	1100°C	1200°C
H	3.3×10^3	4.3×10^3	5.4×10^3	6.5×10^3	7.6×10^3	8.7×10^3
Li	5.8×10^2	8.4×10^2	1.1×10^3	1.5×10^3	1.8×10^3	2.2×10^3
Cu	3.2×10^3	4.0×10^3	4.9×10^3	5.8×10^3	6.7×10^3	7.5×10^3
Ag	1.9	4.6	9.7	1.8×10	3.1×10	4.9×10
B	3.5×10^{-4}	2.2×10^{-3}	1.1×10^{-2}	3.9×10^{-2}	1.2×10^{-1}	3.1×10^{-1}
Al	1.1×10^{-3}	7.4×10^{-3}	3.6×10^{-2}	1.3×10^{-1}	4.1×10^{-1}	1.1
C	7.5×10^{-3}	4.2×10^{-2}	1.8×10^{-1}	5.9×10^{-1}	1.6	4.0
Ge	2.0×10^{-5}	2.7×10^{-4}	2.3×10^{-3}	1.4×10^{-2}	6.6×10^{-1}	2.5×10^{-1}
Ti	3.4×10^{-1}	7.9×10^{-1}	1.6	2.8	4.7	7.2
N	1.6×10^{-3}	1.0×10^{-2}	4.6×10^{-2}	1.7×10^{-1}	5.0×10^{-1}	1.3
P	3.7×10^{-4}	2.9×10^{-3}	1.6×10^{-2}	6.5×10^{-2}	2.2×10^{-1}	6.3×10^{-1}
As	1.9×10^{-4}	1.4×10^{-3}	7.3×10^{-3}	2.9×10^{-2}	9.6×10^{-2}	2.7×10^{-1}
Sb	9.4×10^{-5}	7.1×10^{-4}	3.9×10^{-3}	1.6×10^{-2}	5.4×10^{-2}	1.5×10^{-1}
O	6.5×10^{-2}	2.7×10^{-1}	8.9×10^{-1}	2.4	5.6	1.2×10
Cr	1.6×10^2	2.7×10^2	4.2×10^2	6.2×10^2	8.7×10^2	1.2×10^3
Mn	9.6×10^1	2.0×10^2	3.6×10^2	6.0×10^2	9.2×10^2	1.3×10^3
Fe	3.7×10^2	5.4×10^2	7.4×10^2	9.7×10^2	1.2×10^3	1.5×10^3
Co	9.9	4.7×10	1.7×10^2	5.1×10^2	1.3×10^3	2.9×10^3
Ni	1.6×10^3	2.1×10^3	2.6×10^3	3.1×10^3	3.7×10^3	4.2×10^3

[a] Diffusion length $L = \sqrt{Dt}$ was calculated from Table XIII.

TABLE XV

Interstitial Solubility of Transition Metals at 1100°C[a]

Element	Solubility (cm^{-3})
Fe	8×10^{14}
Co	8×10^{15}
Ni	5×10^{17}
Cu	5×10^{17}

[a] From reference [62]. Additional solubility data are available in references [36, 63] and L. C. Kimerling and J. R. Patel, in "VLSI Electronics Microstructure Science," Vol 12 (N. G. Einspruch and H. R. Huff, eds.). Academic Press, New York (1985).

TABLE XVI

Fractional Diffusivity G_I^S via Self-interstitials for Some Substitutional Group III and Group V Dopants at 1100°C[a,b]

	Group III dopants			Group V dopants		
	B	Ga	Al	P	As	Sb
r_s/r_{Si}	0.75	1.08	1.08	0.94	1.01	1.16
G_I^S	0.8–1.0	0.6–0.7	0.6–0.7	0.5–1.0	0.2–0.5	0.02

[a] From reference [64].
[b] The ratio of the radii r_s of the substitutional dopant and r_{Si} of a silicon atom is also given. Additional detailed data on the fractional diffusivity in silicon continues to be an extremely active area of current research (i.e., Thirteenth International Conference on Defects in Semiconductors, San Diego, 1984). Therefore, great care must be exercised in the use of these data.

TABLE XVII

Concentration C and Diffusion Constant D of Vacancies V and Self-Interstitials I in Silicon[a,b]

Temperature (°C)	Si vacancies		Si self-interstitials	
	C_V (cm^{-3})	D_V (cm$^2 \cdot$ s^{-1})	C_I (cm^{-3})	D_I (cm$^2 \cdot$ s^{-1})
1420	6.5×10^{14}	4.2×10^{-4}	2.0×10^{16}	4.3×10^{-6}
1227	2.5×10^{13}	1.2×10^{-4}	1.4×10^{15}	7.2×10^{-7}
1027	3.3×10^{11}	2.0×10^{-5}	4.3×10^{13}	7.6×10^{-8}
827	8.5×10^{8}	1.8×10^{-6}	3.9×10^{11}	3.7×10^{-9}

[a] From reference [60].
[b] Additional detailed discussions on self-interstitials and vacancies in silicon have been published in references [65,66]. In reference 66, the self-interstitials have been shown to diffuse faster than vacancies. For example, at 1100°C, $D_I \geq 2 \times 10^{-7}$ cm^2/s whereas $D_V = 2 \times 10^9$ cm^2/s. On the other hand, $D_I \simeq 1.8 \times 10^{-9}$ cm^2/s and $C_I \simeq 6.5 \times 10^{16}$ cm^{-3} at 1100°C according to K. Taniguchi et al., Appl. Phys. Lett. **42**, 961 (1983). Therefore, great care must be exercised in the use of these data as well as Figure 24.

TABLE XVIII

Optical Absorption Bands of Several Impurities in Silicon[a]

Source	Wavelength[b] (μm)	Temperature (K)
Oxygen	9.042	298
Carbon	16.53	298
Boron[c]	31.2	4
	32.2	
Aluminum[c]	21.1	4
	22.5	
Gallium[c]	21.2	4
	20.0	
Phosphorus[c]	31.3	—
Arsenic[c]	31.6	—
Antimony[c]	33.9	—
	31.0	—

[a] From references [67,68].

[b] There are several bands from each source; the ones listed are the most intense.

[c] These impurities are ionized at room temperature. When the temperature is reduced to the point at which they are no longer ionized, transitions may occur between their levels and the appropriate valence or conduction band.

TABLE XIX

Commonly Used Diffusion Profile Measurement Techniques[a]

Profile techniques	Characteristics	Ref.
Capacitance–voltage	Carrier concentration at the edge of the depletion layer of a $p-n$ junction; maximum total dopants 2×10^{12} atoms/cm^2	70
Differential conductance and Hall effect	Resistivity and mobility of net electrically active species; requires thin layer removal; 10^{20}–10^{18} atoms/cm^3	71
Spreading resistance	Resistance on angle-beveled sample; good for comparison with known profiles and quick semiquantitative evaluation; depth >1 μm	72
SIMS	High sensitivity on many elements, for B and As the detection limit is 5×10^{15} cm^{-3}; capable of measuring profiles in 1000-Å range; needs standards	73
Radioactive tracer analysis	Total concentration; lower limit 10^{15} cm^{-3}; limited to radioactive elements with suitable half-life times (P, As, Sb, Na, Cu, Au, etc.)	74
Rutherford backscattering	Applicable only for elements heavier than Si	75
Nuclear reaction	Measures total boron through ^{10}B(n, ^{4}He)^{7}Li, or ^{11}B(p, α); needs van de Graff generator	76, 77

[a] From reference [69].

TABLE XX
Lowest Binary Eutectic Temperatures and Resistivities of Several Silicides[a]

Silicide	Lowest binary eutectic temperature (°C)	Specific resistivity ($\mu\Omega \cdot cm$)
$CoSi_2$	1195	18–25
$HfSi_2$	1300	45–50
$MoSi_2$	1410	100
$NbSi_2$	1295	50
$NiSi_2$	966	50–60
Pd_2Si	720	30–35
$PtSi$	830	28–35
$TaSi_2$	1385	35–45
$TiSi_2$	1330	13–25
VSi_2	1385	50–55
WSi_2	1440	70
$ZrSi_2$	1355	35–40

[a] From reference [78].

TABLE XXI
Silicon (Cubic) Interplanar Angles

Crystallographic plane	(100)	(110)	(111)	(211)	(511)
(100)	2 of 90°00'	2 of 45°00' 90°00'	3 of 54°64'	35°16' 2 of 65°54'	15°48' 2 of 78°54'
(110)	4 of 45°00' 2 of 90°00'	4 of 60°00' 90°00'	3 of 35°16' 3 of 90°00'	2 of 30°20' 54°44' 2 of 73°13' 90°00'	2 of 35°16' 2 of 57°01' 74°12' 90°00'
(111)	4 of 54°44'	2 of 35°16' 2 of 90°00'	3 of 70°32'	19°28' 2 of 61°52' 90°00'	38°57' 2 of 56°15' 70°32'
(211)	4 of 35°16' 8 of 65°54'	4 of 30°00' 2 of 54°44' 4 of 73°13' 2 of 90°00'	3 of 19°28' 6 of 61°52' 3 of 90°00'	2 of 33°33' 2 of 48°11' 2 of 60°00' 70°32' 4 of 80°24'	19°28' 2 of 38°13' 3 of 51°03' 2 of 61°53' 2 of 71°41' 2 of 80°58'
(511)	4 of 15°48' 8 of 78°54'	4 of 35°16' 4 of 57°01' 2 of 74°12' 2 of 90°00'	3 of 38°57' 6 of 56°15' 3 of 70°32'	19°28' 2 of 38°13' 3 of 51°03' 2 of 61°53' 2 of 71°41' 2 of 80°58'	2 of 22°11' 31°35' 2 of 65°57' 2 of 70°32' 4 of 87°53'

TABLE XXII

Properties of Silicon Crystal Planes[a]

Plane	Surface energy (J/cm^2)	Area of unit cell	Si atoms in area	Si bonds in area	Bonds available	Bond density ($\times 10^{14}$/cm^2)	Available bond density[b] ($\times 10^{14}$/cm^2)
(100)	2.53	a^2	2	4	2	13.55	6.77
(110)	1.78	$\sqrt{2}a^2$	4	8	4	19.18	9.59
(111)	1.46	$\sqrt{3}a^2/2$	2	4	3	15.68	11.76

[a] From references [79,80].
[b] The available bond density is less than the atomistic bond density due to a variety of geometric related effects referred to as steric hindrances.

TABLE XXIII

Hydrogen Peroxide-Based Immersion Cleaning Procedures for Silicon Product[a]

A. Preliminary cleaning if necessary)
  1. Remove bulk of photoresist film (if present) by plasma oxidation stripping, immersion in organic photoresist stripper, or with a hot 1 : 2 v/v H_2O_2–H_2SO_4 mixture if adequate safety precautions are exercised.
  2. Rinse with water (see note on water purity for entire processing[b]).
  3. Transfer the wafers to a clean Teflon holder. Pick up wafers with Teflon or polypropylene plastic tweezers.

B. Removal of residual organic contaminants and certain metals (SC-1)
  1. Prepare a fresh mixture of H_2O–NH_4OH–H_2O_2(5 : 1 : 1) by measuring the following reagents into a beaker of fused silica (opaque silica-ware is acceptable):
     (a) 5 volumes of water,
     (b) 1 volume of ammonium hydroxide (29%, electronic grade, w/w % based on NH_3),
     (c) 1 volume of hydrogen peroxide (30%, unstabilized electronic grade, w/w %).
  2. Stir the solution with a clean rod of fused quartz.
  3. Submerge holder with wafers in the cold solution and place the beaker on a hotplate.
  4. Heat to 75–80°C. Then reduce heat to maintain the solution at 80°C for an additional 10 min. (The vigorous bubbling is due to oxygen evolution. Make sure not to boil the solution, to prevent rapid decomposition of the H_2O_2 and volatilization of the ammonia.)
  5. Overflow-quench the solution by placing the beaker under running water for about 1 min.
  6. Remove holder with wafers and immediately place it in a cascade water rinse tank for 5 min.

C. Stripping of thin hydrous oxide film (1 : 50 HF–H_2O)
  1. Submerge wafer assembly from step B.6 directly in an agitated mixture of 1 volume hydrofluoric acid (49%, electronic grade) and 50 volumes of water.
  2. Allow wafer assembly to remain in the solution for only 15 s. Exposed silicon (but not Si_2) should repel the HF solution [24]. Use a polypropylene beaker for this step.
  3. Transfer the wafer assembly to a water tank, but rinse it for only 20–30 s with agitation to remove the HF solution (this minimizes regrowth of a hydrous oxide film).
  4. Transfer the wafer assembly immediately, without drying, into the hot SC-2 solution of step D.

TABLE XXIII (Continued)

D. Desorption of remaining atomic and ionic contaminants (SC-2)
  1. Prepare a fresh mixture of H_2O–HCl–H_2O_2 (6:1:1) by measuring the following reagents into a beaker of fused quartz:
     (a) 6 volumes of water,
     (b) 1 volume of hydrochloric acid (37%, electronic grade),
     (c) 1 volume of hydrogen peroxide (30%, unstabilized, electronic grade).
  2. Place the beaker on a hotplate and heat to 75–80°C.
  3. Submerge the still-wet wafers in the holder (after step B.6 or C.3) in the hot solution.
  4. Maintain the solution at 80°C for 10–15 min.
  5. Overflow quench as in step B.5.
  6. Continue the rinsing at this stage for a total of 20 min in a cascade rinsing system.
E. Drying of the wafers
  1. Transfer the holder with the wet wafers into a wafer centrifuge.
  2. Apply a final water rinse during spinning.
  3. Allow to dry while gradually increasing the spinning speed (to avoid aerosol formation from the water droplets).
  4. Remove the wafers by dump transfer for high-temperature processing. If single-wafer handling must be used, handle the wafers only at the edge with plastic tweezers.
F. Storage
  1. Avoid storage of cleaned wafers, preferably by immediate continuation of processing. If storage is unavoidable, store the wafers in closed glass containers cleaned with hot SC-1 solution, followed by water rinsing and over-drying.

[a] From reference [81].
[b] All water used for preparing the reagent mixtures or for rinsing should be thoroughly deionized and ultrafiltered, with a resistivity in the 10–20 MΩ range at 18–23°C. All reagents should be electronic grade, preferably ultrafiltered for freedom from particulate impurities.

TABLE XXIV
Compositions of Commonly Used Concentrated Aqueous Reagents

Reagent	Weight %
HCl	37
HF	49
H_2SO_4	98
H_3PO_4	85
HNO_3	70
$HClO_4$	70
CH_3COOH	99
H_2O_2	30
NH_4OH	29 (as NH_3)

TABLE XXV

Etchants for Crystalline Silicon[a]

Application	Etchant	Chemical composition (volume ratio)	Comment
Defect delineation	Sirtl [83]	$HF:CrO_3(5M) = 1:1$	Best applicable to (111)
	Dash [84]	$HF:HNO_3:CH_3COOH = 1:3:10$	Applicable for both n- and p-type (111) and (100)
			Better for p-type than n-type
	Secco [85]	$HF:K_2Cr_2O_7(0.15M) = 2:1$	Suitable for (100)
			Ultrasonic agitation recommended
	Schimmel-1 [86]	$HF:HNO_3 = 155:1$	Applicable to p-type (100)
	Schimmel-2 [87]	$HF:CrO_3(1M)(:H_2O) = 2:1$ $(:1.5)$	Useful for 0.6–15 $\Omega \cdot cm$ p- and n-types (100)
			Adding 1.5 (H_2O) is recommended for heavily doped (100)
	Wright [88]	$HF:HNO_3:CrO_3(5M):Cu(NO_3)_2 \cdot 3H_2O:CH_3COOH:H_2O =$ 60 ml; 30 ml : 30 ml : (2g) : 60 ml : 60 ml	Widely applicable for p- and n-type (100) and (111)
Polish	CP-4A	$HF:HNO_3:CH_3COOH = 3:5:3$	Can be applied generally
	5-1 etch	$HF:HNO_3 = 1:5$	Can be applied generally
	Caustic etch	KOH or $NaOH:H_2O = (20\ g):(100\ cc)$	70°C–130°C
			Uniform etching develops flat surface
Stain	Staining etch	$HF:HNO_3 = 1000:\sim 1–2$	p–n junction delineation

[a] From reference [15,82].

TABLE XXVI

Etchants for Noncrystalline Films[a]

Material	Etchant	Remark
SiO_2	28 ml HF 170 ml H_2O 113 g NH_4F	BHF, 1000–2500 Å/min at 25°C
	15 ml HF 10 ml HNO_3 300 ml H_2O	P-etch, 128 Å/min at 25°C
	1 ml BHF 7 ml H_2O	800 Å/min

TABLE XXVI (*Continued*)

Material	Etchant	Remark
BSG	1 ml HF 100 ml HNO_3 100 ml H_2O	R-etch, 300 Å/min for 9 mole % B_2O_3, 50 Å/min for SiO_2
	4.4 ml HF 100 ml HNO_3 100 ml H_2O	S-etch, 750 Å/min for 9 mole % B_2O_3, 135 Å/min for SiO_2
PSG	28 ml HF 170 ml H_2O 113 g NH_4F	BHF, 5500 Å/min for 8 mole % P_2O_5
	15 ml HF 10 ml HNO_3 300 ml H_2O	P-etch, 34,000 Å/min for 16 mole % P_2O_5, 110 Å/min for SiO_2
	1 ml BHF 7 ml H_2O	800 Å/min
Si_3N_4	HF	140 Å/min, CVD at 1100°C 750 Å/min, CVD at 900°C 1000 Å/min, CVD at 800°C
	28 ml HF 170 ml H_2O 113 g NH_4F	BHF, 5–10 Å/min
	H_3PO_4	100 Å/min at 180°C
Polysilicon	6 ml HF 100 ml HNO_3 40 ml H_2O	8000 Å/min, smooth edges
	1 ml HF 26 ml HNO_3 33 ml CH_3COOH	1500 Å/min
SIPOS	1 ml HF 6 ml H_2O 10 ml NH_4F (40%)	2000 Å/min for 20% O_2 film
Al	1 ml HCl 2 ml H_2O	80°C, fine line, can be used with gallium arsenide
	4 ml H_3PO_4 1 ml HNO_3 4 ml CH_3COOH 1 ml H_2O	350 Å/min, fine line, will attack gallium arsenide
	16–19 ml H_3PO_4 1 ml HNO_3 0–4 ml H_2O	1500–2500 Å/min, will attack gallium arsenide
	0.1 M $K_2Br_4O_7$ 0.51 M KOH 0.6 M $K_3Fe(CN)_6$	1 µm/min, pH 13.6, no gas evolved during etching
Au	3 ml HCl 1 ml HNO_3	Aqua regia, 25–50 µm/min

TABLE XXVI (*Continued*)

Material	Etchant	Remark
Au	4 g KI 1 g I_2 40 ml H_2O	0.5–1 μm/min, can be used with resist
Ag	1 ml NH_4OH 1 ml H_2O_2 4 ml CH_3OH	3600 Å/min, can be used with resists, must be rinsed rapidly after etching
Cr	1 ml HCl 1 ml glycerine	800 Å/min, needs depassivation
	1 ml HCl 9 ml saturated $CeSO_4$ solution	800 Å/min, needs depassivation
	1 ml, 1 g NaOH in 2 ml H_2O 3 ml, 1 g $K_3Fe(CN)_6$ in 3 ml H_2O	250–1000 Å/min, no depassivation, resist mask can be used
Mo	5 ml H_3PO_4 2 ml HNO_3 4 ml CH_3COOH 150 ml H_2O	0.5 μm/min, resist mask can be used
	5 ml H_3PO_4 3 ml HNO_3 2 ml H_2O	Polishing etch
	11 g $K_3Fe(CN)_6$ 10 g KOH 150 ml H_2O	1 μm/min
W	34 g KH_2PO_4 13.4 g KOH 33 g $K_3Fe(CN)_6$ H_2O to make 1 L	1600 Å/min, high resolution, resist mask can be used
Pt	3 ml HCl 1 ml HNO_3	Aqua regia, 20 μm/min, precede by a 30-s immersion in HF
	7 ml HCl 1 ml HNO_3 8 ml H_2O	400–500 Å/min, 85°C
Pd	1 ml HCl 10 ml HNO_3 10 ml CH_3COOH	1000 Å/min
	4 g KI 1 g I_2 40 ml H_2O	1 μm/min, opaque, must be rinsed before visual inspection

[a] From reference [15].

15. VLSI Silicon Material Criteria

TABLE XXVII

Color Chart for Thermally Grown SiO_2 Films Observed Perpendicularly under Daylight Fluorescent Lighting[a]

Film thickness (μm)	Color and comments	Film thickness (μm)	Color and comments
0.05	Tan	0.63	Violet red
0.07	Brown	0.68	Bluish (not blue but borderline between violet and blue green; it appears more like a mixture of violet red and blue green and looks grayish)
0.10	Dark violet to red violet		
0.12	Royal blue		
0.15	Light blue to metallic blue		
0.17	Metallic to very light yellow green		
0.20	Light gold or yellow, slightly metallic	0.72	Blue green to green (quite broad)
0.22	Gold with slight yellow orange	0.77	Yellowish
0.25	Orange to melon	0.80	Orange (rather broad for orange)
0.27	Red violet		
0.30	Blue to violet blue	0.82	Salmon
0.31	Blue	0.85	Dull, light red violet
0.32	Blue to blue green	0.86	Violet
0.34	Light green	0.87	Blue violet
0.35	Green to yellow green	0.89	Blue
0.36	Yellow green	0.92	Blue green
0.37	Green yellow	0.95	Dull yellow green
0.39	Yellow	0.97	Yellow to yellowish
0.41	Light orange	0.99	Orange
0.42	Carnation pink	1.00	Carnation pink
0.44	Violet red	1.02	Violet red
0.46	Red violet	1.05	Red violet
0.47	Violet	1.06	Violet
0.48	Blue violet	1.07	Blue violet
0.49	Blue	1.10	Green
0.50	Blue green	1.11	Yellow green
0.52	Green (broad)	1.12	Green
0.54	Yellow green	1.18	Violet
0.56	Green yellow	1.19	Red violet
0.57	Yellow to yellowish (not yellow but is in the position where yellow is expected; at times it appears to be light creamy gray or metallic)	1.21	Violet red
		1.24	Carnation pink to salmon
		1.25	Orange
		1.28	Yellowish
		1.32	Sky blue to green blue
		1.40	Orange
0.58	Light orange or yellow to pink borderline	1.45	Violet
		1.46	Blue violet
0.60	Carnation pink	1.50	Blue
		1.54	Dull yellow green

[a] From reference [89].

TABLE XXVIII
Diffusivities of Elements in SiO_2[a]

Element	D at 1100°C (cm^2/s)	D at 1200°C (cm^2/s)
B	3×10^{-17}–2×10^{-14}	2×10^{-16}–5×10^{-14}
Ga	5.3×10^{-11}	5×10^{-8}
P	2.9×10^{-16}–2×10^{-13}	2×10^{-15}–7.6×10^{-13}
Sb	9.9×10^{-17}	1.5×10^{-14}
As	1.2×10^{-16}–3.5×10^{-15}	2×10^{-15}–2.4×10^{-14}

[a] From reference [90].

TABLE XXIX
Properties of Common Gases Used in CVD[a]

Gas	Properties
Silane	Toxic, flammable, pyrophoric
Dichlorosilane	Toxic, flammable, corrosive
Phosphine	Very toxic, flammable
Diborane	Very toxic, flammable
Arsine	Very toxic, flammable
Hydrogen chloride	Toxic, corrosive
Ammonia	Toxic, corrosive
Hydrogen	Nontoxic, flammable
Oxygen	Nontoxic, supports combustion
Nitrous oxide	Nontoxic, nonflammable
Nitrogen	Usually inert
Argon	Inert

[a] From reference [91].

TABLE XXX
Typical Reactions for Depositing Dielectrics and Polysilicon[a]

Product	Reactants	Deposition temperature (°C)
Silicon dioxide	$SiH_4 + CO_2 + H_2$	850–950
	$SiCl_2H_2 + N_2O$	850–900
	$SiH_4 + N_2O$	750–850
	$SiH_4 + NO$	650–750
	$Si(OC_2H_5)_4$	650–750
	$SiH_4 + O_2$	400–450
Silicon nitride	$SiH_4 + NH_3$	700–900
	$SiCl_2H_2 + NH_3$	650–750
Plasma silicon nitride	$SiH_4 + NH_3$	200–350
	$SiH_4 + N_2$	200–350
Plasma silicon dioxide	$SiH_4 + N_2O$	200–350
Polysilicon	SiH_4	600–650

[a] From reference [91].

TABLE XXXI
Comparison of Different Deposition Methods[a]

Deposition properties	Atmospheric-pressure CVD	Low-temperature LPCVD	Medium-temperature LPCVD	Plasma-assisted CVD
Temperature (°C)	300–500	300–500	500–900	100–350
Materials	SiO_2	SiO_2	Poly-Si	SiN
	P-glass	P-glass	SiO_2	SiO_2
			P-glass	
			Si_3N_4	
Uses	Passivation, insulation	Passivation, insulation	Gate metal, insulation, passivation	Passivation, insulation
Throughput	High	High	High	Low
Step coverage	Poor	Poor	Conformal	Poor
Particles	Many	Few	Few	Many
Film properties	Good	Good	Excellent	Poor
Low temperature	Yes	Yes	No	Yes

[a] From reference [91].

TABLE XXXII
Properties of Deposited Silicon Dioxide[a]

Deposition	Plasma	$SiH_4 + O_2$	TEOS	$SiCl_2H_2 + N_2O$
Temperature (°C)	200	450	700	900
Composition	$SiO_{1.9}(H)$	$SiO_2(H)$	SiO_2	$SiO_2(Cl)$
Step coverage	Nonconformal	Nonconformal	Conformal	Conformal
Thermal stability	Looses H	Densifies	Stable	Losses Cl
Density (g/cm³)	2.3	2.1	2.2	2.2
Refractive index	1.47	1.44	1.46	1.46
Stress (10^9 dynes/cm²)	3C–3T	3T	1C	3C
Dielectric strength (10^6 V/cm)	3–6	8	10	10
Etch rate (Å/min) (100:1 H_2O:HF)	400	60	30	30

[a] From reference [91].

TABLE XXXIII
Properties of Silicon Nitride[a]

Deposition	LPCVD	Plasma
Temperature (°C)	700–800	250–350
Composition	$Si_3N_4(H)$	SiN_xH_y
Si/N Ratio	0.75	0.8–1.2
At. % H	4–8	20–25
Refractive index	2.01	1.8–2.5
Density (g/cm³)	2.9–3.1	2.4–2.8
Dielectric constant	6–7	6–9
Resistivity (Ω·cm)	10^{16}	10^6–10^{15}
Dielectric strength (10^6 V/cm)	10	5
Energy gap (eV)	5	4–5
Stress (10^9 dynes/cm²)	10T	2C–5T

[a] From reference [91].

TABLE XXXIV
Single Bond Energy of Relevant Elements[a]

Bond	(eV/bond)	Bond	(eV/bond)
H—H	4.51	Cl—Cl	2.52
C—C	3.60	Si—H	3.05
Si—Si	1.83	Si—C	3.01
Ge—Ge	1.63	Si—O	3.83
N—N	1.67	Si—S	2.35
P—P	2.23	Si—F	5.61
As—As	1.39	Si—Cl	3.72
Sb—Sb	1.31	Si—Br	3.00
O—O	1.44	Si—I	2.21

[a] From reference [39].

15. VLSI Silicon Material Criteria

TABLE XXXV

Properties of Elements Used in Ion Implantation[a]

Element	Atomic number	Atomic weight	Density (g/cm^3)
Al	13	26.98	2.70
Sb	51	121.75	6.62
As	33	74.92	5.72
Be	4	9.01	1.85
B	5	10.81	2.34
Cd	48	112.40	8.65
Ga	31	69.72	5.91
Ge	32	72.59	5.32
Mg	12	24.31	1.74
N	7	14.01	1.25[b]
O	8	16	1.43[b]
P	15	30.97	1.83
Se	34	78.96	4.79
Si	14	28.09	2.33
S	16	32.06	2.07
Te	52	127.6	6.24
Sn	50	118.69	7.30
Zn	30	65.37	7.13

[a] From reference [15].
[b] g/liter

TABLE XXXVI

Thermomechanical Properties of Crystalline Silicon[a]

Hardness[b]	7 Moh, 1000 Vickers, 950–1150 Knoop
Elastic constants	C_{11}: 1.6740×10^{12} dynes/cm^2
	C_{12}: 0.6523×10^{12} dynes/cm^2
	C_{44}: 0.7959×10^{12} dynes/cm^2
Temperature coefficients of elastic constants	$K_{C_{11}}$: -75×10^{-6}/°C
	$K_{C_{12}}$: -24.5×10^{-6}/°C
	$K_{C_{44}}$: -55.5×10^{-6}/°C
Young's modulus	1.9×10^{12} dynes/cm^2, $\langle 111 \rangle$ direction
Bulk modulus	7.7×10^{11} dynes/cm^2
Modulus of rupture (in bending)	700–3500 kg/cm^2
Breaking strength (in compression)	4900–5600 kg/cm^2
Linear thermal coefficient of expansion	2.33×10^{-6}/°C
Surface tension	720 dynes/cm (freezing point)

[a] From reference [7].
[b] See also Table XXXVII.

TABLE XXXVII

Effect of Crystal Plane on Hardness[a] (Knoop Indenter, 100-g Load)

Crystallographic plane	Average (Knoop)	Range (depending on direction of long axis of indenter) (Knoop)
(100)	964	950–980
(110)	964	940–980
(111)	948	935–970

[a] From reference [92].

TABLE XXXVIII

Properties of Liquid Silicon[a]

Property	At 1420°C	At 1500°C
Thermal conductivity (W/m·K)	41.84	—
Dynamic viscosity (mPa·s) (= cP)	0.88	0.7
Kinematic viscosity (mm^2/s) (= cSt)	0.347	0.28
Surface tension (mN/m) (= dynes/cm)	736.00	720.00
Heat capacity (J/kg·K)	0.16	6.84
Density (g/cm^3)	2.533	2.50
Electrical resistivity (μ·ohm·cm)	80.00	100.00
Total optical emissivity	0.33	0.33
Reflectivity at 633 nm (%)	72.00	70.00

[a] From reference [93].

15. VLSI Silicon Material Criteria

TABLE XXXIX

Selected VLSI Silicon Material Target Values[a]

Material property	Trend
Electrical	
Oxide breakdown voltage	≤1% failure for electric fields $\leq 5 \times 10^6$ V/cm
Flat-band voltage shift	≤0.2 V (specifically defined metal gate test)
Generation lifetime	~300–1000 μs
Resistivity variation	≤10%
Chemical	
Cleanliness	≤0.03/cm^2 particles (≤0.5 μm) wafer front surface; ≤0.05/cm^2 particles (≤1 μm) wafer back surface; no wafer backsurface stain
Oxygen concentration	Customer specified; ±2 ppma
Oxygen radial gradient	≤3%
Carbon concentration	≤0.3 ppma
Metals	
Bulk	≤0.01 ppba for specific metals
Surface	≤10^{10} cm^{-2}
Structural	
Grown-in dislocation (etch pits)	0 cm^{-2}
Oxidation-induced stacking faults	≤3/cm^2 (specifically defined oxidation test such as ASTM F-416)
Mechanical	
Diameter	>150 mm
Tolerance	≤0.2 mm
Thickness	625, 675 μm
Tolerance	≤10 μm
Orientation flat	
Tolerance	≤1.5 mm
Total thickness variation	≤10 μm
Global flatness	≤3 μm
Bow	≤10 μm
Warp	≤10 μm
Local site flatness	≤1.0 μm/20 × 20 mm^2 field
Wafer curvature (polished surface)	Convex or concave specified by customer
Edge contour	Chip-free

[a] These data are based on a variety of sources including several published papers [132–134], and H. R. Huff and F. Shimura, *Solid State Technology*, **3**, 103–118 (1985). "Silicon Material Criteria for VLSI Electronics." The data also form the basis for VLSI target values.

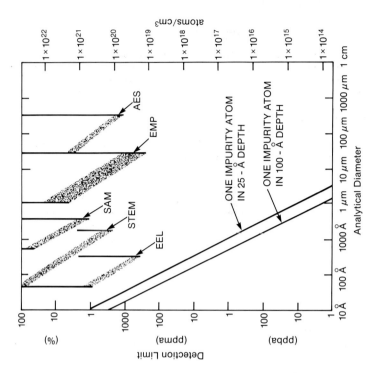

Fig. 2. Comparison of detection limit of electron beam techniques as a function of analytical diameter. Notice the two lines at the lower left that define the concentration and diameter where only one impurity is present in analytical depths of 25 and 100 Å [13].

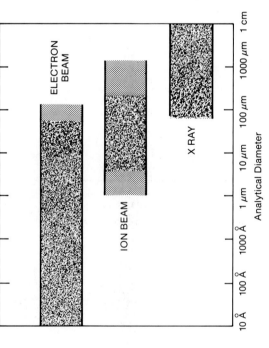

Fig. 1. Analytical diameter capabilities of electron, ion, and X-ray beam characterization techniques (based on Reference [13]).

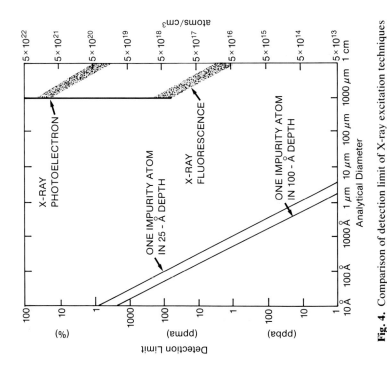

Fig. 4. Comparison of detection limit of X-ray excitation techniques as a function of analytical diameter. Notice the two lines at the lower left that define the concentration and diameter where only one impurity is present in analytical depths of 25 and 100 Å [13].

Fig. 3. Comparison of detection limit of ion-beam techniques as a function of analytical diameter. Notice the two lines at the lower left that define the concentration and diameter where only one impurity is present in analytical depths of 25 and 100 Å [13].

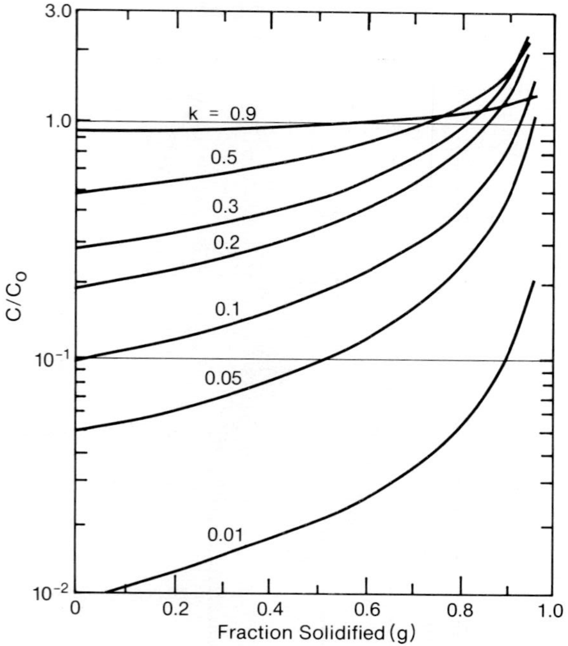

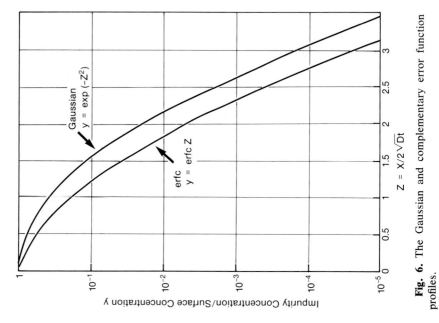

Fig. 6. The Gaussian and complementary error function profiles.

Fig. 5. Distribution of impurities during normal freezing.

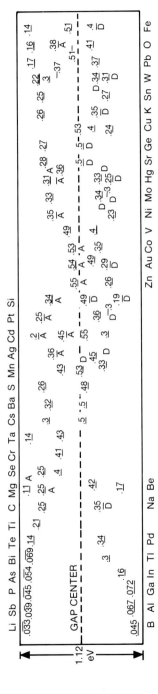

Fig. 7. Measured ionization energies of various impurities in Si. The levels below the gap center are measured from the top of the valence band and are acceptor levels unless indicated by D for donor level. The levels above the gap center are measured from the bottom of the conduction band and are donor levels unless indicated by A for acceptor level [94,95]. Additional energy levels associated with various dopant–impurity structural complexes are tabulated by J. L. Benton and L. C. Kimerling, *J. Electrochem. Soc. 129*, 2098 (1982).

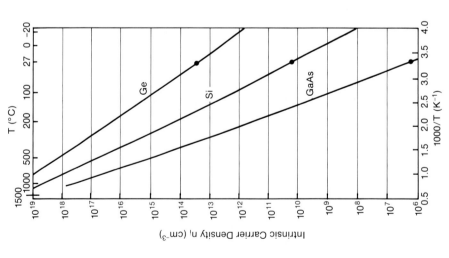

Fig. 8. Intrinsic carrier densities of Ge, Si, and GaAs as a function of reciprocal absolute

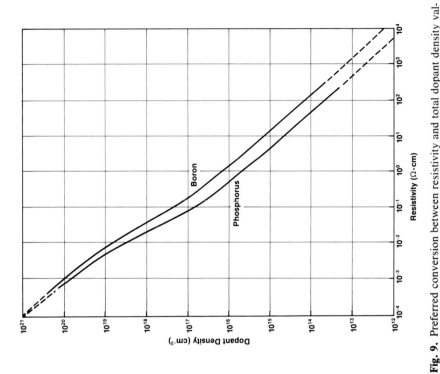

Fig. 9. Preferred conversion between resistivity and total dopant density values for boron- and phosphorus-doped silicon. Solid lines show range of data. Dashed lines show regions of extrapolation from data [9].

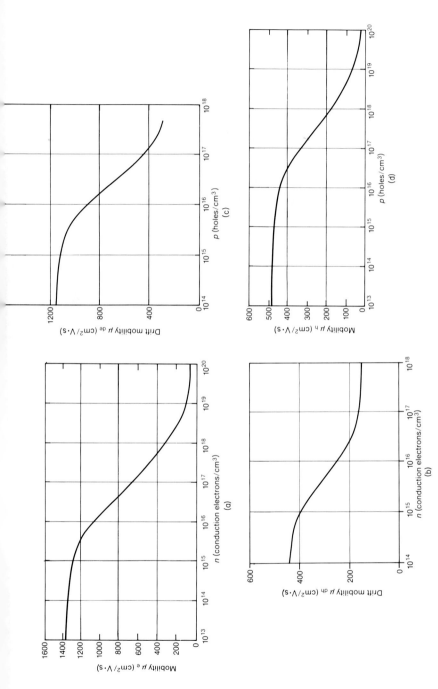

Fig. 10. Majority and minority carrier mobilities in silicon at 300 K as a function of the majority carrier concentration. (a) Electrons in n-type silicon. (b) Holes in n-type silicon [95]. (c) Electrons in p-type silicon. (d) Holes in p-type silicon.

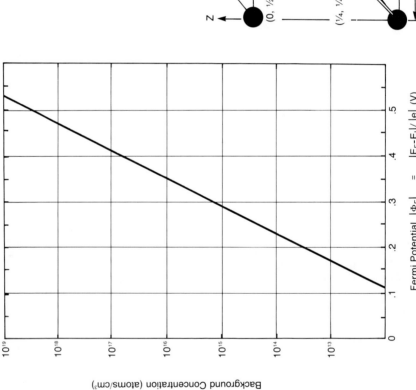

Fig. 11. Fermi potential as a function of the background doping concentration in p- or n-type silicon at 300 K ($\phi_F > 0$ for p-type; $\phi_F < 0$ for n-type) [98].

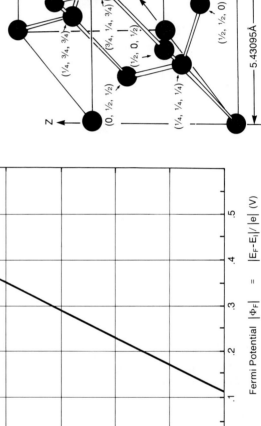

Fig. 12. Silicon crystal structure.

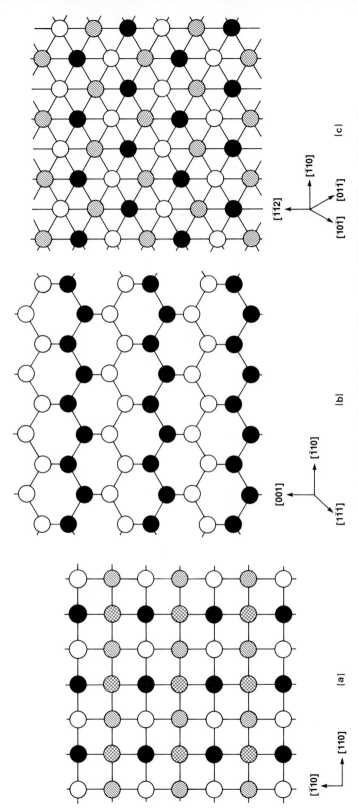

Fig. 13. Channeling in a silicon crystal. (a) Viewed along [100] axis. (b) Viewed along [110] axis. (c) Viewed along [111] axis. Different shading refers to different atomic layers.

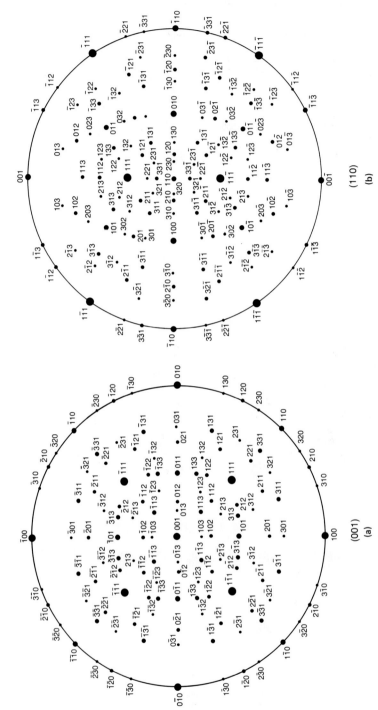

Fig. 14. Standard stereographic projections for a face-centered cubic crystal, showing (a) (001) projection, (b) (110) projection, and (c) (111) projection.

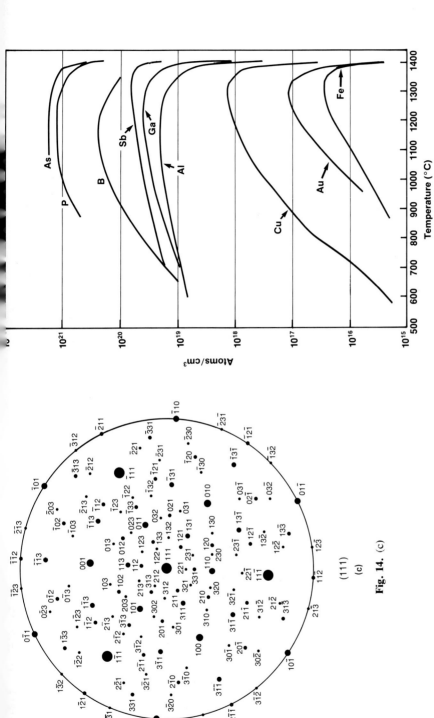

Fig. 15. Solid solubility of impurities in silicon [36,99]. Additional solubility data are summarized in Table XV.

Fig. 14. (c)

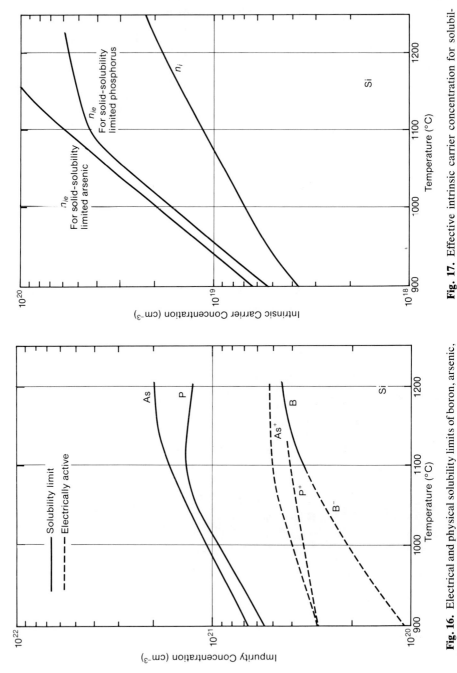

Fig. 16. Electrical and physical solubility limits of boron, arsenic, and phosphorus in silicon [46].

Fig. 17. Effective intrinsic carrier concentration for solubility-limited doping [100].

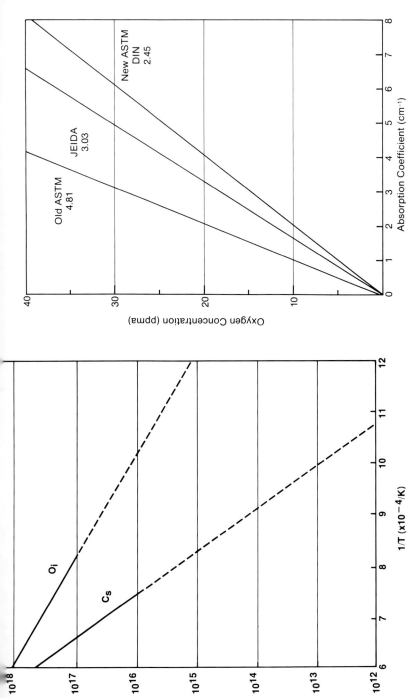

Fig. 18. Solubility of substitutional carbon and interstitial oxygen in silicon. Solubility of interstitial oxygen is given by $1.53 \times 10^{21} \exp(-1.03\ eV/kT)$ cm^{-3} [101] and for substitutional carbon by $4 \times 10^{24} \exp(-2.3\ eV/kT)$ cm^{-3} [102].

Fig. 19. Oxygen concentration versus infrared absorption coefficient at 9 μm. Three conversion factors, 2.45 (new ASTM [103] and DIN [104]), 3.03 (JEIDA [105]), and 4.81 (old ASTM [106]) are given.

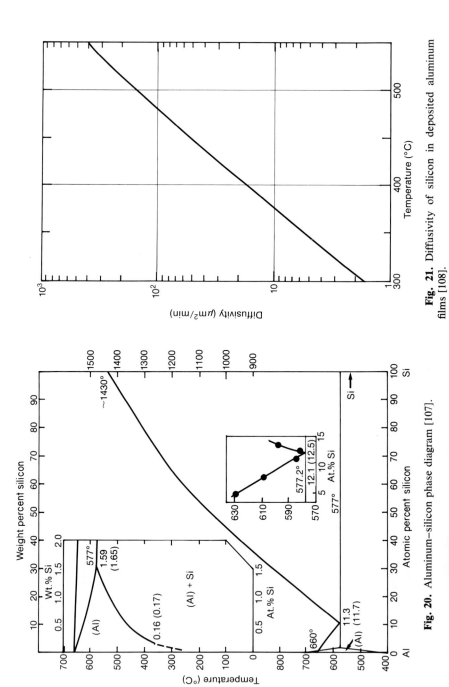

Fig. 21. Diffusivity of silicon in deposited aluminum films [108].

Fig. 20. Aluminum–silicon phase diagram [107].

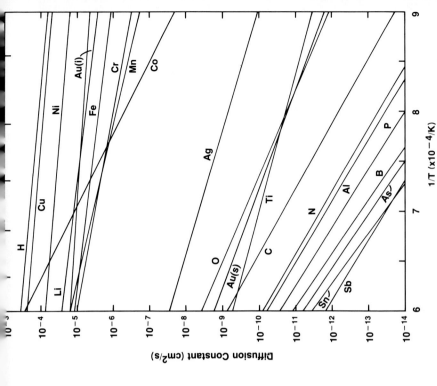

Fig. 23. Diffusion constants of impurities in silicon following Table XIII. Additional useful surveys of diffusion constants in silicon have been published in a number of references [38,59–63].

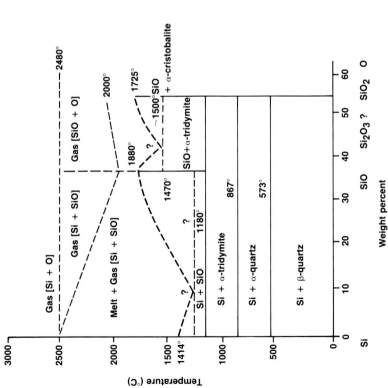

Fig. 22. Oxygen–silicon phase diagram (estimated) [109].

243

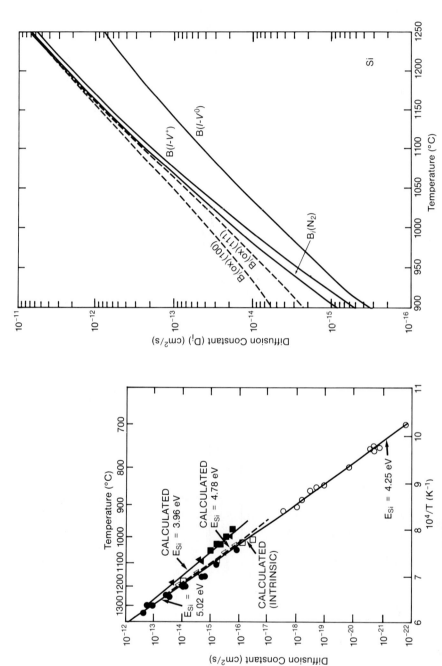

Fig. 25. Diffusion constants for boron in silicon as a function of temperature, ambient, and orientation for several microscopic models [15, 100].

Fig. 24. Silicon self-diffusivity as a function of temperature. ● and □, intrinsic silicon; ▲, boron doped to 2.5×10^{19} cm^{-3}; △, arsenic doped to 8×10^{19} cm^{-3}; ■, phosphorus doped; ○, nickel doped in intrinsic silicon [110]. See

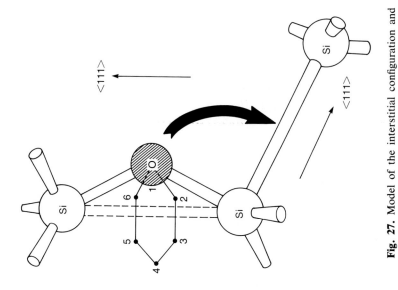

Fig. 27. Model of the interstitial configuration and diffusion jump processes of oxygen in silicon. Six equivalent positions are also shown [61].

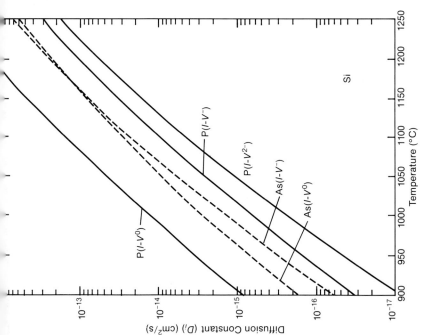

Fig. 26. Diffusion constants for arsenic and phosphorus in silicon as a function of temperature for several microscopic models [15,100].

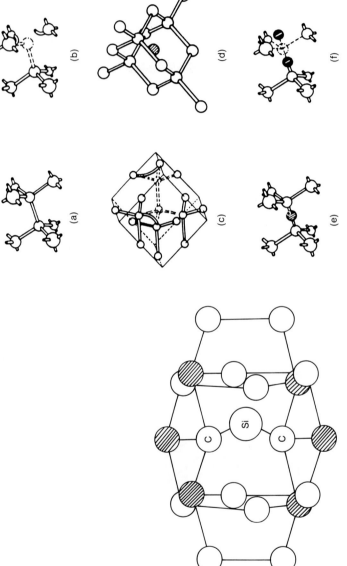

Fig. 28. Defect consisting of a silicon interstitial bound to two substitutional carbon atoms [111].

Fig. 29. Geometrical configurations of vacancy and interstitial point defects. (a) Eight Si atoms form two adjacent tetrahedral bonds. (b) A simple vacancy. (c) Divacancy. (d) A simple tetrahedral interstitial. (e) A bond centered interstitial. (f) A (100) split interstitial [69].

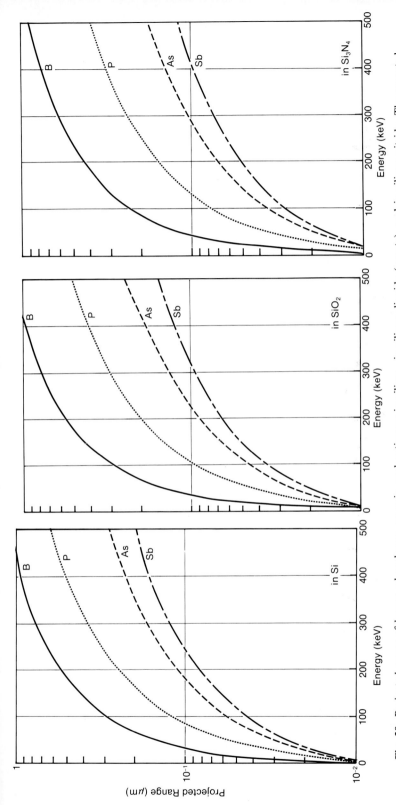

Fig. 30. Projected range of boron, phosphorus, arsenic, and antimony in silicon, in silicon dioxide (quartz), and in silicon nitride. The reported densities for silicon dioxide and silicon nitride vary over a wide range. Therefore, atomic mass units of $(28.09 \times 2.66 \times 10^{22}) + (16 \times 5.33 \times 10^{22})$ and $(28.09 \times 4.43 \times 10^{22}) + (14 \times 5.98 \times 10^{22})$ were used for silicon dioxide (quartz) and silicon nitride, respectively, for Figs. 30 through 32 [112]. The importance of higher-order moments such as skewness, especially required for boron implantations, is discussed in reference [112].

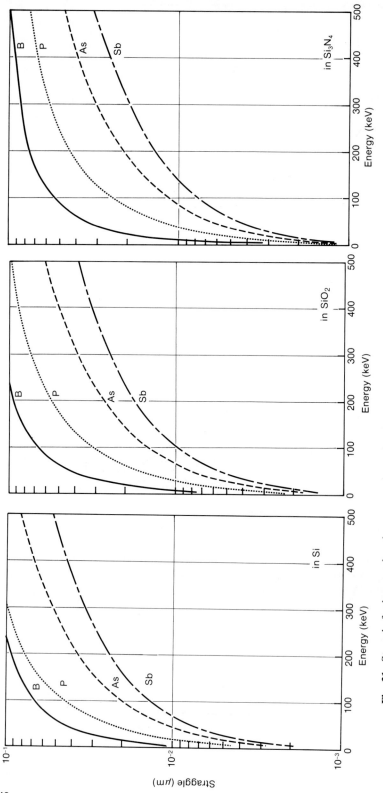

Fig. 31. Straggle for boron, phosphorus, arsenic, and antimony in silicon, in silicon dioxide (quartz), and in silicon nitride [112].

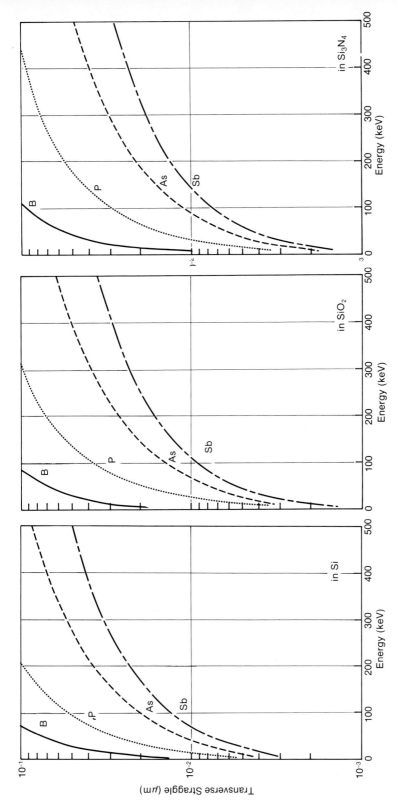

Fig. 32. Transverse straggle for boron, arsenic, phosphorus, and antimony in silicon, in silicon dioxide (quartz), and in silicon nitride [112].

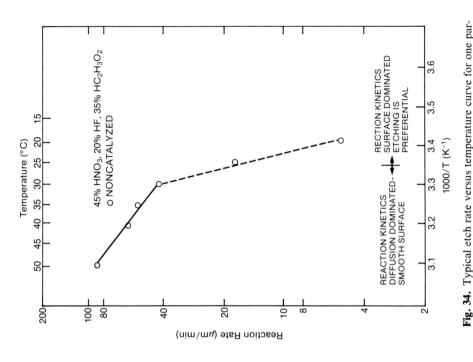

Fig. 34. Typical etch rate versus temperature curve for one particular mixture of HF, HNO_3, and $HC_2H_3O_2$ acids. Note the surface- and diffusion-limited regimes [114].

Fig. 33. Isoetch curves for silicon (HF : HNO_3 : diluent system) [113].

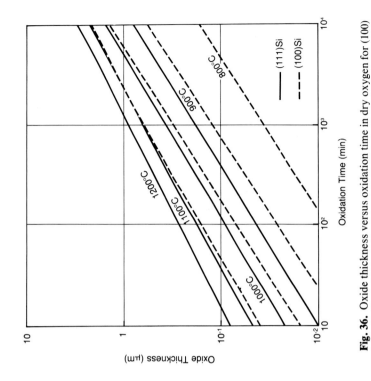

Fig. 36. Oxide thickness versus oxidation time in dry oxygen for (100) and (111) silicon [116,117].

Fig. 35. Comparative etch rates of silicon, silicon dioxide, and silicon nitride in phosphoric acid [115].

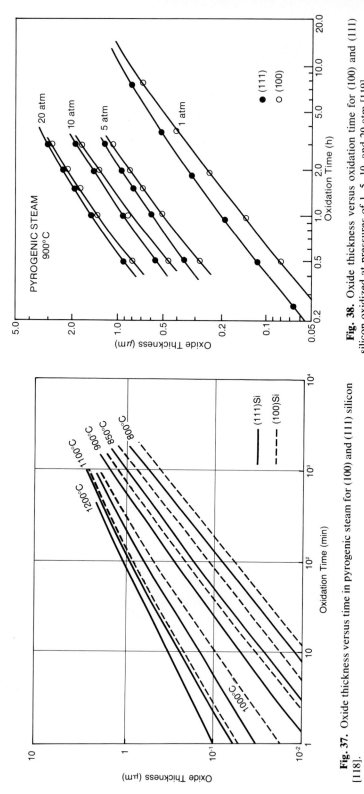

Fig. 37. Oxide thickness versus time in pyrogenic steam for (100) and (111) silicon [118].

Fig. 38. Oxide thickness versus oxidation time for (100) and (111) silicon oxidized at pressures of 1, 5, 10, and 20 atm [119].

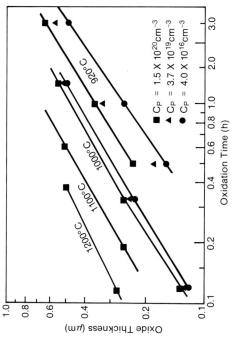

Fig. 40. Oxide thickness versus oxidation time for phosphorus-doped silicon in wet oxygen (95°C, H_2O) as a function of temperature and concentration [120].

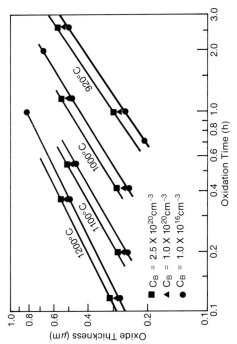

Fig. 39. Oxide thickness versus oxidation time for boron-doped silicon in wet oxygen (95°C, H_2O) as a function of temperature and concentration [120].

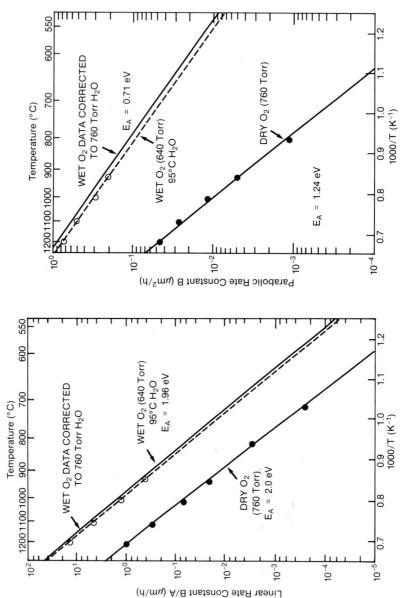

Fig. 41. The linear rate constant versus reciprocal absolute temperature in dry and wet oxygen [121].

Fig. 42. The parabolic rate constant versus reciprocal absolute temperature in dry and wet oxygen [121].

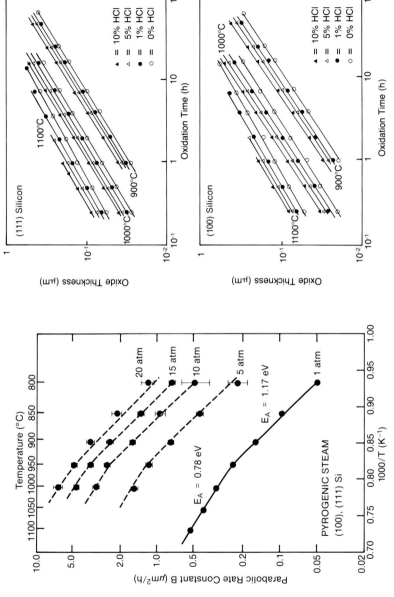

Fig. 43. The parabolic rate constant versus reciprocal absolute temperature for (100) and (111) silicon oxidized at pressures of 1, 5, 10, 15, and 20 atm in pyrogenic steam [119].

Fig. 44. Oxide thickness versus oxidation time in dry oxygen with HCl for (100) and (111) silicon [116,117].

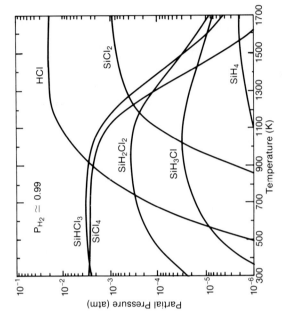

Fig. 46. Temperature and partial pressure variation of the equilibrium gas phase composition at approximately 1 atm H_2 total pressure and Cl/H partial pressure of 0.01 [122].

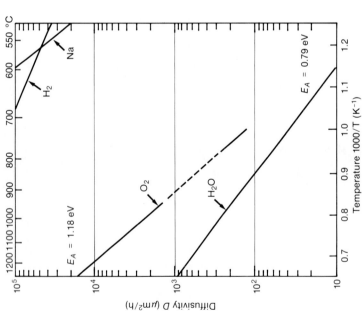

Fig. 45. Diffusivities of hydrogen, oxygen, sodium, and water vapor in silica glass [15].

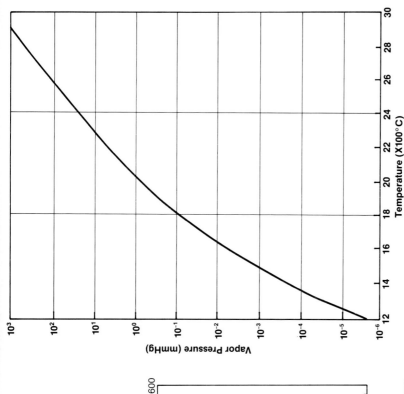

Fig. 47. Temperature dependence of the epitaxial growth rate for various silicon sources. In region A, the process can be characterized as reaction-rate or kinetic limited, whereas region B is transport limited [123].

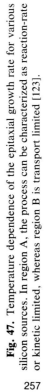

Fig. 48. Vapor pressure as a function of temperature for silicon [124].

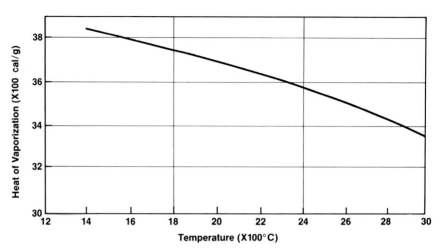

Fig. 49. Heat of vaporization as a function of temperature for silicon [124].

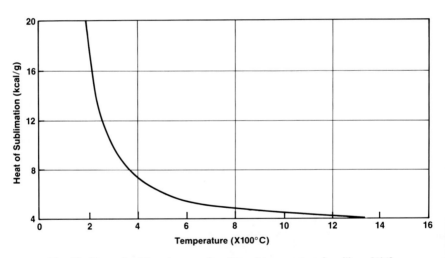

Fig. 50. Heat of sublimation as a function of temperature for silicon [124].

15. VLSI Silicon Material Criteria

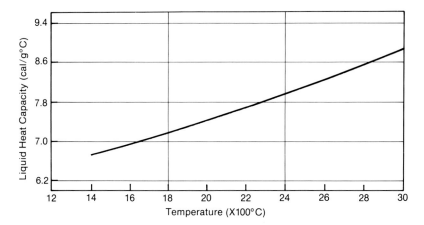

Fig. 51. Liquid heat capacity as a function of temperature for silicon [124].

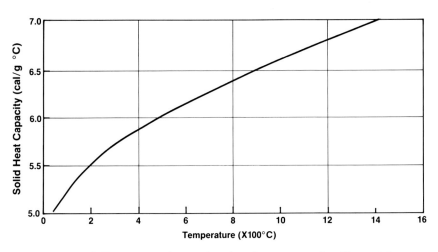

Fig. 52. Solid heat capacity as a function of temperature for silicon [124].

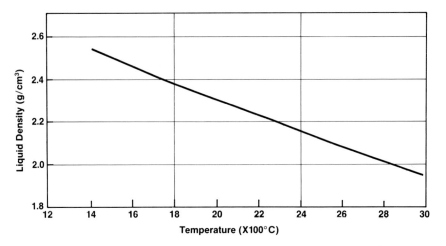

Fig. 53. Liquid density as a function of temperature for silicon [124].

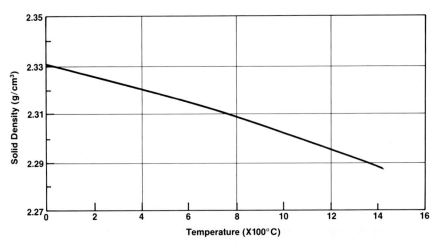

Fig. 54. Solid density as a function of temperature for silicon [124].

15. VLSI Silicon Material Criteria 261

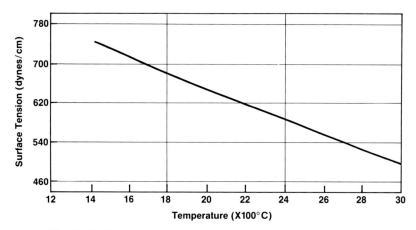

Fig. 55. Surface tension as a function of temperature for silicon [124].

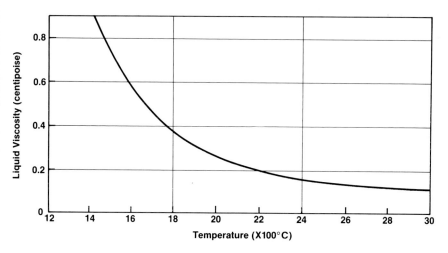

Fig. 56. Liquid viscosity as a function of temperature for silicon [124].

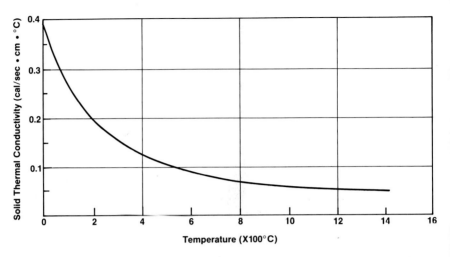

Fig. 57. Solid thermal conductivity as a function of temperature for silicon [124].

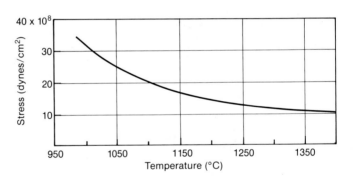

Fig. 58. Ultimate tensile stress as a function of temperature for silicon [125].

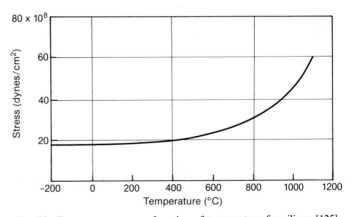

Fig. 59. Fracture stress as a function of temperature for silicon [125].

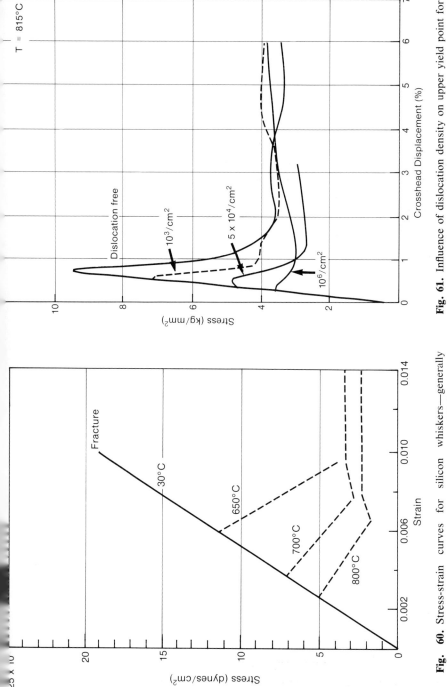

Fig. 60. Stress-strain curves for silicon whiskers—generally stronger than silicon crystals—at various temperatures [126].

Fig. 61. Influence of dislocation density on upper yield point for silicon [127]. Detailed thermomechanical discussions related to plastic deformation are given in references [128,129].

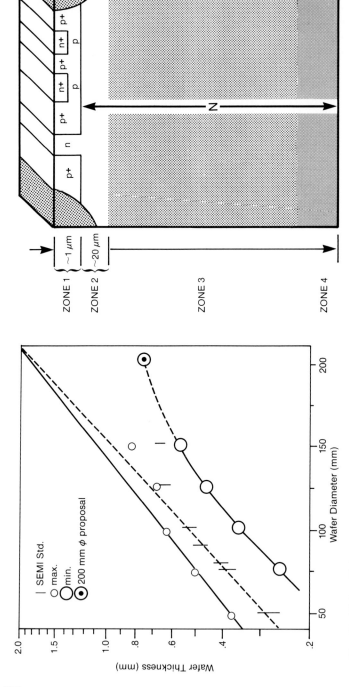

Fig. 63. Recommended multizoned structure for silicon product. Zones 1 and 2 contain VLSI AEGs in a region essentially free of unwanted point defects and structural imperfections. Zones 3 and 4 have processed-in large numbers of structural imperfections designed to function as an "infinite sink" for unwanted point defects [5].

Fig. 62. Wafer thickness versus wafer diameter, after proposed SEMI standard; recommended maximum and minimum thicknesses for commercial wafers [134].

ACKNOWLEDGMENTS

The authors especially wish to thank Jean Collins, Linda Haggett, and Lori Heintz for typing the manuscript.

REFERENCES

1. T. Abe, *in* "VLSI Electronics: Microstructure Science," vol. 12 (N. G. Einspruch and H. R. Huff, eds.). Academic Press, New York, 1984.
2. J. A. Moreland, *in* "VLSI Electronics: Microstructure Science," vol. 12 (N. G. Einspruch and H. R. Huff, eds.). Academic Press, New York, 1984.
3. S. M. Hu, *J. Vac. Sci. Technol.* **14,** 17 (1977).
4. K. V. Ravi, "Imperfections and Impurities in Semiconductor Silicon." Wiley, New York, 1981.
5. J. E. Lawrence and H. R. Huff, *in* "VLSI Electronics: Microstructure Science," vol. 5 (N. G. Einspruch, ed.), p. 51. Academic Press, New York, 1982.
6. B. O. Kolbesen and H. Strunk, *in* "VLSI Electronics: Microstructure Science," vol. 12 (N. G. Einspruch and H. R. Huff, eds.). Academic Press, New York, 1985.
7. W. R. Runyan, "Silicon Semiconductor Technology." McGraw-Hill, New York, 1965.
8. H. F. Wolf, "Silicon Semiconductor Data." Pergamon Press, New York, 1969.
9. H. F. Wolf, "Semiconductors." Wiley, New York, 1971.
10. "Annual Book of ASTM Standards." American Society for Testing and Materials, Philadelphia, 1981.
11. Following G. B. Larrabee, *in* "VLSI Electronics; Microstructure Science," vol. 2 (N. G. Einspruch, ed.), p. 37. Academic Press, New York, 1981. R. A. Craven, F. Shimura, R. S. Hockett, L. W. Shive, P. B. Fraundorf, and G. K. Fraundorf, *in* "VLSI Science and Technology/1984, PV 84-7" (K. E. Bean and G. A. Rozgonyi, eds.), p. 20. The Electrochemical Society, Pennington, New Jersey, 1984.
12. Following J. Matsui and K. Nishida, *Oyo Butsuri* **51,** 827 (1982). In Japanese.
13. G. B. Larrabee, *in* "VLSI Electronics: Microstructure Science," vol. 2 (N. G. Einspruch, ed.) p. 37. Academic Press, New York, 1981.
14. S. M. Sze (ed.), "VLSI Technology." McGraw-Hill, New York, 1983.
15. S. K. Ghandi, "VLSI Fabrication Principles—Silicon and Gallium Arsenide." Wiley, New York, 1983.
16. Mainly following S. M. Sze, "Physics of Semiconductor Devices," Second ed. Wiley, New York, 1981.
17. J. F. Nye, "Physical Properties of Crystals." Oxford University Press, New York, 1960.
18. W. R. Runyan, "Semiconductor Measurements and Instrumentation." McGraw-Hill, New York, 1975.
19. W. G. Pfann, *Trans. AIME* **194,** 747 (1952).
20. J. A. Burton, R. C. Prim, and W. P. Slichte, *J. Chem. Phys.* **21,** 1987 (1953).
21. K. Morizane, A. F. Witt, and H. C. Gatos, *J. Electrochem. Soc.* **113,** 51 (1966).
22. A. F. Witt and H. C. Gatos, *J. Electrochem. Soc.* **113,** 808 (1966).
23. K. Morizane, A. Witt, and H. C. Gatos, *J. Electrochem. Soc.* **114,** 738 (1967).
24. S. M. Hu, *J. Appl. Phys.* **40,** 4413 (1969).
25. D. J. Dumin, *J. Appl. Phys.* **36,** 2700 (1965).
26. H. Alexander and P. Haasen, *in* "Solid State Physics," vol. 22 (F. Seitz, D. Turnbull, and H. Ehenreich, eds.), Academic Press, New York, 1968.

27. J. W. Christian, "The Theory of Transformations in Metals and Alloys." Pergamon Press, New York, 1965.
28. L. J. Van der Pauw, *Philips Tech. Rev.* **20,** 220 (1959).
29. "Annual Book of ASTM Standards," F 391-78, p. 795. American Society for Testing and Materials, Philadelphia, 1981.
30. R. P. Heiman, *IEEE Trans. Electron Devices* **ED-14,** 781 (1967).
31. Private communication. P. Chatterjee, A. Tasch, Jr., and H.-S. Fu.
32. R. F. Pierret and D. W. Small, *IEEE Trans. Electron Devices* **ED-22,** 1051 (1975).
33. W. D. Eades, J. P. Schott, and R. M. Swanson, *IEEE Trans. Electron Devices* **ED-30,** 1274 (1983).
34. M. Zerbst, *Z. Agnew. Phys.* **22,** 30 (1966). In German.
35. A. Many, Y. Goldstein, and B. Grover, "Semiconductor Surfaces." North-Holland, Amsterdam, 1965.
36. F. A. Trumbore, *Bell Sys. Tech. J.* **39,** 205 (1960).
37. Y. Yatsurugi, N. Akiyama, Y. Endo, and T. Nozaki, *J. Electrochem. Soc.* **120,** 975 (1973).
38. J. A. Keenan and G. B. Larrabee, *in* "VLSI Electronics: Microstructure Science," vol. 6 (N. G. Einspruch, ed.) p. 1. Academic Press, New York, 1983.
39. L. Pauling, "The Nature of the Chemical Bond." Cornell University Press, New York, 1960.
40. A. Van Wieringen and N. Warmoltz, *Physica* **22,** 849 (1956).
41. E. M. Pell, *Phys. Rev.* **119,** 1014 (1960).
42. R. N. Hall and J. H. Racette, *J. Appl. Phys.* **35,** 379 (1964).
43. B. I. Boltaks and H. Shih-gin, *Sov. Phys.-Solid State* **2,** 2383 (1961).
44. W. R. Wilcox and T. J. LaChapelle, *J. Appl. Phys.* **35,** 240 (1964).
45. M. K. Bakhadryrkhanov, B. I. Boltaks, G. S. Kulikov, and E. M. Pedyach, *Sov. Phys.-Semicond.* **4,** 739 (1970).
46. R. B. Fair, *in* "Semiconductor Silicon 1977, PV 77-2" (H. R. Huff and E. Sirtl, eds.), p. 968. The Electrochemical Society, Princeton, 1977.
47. R. C. Newman and J. Wakefield, *J. Phys. Chem. Solids* **19,** 230 (1961).
48. G. L. McVay and A. R. Ducharme, *J. Appl. Phys.* **44,** 1409 (1973).
49. T. H. Yeh, S. M. Hu, and R. J. Kastl, *J. Appl. Phys.* **39,** 4266 (1968).
50. V. P. Boldyrev, *Sov. Phys.-Semicond.* **11,** 709 (1977).
51. P. V. Pavlov, E. I. Zorin, D. I. Tetelbaum, and A. F. Khokhlov, *Phys. Status Solidi (a)* **35,** 11 (1976).
52. G. D. Watkins, J. W. Corbett, and R. S. McDonald, *J. Appl. Phys.* **53,** 7097 (1982).
53. R. O. Carlson, R. N. Hall, and E. M. Pell, *J. Phys. Chem. Solids* **8,** 81 (1959).
54. N. T. Bendik, V. S. Garnyk, and L. S. Mileuskii, *Sov. Phys.-Solid State* **12,** 150 (1970).
55. M. K. Bakhadyrkhanov, B. I. Boltaks, and G. S. Kulikov, *Sov. Phys.-Solid State* **14,** 1441 (1972).
56. E. R. Weber, *Appl. Phys. A* **30,** 1 (1983).
57. H. Kitagawa and K. Hashimoto, *Jpn. J. Appl. Phys.* **16,** 173 (1977).
58. M. K. Bakhadyrkhanov, S. Zainabidinov, and A. Khamidov, *Sov. Phys.-Semicond.* **14,** 243 (1980).
59. D. L. Kendall and D. B. DeVries, *in* "Semiconductor Silicon" (R. R. Haberecht and E. L. Kern, eds.), p. 358. The Electrochemical Society, New York, 1969.
60. W. Zulehner and D. Huber, *in* "Crystals 8: Silicon·Chemical Etching" (H. C. Freyhardt, ed.), p. 1. Springer-Verlag, Berlin, 1982.
61. W. Frank, U. Gösele, H. Mehrer, and A. Seeger, *in* "Diffusion in Solids II." (A. S. Nowick and G. Murch, eds.). Academic Press, New York, 1984.
62. W. T. Stacy, D. F. Allison, and T.-C. Wu, *in* "Semiconductor Silicon 1981, PV 81-5"

15. VLSI Silicon Material Criteria

(H. R. Huff, R. J. Kriegler, and Y. Takeishi, eds.), p. 344. The Electrochemical Society, Pennington, New Jersey, 1981.
63. K. Graff, in "Aggregation Phenomena of Point Defects in Silicon, PV 83-4" (E. Sirtl and J. Goorissen, eds.), p. 121. The Electrochemical Society, Pennington, New Jersey, 1983.
64. U. Gösele and T. Y. Tan, in "Aggregation Phenomena of Point Defects in Silicon, PV 83-4" (E. Sirtl and J. Goorissen, eds.), p. 17. The Electrochemical Society, Pennington, New Jersey, 1983.
65. A. Seeger and M. L. Swanson, in "Lattice Defects in Semiconductors" (R. R. Hasiguti, ed.), p. 93. University of Tokyo Press, Tokyo, 1968.
66. T. Y. Tan, F. Morehead, and U. Gosele, in "Defects in Silicon, PV83-9" (M. W. Bullis and L. C. Kimerling, eds.), p. 325. The Electrochemical Society, Pennington, New Jersey, 1983.
67. T. S. Moss, "Optical Properties of Semiconductors." Academic Press, New York, 1955.
68. J. I. Pankove, "Optical Processes in Semiconductors." Dover, New York, 1975.
69. J. C. C. Tsai, in "VLSI Technology" (S. M. Sze, ed.), p. 169. McGraw Hill, New York, 1983.
70. C. P. Wu, E. C. Douglas, and C. W. Mueller, *IEEE Trans. Electron Devices* **ED-22,** 319 (1975).
71. E. Tannenbaum, *Solid State Electron.* **3,** 123 (1961).
72. R. G. Mazur and D. H. Dickey, *J. Electrochem. Soc.* **113,** 255 (1966).
73. W. K. Hofker, *Philips Res. Rep. Suppl.* **8,** 1 (1975).
74. P. F. Kane and G. B. Larrabee, "Characterization of Semiconductor Materials." McGraw-Hill, New York, 1970.
75. W. K. Chu, J. W. Mayer, M.-A. Nicolet, T. M. Buck, G. Amsel, and F. Eisen, in "Semiconductor Silicon 1973." (H. R. Huff and R. R. Burgess, eds.), p. 416. The Electrochemical Society, Pennington, New Jersey, 1973.
76. J. F. Ziegler, G. W. Cole, and J. E. E. Baglin, *J. Appl. Phys.* **43,** 3809 (1972).
77. J. L. Combasson, J. Bernard, G. Guernet, N. Hilleret, and M. Bruel, in "Ion Implantation in Semiconductors and Other Materials" (B. L. Crowder, ed.), p. 285. Plenum, New York, 1973.
78. S. P. Murarka, *J. Vac. Sci. Technol.* **17,** 775 (1980).
79. J. R. Ligenza, *J. Phys. Chem.* **65,** 2011 (1961).
80. J. E. Sinclair and B. R. Lawn, *Proc. Roy. Soc. London* **A329,** 83 (1972).
81. W. Kern, *RCA Engineer* **28-4,** 99 (1983).
82. R. B. Heiman, *In* "Crystals 8: Silicon · Chemical Etching" (H. C. Freyhardt, ed.), p. 173. Springer-Verlag, Berlin, 1982.
83. E. Sirtl and A. Adler, *Z. Metallk* **52,** 529 (1961). In German.
84. W. C. Dash, *J. Appl. Phys.* **27,** 1193 (1956).
85. F. Secco d'Aragona, *J. Electrochem. Soc.* **119,** 948 (1972).
86. D. G. Schimmel, *J. Electrochem. Soc.* **123,** 734 (1976).
87. D. G. Schimmel, *J. Electrochem. Soc.* **126,** 479 (1979).
88. M. W. Jenkins, *J. Electrochem. Soc.* **124,** 757 (1977).
89. W. A. Pliskin, and E. E. Conrad, *IBM J. Res. Dev.* **8,** 43 (1964).
90. M. Ghezzo and D. M. Brown, *J. Electrochem. Soc.* **120,** 146 (1973).
91. A. C. Adams, in "VLSI Technology" (S. M. Sze, ed.), p. 93. McGraw-Hill, New York, 1983.
92. A. A. Giardini, *Am. Mineralogist* **43,** 957 (1958).
93. W. R. Runyan, in "Encyclopedia of Chemical Technology," vol. 20, 3rd ed. (Kirk-Ostner, ed.), p. 826. Wiley, New York, 1982.
94. S. M. Sze, "Physics of Semiconductor Device," 2nd ed. Wiley, New York, 1981.

95. E. M. Conwell, *Proc. IRE* **46,** 1281 (1958).
96. C. D. Thurmond, *J. Electrochem. Soc.* **122,** 1133 (1975).
97. "Annual Book of ASTM Standards," F723-81, p. 1265. American Society for Testing and Materials, Philadelphia, 1981.
98. A. S. Grove, "Physics and Technology of Semiconductor Devices." Wiley, New York, 1967.
99. G. L. Vick and K. M. Whittle, *J. Electrochem. Soc.* **116,** 1142 (1969).
100. R. A. Colclaser, "Microelectronics: Processing and Device Design." Wiley, New York, 1980.
101. R. A. Craven, *in* "Semiconductor Silicon 1981, PV 81-5" (H. R. Huff, R. J. Kriegler, and Y. Takeishi, eds.), p. 254. The Electrochemical Society, Pennington, New Jersey, 1981.
102. A. R. Bean and R. C. Newman, *J. Phys. Chem. Solids* **32,** 1211 (1971).
103. "Annual Book of ASTM Standards," F121-80, p. 538. American Society for Testing and Materials, Philadelphia, 1981.
104. Deutsche Norman DIN 50 438/1, p. 1. Beuth Verlag GmbH, Berlin, 1978. In German.
105. T. Iizuka, S. Takasu, M. Tajima, T. Arai, M. Nozaki, N. Inoue, and M. Watanabe, *in* "Defects in Silicon, PV 83-9" (W. M. Bullis and L. C. Kimerling, eds.), p. 265. The Electrochemical Society, Pennington, New Jersey, 1983.
106. "Annual Book of ASTM Standards," F121-76, p 518. American Society for Testing and Materials, Philadelphia, 1976.
107. M. Hansen and K. Anderko, "Constitution of Binary Alloys." McGraw-Hill, New York, 1958.
108. J. O. McCaldin and H. Sankur, *Appl. Phys. Lett.* **19,** 524 (1971).
109. A. S. Berezhnoi, "Silicon and Its Binary Systems." Consultants Beaureau, 1960.
110. R. B. Fair, *in* "Impurity Doping Processes in Silicon" (F. F. Y. Wang, ed.), North-Holland, New York, 1981.
111. K. P. O'Donnel, K. M. Lee, and G. D. Watkins, *in* "Proc. 12th Int. Cont. Defects in Semiconductors," 1982.
112. J. F. Gibbons, W. S. Johnson, and S. W. Mylroie, (1975). "Projected Range Statistics in Semiconductors and Related Materials," 2nd ed. Dowden, Hutchinson & Ross, New York, 1975.
113. H. Robbins and B. Schwartz, *J. Electrochem. Soc.* **106,** 505 (1959).
114. H. Robbins and B. Schwartz, *J. Electrochem. Soc.* **108,** 365 (1961).
115. W. van Gelder and V. E. Hauser, *J. Electrochem. Soc.* **144,** 869 (1967).
116. D. Hess and B. E. Deal, *J. Electrochem. Soc.* **124,** 735 (1977).
117. J. J. Barnes, J. M. DeBlasi, and B. E. Deal, *J. Electrochem. Soc.* **126,** 1779 (1979).
118. B. E. Deal, *J. Electrochem. Soc.* **125,** 576 (1978).
119. R. R. Razouk, L. N. Lie, and B. E. Deal, *J. Electrochem. Soc.* **128,** 2214 (1981).
120. B. E. Deal and M. Sklar, *J. Electrochem. Soc.* **112,** 430 (1965).
121. B. E. Deal and A. S. Grove, *J. Appl. Phys.* **36,** 3770 (1965).
122. E. Sirtl, L. P. Hunt, and D. H. Sawyer, *J. Electrochem. Soc.* **121,** 919 (1974).
123. F. C. Eversteyn, *Philips Res. Rep.* **29,** 45 (1974).
124. C. L. Yaws, L. L. Dickens, R. Lutwack, and G. Hsu, *Solid State Technol.* Jan., 87.
125. W. D. Sylwestrowicz, *Philos. Mag.* **7,** 1825 (1962).
126. G. L. Pearson, W. T. Read, and W. L. Feldman, *Acta Met.* **5,** 181 (1957).
127. J. R. Patel and A. R. Chaudhuri, *J. Appl. Phys.* **34,** 2788 (1963).
128. K. Sumino, *in* "Semiconductor Silicon 1981, PV 81-5" (H. R. Huff, R. J. Kriegler, and Y. Takeishi, eds.), p. 208. The Electrochemical Society, Pennington, New Jersey, 1981.
129. Y. Kondo, *in* "Semiconductor Silicon 1981, PV 81-5" (H. R. Huff, R. J. Kriegler, and

Y. Takeishi, eds.), p. 220. The Electrochemical Society, Pennington, New Jersey, 1981.
130. F. Shimura, *in* "VLSI Science and Technology/1982" (C. J. Dell'Oca and W. M. Bullis, eds.), p. 17. 1982.
131. G. B. Larrabee and T. J. Shaffner, *in* "VLSI Handbook." (N. G. Einspruch, ed.). Academic Press, New York, (1983).
132. S. Takasu, *Denshi Zairyo, extra number,* November, 13 (1983). In Japanese.
133. S. Takasu, *in* "VLSI Science and Technology/1984, PV 84-7" (K. E. Bean and G. A. Rozgonyi, eds.), p. 490. The Electrochemical Society, Pennington, New Jersey, 1984.
134. S. Takasu, North-Holland, (to be published).
135. D. C. Ahlgren, A. G. Domenicucci, R. Karcher, S. R. Mader, and M. R. Poponiak, *in* "Defect in Silicon, PV 83-9" (W. M. Bullis and L. C. Kimerling, eds.), p. 472. The Electrochemical Society, Pennington, New Jersey, 1983.
136. J. D. Meindl, *IEEE Trans. Electron Devices,* **ED-31,** 1555 (1984).

Chapter 16
Characterization of Bulk Silicon Materials

GRAYDON B. LARRABEE
THOMAS J. SHAFFNER

Texas Instruments Incorporated
Dallas, Texas

I. Polycrystalline Silicon	271
II. Bulk Single Crystal Silicon	272
III. Single Crystal Silicon Wafers	272
References	283

Semiconductor grade silicon is synthesized by the reaction of chlorosilanes with hydrogen to produce a very high-purity polycrystalline product. This material is grown into single crystal boules using either a Czochralski or floating-zone technique. The single crystal boules are then cut into wafers of approximately 500 μm in thickness, lapped and polished, and introduced into the device fabrication process. Characterization of these bulk materials must be carried out in a systematic manner. For the purposes of this chapter, bulk silicon materials are defined as polycrystalline, single crystal bulk, and single crystal slices. Epitaxial silicon is not covered.

POLYCRYSTALLINE SILICON I

The characterization of polycrystalline silicon is significantly different from that of single crystal bulk and wafers, due to the polycrystalline nature of the material. Polycrystalline silicon is usually characterized for dopants, oxygen/carbon, metallic impurities, and electrical properties. Normally it is necessary to zone melt a rod of the material to obtain large single crystal regions [1]. Once this is done it is possible to obtain meaningful values for the oxygen, carbon, and dopant impurities using Fourier transform infrared spectroscopy (FTIR) (Tables I and II). Electrical measurements are not often performed, because the measurement of each dopant concentration using FTIR provides the resistivity as well as a measure of the amounts of compensating impurities. When electrical measurements are performed, generally only two- and four-point probe techniques are employed (Table IV). Metallic impurity determinations are performed on the polycrystalline material directly, using the techniques shown in Table III with typical detection limits shown in Table VII.

II BULK SINGLE CRYSTAL SILICON

The characterization of bulk single crystal silicon involves the measurements of dopants, oxygen/carbon, metallic impurities, electrical properties, and physical defect levels. Unlike polycrystalline silicon, the samples are cut directly from the single crystal boule to the desired shape and size and then analyzed. The characterization techniques used for dopants, oxygen/carbon and metallic impurities are shown in Tables I, II, and III. A diversity of electrical techniques are used for single crystal material, and the reader is referred to the review by Runyan [2]. Table IV lists the most frequently used electrical techniques. Single crystal silicon that has been grown by the Czochralski technique is essentially free of physical defects. The normal defect characterization techniques of x-ray topography and transmission electron microscopy do not detect defects in this material (Table V).

III SINGLE CRYSTAL SILICON WAFERS

Single crystal wafers are prepared for device processing by sawing, lapping, and frontside polishing. Frequently there is deliberately introduced damage on the backside of the wafer. This backside damage is for gettering during the device process. The wafers also receive a *resistivity stabilization* heat treatment which makes the oxygen in Czochralski silicon electrically inactive. It is during this heat treatment and subsequent device thermal cycles that physical defects are introduced into the silicon material. An oxygen depleted region, or denuded zone, is introduced into the outer 5–50 μm of the front surface by the initial thermal cycles.

Wafers are characterized for dopants, oxygen/carbon, metallic impurities, electrical properties, physical defects, and surface contamination.

16. Characterization of Bulk Silicon Materials

(See Shaffner [3] for more detailed information on surface characterization.) Physical parameters of slices are performed by methods described by ASTM [4]. The characterization of backside damage and denuded zones involves chemical etching and microscopy techniques, which have not been standardized. The characterization techniques most frequently used are shown in Tables I–VI. Table VII is a list of detection limits for elements in silicon. Impurity energy levels are usually determined by deep-level transient spectroscopy (DLTS). A list of energy levels of impurities and dopants in silicon is given in Tables VIII and IX as a function of energy and is arranged alphabetically by element in Tables X and XI.

TABLE I

Dopants

	Fourier transform infrared	Photo-luminescence	Neutron activation analysis	Functional temperature Hall
Depth analyzed	1–10 mm	1–3 μm	>1 μm	1 mm
Diameter of analysis region	2 mm	>5 μm	>1 cm	1 cm
Detection limit (atoms/cm^3)	1×10^{11}	1×10^{11}	1×10^{11}	1×10^{11}
Detection limit (ppma)	2×10^{-6}	2×10^{-6}	2×10^{-6a}	2×10^{-6}
In-depth profiling resolution (μm)	None	None	1	None
Time for analysis	2 h	2 h	2 days	1–2 days
Comments	Performed at ~10–15 K	Performed at ~4 K	In-depth profiling achieved by chemical etching	Performed at 4–300 K

[a] See Table VII.

TABLE II
Carbon and Oxygen

	Fourier transform infrared	Charged particle activation	Ion microprobe (Cs^+)	Photoneutron activation
Depth analyzed	2–5 mm	300 μm	50 Å	0.5 cm
Diameter of analysis region	2 mm	5 mm	100 μm	0.5 cm
Detection limit (atoms/cm³)	5×10^{15}	5×10^{13}	5×10^{17a}	5×10^{15}
Detection limit (ppma)	0.1	0.001	10	0.1
In-depth profiling resolution	None	25 μm	50 Å	None
Time for analysis	10 min	2 h	1 h	2 h
Comments	Bulk measurement only	In-depth profiling achieved by chemical etching	First 100 Å unreliable	Bulk measurement only

[a] Oxygen only.

TABLE III
Metallic Impurities

	Emission spectroscopy	Neutron activation	Ion microprobe (Cs^+ and O_2^-)	Spark source spectroscopy
Depth analyzed	None	1 μm	50 Å	>10 μm
Diameter of analysis region	None	>1 cm	100 μm	>1 mm
Detection limit (atoms/cm³)	$5 \times 10^{15} - 5 \times 10^{18}$	$5 \times 10^{12} - 5 \times 10^{18}$	$5 \times 10^{15} - 5 \times 10^{18}$	$5 \times 10^{14} - 5 \times 10^{17}$
Detection limit (ppma)	$0.1-100^a$	$0.001-100^a$	$0.1-100^a$	$0.01-10^a$
In-depth profiling resolution	None	1 μm	50 Å	None
Time for analysis	30 min	2 days	1 h	2–8 h
Comments	Bulk measurement only	In-depth profiling achieved by chemical etching	First 100 Å unreliable	Bulk measurement only

[a] See Table VII.

TABLE IV
Electrical Properties: Resistivity

	2-Point probe	4-Point probe	Spreading resistance	Hall van der Pauw
Probe spacings	5–10 mm	1.6 mm	25–100 μm	2–10 mm
Probe diameter	50–100 μm	50–100 μm	10–50 μm	Soldered contacts
Resistivity range (Ω cm)	0.00009–3000	0.0008–6000	0.001–100	Up to 10,000
In-depth profiling resolution	None	1 μm	0.2 μm	None
Time for analysis	5 min	2 min	1–2 h	1–2 h
Comments	Must know geometry	In-depth profiling achieved by chemical etching	In-depth profiling achieved by angle lap	Obtains net carrier concentration

TABLE V
Physical Defects

	Chemical etching	Transmission electron microscopy	Lang x-ray topography	Double crystal x-ray topography
Depth analyzed	10–50 μm	1000 Å	500 μm	5–100 μm
Diameter of analysis region	>1 mm	10 μm	>1 cm	>1 cm
Lattice distortion ($\Delta d/d$)	1×10^{-3}	Image defects	1×10^{-3}	1×10^{-7}
Spatial resolution	1 μm	2–5 Å	1–10 μm	1–10 μm
In-depth profiling resolution	1 μm	Stereo microscopy	Stereo topography	None
Time for analysis	20 min	1 day	1 h	4 h
Comments	Whole slice survey	3 mm sample	Whole slice survey	Whole slice survey

TABLE VI
Contamination

	Auger electron spectroscopy	x-ray photoelectron spectroscopy	Electron microprobe (EDS & WDS)	Ion microprobe (Cs^+ & O_2^-)	Raman microprobe	High energy electron diffraction	x-ray powder diffraction	Seeman Bohlin x-ray diffraction
Depth analyzed	20 Å	20 Å	1 μm	50 Å	1 μm	50 Å	15 μm	0.1–1 μm
Diameter of analysis region	100 μm [1000 Å]	5 mm	1 μm	100 μm	1 μm	0.1–10 μm	3 mm	3–5 mm
Detection limit (atoms/cm³)	5×10^{19} [1×10^{21}]	2×10^{20}	5×10^{19}	5×10^{17}	5×10^{19}	3×10^{20}	1×10^{20}	5×10^{20}
Detection limit (ppma)	1000 [20,000]	5000	1000	10[a]	1000	5000	2000	10,000
In-depth profiling resolution	20 Å	20 Å	None	50 Å	None	None	None	None
Time for analysis	30 min	45 min	3–45 min	1 h	15 min	30 min	30 min	45 min
Comments	Brackets [] denote scanning Auger	In-depth profiling achieved by ion sputtering	Wavelength dispersive takes longer than energy dispersive	First 100 Å unreliable	Identification or organics possible	Must be crystalline	Must be crystalline	Must be crystalline

[a] See Table VII.

Interference-Free Detection Limits in Bulk Silicon[a,b,c]

	Emission spectroscopy	Neutron activation	Electron microprobe	O$^+$-ion microprobe	Cs$^+$-ion microprobe	X-ray fluorescence	Spark source mass spectroscopy
Aluminum	0.2	—	110	0.005	3	—	0.003
Antimony	0.2	0.000003	2000	1	2	10	0.03
Arsenic	0.4	0.000005	1200	10	0.1	0.5	0.02
Barium	1	0.006	750	0.05	>1000	5	0.03
Beryllium	0.002	—	—	0.05	3	—	0.02
Bismuth	0.05	—	2200	1	4	15	0.02
Boron	0.4	—	—	0.01	0.1	—	0.03
Bromine	—	0.000008	2700	0.1	<0.05	0.4	0.03
Cadmium	5	0.01	1600	1	>100	10	0.15
Calcium	4	—	220	0.1	0.9	0.7	0.02
Carbon	—	—	1100	100	0.006	—	0.30
Cerium	200	0.00019	1500	1	>10	10	0.15
Cesium	50	0.0005	—	0.1		5	0.02
Chlorine	—	—	230	0.1	<0.05	2	0.03
Chromium	2	0.001	170	0.1	6	0.4	0.02
Cobalt	0.04	0.0003	300	1	2	0.2	2.00
Copper	0.03	0.000008	290	0.1	0.8	0.1	0.02
Dysprosium	200	30	1500	1	>10	10	0.02
Erbium	100	—	1500	1	>10	10	0.02
Europium	100	0.000012	1500	1	>10	10	0.02
Fluorine	—	—	3600	0.1	<0.1	—	0.02
Gadolinium	300	0.001	1500	1	>10	10	0.03
Gallium	0.02	0.00006	490	0.05	70	0.6	0.03
Germanium	—	0.1	710	1	0.09	0.6	0.06
Gold	0.07	0.0000016	1400	100	0.06	10	0.10
Hafnium	3	0.00003	560	0.1	>10	5	0.1
Holmium	300	0.0001	1500	1	>10	10	0.01
Hydrogen	—	—	100	—	≤0.1	—	—

(*Continued*)

TABLE VII (*Continued*)

	Emission spectroscopy	Neutron activation	Electron microprobe	O$^+$-ion microprobe	Cs$^+$-ion microprobe	X-ray fluorescence	Spark source mass spectroscopy
Indium	0.004	0.00022	2000	0.05	200	10	15
Iodine	—	—	—	1	~0.1	10	0.02
Iridium	0.8	0.0000007	1000	5	0.07	8	0.03
Iron	0.1	0.1	260	50	50	0.3	0.3
Lanthanum	50	0.000003	560	1	>10	5	0.02
Lead	0.02	—	2300	0.5	~3	15	0.02
Lithium	0.4	—	—	0.1	~1	—	0.3
Lutetium	—	0.000004	1500	1.0	>10	10	0.02
Magnesium	0.0004	—	1300	0.01	>10	—	0.2
Manganese	0.04	0.28	230	0.1	>1000	0.3	0.02
Mercury	0.3	0.00008	1900	—	—	12	0.1
Molybdenum	0.07	0.00026	1000	0.1	9	1.5	0.1
Neodymium	300	0.001	1500	1	>10	10	0.01
Nickel	0.3	0.001	340	10	10	0.2	0.6
Niobium	—	—	950	0.1	0.7	0.7	0.02
Nitrogen	—	—	—	—	>1000	—	0.6
Osmium	300	0.0002	1000	1	0.1	8	0.06
Oxygen	—	—	10000	—	0.1	—	1.0
Palladium	0.003	0.001	1200	100	8	5	0.1
Phosphorous	0.9	—	170	10	5	—	0.02
Platinum	0.007	0.0009	1100	100	0.07	10	0.06
Potassium	0.1	0.05	250	0.01	~1	1	0.02
Praseodymium	300	0.001	1500	1	>10	10	0.1
Rhenium	0.3	0.0001	1000	1	40	8	0.3
Rhodium	0.2	—	1100	1	2	8	0.02
Rubidium	0.1	0.01	1100	1	~1	8	15
Ruthenium	0.6	—	1000	1	6	8	0.06

Element	Emission spectroscopy	Neutron activation	Electron microprobe	Ion microprobe	X-ray fluorescence	Spark source mass spectroscopy
Scandium	—	0.00016	200	—	>1000	0.02
Selenium	50	0.01	2700	10	0.07	0.03
Silver	0.002	0.0008	1200	1	3	0.03
Sodium	0.1	0.00009	2200	0.01	2	0.02
Strontium	0.06	0.25	1000	100	>1000	20
Sulfur	—	—	220	1	0.04	0.6
Tantalum	50	0.00006	560	1	500	1
Tellurium	0.6	10	2600	10	0.3	0.06
Terbium	300	0.001	1500	—	>10	0.01
Thallium	50	—	2000	1	>10	0.03
Thulium	—	0.0002	1500	1	>10	1
Tin	0.2	0.028	2200	0.5	1	0.06
Titanium	0.07	—	100	0.02	1	0.2
Tungsten	100	0.000005	600	1	0.8	0.06
Vanadium	0.6	—	150	0.05	0.5	0.02
Ytterbium	100	0.000009	1500	—	10	0.1
Zinc	0.4	0.010	390	1	>1000	0.1
Zirconium	0.5	0.011	900	0.02	7	0.1

[a] Detection limits are in ppma.

[b] The Table is a best estimate of the detection limits for trace impurities in bulk silicon for six analytical techniques. Each technique is subject to interferences, and depending on the sample, detection limits will vary. A brief description of the analytical procedure for each technique follows.

Emission spectroscopy. A 30-mg sample of silicon is burned in a dc arc using an enclosed Stallwood jet. The spectrum in the 2400–3600-Å range is recorded. When high-sensitivity alkali metal analyses are required, the spectrum in the 5000–8000-Å range is recorded.

Neutron activation. A 1–2-g sample of silicon is irradiated at a flux of 1×10^{13} $n/s\,cm^2$ for 14 h, allowed to decay for 24–36 h, and analyzed using gamma ray spectroscopy.

Electron microprobe. A 1–2-μm, 5–40-keV beam of electrons is used to excite the silicon, and the resultant x rays are analyzed. The volume analyzed is a sphere typically 2–25 μm in diameter.

Ion microprobe. A 20-keV, 5-μm diameter beam of oxygen or cesium ions is used to bombard the sample, and the sputtered secondary ions are mass analyzed. Typically, areas of 125×100 μm are analyzed with depths of 100–100 Å.

X-ray fluorescence. X rays are used to excite a 2-cm diameter spot on the silicon, and the resulting x rays are energy analyzed. A depth of 10–50 μm is analyzed.

Spark source mass spectroscopy. Two electrodes of silicon are sparked against the other at 20 keV and at 800 kHz. The sputtered secondary ions are mass analyzed. Typically, milligram amounts of materials are consumed.

[c] Table and procedures are from "VLSI Electronics: Microstructure Science" (N. Einspruch and G. B. Larrabee, eds.), vol. 6, pp. 10–13, Academic Press, New York, 1983.

TABLE VIII

Energy Levels below the Conduction Band Arranged by Energy

$E_c - 0.017$	Nitrogen	$E_c - 0.30$	Sulfur
$E_c - 0.032$	Barium	$E_c - 0.30$	Tungsten
$E_c - 0.032$	Sodium	$E_c - 0.31$	Mercury
$E_c - 0.033$	Lithium	$E_c - 0.33$	Molybdenum
$E_c - 0.039$	Antimony	$E_c - 0.33$	Neodymium
$E_c - 0.04$	Potassium	$E_c - 0.35$	Scandium
$E_c - 0.045$	Phosphorus	$E_c - 0.353$	Rhodium
$E_c - 0.049$	Arsenic	$E_c - 0.36$	Cobalt
$E_c - 0.069$	Bismuth	$E_c - 0.36$	Mercury
$E_c - 0.07$	Zirconium	$E_c - 0.37$	Nickel
$E_c - 0.073$	Potassium	$E_c - 0.37$	Tungsten
$E_c - 0.1$	Potassium	$E_c - 0.385$	Iridium
$E_c - 0.115$	Magnesium	$E_c - 0.39$	Manganese
$E_c - 0.16$	Oxygen	$E_c - 0.40$	Chromium
$E_c - 0.17$	Lead	$E_c - 0.40$	Magnesium
$E_c - 0.17$	Rhenium	$E_c - 0.44$	Cobalt
$E_c - 0.170$	Vanadium	$E_c - 0.446$	Vanadium
$E_c - 0.18$	Osmium	$E_c - 0.45$	Cadmium
$E_c - 0.2$	Tellurium	$E_c - 0.45$	Ruthenium
$E_c - 0.20$	Sulfur	$E_c - 0.472$	Tantalum
$E_c - 0.21$	Manganese	$E_c - 0.5$	Rhenium
$E_c - 0.22$	Palladium	$E_c - 0.5$	Scandium
$E_c - 0.22$	Tungsten	$E_c - 0.5$	Zirconium
$E_c - 0.227$	Magnesium	$E_c - 0.52$	Selenium
$E_c - 0.23$	Chromium	$E_c - 0.53$	Manganese
$E_c - 0.23$	Platinum	$E_c - 0.53$	Osmium
$E_c - 0.232$	Tantalum	$E_c - 0.535$	Cobalt
$E_c - 0.24$	Ruthenium	$E_c - 0.54$	Gold
$E_c - 0.26$	Titanium	$E_c - 0.55$	Iron
$E_c - 0.27$	Iron	$E_c - 0.55$	Scandium
$E_c - 0.27$	Scandium	$E_c - 0.55$	Zinc
$E_c - 0.29$	Silver	$E_c - 0.59$	Sulfur
$E_c - 0.29$	Thulium	$E_c - 0.591$	Rhodium
$E_c - 0.3$	Rhenium	$E_c - 0.629$	Iridium
$E_c - 0.30$	Cesium		

TABLE IX
Energy Levels below the Conduction Band
Arranged Alphabetically

Ec − 0.039	Antimony	Ec − 0.045	Phosphorus
Ec − 0.049	Arsenic	Ec − 0.073	Potassium
Ec − 0.032	Barium	Ec − 0.1	Potassium
Ec − 0.069	Bismuth	Ec − 0.04	Potassium
Ec − 0.45	Cadmium	Ec − 0.5	Rhenium
Ec − 0.30	Cesium	Ec − 0.3	Rhenium
Ec − 0.23	Chromium	Ec − 0.17	Rhenium
Ec − 0.40	Chromium	Ec − 0.353	Rhodium
Ec − 0.535	Cobalt	Ec − 0.591	Rhodium
Ec − 0.36	Cobalt	Ec − 0.24	Ruthenium
Ec − 0.44	Cobalt	Ec − 0.45	Ruthenium
Ec − 0.54	Gold	Ec − 0.27	Scandium
Ec − 0.385	Iridium	Ec − 0.35	Scandium
Ec − 0.629	Iridium	Ec − 0.5	Scandium
Ec − 0.27	Iron	Ec − 0.55	Scandium
Ec − 0.55	Iron	Ec − 0.52	Selenium
Ec − 0.17	Lead	Ec − 0.29	Silver
Ec − 0.033	Lithium	Ec − 0.032	Sodium
Ec − 0.115	Magnesium	Ec − 0.20	Sulfur
Ec − 0.40	Magnesium	Ec − 0.30	Sulfur
Ec − 0.227	Magnesium	Ec − 0.59	Sulfur
Ec − 0.53	Manganese	Ec − 0.232	Tantalum
Ec − 0.39	Manganese	Ec − 0.472	Tantalum
Ec − 0.21	Manganese	Ec − 0.2	Tellurium
Ec − 0.31	Mercury	Ec − 0.29	Thulium
Ec − 0.36	Mercury	Ec − 0.26	Titanium
Ec − 0.33	Molybdenum	Ec − 0.22	Tungsten
Ec − 0.33	Neodymium	Ec − 0.30	Tungsten
Ec − 0.37	Nickel	Ec − 0.37	Tungsten
Ec − 0.017	Nitrogen	Ec − 0.446	Vanadium
Ec − 0.18	Osium	Ec − 0.170	Vanadium
Ec − 0.53	Osium	Ec − 0.5	Zirconium
Ec − 0.16	Oxygen	Ec − 0.07	Zirconium
Ec − 0.22	Palladium	Ec − 0.55	Zinc
Ec − 0.23	Platinum		

TABLE X

Energy Levels above the Valence Band Arranged by Energy

Ev + 0.0394	Lithium	Ev + 0.30	Molybdenum
Ev + 0.04	Samarium	Ev + 0.31	Tungsten
Ev + 0.045	Boron	Ev + 0.31	Zinc
Ev + 0.0563	Aluminum	Ev + 0.32	Palladium
Ev + 0.057	Aluminum	Ev + 0.32	Platinum
Ev + 0.065	Gallium	Ev + 0.33	Manganese
Ev + 0.0685	Aluminum	Ev + 0.33	Mercury
Ev + 0.07	Barium	Ev + 0.33	Palladium
Ev + 0.11	Chromium	Ev + 0.34	Molybdenum
Ev + 0.111	Indium	Ev + 0.34	Tungsten
Ev + 0.12	Niobium	Ev + 0.35	Vanadium
Ev + 0.145	Beryllium	Ev + 0.37	Copper
Ev + 0.16	Indium	Ev + 0.37	Lead
Ev + 0.18	Rhenium	Ev + 0.384	Cobalt
Ev + 0.18	Osmium	Ev + 0.395	Silver
Ev + 0.19	Beryllium	Ev + 0.4	Rhenium
Ev + 0.22	Cobalt	Ev + 0.40	Cobalt
Ev + 0.24	Copper	Ev + 0.40	Iron
Ev + 0.24	Nickel	Ev + 0.43	Beryllium
Ev + 0.246	Thallium	Ev + 0.44	Manganese
Ev + 0.25	Mercury	Ev + 0.45	Cobalt
Ev + 0.26	Thallium	Ev + 0.50	Barium
Ev + 0.29	Cobalt	Ev + 0.51	Manganese
Ev + 0.29	Titanium	Ev + 0.52	Copper
Ev + 0.33	Gold	Ev + 0.55	Cadmium
Ev + 0.30	Cadmium	Ev + 0.62	Selenium

TABLE XI
Energy Levels above the Valence Band Arranged Alphabetically

Ev + 0.0563	Aluminum	Ev + 0.0394	Lithium
Ev + 0.057	Aluminum	Ev + 0.33	Manganese
Ev + 0.0685	Aluminum	Ev + 0.44	Manganese
Ev + 0.07	Barium	Ev + 0.51	Manganese
Ev + 0.50	Barium	Ev + 0.25	Mercury
Ev + 0.145	Beryllium	Ev + 0.33	Mercury
Ev + 0.19	Beryllium	Ev + 0.30	Molybdenum
Ev + 0.43	Beryllium	Ev + 0.34	Molybdenum
Ev + 0.045	Boron	Ev + 0.24	Nickel
Ev + 0.30	Cadmium	Ev + 0.12	Niobium
Ev + 0.55	Cadmium	Ev + 0.18	Osmium
Ev + 0.11	Chromium	Ev + 0.32	Palladium
Ev + 0.22	Cobalt	Ev + 0.33	Palladium
Ev + 0.29	Cobalt	Ev + 0.32	Platinum
Ev + 0.384	Cobalt	Ev + 0.18	Rhenium
Ev + 0.40	Cobalt	Ev + 0.4	Rhenium
Ev + 0.45	Cobalt	Ev + 0.04	Samarium
Ev + 0.24	Copper	Ev + 0.62	Selenium
Ev + 0.37	Copper	Ev + 0.395	Silver
Ev + 0.52	Copper	Ev + 0.246	Thallium
Ev + 0.065	Gallium	Ev + 0.26	Thallium
Ev + 0.33	Gold	Ev + 0.29	Titanium
Ev + 0.111	Indium	Ev + 0.31	Tungsten
Ev + 0.16	Indium	Ev + 0.34	Tungsten
Ev + 0.40	Iron	Ev + 0.35	Vanadium
Ev + 0.37	Lead	Ev + 0.31	Zinc

REFERENCES

1. J. A. Keenan and G. B. Larrabee, "VLSI Electronics: Microstructure Science" (N. G. Einspruch and G. B. Larrabee, eds.), vol. 6, p. 1. Academic Press, New York, 1983.
2. W. R. Runyan, "Semiconductor Measurements and Instrumentation," p. 65. McGraw-Hill, New York, 1975.
3. T. J. Shaffner, "VLSI Electronics: Microstructure Science" (N. G. Einspruch and G. B. Larrabee, eds.), vol. 6, p. 497. Academic Press, New York, 1983.
4. "Annual Book for ASTM Standards," part 43, Electronics. American Society for Testing and Materials, Philadelphia, Pennsylvania, 1983.

Chapter 17

Growth of Epitaxial Films for VLSI Applications

T. L. CHU

Southern Methodist University
Dallas, Texas

I. Introduction	285
A. Definition of Epitaxy	285
B. Development of Epitaxial Films	286
C. Types of Epitaxial Films	286
II. Epitaxial Growth of Silicon Films	286
A. Fabrication Steps	287
B. Buried Layers	288
C. Chemical Processes	289
D. Doping and Autodoping	292
E. Equipment	293
F. Procedure	294
III. Characterization of Epitaxial Silicon Films	296
A. Standards	296
B. Structural Defects	296
C. Electrical Properties	297
IV. Epitaxial Growth of Gallium Arsenide Films	301
A. Liquid Phase Epitaxy	301
B. Chemical Vapor Epitaxy	302
References	303

INTRODUCTION   I

Definition of Epitaxy   A

Epitaxial growth, single crystal overgrowth of one material on a single crystal substrate of the same (homoepitaxy) or different (heteroepitaxy) chemical composition, is an essential part of the VLSI technology. Heteroepitaxy can be achieved if the substrate and the grown film are similar in crystal structure and lattice parameters, for example, silicon on sapphire or gallium aluminum arsenide on gallium arsenide. The major advantage of epitaxial growth is flexibility in controlling the concentrations and distribution of dopants in a semiconductor film of high chemical purity and structural perfection.

B Development of Epitaxial Films

The importance of silicon epitaxial growth in device technology was first demonstrated in the late fifties. A chemical vapor epitaxial (CVE) silicon film of about 20 μm thickness and several ohm-cm resistivity deposited on a low-resistivity substrate improved the frequency response of bipolar transistors. The flexibility of CVE was subsequently applied to the fabrication of monolithic ICs with equal success, and CVE has since become an important process in the silicon integrated circuit technology.

C Types of Epitaxial Films

Epitaxial films of semiconductors can be grown from the vapor phase (VPE) using chemical vapor growth (CVE) or physical vapor growth such as molecular beam epitaxy, MBE, or from solution (liquid phase epitaxy, LPE). Each technique has its own merits, summarized in Table I. At present, silicon ICs dominate the commercial market. Gallium arsenide technology is behind the silicon technology, and gallium arsenide circuits are limited mainly to military applications. Thus, this chapter is concerned mainly with the epitaxial growth and characterization of silicon films with a very brief summary on the epitaxial growth of gallium arsenide films.

II EPITAXIAL GROWTH OF SILICON FILMS[†]

Epitaxial structures are used in bipolar, CMOS, and other ICs. The requirements for the properties of epitaxial silicon films for VLSI applications are more stringent and more difficult to achieve than those for discrete devices. For the fabrication of VLSI circuits, the crystal perfection,

† See references [1,2].

TABLE I

Epitaxial Growth Processes

Process	Requirements	Advantages	Disadvantages
LPE	Electrically inactive solvent Low solubility of solvent in semiconductor	High purity films obtainable Dissociation of compound semiconductors minimized	Segregation of dopants Complex equipment required for multilayer structures
CVE	Thermochemically and kinetically feasible chemical reactions No undesirable by-products	Flexibility in controlling the dopant concentration and distribution Simplicity in operation	Enhanced diffusion and autodoping due to high deposition temperature
MBE	High vacuum system to attain 10^{-10} Torr consistently Materials used for equipment construction have low vapor pressures and low sticking coefficients	Precise control of very thin films (50 Å) achievable Versatile in growing high quality multilayer structures at relatively low temperatures Abrupt changes in doping or in composition obtainable	Sophisticated high vacuum equipment required Low capacity

thickness and thickness uniformity, and dopant concentration and distribution must be highly uniform over large areas. The importance of these properties, major factors affecting them, and the control techniques are summarized in Table II.

Fabrication Steps   A

In *bipolar circuits,* for example, *n*-type epitaxial films of 1–5 ohm-cm resistivity and 5–10 μm thickness are deposited on *p*-type substrates of 10–20 ohm-cm resistivity; then *p*-type isolation moats are made by diffusion from the surface of the epitaxial film. Transistors and other components are fabricated in these isolated regions of the epitaxial film. Double diffusion can be used for the fabrication of these transistors, but such transistors have a high emitter–collector resistance.

To reduce this parasitic resistance, a low resistivity region is formed between the collector of the transistor and the substrate prior to the epitaxial growth of *n*-type film. This *buried layer* can be formed in selected regions of the substrate by using oxidation and photolithography,

TABLE II

Properties of CVE Silicon Layers for VLSI Circuits

Property	Importance	Determining factors	Control techniques
Structural perfection	Good structural perfection essential for reasonable carrier mobility and lifetime in epitaxial films	Structure perfection of substrates Cleanliness of substrate surface Purity of reactants and deposition equipment	Use substrates of low defect density In situ etching of substrates at 1150–1200°C using a $H_2:HCl$ mixture
Thickness and thickness uniformity	—	Rate and duration of deposition Reactant flux over the substrates surface Temperature uniformity	Uniform reactant flux over substrate surface through equipment design
Dopant concentration and distribution	Voltage capability, frequency, and other properties of ICs determined by dopant concentration and distribution	Concentration of dopant in reactant mixture Reactant flux over the substrate surface Impurity content in the reactant and deposition chamber Temperature uniformity	Programmed introduction of dopant into the reactant mixture Thorough mixing in the reactant mixture Uniform reactant flux over substrate surface Minimize concentration of background impurities

followed by diffusion or ion implantation. Subsequent to the buried layer formation, epitaxial growth and isolation diffusion are carried out in the usual manner. The configuration of the resulting transistors is shown schematically in Fig. 1, in which the reduction of collector resistance associated with the flow of emitter–collector current through the buried layer is apparent.

B Buried Layers

The use of buried layers has several problems: autodoping of epitaxial films during subsequent processing (see Section IID), pattern shift, and pattern distortion. In the formation of the buried layer a gap of a few

17. Growth of Epitaxial Films for VLSI Applications

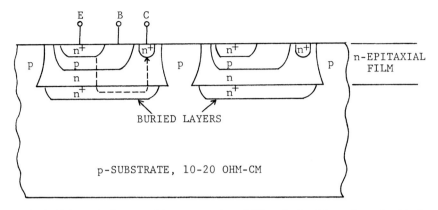

Fig. 1. Schematic diagram of the cross section of diffused epitaxial transistors in a bipolar IC showing the buried layers under the collectors.

hundred angstroms is left around its periphery to mark its location, because subsequent masking steps must be aligned with the buried layer. However, the epitaxial growth process shifts and distorts the buried layer pattern (Fig. 2). The extent of pattern shift and distortion is related to the substrate orientation, substrate temperature, growth rate, epitaxial film thickness, the nature of the chemical reaction, and so on. In general, the use of $\langle 111 \rangle$ substrates oriented 2–3 degrees toward a $\langle 110 \rangle$ direction, high deposition temperature, low growth rate, and a chemically irreversible deposition process reduces the pattern shift. The pattern and distortion must be minimized for the fabrication of high density circuits.

Chemical Processes C

In the CVE process, a reaction mixture of gases containing a silicon compound is passed over the surface of heated silicon substrates in a

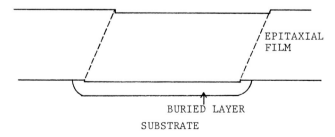

Fig. 2. Schematic of the shift of a buried layer pattern after the growth of an epitaxial film.

deposition chamber, where the chemical reaction takes place on the substrate surface, depositing silicon.

1 Silane Reactions

The thermal reduction of chlorosilanes ($SiCl_4$ and $SiHCl_3$) with hydrogen and the thermal decomposition of dichlorosilane (SiH_2Cl_2) and monosilane (SiH_4) in a hydrogen atmosphere have all been used for the epitaxial growth of device-quality silicon films in a gas flow system. The chemical reactions, their equilibrium constants at 1350 K ($K_{1350 K}$), the temperature range used for the deposition process, and the deposition rate used in manufacturing processes are summarized in Table III.

Chemical equilibrium is not established during the deposition process; however, the K values indicate the relative ease of these reactions. As the chlorine content in chlorosilanes decreases, the reaction becomes more thermochemically feasible, and lower deposition temperature can be used. However, the tolerable concentration of oxygen or water in the reaction mixture also decreases with decreasing substrate temperature: 5–10 ppm at 1200°C and <2 ppm at 1000°C.

2 Deposition Rates

In a given deposition equipment, the deposition rate is determined mainly by the substrate temperature and the composition and flow rate of the reaction mixture. By adjusting these parameters, one can obtain the deposition rate in the range shown in Table III. For a given chemical reaction, higher substrate temperature is required for higher deposition rates. Note in Table III that the deposition rate of device-quality silicon film from the silane process is considerably lower than that from the chlorosilane processes because of *gas-phase nucleation* at high concen-

TABLE III

Chemical Vapor Epitaxial Growth of Silicon Films[a]

Chemical reaction	$K_{1350 K}$	Temperature range, (°C)	Growth rate obtainable (μm/min)
$SiCl_4(g) + 2H_2(g) \rightleftarrows Si(s) + 4HCl(g)$	2.2×10^{-3}	1150–1250	0.4–15.0
$SiCl_3(g) + H_2(g) \rightleftarrows Si(s) + 3HCl(g)$	0.27	1100–1200	0.4–2.0
$SiH_2Cl_2(g) \rightleftarrows Si(s) + 2HCl(g)$	4.0	1050–1150	0.4–3.0
$SiH_4(g) \rightarrow Si(s) + 2H_2(g)$[a]	10^6	950–1050	0.2–0.3

[a] Growth rate is significantly enhanced in a helium atmosphere.

17. Growth of Epitaxial Films for VLSI Applications

trations of silanes. This gas-phase nucleation is associated with the thermal stability of silane and can be minimized by using reduced pressures or by adding hydrogen chloride to the reactant mixture.

The CVE growth of silicon takes place in five steps:

(1) transport of reactant species to the substrate surface,
(2) absorption of reactants onto the surface
(3) chemical reaction on the surface,
(4) desorption of product gases from the surface, and
(5) transport of product gases away from the surface.

The importance of these steps can be deduced from the *temperature dependence* of deposition rates. Figure 3 shows a plot of the logarithm of deposition rate versus reciprocal temperature for the four chemical processes described. In the low temperature range, the deposition rate increases exponentially with increasing temperature, indicating that the rate of chemical reaction (step 3), is rate determining. At higher temperatures, the deposition rate is relatively insensitive to temperature, indicating that the transport processes are rate determining. Note that the transition from the kinetic regime to the *transport regime* occurs at high temperatures as

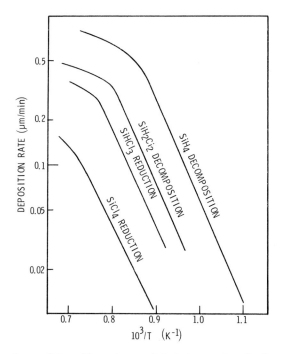

Fig. 3. Dependence of deposition rate on substrate temperature for the epitaxial growth of silicon films by various chemical reactions.

the chlorine content in the silicon compound increases. The deposition process is usually carried out in the transport regime to minimize the effect of temperature variations.

3 Choice of Process

The chemical process chosen depends on the application of the epitaxial silicon film. One important consideration is that the reactions involving the chlorosilanes are chemically reversible, whereas the thermal decomposition of monosilane is essentially irreversible. The reversible chemical reactions may contribute to increased *autodoping* (see Section IID) of the epitaxial silicon film. Further, the silane process requires lower temperature, which minimizes autodoping due to diffusion. Dichlorosilane is widely used as the silicon source because of the relatively high growth rates at relatively low temperatures. In view of the advantages of using monosilane for the epitaxial growth of silicon and the use of monosilane for the deposition of silicon oxide and silicon nitride films, industrial gas manufacturers have significantly improved its purity and reproducibility. It is expected that the thermal decomposition of silane will become the most commonly used process for silicon epitaxial growth.

D Doping and Autodoping

To control the conductivity type and electrical resistivity of epitaxial silicon films, dopants are introduced into the reaction mixture during the growth process. The incorporation of dopant atoms into the epitaxial film involves steps similar to those of the deposition of silicon. The choice of gaseous dopants is limited; those commonly used are in the form of hydrides, such as B_2H_6, PH_3, and AsH_3, supplied as dilute mixtures (10–500 ppm) in hydrogen or inert gas. They are thermochemically unstable at room temperature and above. The degree of incorporation of dopant atoms depends on the substrate temperature, deposition rate, the concentration of dopant in the reaction mixture, geometry of the deposition chamber, and so on. Thus, the dopant/Si ratio in the solid phase is different from that in the gas phase and must be determined experimentally for each set of conditions. Epitaxial films with net carrier concentrations in the range of 10^{14}–10^{20}/cm^3 can be readily obtained.

Unintentional doping also occurs. For example, impurities from the substrate may incorporate into the epitaxial film by diffusion or chemical transport if the deposition process is chemically reversible. This autodoping effect can cause the change in dopant concentration at the substrate–

film interface region to become gradual, which is highly undesirable. One example of autodoping is the epitaxial growth of a lightly doped film over the buried layer. The diffusion into the epitaxial film of impurities from the buried layer, which behaves as an infinite source, follows the complementary error function. In addition to diffusion during growth, the dopant profile in epitaxial films may also change due to diffusion during subsequent processing.

Equipment E

The equipment for the epitaxial growth of silicon consists essentially of three parts:

(1) gas flow control,
(2) deposition chamber, and
(3) exhaust gas scrubber.

There must be appropriate interlocks to minimize accidents.

In the *deposition chamber,* the substrates are usually supported on a silicon carbide-coated graphite *susceptor,* which can be heated resistively or inductively. The substrate–susceptor combination can also be heated by radiation. Several designs of the deposition chamber have been used for manufacturing purposes. For example, the substrates are placed on a susceptor (30 cm × 60 cm in area, or larger) in a fused silica tube of rectangular cross section, and the susceptor is heated externally by an rf generator.

The most widely used system at present is the radiantly heated epitaxial reactor (Figs. 4 and 5). A microprocessor-based control system provides complete operating information with minimal operator effort. The deposition chamber can accommodate twenty-four 10 cm or twelve 12.5 cm diameter substrates. Quartz lamps heat the substrates and susceptor simultaneously. The use of rf heating creates a temperature gradient in the thickness direction of the substrate, resulting in warpage and dislocations in large diameter substrates. The radiant heating system uniformly heats the substrates with virtually no slip lines. Epitaxial films with thickness and resistivity uniformities of ±5% are routinely produced, with variations between runs of ±5% or better.

The epitaxial growth of silicon films can be carried out under atmospheric or reduced pressure (100 Torr, for example). Due to the increased diffusion rate, deposition under reduced pressure provides better uniformity of epitaxial films, reduced autodoping with more abrupt transition of dopant concentration, and reduced pattern shift and distortion.

Fig. 4. Manufacturing equipment for the epitaxial growth of silicon films by radiant heating. (Courtesy of Applied Materials, Inc., Santa Clara, California.)

F Procedure

A typical procedure for the epitaxial growth of silicon film is as follows:

(1) Substrate preparation. Commercially available silicon slices are usually of a ⟨111⟩ (off 2–3 deg toward ⟨110⟩) or ⟨100⟩ orientation with one main face chem-mechanically polished. If the substrates have not been

17. Growth of Epitaxial Films for VLSI Applications

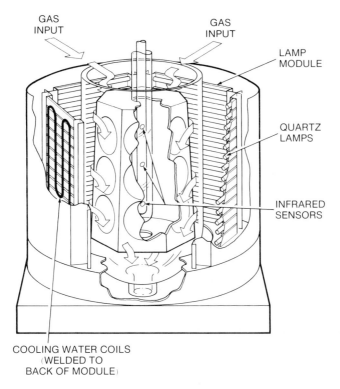

Fig. 5. Schematic diagram illustrating the interior of the deposition chamber of the epitaxial equipment shown in Fig. 4. (Courtesy of Applied Materials, Inc., Santa Clara, California.)

cleaned and packed under proper conditions by the manufacturer, they should be ultrasonically cleaned with perchloroethylene, isopropyl alcohol, and then dried in a laminar-flow work station.

(2) Assembly. The substrates are placed on the susceptor in the deposition chamber, and the chamber is purged with an inert gas (argon, helium, or nitrogen), followed by hydrogen.

(3) In situ etching of substrates. The substrates are heated slowly to 1150–1200°C in hydrogen (the flow rate of hydrogen is usually recommended by the equipment manufacturer), and anhydrous hydrogen chloride is introduced. The concentration of hydrogen chloride in hydrogen should be 1–2%, and approximately 1–2 μm of silicon should be etched from the surface (<0.5 μm when the substrate contains buried layers). Hydrogen chloride flow is then turned off.

(4) Deposition of silicon. Subsequent to in situ etching, the substrate temperature is reduced for the deposition process to a temperature that depends on the nature of the chemical reaction chosen. Dopants are introduced to obtain silicon films of desired resistivity.

(5) Conclusion of process. At the conclusion of the deposition process, the flow of dopant and silicon compound is turned off, and the substrate temperature is reduced gradually to room temperature. The hydrogen flow is replaced with an inert gas, and the epitaxial wafers are removed from the deposition chamber.

IV CHARACTERIZATION OF EPITAXIAL SILICON FILMS[†]

A Standards

The performance of VLSI circuits are affected by the thickness, structural perfection, and electrical properties of epitaxial silicon films. Many techniques for evaluating these properties have been established by the American Society for Testing and Materials and approved by the American National Standards Institute as the American national standard. They are summarized in Table IV.

B Structural Defects

Epitaxial silicon films that are prepared under the most optimum conditions are similar in crystalline perfection to their substrate; that is, essentially all structural defects in the substrate propagate into the epitaxial film. Additional defects may be generated at the substrate–film interface or during the growth process. The most frequently observed defects (e.g., stacking faults, dislocations, precipitates, and inclusions of random orientations) and the principal causes, important effects, and seriousness of these defects are summarized in Table V.

Structural defects can be detected and identified by optical microscopy, transmission electron microscopy, and x-ray diffraction microscopy.

Optical microscopy is the simplest method for detecting structural defects in epitaxial silicon films. The intersections of the line and plane defects with the grown surface are of higher energy than the adjacent crystal and thus exhibit faster etch rates. Etching the grown surface with an appropriate etchant, such as the *Sirtl etch* or the *Dash etch* (3 HNO_3 + 12 CH_3COOH + 1 HF), reveals the defects as pits, grooves, or etch figures of various geometries. Figure 6 shows some examples of defects in epitaxial silicon films revealed by chemical etching and by transmission electron microscopy.

[†] See references [1,2].

TABLE IV

Characterization Techniques for Epitaxial Silicon Films

Property	Sample configuration	Technique	Reference [4]
Thickness	Epitaxial film and substrate of different carrier concentration or conductivity type	Angle-lapping and staining to reveal the substrate–epitaxial film interface for direct measurement	ANSI–ASTM F110
	n/n^+ (substrate) or p/p^+ (substrate)	Measurement of the infrared reflectance spectrum	ANSI–ASTM F95
	Any configuration with stacking faults	Measurement of dimension of stacking fault nucleated at the substrate surface	ANSI–ASTM F143
Structural defects	—	A chromic–hydrofluoric acid mixture preferentially etches structural defects in epitaxial silicon	ANSI–ASTM F80
		Interference contrast microscopy for measurement of stacking fault density	ANSI–ASTM F522
Electrical properties	Epitaxial film and substrate of opposite conductivity type	Radial resistivity variations on epitaxial silicon film by the four-probe method	ANSI–ASTM F81
		Resistivity of epitaxial silicon films by the three-probe voltage breakdown method	ANSI–ASTM F108
		Sheet resistance of epitaxial silicon films by four-probe method	ANSI–ASTM F374
		Net carrier density in epitaxial silicon by voltage–capacitance measurement of p–n junction or Schottky diode	ANSI–ASTM F419
		Dopant profile by a spreading resistance probe	ANSI–ASTM F525

Electrical Properties C

The important electrical properties of epitaxial silane films include electrical resistivity, majority carrier mobility, minority carrier lifetime, and dopant profile.

TABLE V

Structural Defects in Epitaxial Silicon Films

Defect	Principal cause	Important effects	Seriousness
Stacking faults	Defects at substrate surface Impurities on substrate surface Stress during growth	Transformation to other defects during diffusion No direct effects on junction properties	Probably not serious at concentrations $<10^3$ cm^{-3} Secondary effects at high concentration can be serious Relatively easy to eliminate
Dislocations	Propagation from substrate Deformation of substrate Contaminations on substrate surface	Decrease of lifetime, mobility, and breakdown voltage Enhancement of dopant diffusion Preferential precipitation of impurities	Concentrations $>10^4$ cm^{-2} usually have noticeable effects on devices, directly or through secondary effects Relatively easy to control
Polycrystalline inclusion	Impurities on substrate surface	Harmful because of grain boundaries Soft, low-voltage junction	The most serious defects in epitaxial silicon Can be eliminated
Precipitates	Exceeding solubility limits Chemical interaction Impurities in reactants and etchants used for surface preparation and epitaxial growth	Local shorting Sites for second breakdown General degradation of junction characteristics Change in diffusion profile	Serious defect of varied origin Can be minimized by using pure reactants and proper handling techniques

1 Carrier Mobility

The carrier mobility can be determined by Hall-type measurements, provided that the epitaxial film and the substrate are of opposite conductivity.

2 Minority Carrier Lifetime

The minority carrier lifetime (or diffusion length) can be measured by open-circuit voltage decay (for a lightly-doped film on a heavily doped

17. Growth of Epitaxial Films for VLSI Applications

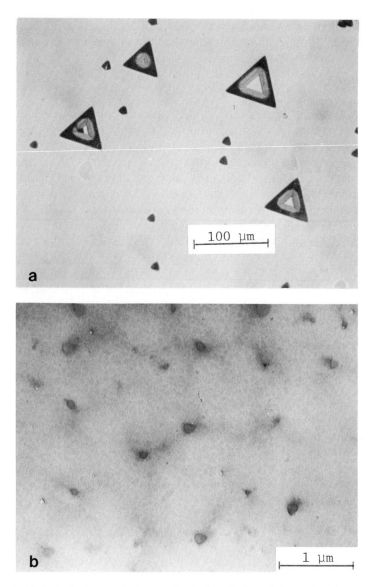

Fig. 6. (a) Optical micrograph showing the stacking faults and dislocations in an epitaxial film revealed by chemical etching. (b) Transmission electron micrograph showing precipitates in an epitaxial film.

substrate of the opposite conductivity) or steady-state surface voltage method (for an epitaxial film on a substrate of the same conductivity and film thickness greater than four times the diffusion length).

3 Dopant Profile

The most useful high-resolution technique for the measurement of the dopant concentration and distribution in epitaxial structures is the *spreading resistance technique* [3]. This technique is based on the resistance measurement of the pressure contact over a small area between a hemispherically surfaced metal and a flat semiconductor surface. The contact, or spreading, resistance R_s is given by $R_s = K\rho/4a$, where ρ is the resistivity of the semiconductor, a is the contact radius, and K is a parameter related to the *zero bias barrier resistance* at the metal–semiconductor interface. K depends on the conductivity type and resistivity of the material and can be evaluated by calibration. Most of the resistance occurs in a volume of about $(2a)^3$ immediately beneath the contact, and the spatial resolution of the spreading resistance technique is thus limited by $(2a)^3$. The radial resistivity profiles in epitaxial silicon films can be obtained by applying a spreading resistance probe along a diameter of the specimen

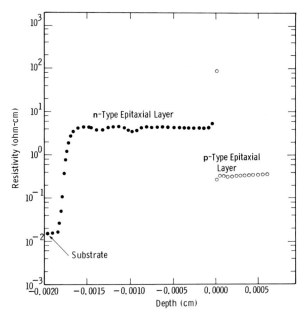

Fig. 7. Resistivity profile of *n*- and *p*-type epitaxial films deposited successively on low resistivity *n*-type substrate by the thermal decomposition of silane at 1050°C. ● *n*-type; ○ *p*-type.

surface and measuring the resistance between this probe and a low resistance contact on the specimen. Since the contact radius of the probe is about 4 μm, the resolution of this technique is about 10 μm. The resistivity profile along the thickness of the specimen is usually measured on a beveled surface to get better resolution. The beveled specimen is mounted on the stage of a micromanipulator, and one or two spreading resistance probes are set transverse to the bevel to measure the spreading resistance at the desired lateral increments. An example of the resistivity profile in an epitaxial $p/n/n^+$ (substrate) structure is shown in Fig. 7. Various aspects of the spreading resistance technique have been described in a number of publications.

IV. EPITAXIAL GROWTH OF GALLIUM ARSENIDE FILMS[†]

Gallium arsenide devices and circuits are superior to their silicon counterparts in operating frequency, operating temperature, and radiation resistance. These features have motivated the development of a variety of gallium arsenide ICs for signal processing, communication, and other applications. Metal–gallium arsenide field-effect transistors (MESFETs) are used as building blocks for several types of ICs such as multipliers and RAMs. They are fabricated from bulk gallium arsenide crystals by ion implantation and high-resolution photolithography. However, the fabrication of other devices, such as high electron-mobility transistors that are particularly suited for high-performance VLSI, requires epitaxial films of gallium arsenide and gallium aluminum arsenide with precisely controlled properties. The control of dopant concentration during the material preparation process is far superior to doping by diffusion or ion implantation. The epitaxial growth of gallium arsenide films has been carried out by several techniques such as MBE, LPE, and CVE.

A. Liquid Phase Epitaxy

The growth of gallium arsenide films by LPE is based on the fact that the solubility of gallium arsenide in gallium or other metallic solvents increases with increasing temperature. In practice, a solution of gallium arsenide in gallium, saturated at 750–950°C, is allowed to come in contact with a substrate and is then cooled at a predetermined rate. Using inert,

† See reference [5].

pure containers and high-purity reagents, epitaxial gallium arsenide layers with electron concentrations of 10^{13}–10^{14} cm^{-3} and room temperature electron mobilities up to 9000 cm^2 V^{-1} s^{-1} have been obtained. The superior quality of the gallium arsenide that is grown in this manner is apparently due to the low segregation coefficients of impurities from the gallium solution. The carrier concentration in solution-grown gallium arsenide can also be controlled by using proper dopants to yield p–n junctions.

B Chemical Vapor Epitaxy

The growth of gallium arsenide by CVE has been carried out with halides or metallo-organic compounds. The halide process uses the reaction of arsenic chloride with gallium in a hydrogen flow system. Two stages can be identified in the process: the formation of gallium monochloride and arsenic in the gallium zone at about 800°C and the deposition of gallium arsenide on the substrate surface at about 750°C, according to the equation:

$$6\ \text{GaCl(g)} + \text{As}_4\text{(g)} = 4\ \text{GaAs(s)} + 2\ \text{GaCl}_3\text{(g)}.$$

Without intentional doping, the deposited material usually has an electron concentration of 10^{14}–10^{15} cm^{-3} and a room temperature electron mobility of 6000–8000 cm^2 V^{-1} s^{-1}. The epitaxial growth of gallium arsenide has also been achieved by using arsine instead of arsenic chloride. The carrier concentration in the epitaxial layer can be controlled by introducing n-type (such as hydrogen selenide) and p-type (such as dimethyl zinc) dopants into the reactant mixture. Thus, the p–n junctions can be produced during the growth process.

The use of metallo-organic compounds was developed in the late sixties. For example, the reaction of trimethylgallium, $(\text{CH}_3)_3\text{Ga}$, and arsine, AsH_3, on the surface of gallium arsenide substrates at about 700°C was used in a gas flow system to produce epitaxial layers of gallium arsenide. Similarly, the reaction of trimethylgallium, trimethylaluminum, and arsine has been used for the deposition of epitaxial layers of gallium aluminum arsenide. In fact, this process is the only technique, scalable to high volume, for the deposition of gallium aluminum arsenide layers of controlled thicknesses and carrier concentrations. A microprocessor controlled metallo-organic deposition system is shown in Fig. 8. The importance of metallo-organic reactions was not recognized until the new material requirements, such as submicron layers and high-quality heterojunctions, developed during the past few years. For example, the technologically important gallium aluminum arsenide, $\text{Ga}_x\text{Al}_{1-x}\text{As}$ and gallium

17. Growth of Epitaxial Films for VLSI Applications

Fig. 8. Microprocessor-controlled equipment for the epitaxial growth of III–V compounds and their solid solutions by the metallo-organic process. (Courtesy of Cambridge Instruments, Inc., Monsey, New York.)

arsenide phosphide, $GaAs_xP_{1-x}$, can be conveniently deposited over a wide composition range. An important feature of the metallo-organic process is that multilayer, multicomponent growths can be accomplished in a single growth experiment. Numerous multilayer structures have led to devices with characteristics equal to or better than comparable devices grown by other materials technologies. For example, $Ga_xAl_{1-x}As/GaAs$ structures have been fabricated for heterojunction transistors for high-speed low-power digital logic circuits. Because of the similar lattice parameters of GaAs and AlAs, these heterojunctions exhibit a high degree of crystal perfection.

REFERENCES

1. "Semiconductor Silicon" (R. R. Haberecht and E. L. Kern, eds.). The Electrochemical Society Inc., 1969.

2. "Silicon Device Processing" (Charles P. Marsden, ed.). NBS Special Publication 337, November, 1970.
3. "Spreading Resistance Symposium" (James R. Ehrstein, ed.). NBS Special Publication 400-10, December, 1974.
4. "1983 Annual Book of ASTM Standards," vols. 10.04 and 10.05. American Society for Testing and Materials, 1983.
5. "III-V Opto-Electronics Epitaxy and Device Related Processes" (V. G. Keramidus and S. Mahajan, eds.). Electrochemical Society Proceedings Volume 83-13, 1983.

Chapter 18
Epitaxial Silicon: Material Characterization

S. B. KULKARNI

IBM East Fishkill Facility
Hopewell Junction, New York

I. Introduction	305
II. Growth Characterization	306
A. General Classification	306
B. Growth Techniques	306
C. Impurity Transport	307
III. Electrical Characterization	310
A. Resistivity Measurements	312
B. Dopant Profile Measurements	312
C. Measurement of Electrically Active Impurities	314
IV. Physical and Optical Characterization	316
A. Conventional Thickness Measurements	316
B. Weighing Technique	316
C. Infrared Reflectance Technique	316
D. Fourier Transform Infrared Spectrometry	320
V. Epitaxial Defect Characterization	321
A. Electrical Properties of Defects	321
B. Process Induced Defects	324
C. Defect Prevention	324
VI. Epitaxial Defect Measurements	324
References	325

INTRODUCTION   I

Epitaxial growth is a process whereby a crystallographically oriented film is grown as an extension onto an existing crystal (usually referred to as *substrate*). In VLSI circuit technology, substrates usually include buried diffused islands, separated by junctions or dielectric isolation areas.

Transistors are often fabricated in the epitaxial films by various device isolation schemes so that characterization of epitaxial films plays a key role in successful fabrication of VLSI circuits.

The various process and material characterization techniques that are described in this chapter represent the state of the art. For detailed information, a reader is encouraged to read recently published review articles [1,2]. Although the discussion is limited to silicon technology, the characterization techniques can be applied to other semiconductor materials with minor modifications.

II GROWTH CHARACTERIZATION

A General Classification

Epitaxial growth should be considered in terms of two basic categories: homoepitaxy and heteroepitaxy. The term *homoepitaxy* refers to the growth of a particular chemical species onto a substrate that has the same chemical formula (e.g., silicon on silicon). The electrical properties of the films may differ from those of their substrates. The term *heteroepitaxy* refers to the growth of a film that has a chemical formula different from that of the substrate (e.g., silicon on Al_2O_3 substrate).

B Growth Techniques

Any film growth involves a material transport mechanism; therefore, all epitaxial growth techniques can be classified accordingly (Table I). Techniques based on vapor phase growth are commonly used in the industry today. The CVD and MBE techniques are described here, and other techniques can be found in various articles published in the literature.

1 Chemical Vapor Deposition

In the chemical vapor deposition (CVD) technique a suitable vapor source of silicon is reduced by hydrogen at high temperature, and the reduction byproduct is adsorbed on a heated silicon wafer. Overall simplified chemical reactions are the following.

$$SiCl_4(v) + 2\ H_2(g) \rightleftharpoons Si(s) + 4\ HCl(v)$$

$$SiHCl_3(v) + \tfrac{3}{2} H_2(g) \rightleftharpoons Si(s) + 3\ HCl(v)$$

$$SiH_2Cl_2(v) \rightleftharpoons Si(s) + 2\ HCl(v)$$

TABLE I

Classification of Epitaxial Growth Techniques

Growth type	Example	Growth conditions	Comments and key references
Solid phase: solid ⇌ solid	Epitaxial regrowth of amorphous Si by laser annealing	Temperature >600°C Pressure 1 atm Damaged layer required as substrate	Technique being explored for VLSI circuits in the research environment
Liquid phase: liquid ⇌ solid	Liquid phase epitaxy	Temperature near liquidus temperature Rotating substrate Pressure 1 atm	[2]
Gas phase: vapor ⇌ solid	Chemical vapor deposition (CVD)	Temperature >900°C Pressure ≤1 atm	Currently used in the semiconductor industry [2]
	Molecular beam epitaxy (MBE)	Temperature >700°C Pressure UHV	Being explored in the research environment [2]
	Sputter epitaxy	Temperature >700°C Pressure UHV	Technique in the research stage

For controling the resistivity of the film, arsine gas is usually used during growth. Arsine atoms that are adsorbed on the surface occupy substitutional silicon lattice sites. A typical reactor used for the film growth is sketched in Fig. 1.

2 Molecular Beam Epitaxy

Molecular beam epitaxy (MBE) is basically a technique of vacuum evaporation. A variety of source materials can be used that can be conventionally transformed by heating into their gaseous forms, which then expand into the vacuum space and condense on the substrates. Ultrahigh vacuum systems are required. Substrate cleaning, prior to the continuation of growth, and the control of dopant evaporation are the two most important parts of the growth process.

C Impurity Transport

Transport of impurity atoms during the vapor phase epitaxial growth process controls various electrical properties of the devices fabricated.

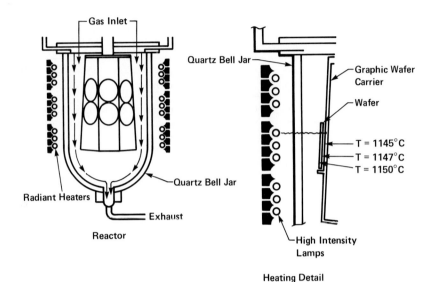

Fig. 1. Radiant-heated barrel reactor. (From Pogge [2]. Reprinted with permission of North-Holland, Amsterdam.)

Transport phenomena can be divided into two major groups: solid state diffusion and vapor phase autodoping.

Solid state diffusion of atoms from the buried layers (uniform doping N_s) can be described by the impurity diffusion profile:

$$N(x,t) = \tfrac{1}{2} N_s \operatorname{erfc} x/2\sqrt{Dt}. \tag{1}$$

An assumption of high growth rate is made such that $vt \gg Dt$, where v is the film growth rate.

In a typical CVD growth process some impurity atoms from heavily doped substrates are transferred from solid phase to vapor phase because of their vapor pressure. A fraction of these atoms are incorporated into the epitaxial film. This phenomenon is called *autodoping* (Fig. 2). The vertical autodoping can be calculated using the gas phase autodoping model developed by Kühne [3]. It is given by

$$\log N_{AS} = A - B \left[\frac{C}{V_{\text{epi}}} + 1 \right] \Delta X_{\text{epi}} \tag{2}$$

where A, B, and C are factors involving temperature and the boundary layer thickness; X_{epi} is the distance from the substrate–epitaxial interface, and V_{epi} is the growth rate.

The overall dopant profile in an epitaxial film consists of solid state diffusion from the substrate and the vertical autodoping (Fig. 3). The

18. Epitaxial Silicon: Material Characterization   309

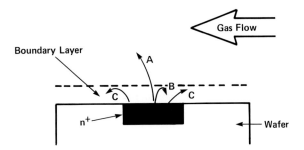

Fig. 2. Conditions for vertical and lateral autodoping. (a) Impurity is picked up by gas stream; no autodoping. (b) Vertical autodoping. (c) Lateral autodoping. (From Pogge [2]. Reprinted with permission of North-Holland, Amsterdam.)

effect of process parameters such as temperature, growth rate, and pressure on the epitaxial profiles can be calculated using Eq. (2). Other parameters affecting gas phase autodoping are listed in Table II.

The influence of the above transport phenomena on various device parameters is schematically shown in Fig. 4.

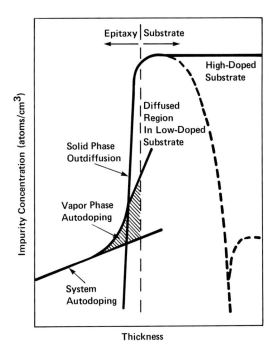

Fig. 3. Relative dopant profile in an epitaxial film indicating various contributions. (From Pogge [2]. Reprinted with permission of North-Holland, Amsterdam.)

TABLE II

Parameters Controlling Gas Phase Autodoping

Parameter	How to reduce autodoping
Substrate	
Dopant type	Select dopant with low vapor pressure
Dopant concentration	Minimize surface concentration
Surface area	Minimize the area
Process	
Prebake	Long prebake at high temperature
Pressure (prebake and deposition)	Low for n type and high for p type impurity
Deposition temperature	High for n type impurity and low for p type impurity
Growth	Interrupt growth and flush system with high gas flows

III ELECTRICAL CHARACTERIZATION

Electrical characterization of epitaxial films is required to measure and monitor electrically active impurity concentrations, which vary according to the transport phenomena previously described. Various characterization techniques, differing in both detection limits and resolution, are used to obtain information about the bulk as well as the local electrical activity of the epitaxial layers. Their relative merits are listed in Table III.

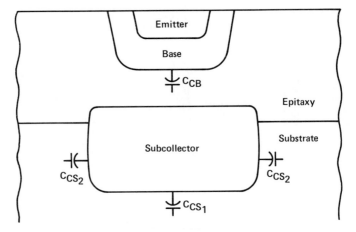

Fig. 4. A transistor schematic showing various capacitance components. Collector-base capacitance C_{CB} is determined by vertical autodoping, and collector–substrate C_{CS} capacitance is determined by lateral autodoping phenomenon.

TABLE III

Comparison of Electrical Characterization Techniques

Technique	Sample preparation	Applicable concentration range (atoms/cm^3)	Advantages	Disadvantages	Accuracy	Comments
Four-point probe	n/p or n/p	10^{14}–10^{18}	Speed	Separation of surface from bulk properties	±10%	Poor accuracy with submicron layers
Junction C–V	Diffused diode	10^{14}–10^{16}	On-product measurement	Correction required for diffused diodes; limited depth of penetration	±15%	Good for asymmetrical diodes only
Pulsed C–V	MOS device	10^{14}–10^{16}	No corrections required to test structure	Limited depth penetration; sensitive to epitaxial layer quality and unwanted autodoping impurities	±7%	Moderately slow; limited by sample preparation
Metal-Schottky C–V	Metal-Schottky device	10^{14}–10^{16}	On product; final profile determination	Limited depth of penetration	±7%	Good in-line characterization technique
Hg-Schottky	Hg-Schottky with backside-substrate contact	10^{14}–10^{16}	Speed of measurement; simple sample preparation	Inaccuracy due to error in contact area measurement	±15%	Fast for p^- surfaces; n^- surfaces require thin oxide formation
Spreading resistance probe	Small-angle bevel	10^{14}–10^{20}	Dynamic dopant concentration range; accurate junction depth	Not reliable for profile determination in epitaxy	±15%	Good technique for failure analysis

A Resistivity Measurements

A high resistivity value of the epitaxial film can be detrimental to the performance of a metal Schottky device, and too low a resistivity value is undesirable due to high collector-to-base capacitance. For convenience, the resistivity value ρ is reported as a sheet resistance R_s through a relation

$$\rho = R_s h F, \qquad (3)$$

where h is the thickness of the film and F a geometrical correction factor. R_s is measured using a collinear four-point probe array, where the two outermost probes (1 and 2 in Fig. 5) carry a known constant amount of current, and a voltage difference is measured across the other probes (3 and 4 in Fig. 5). The geometrical factor depends on the ratio of film thickness to the probe separation distance.

B Dopant Profile Measurements

Accurate dopant profile measurements are essential where dopant distribution determines such device characteristics as capacitance and series resistance as a function of voltage, the reverse recovery and transition times, and the avalanche breakdown voltage.

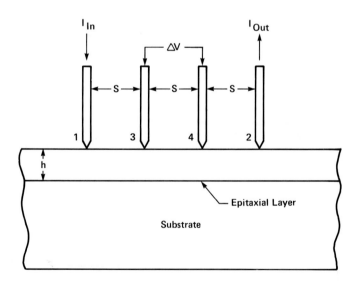

Fig. 5. A conventional four-point probe array. $R_s = \Delta V/I$.

Spreading Resistance Probe

The spreading resistance of a reproducibly formed point contact between the semiconductor surface and metal probe is measured.

If a flat circular voltage probe of radius a is brought into contact with a thin conducting epitaxial film of resistivity ρ the spreading resistance R_s of the contact is

$$R_s = n\rho_{cor} CF/4a, \qquad (4)$$

where n is number of current carrying probes, CF is the correction factor, and ρ_{cor} is the corrected resistivity. The value of CF depends on (i) relative resistivities and the type of epitaxial film and substrate and (ii) the radius h/a, where h is the epitaxial layer thickness. Values of CF are available in the literature [4].

When an in-depth resistivity profile is desired, the sample is angle lapped using the mechanical polishing technique. A set of known resistivity standards is measured to calibrate the spreading resistance probe (SRP). Variations of R_s down the bevel are transformed into the depth coordinate Z, after accurately measuring the bevel angle. From the $R_s(Z)$ function, $\rho(Z)$ and therefore $N(Z)$ are obtained from a known ρ versus N relation. Commercial tools are available for routine measurements. A software program is required for calculating the entire depth profile $N(Z)$.

Capacitance–Voltage (C–V) Technique

The capacitance–voltage (C–V) technique requires fabrication of a diode in the epitaxial film, the capacitance of which is determined as a function of voltage applied. The doping concentration N is determined from the slope of $1/C^2$ versus voltage V using the relation

$$N = \frac{2}{q\varepsilon A^2 \, d(1/C^2)/dV}, \qquad (5)$$

where q is charge; ε, dielectric constant; A, diode area; C, capacitance; and V, voltage. A depletion width W is given by

$$W = \varepsilon A/C. \qquad (6)$$

The depth at which the density is determined is given by

$$d = X_j + W, \qquad (7)$$

where X_j is the depth of the metallurgical junction.

The diode can be fabricated in the epitaxial film by either diffusion (junction diode) or metal evaporation and sintering (metal Schottky). The latter technique has the advantage that the junction is asymmetrical, and no corrections are required to Eq. (5). The metal Schottky diode can also be formed simply by bringing a mercury probe in contact with the epitaxial film that has a thin film of oxide (~50 Å).

3 Pulsed Capacitance–Voltage Technique

In this technique, a MOS device is reverse biased instead of a junction or Schottky diode. When a voltage pulse of a duration that is short compared with the time constant of thermal generation in silicon is applied to a MOS capacitor, no inversion layer can form. Using this technique one can penetrate into the silicon, and using the same equations as for junction diodes, one can calculate the dopant profiles.

The MOS device fabrication involves thermal oxidation (1000–3000 Å) followed by aluminum dots (100–200 μm) that are sintered at 450°C.

Figure 6 shows application of the metal Schottky technique for obtaining profiles in the epitaxial layers.

C Measurement of Electrically Active Impurities

Typical doping levels of epitaxial layers are in the range 10^{14} to 5×10^{16} atoms/cm^3. For direct measurements of impurities at such low levels, a number of analytical techniques listed in Table IV are employed. The NAA and SIMS techniques, though time-consuming, have the ability to profile epitaxial layers. Other techniques, when used in transmission, give

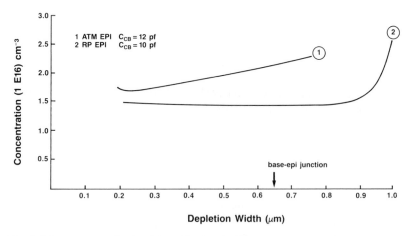

Fig. 6. Measurement of epitaxial profiles by platinum Schottky C–V technique. Reduction in C_{CB} by RP epi process is due to reduction in vertical autodoping.

TABLE IV
Analytical Techniques for Detecting Chemical Impurities

Technique	Sample preparation	Minimum detectable concentration (atoms/cm^3)	Accuracy	Comments
Neutron activation analysis (NAA)	Neutron irradiation; chemical sectioning	10^{16} for Sb	Poor at low concentration	Good depth precision; profile dependent on irradiation time; time-consuming technique
Secondary ion mass spectrometry (SIMS)	Relatively easy	10^{16} for Sb 10^{15} for B 5×10^{16} for As	Poor at low concentration	Sensitivity dependent on secondary ion efficiency
Fourier Transform Infrared (FTIR) spectroscopy	Relatively easy	10^{13}–10^{14} for B and P	Good	Needs liquid helium temperature for measurements; tool available commercially for transmission
Raman spectroscopy	Relatively easy	10^{16} for P 10^{17} for As	Moderate	Raman probe available commercially
Photoluminescence (PL)	Relatively easy	10^{13} for P and As	Good	Needs liquid helium temperature for measurements

a contribution from the substrate and, therefore, are limited in accuracy when a graded dopant profile is encountered. However, when used in the reflection mode, they can be accurate. In particular, the FTIR (or PL) technique is capable of detecting very low levels of dopants in silicon.

IV PHYSICAL AND OPTICAL CHARACTERIZATION

Epitaxial thickness measurement and control are required to determine device parameters such as breakdown voltage, capacitance, ac performance, transistor gain, switching speed, Schottky series resistance, and forward voltage. Table V lists and compares all the known techniques of measuring epitaxial thickness.

A Conventional Thickness Measurements

Techniques such as bevel and stain, SRP, and electrochemical junction decoration, although simple, have poor accuracy and precision. Therefore, such techniques are inadequate for measuring submicron films.

B Weighing Technique

The weighing technique is very simple and has high accuracy and precision. It assumes a uniformly thick film and calculates the thickness from the weight gained, divided by the product of area and the film density. Such a measurement correlates with the SRP thickness calculated from the peak concentration of lateral autodoping (Fig. 7). The difference between the two measurements depends on the epitaxial process used for the film deposition and it may be related to the delay in the dopant incorporation that normally occurs [5].

C Infrared Reflectance Technique

Electromagnetic radiation incident on the surface of the film is partially reflected from the film–air interface, transmitted through the film, and reflected from the substrate–film interface because of a change in optical constants between the two (for heavily doped substrates). This reflected radiation combines with the originally reflected radiation from the film–air

TABLE V
Epitaxial Thickness Measurement Techniques

Technique	Sample preparation	Advantage	Disadvantage	Accuracy
Bevel and stain	Diode, bevel at an angle followed by chemical staining	Fast; simple	Destructive; poor accuracy	±10%
Weighing	Weight of substrate required	Accurate; nondestructive; gives physical thickness	Average thickness; lacks spatial resolution	±1%
Spreading resistance probe (SRP)	Same as for SRP technique	Simple	Destructive	±10%
Electrochemical	n^-/n^+ junction decoration by preferential etch on a beveled sample	Measures device-related thickness	Destructive	±5%
Ellipsometry	n^-/n^+ or p^-/p^+	Nondestructive	Tool development required for submission layers	±5%
IR-reflection	n^-/n^+ or p^-/p^+	Nondestructive	Depends on substrate doping profile	±5%
(a) ASTM	n^-/n^+ or p^-/p^+	—	—	—
(b) Fourier transform infrared (FTIR) spectrometry		Nondestructive; very fast; product measurements with reproducibility ±1%		±1%

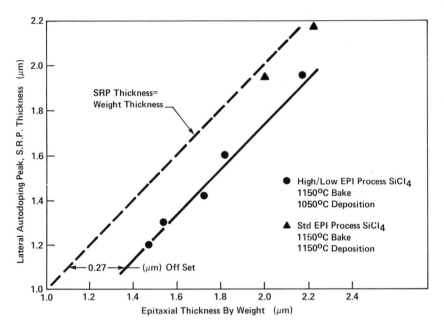

Fig. 7. Correlation of spreading resistance probe (SRP) measurement with the weight measurement. (From Kulkarni [1]. Reprinted with permission of Academic Press, New York.)

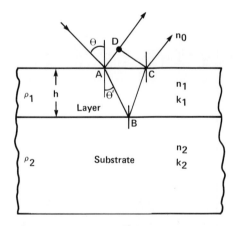

Fig. 8. Geometry of infrared interference method of measuring epitaxial layer thickness. (From Schumann [6]. Reprinted with permission of Academic Press, New York.)

18. Epitaxial Silicon: Material Characterization

interface (Fig. 8). The resulting interaction, called *interference*, is an intensity variation as a function of wavelength of radiation. Interference fringes and their periodicity in the spectrum are used to characterize film thickness.

According to the ASTM method [6] any order of reflection P (maximum or minimum) at a wavelength can be calculated from

$$P_2 = \frac{m\lambda_1}{\lambda_1 - \lambda_2}, \tag{8}$$

where $P_2 = P_1 + m$, $m = 0, 1, 2, 3, \ldots$, and the thickness is given by

$$h = \frac{m\lambda_1\lambda_2}{2(n_1^2 - \sin^2\theta)^{1/2}(\lambda_1 - \lambda_2)}\left[1 - \frac{\Phi_{21} - \Phi_{22}}{2\pi m}\right], \tag{9}$$

where Φ_{21} and Φ_{22} are phase corrections at the epitaxial substrate interface at wavelengths λ_1 and λ_2, respectively. The phase corrections depend on the doping level of the substrate and can be found in the literature.

To illustrate the point that infrared (IR) thickness depends on the doping level of the substrate and the out diffusion, a correlation between the weight thickness and the IR thickness has been established (Fig. 9). Deposition at higher temperatures (1150°C versus 1050°C) then reduces the IR thickness for the same weight thickness by about 0.05 μm.

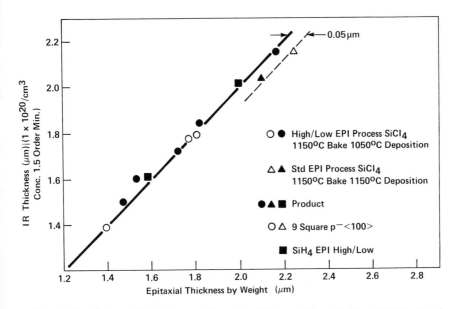

Fig. 9. Correlation of IR thickness with weight thickness. (From Kulkarni [1]. Reprinted with permission of Academic Press, New York.)

D Fourier Transform Infrared Spectrometry

Fourier transform infrared (FTIR) spectrometers have been proven far superior to grating or prism spectrometers. The advantages are lack of mechanical gears, throughput, high signal-to-noise ratio, accuracy, and speed. The technique uses a Michelson interferometer consisting of a fixed and a movable mirror, the motion of which is accurately calculated by a laser interferometer. The IR beam travels through the interferometer and then reflects in the same fashion as in Fig. 8. The movable mirror produces an interferogram, a variable amplitude plot as a function of the mirror position. This, when Fourier transformed allows one to separate the amplitudes per wavenumber, is called the *spectrum*. The calculation is done by a fast computer that plots the entire spectrum in a few seconds. The technique can be used as a regular ASTM method, once the spectrum is obtained; or from the interferogram alone one can calculate the thick-

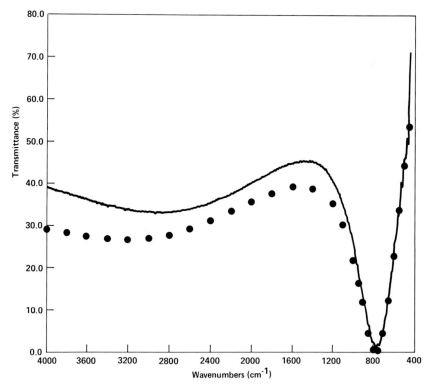

Fig. 10. Comparison of theoretical and measured reflectivity curves for a 0.5 μm epitaxial layer using FTIR spectrometry. (From Kulkarni [7]. Reprinted with permission of Academic Press, New York.)

18. Epitaxial Silicon: Material Characterization

ness after convolution, when substrate doping levels are taken into consideration.

Recently, a correlation technique has been developed using FTIR Spectrometry [7], in which theoretical reflectivity spectra have been generated for various thicknesses and different substrate doping levels using optical physics. The sample spectrum is correlated to the stored data and the best correlation gives accurate thickness. Figure 10 shows the comparison.

V. EPITAXIAL DEFECT CHARACTERIZATION

A. Electrical Properties of Defects

Defects in epitaxial films can result from improper growth processes or can be extensions of defects from substrates. Crystallographic defects are associated with electrical activity. For example, a 60-deg dislocation in a

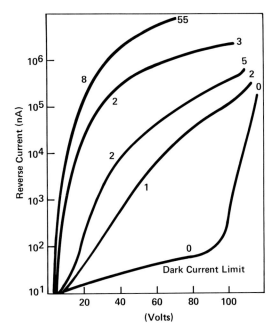

Fig. 11. I–V characteristics of five diodes illustrating the effects of electrically active stacking faults on excess reverse currents. The total number of faults within the perimeter of the diode is shown at the ends of the curves, and the total number of electrically active faults is indicated at the center. (From Ravi *et al.*, 1973. Reprinted with permission of Electrochemical Society, New Jersey.)

TABLE VI

Epitaxial Films: Substrate Induced Defects

Nature of defects	Probable cause	Remedy
Substrate		
Oxidation induced stacking faults in both epi and substrate	High oxygen or excessive precipitation due to long low temperature heat cycles	Substrate preparation for defect free zone; controlled precipitation
Stacking faults and dislocations	Incomplete annealing after implantation; improper cleaning	Proper annealing of damage layer prior to epitaxial growth
Hillocks	Improper cleaning of substrates	Improved cleaning procedures for substrates
Localized area containing dislocations or dislocations with stacking faults	Metal contamination; improper annealing of damaged layer after implantation	Clean processing free of metal contamination; improved annealing of damaged layer
Growth Process		
Stacking faults all over the wafers	Improper cleaning; moisture in reactor or gas lines	High-temperature bake in hydrogen; leak check system; replace quartzware and graphite parts
Localized area of dislocation network detected after W.J. etch	Metal contamination in the reactor or precleaning process	Change of reactor parts, gas line, etc.
Polysilicon islands	Residual oxide islands left on wafers; improper cleaning	Ensure complete oxide removal by 10:1 HF treatment
Orange peel surface or surface pitting	Improper cleaning or air leak	Improved cleaning; check leaks in the system
Mounds	Deposits from quartzware	Clean or replace quartzware

TABLE VII

Parameters Affecting Bipolar Device Yields

Parameter	Recently observed trends to improve yields
Wafer precleaning	(i) 150°C H_2SO_4 followed by 10:1 HF (ii) RCA cleaning
Substrates	Low dose of ion implantation
Epitaxial reactors	Vertical barrel reactors; radiant heating
Prebake	High temperature in hydrogen ambient at low pressures
Deposition	Temperature in mass controlled region; used of dichlorosilane at low pressure and growth rates

TABLE VIII

Epitaxial Defect Measurement Techniques

Technique	Sample preparation	Detects following	Advantages, disadvantages, and comments
Deep level transient spectroscopy (DLTS)	p/n junction or Schottky diode	Trap levels in band gap due to low level metallics or point defects or clusters	Below $10^{15}/cm^3$, indirect measurements without direct evidence; ideal for low level metallic contaminations
Optical microscopy	Etch sample in chemical solutions	Stacking faults dislocations and hillocks	Simple, with good resolution for failure analysis; destructive technique
X-ray topography	Thin wafer samples	Crystal defects, slip, clusters of point defects, precipitation in substrates	Wafer map available; long measurement time
Scanning electron microscope (SEM)	p/n junction or Schottky diode	Stacking fault or dislocation or slip	Surface technique with good resolution; destructive technique
Transmission electron microscopy (TEM)	Thin foil	All crystallographic defects; precipitates of metal	Thorough defect identification; requires large defect concentration in a local area; excellent for failure analysis
Scanned surface photovoltage [8]	Oxide on epi or epi on substrate	All crystallographic defects; slip lines; substrate defects	Nondestructive contactless surface technique with good resolution; interpretation of images requires confirmation by other techniques

semiconductor lattice distorts the energy bands because of deformation potential and adds bound states near the band edges.

Stacking faults, being bound by the dislocations, also have electrical activity; therefore, the reverse-biased characteristic of a diode is affected by their presence (Fig. 11). A leaky diode is therefore associated with crystallographic defects.

B Process Induced Defects

Epitaxial defects are induced by the growth process and substrate preparations are listed in Table VI, with the cause and remedy for each.

C Defect Prevention

Growing defect-free epitaxial films that enhance device yields is a difficult task. Some of the important process parameters are listed in Table VII.

VI EPITAXIAL DEFECT MEASUREMENTS

Epitaxial defects are detected by a number of techniques that are capable of identifying them by their electrical activity, morphology, dimensions, and the strain associated with them. The techniques are summarized in Table VIII.

The most widely used technique is optical microscopy using Nomarski interference contrast. The sample is chemically etched by any one of the etchants listed in Table IX. Among nondestructive techniques, scanned

TABLE IX

Chemical Etches for Epitaxial Silicon

Etch	Formula	Comments
Sirtl and Adler	1 part conc. HF, 1 part CrO_3 (5 M), 500 g/L solution	3.5 μm/min etch rate, good for {111}, poor on {100}, faceted pits
Secco–D'Aragona	2 parts conc HF, 1 part $K_2Cr_2O_7$ (0.15 M), 44 g/L solution	1.5 μm/min etch rate; ultrasonic agitation; noncrystallographic pit; typical dislocation pit appears elliptical in shape
Wright Jenkins	2 parts conc. HF, 2 parts conc. CH_3COOH, 1 part conc. HNO_3, 1 part CrO_3 (4 M), 400 g/L solution, 2 $Cu(NO_3)_2 3H_2O$ (0.14 M), 33 g/L solution	1.7 μm/min etch rate, good on all orientations, faceted pit; typical dislocation pit on {100} surface appears like a kite figure with four sides, stacking faults on {100} show square pits

18. Epitaxial Silicon: Material Characterization

surface photovoltage [8] has the advantage that the surface examination can be limited to part of the epitaxial film, with good spatial resolution.

REFERENCES

1. S. B. Kulkarni, *in* "VLSI Electronics, Microstructure Science" (N. Einspruch, ed.), pp. 73–145. Academic Press, New York, 1982.
2. S. P. Keller, "Handbook on Semiconductors, vol. 3, Materials, Properties and Preparation." North Holland, Amsterdam, 1980.
3. H. Kühne, *Crystal Res. Technol.* **17,** 1097 (1982).
4. D. M. Dickey and J. R. Ehrstein, *National Bureau of Standards Special Publication 400-48* (1979).
5. R. Reif and R. W. Dutton, *J. Electrochem. Soc.* **128,** 909 (1981).
6. Annual Book of ASTM Standards, vol. 10.05, F95-76, pp. 207–214. American Society for Testing and Materials, Philadelphia, Pennsylvania, 1983.
7. S. B. Kulkarni, "Fournier Transform Infrared Spectroscopy Application Brief." IBM Instruments Inc., Danbury, Connecticut, 1983.
8. T. H. Distefano, *in* "Nondestructive Evaluation of Semiconductor Materials and Devices" (J. N. Zemel, ed.), pp. 457–514. Plenum, Press, New York, 1979.

Chapter 19
Resist Technology in VLSI Device Processing

RONALD C. BRACKEN
Solid State Technology Center
RCA Corporation
Somerville, New Jersey

SYED A. RIZVI
United Technologies
Mostek
Carrollton, Texas

I. Introduction	328
II. Optical Patterning	328
A. Negative Resists	331
B. Positive Resists	333
C. Deep Ultraviolet Radiation (DUV) Resists	338
III. Multilevel Resists	339
A. Reason for Developing Multilevel Resists	339
B. Bilevel Resist Systems	341
C. Trilevel Resist Processes	343
IV. Electron Beam Patterning	343
A. Negative E-Beam Resists	344
B. Positive E-Beam Resists	345
C. Proximity Effect in E-Beam Lithography	346
D. Multilevel E-Beam Resists	347
V. X-Ray Resists	347
A. Resist System	347
B. Plasma Developable Resists	349
C. Inorganic Resists	349
VI. Conclusions	349
References	350

I INTRODUCTION

The purpose of this chapter is to acquaint the engineer with some of the principles of microlithographic patterning and to discuss how these principles relate to the currently perceived design rules of VLSI capabilities.

Table I shows the authors' estimate of current practice in VLSI patterning. Several trends are apparent. As the design rules decrease, the use of positive resist becomes prominent and then gives way to multilevel resist systems that usually involve positive resist. Also, as design rules decrease, patterning equipment changes. In our discussion we shall examine the principles involved in this evolution and why it has occurred as it has.

II OPTICAL PATTERNING

There are three components to the process of wafer patterning: the resist chemistry, the patterning equipment, and the wafers themselves. A few optical definitions will be needed before we can proceed.

TABLE I

Estimate of Current Practice in VLSI Patterning

Design rule (μm)	D_c (μm)[a]	Environ[b]	Mask D_o/M[c]	Resist	Patterning	Die area (Kmil²)	D_T (d/in²)[d]
≥7.0	3.0	10 K + VHF	4.0/SL	Neg	Proximity	≤25	160
≥5.0	2.0	10 K + VHF	4.0/LE	Neg	Contact	≤25	250
≥5.0	2.0	10 K + VHF	1.0/LE	Neg	Contact	≤25	60
≥2.5	2.0	100	1.0/Q	Neg/Pos	Projected	<25	100
≥2.5	1.0	100 + VHF	1.0/Q	Pos	Projected	≤50	50
≥1.25	0.5	10	—	Pos	Stepper	≤100	40
≥0.5	0.25	"1"	—	Multi	E-beam[e]	≤200	40
≥0.5	0.25	10	—	Multi	X-ray	≤200	40

[a] D_c is the linear dimension of a critical defect or the size setting for defect scanner.

[b] Environ is the environmental quality that is appropriate for these design rules.

[c] Mask Do is the accepted criteria for the mask in defective dies per in.² M is mask material. SL soda lime. LE = Borosilicate. Q = quartz.

[d] D_T is the total effective defect density that might be experienced for a die of the size shown that produced in a wafer fabrication area under effective particulate control. D_T is derived using $Y = 1/(1 D_T A)$ where Y = probe yield/100, and A is die area.

[e] For the looser design rule circuits, multilevel resist are neither required nor used.

19. Resist Technology in VLSI Device Processing

Contrast. This quantity relates the radiation in the area of intended exposure I_1 to that in the area of intended unexposure I_0.

$$\text{contrast} = (I_1 - I_0)/(I_1 + I_0)$$

Modulation. This quantity describes how image contrast changes in its transfer from masks (image) to wafer (object)

$$M = C_{\text{image}}/C_{\text{object}}$$

Modulation Transfer Function. This quantity describes how the modulation varies as a function of the dimensions of the pattern that is being transferred from the mask to the wafer. For contact printing (shadow casting) under the conditions of perfect wafer–mask intimacy, the modulation transfer function (MTF) is independent of the dimension and is unity. With proximity printing, there is a diffraction of light into the area of intended unexposure, so the MTF is less than unity. For other systems of optical transfer such as lenses or mirrors, the MTF is more complicated. In the usual case, the smallest useful resolution of a lens is the point at which 60% contrast is obtained. For a diffraction limited lens, the MTF is related to the Rayleigh limit. Figure 1 shows the MTF curve for an optical lens, the 10-77-82 Zeiss lens, used in many of the commercial step and repeat cameras [1].

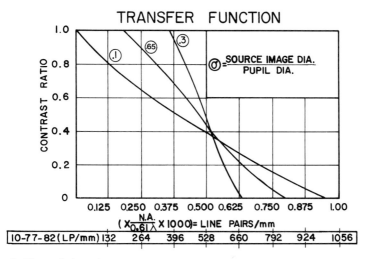

Fig. 1. The variation of contrast or modulation as a function of line-space density per millimeter. The influence of partial coherence of the illumination is seen in the shape of the MTF curve. At $\sigma = 1$ is effective incoherence. At $\sigma = 0$ is complete coherence and a step function cutoff at 0.5 normalized frequency occurs. (From Roussel [1].)

Rayleigh Limit. This quantity is the minimum distance of separation at which two objects that are imaged by an optical system can be said to be distinct.

$$D = 0.61\lambda/NA,$$

where λ = illuminating radiation, and NA is the numerical aperture of the lens. Usually three times the Rayleigh limit is used as a practical resolution.

Numerical Aperture. This quantity is a figure of merit for an optical system.

$$NA = n \sin(\theta/2),$$

where n = index of refraction of the medium and $\theta/2$ is the half-angle of the cone described by the lens and its focus.

Edge Gradient. This quantity describes the variation of radiation intensity with distance in the direction normal to the edge of a geometry. For shadow casting, this gradient is steep; for imaging at a modulation of 60%, this gradient is less steep.

Gamma (or contrast). This quantity measures the response of a photosensitive material, such as a photographic film, to incident radiation.

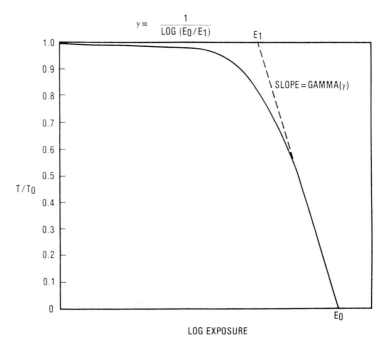

Fig. 2. The variation of positive resist film thickness as a function of exposure energy. The contrast (or gamma) and sensitivity E_0 are the quantities indicated on the figure.

Mathematically, gamma is the slope of the optical density versus exposure energy. By analog, for photoresist, the optical density becomes the resist thickness after exposure and development.

Sensitivity. This quantity measures the exposure needed to achieve either 50% film thickness development for negative resists or complete removal for positive resists. Both the gamma and the sensitivity of a resist depend on the specific application. Such parameters as developer concentration, developer temperature, and the type of substrate being exposed can have a large effect on the sensitivity curve. A typical sensitivity curve for positive resist is shown in Fig. 2.

Standing Wave Effect. This quantity refers to a periodic variation of light intensity in the photoresist film. These variations arise from interference between incident and reflected light. The reflection occurs at an interface at which the index of refraction has a large variation. The period of the variation is $\lambda/2n$. The index of refraction of SiO_2 is so near that of resist that these two films behave optically as a single film. In that case, the reflection of significance occurs at the SiO_2–Si interface.

Negative Resists A

Negative resists form patterns by protecting the wafer areas exposed to UV radiation. They reverse the tone of the mask and yield the negative of the mask pattern, hence their name.

Polymer: *cis*-1,4-polyisoprene (synthetic rubber)
Solvent: xylene, toluene
Activator: aromatic *bis*-diazides (2–4% concentration)
Exposure wavelengths: 365 nm
Developers: organic solvents (xylene, toluene, Stoddard solvent—proprietary brands)
Rinse: *n*-butyl acetate, alcohol solutions
Soft bake: 65°–85°C (type dependent)
Hard bake: 130°–170°C (type dependent)
Vendors: Kodak, Hunt, Tokyo Okha
Comments: Oxygen sensitivity tends to limit most of the applications to contact or proximity printing; negative resist is less expensive than positive resist; it has excellent cleanliness and shelf life but is sensitive to low temperature; do not expose bottles to temperatures around 32°F or below.

Adhesion to clean dry oxides is excellent. PSG oxides need adhesion promoters if more than 60 min elapse between furnace and resist application. Developer tends to swell resist, causing pattern distortion and webbing in small geometries.

Almost all the commonly used negative resists are based on xylene solutions of *cis*-1,4 polyisoprene, which is a synthetic rubber material. This polymer, as used, is of the order of 2×10^6 in molecular weight. The end-use properties of the polymer depend on the degree of cyclization and unsaturation that have been induced into the linear polyisoprene starting material. The unsaturation in the linear polymer adds beneficial adhesion properties but poor thermal characteristics (low m.p.). Unsaturation can be decreased by internal cyclization, which increases the polymer thermal flow point but decreases adhesion. The commercial product is an optimization of these properties.

The source of *UV illumination* for nearly all exposure systems is a mercury arc lamp which has a discrete emission spectrum. The most commonly used spectral wavelengths are 300, 340, 365, 405, and 436 nm.

The basis for all photoresists is a *photolytic reaction* in the film that changes the solubility of the exposed material. In the case of negative resists, the exposed material becomes less soluble than the unexposed material.

The polyisoprene itself is transparent to UV above 300 nm. There is an *activator* compound dissolved in the polymer solution that absorbs UV above 300 nm. This activator is usually an aromatic *bis*-diazide. Typically, 2–4% of such a compound is used in the negative resist. These compounds are capable of a photoreaction when illuminated with 360-nm wavelength UV radiation.

$$N_3\text{—R—}N_3 \xrightarrow{UV} N_3\text{—R—N}\cdot + N_2 \uparrow$$

The reaction product is a *nitrogen free-radical* that can extract hydrogen from the polyisoprene and initiate a cross-linking reaction.

$$R\text{—N}\cdot + H\text{—C—} \longrightarrow R\text{—N—H} + C\cdot$$

The *carbon free-radical* can then cross-link with the other polyisoprene molecules. As the molecular weight is increased by cross-linking, the polymer solubility drops, and a protective film is formed.

An undesirable side reaction is the *oxidation* of the nitro radical, which is favored by a factor of about 10^4; so the presence of oxygen during the UV exposure is unfavorable to the polymer cross-linking. The effect of oxygen is to decrease the differential solubility and so increase the tendency for pinholes to form in the film. This effect also increases the critical dimension (CD) variation across the wafer. The usual way to avoid oxygen exposure is to assure a tight seal to the vacuum chuck that clamps the wafer to the mask and to blanket the wafer–chuck periphery with nitrogen. This oxygen effect places some severe limitations on the use of negative resist in noncontact printing applications [2].

Another limitation on the application of negative resists derives from their very nature. Low levels of light can initiate some *polymerization*, which gives rise to a thin but partially protective film. Such situations arise when projection printing is used for pattern definition. The contrast between exposed and unexposed regions is incomplete. To compensate for this problem, the overall exposure must be reduced and the developer made less vigorous to preserve the film in the areas of exposure.

In addition, small geometries which have decreased contrast due to MTF become increasingly difficult to pattern. The protective properties of negative resist films begin to decrease as small-geometry resolution is attempted. The widely accepted rule is that negative resist does not serve effectively for projection patterning much below 5.0 µm resolution and that the aspect ratio (the ratio of the width of the pattern being resolved to the resist thickness) is about 3 for negative resists.

Positive Resists B

Positive resists are completely different chemical systems from negative resists. They become soluble when exposed to UV radiation. The protective film is that which remains after the exposed portions of the film are dissolved. These resists preserve the tone of the mask. They form a positive image of the mask; hence they are called positive photoresists. Some commercially available resists are listed in Table II.

Polymer: Novolac resin
Solvent: methyl cellosolve acetate
Activator (inhibitor): diazoquinones (30–40%)
Exposure wavelengths: 300–436 nm
Developers: aqueous alkaline solutions
Rinse: water
Soft bake: 90°–105°C for 30 min
Hard bake: 125°–180°C for 30 min (type dependent)
Vendors: Shipley, Hunt, Azoplate, (American Hoechst) Kodak, Allied
Comments: Excellent adhesion to metals; adhesion promoters are recommended with oxides, nitrides, polysilicon, and polycides; store at cool temperature, preferably below 60°F; decomposition begins above 85°F; bottles can explode if N_2 pressure becomes excessive

Positive resists are based on *Novolac resins,* which are the condensation polymers of substituted phenols and formaldehyde. As phenolic materials, they are soluble in alkali solutions. By themselves, these materials are neither photosensitive nor particularly effective as etch resists.

Both photosensitivity and etch-resistance characteristics are established by the addition of a *photosensitive material, substituted diazo-*

TABLE II
Commercially Available Resists

Vendor	Solids range available (%)	T_S (°C)[a]	T_H (°C)[b,c]	Comments
American Hoechst Somerville, New Jersey				
AZ 4000	17–42	85	120	Available in a range of solids content; said to be able to withstand T_H 180°C without flow; useful in wet and plasma etch applications; striation free
AZ 1300 AZ 1300 SF	17–37	85	120	Older product, very widely used; formerly AZ1350; was associated with Shipley product; striation free
AZ 2400	17–37	85	120	Higher speed than 1300 resist; becoming used in E-beam photoplate manufacture
Shipley Co. Newton, Massachusetts				
Micro Posit 111	—	90	90	Thick film applications
Micro Posit 1300	15–35	90	120	Very widely used resist in mask making, flow, $T \sim 130°C$
Micro Posit 1400	15–35	90	120	Very similar to 1300 series; striation free
Micro Posit 2400	15–35	90	100	Similar comments to AZ 2400 resist above
Micro Posit 23	(NA)	90	100	Higher temperature stable resist; useful for plasma applications
Philip A. Hunt Chemical Co. Pallisades Park, New Jersey				
Waycoat MPR	20	90	115	Designed for a mask application
Waycoat HPR				
204	28	100	125	T_H can be as high as 180°C to improve adhesion
206	33			
WX118	NA	—	—	Stepper resist; high resolution
WX159	NA			DUV resist 240–250 nm
Kodak 809	NA	100	120	Good plasma etch applications; can withstand T_H 150°C; Striation free
Kodak 820	NA	100	130	
Dynachem Corp Santa Anna, California Tokyo Okha, Kawasaki, Japan				
OFPR 800	NA	90	130–140	High T stable resist; plasma applications stable to 180°C

[a] T_S = Temperature of the soft bake, or pre-exposure bake
[b] T_H = Temperature of the hard bake, or post-exposure bake
[c] In many applications a hard bake is unnecessary for positive resists. In mask making the post exposure bake is seldom used, as it contributes potential contamination and confers little benefit.

quinones. These compounds decompose on exposure to UV radiation. These materials are hydrophobic, and their presence decreases the Novolac resin's solubility in alkali solutions.

These resists absorb broadly in the region of 300 to 436 nm. The *reaction products* are nitrogen gas and a ketene. If this radiation occurs in a nondehydrated film, the ketene takes up water to become a carboxylic acid. By this conversion, the solubility of the phenolic film is restored, and the solubility-inhibiting diazoquinone becomes very alkali soluble. The result is that the exposed region becomes much more soluble than the unexposed region.

The diazoquinone (inhibitor) can also decompose thermally. At higher temperatures, water is absent from the film, and the ketene can condense with the phenolic hydroxide of the Novolic resin. The ester that is produced is alkali insoluble, so the effect of the reaction is to decrease the sensitivity of the resist [3]. At temperatures above 110°C, development becomes difficult; after a bake of 30 min at 130°, the film is almost completely insoluble. This thermal reaction makes the *pre-exposure bake* of the positive resist very critical to the exposure characteristics of the resist film. The principal characteristic that is affected by the bake is the dimensional control (CD). It is essential that this bake be uniform over each wafer, between wafers within a lot, and from one wafer lot to another.

The chemical reactions involved in positive resist processing admit of *no oxygen effect*. This absence is a great advantage when projection printing is attempted. In projection printing, the radiation gradient at the pattern edge is not steep, and there is some radiation in the region of intended unexposure. This partial exposure causes scumming in the clear areas when negative resist is used. With positive resist, this partial exposure works from the opposite end of the exposure–thickness curve; it causes some thinning of the resist, but no scumming.

Positive Resist Developers

The developers for positive resists either contain metallic ions, usually sodium or potassium, or are classified as metal-ion free (MF). The latter developers primarily contain tetramethyl ammonium hydroxide. In Table III are the results obtained by Hinsberg relating to developer composition [4].

The metal-ion-free developers are becoming ever more widely accepted. A primary reason is cleanliness. The silicate or borate salts of the metal-ion developers cause particulate contamination around the processing equipment when residue dries. In addition, the sodium or potassium can contaminate oxide films with mobile ions. The MF developers are usually less selective than the metal-ion developers, which tends to make

TABLE III

Compositions of Commonly Used Positive Resist Developers

Developer	Composition	Molar concentration	Reference
AZ developer	Sodium metasilicate	0.22	[4]
	Sodium phosphate	0.12	
AZ 351	Sodium borate	0.95	[4]
	Sodium hydroxide	0.44	
AZ 303A	Sodium hydroxide	1.7	[4]
	Allylaryl sulfonate	0.08	
AZ 2401	Potassium hydroxide surfactant	1.15	[4]
Hunt LSF metal free	TMAH[a]	—	[5]
Shipley MF312	Choline[b]	—	[5]

[a] TMAH = tetramethyl ammonium hydroxide
[b] Choline = N-(2 hydroxyethyl)-N,N,N trimethyl ammonium hydroxide

them less forgiving of processing variations. The MF developers become more active as the temperature decreases.

Table IV shows a comparison of the activities of various developers at different temperatures and concentrations. Note the variation of selectivity with temperature [5]. These values can be taken as the relative activities of the various developers when used with the resists in the situations described. The really significant trends are the following.

(1) Metal-ion developers increase the rate of resist development with temperature but become less selective.

(2) Metal-ion-free developers decrease the rate of resist development with increasing temperature and become more selective. At room temperature there is little difference in selectivity between metal and MF developers at the concentrations in Table IV.

(3) Diluted developers show enhanced selectivity over undiluted developers. Most developers are diluted in their applications.

(4) The impact of spray on diluted developers is to enhance the development rate.

2 Puddle Develop

In this application, developer is dispensed onto a slowly rotating wafer. The dispense cycle ceases; then the wafer is allowed to develop, is rinsed, and is spun dry. This application has the advantages of economy, because less developer is used, and of processing latitude, because the physical details of the nozzle become unimportant to the develop cycle.

TABLE IV
Effect of Temperature on Resist–Developer Systems[a]

Developer	Resist	10°C			15°C			20°C			25°C			Developer type	Reference
		U	E	R	U	E	R	U	E	R	U	E	R		
Hunt LST MF	Hunt 204	1350	8084	6.0	783	6403	8.2	603	5394	9.0	361	3649	10.1	TMAH	[5]
Hunt LST Type 2	Hunt 204	164	2213	13.5	310	3150	10.2	601	5378	8.9	1162	7120	6.1	NaOH	[5]
20% LST Type 2	Hunt 204	—	—	—	101	4892	48.0	248	7545	30.0	313	7716	24.0	SPRAY	[5]
Shipley 350	1350J	93	1694	18.2	130	2005	15.4	362	3878	10.7	652	4594	7.0	NaOH	[5]
Shipley 351	1350J	160	3189	19.9	291	4333	14.9	429	5268	12.3	782	7771	9.9	NaOH	[5]
Shipley MF312	1350J	1460	8750	6.0	763	6626	8.9	512	5018	9.8	303	3728	12.3	Choline	[5]
40% MF312	1350J	—	—	—	378	7560	20.0	220	7008	32.0	102	4452	43.0	SPRAY	[5]
Shipley MF314	1350J	991	9884	10.0	757	8260	10.9	461	6117	13.3	340	4892	14.4	TMAH	[5]
OFPR	OFPR	267	6821	25.5	471	9896	21.0	1038	13982	13.5	—	—	—	Choline	[5]
Kodak LMI	K809	49	890	18.2	110	1792	16.3	251	3011	12.0	491	4965	10.1	Choline	[5]
Shipley AZ 351 50%	1350J	—	—	—	—	—	—	—	—	343.0	—	—	—	Na Salts	[6]
Shipley AZ 351 50%	1350J	—	—	—	—	—	—	—	—	27.5	—	—	—	NaOH	[6]
Shipley AZ 351 25%	1350J	—	—	—	—	—	—	—	—	250.0	—	—	—	NaOH	[6]

[a] Wafers: 600 nm oxide HMDS (100%); immersion developed, except where noted, after a flood exposure.
Resist: 1.5 μm thickness; 4 min. softbake GCA IR at 100°C.
Develop time: 20 s.
Entries: resist thickness loss in Å for unexposed resist (U) and exposed resist (E); R = E/U.

C Deep Ultraviolet Radiation (DUV) Resists

1 Advantages

One way to improve resolution without using multilevel resist systems is simply to use shorter wavelengths: The diffraction limited resolution is $d = 0.61\lambda/\text{NA}$. Decreasing the wavelength from 400 to 240 nm proportionately improves resolution. This approach has an inherent advantage over simply changing optics. Increasing the NA similarly improves resolution, but the depth of focus is $FD = (\lambda/2)/(\text{NA})^2$. So depth of focus is lost as NA increases. Decreasing the depth of focus makes the system increasingly sensitive to wafer and mask flatness, whereas changing the wavelength does not.

2 Problems

The problems with use of DUV resists relate to having a system corrected for DUV, providing DUV radiation sources, and providing DUV resists.

3 DUV Systems

Proximity and *contact* systems are easily adapted to DUV. The *reflective* systems (e.g., the PE-300) have also been adapted to DUV. However, the *refractive* systems, such as the step and repeat reduction cameras, typically absorb strongly in the 200–240 nm region and are not adaptable to DUV exposure.

Hg–Xe sources are rich in DUV and are acceptable sources. Kameko et al. [7] have used a Canon PLA 520 F proximity printer with a Xe–Hg source at 250 nm. They report resolving 1.5 μm line-space pairs in PMMA and PMIPK.

The DUV resists that are used are the two mentioned above and those in the following tabulation:

Resist	Thickness	Resolution	Reference
PMMA	2.0 μm	1.5 μm	[7]
PMMA	1.8 μm	0.5 μm	[8]
Hunt WX 159	1.5 μm	1.0–5 μm	[9]
AZ 2400		1.5 μm	

III MULTILEVEL RESISTS

A Reason for Developing Multilevel Resists

As progressively finer geometries are attempted, the positive resists increasingly become the material of choice, because of the absence of an oxygen effect and the sensitivity characteristics of the resist.

Finer geometries force the use of thinner resists. Typically, a minimum aspect ratio of one is observed. The fine dimensions carry a problem with them, however. Due to a capillary effect, the variation of resist thickness over a step and into an etched pattern gets greater as the pattern size gets less. The work of White [10] illustrates this tendency. We see this tendency illustrated in Tables V and VI, in which the step height in the polymer is shown as a polysilicon step is traversed. Different polymers give different degrees of step leveling.

This variation means that the dimensions of any pattern being imaged in this resist will vary as the step geometry is traversed. Variations are shown in Figs. 3 and 4, in which simulated resist profiles have been calculated for the resist thicknesses encountered at a step [11]. It has been found that the dimensional change is approximately linear with the thickness change, and that it can range from 40 to 80% of the thickness change for mild developers and harsh developers, respectively [12].

Therefore, though patterns can be imaged and resist structures formed

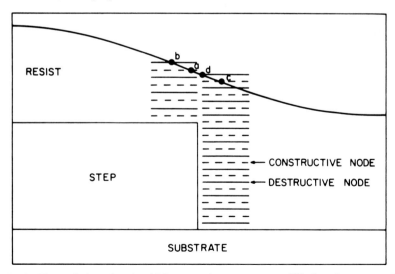

Fig. 3. The variation of resist thickness as it passes over a diffusion step on a wafer. (From Neureuther *et al.* [11].)

TABLE V
Step Height in Resist or Polymer Measured over Polysilicon Step of 1.0 μm Height

Resist	Polymer step height (μm)	Polymer thickness change (%)
HPR 204	0.6	40
Kodak 809 (32)	0.6	40
KTI PMMA (496 K)	0.8	20
Polyimide	0.8	20

TABLE VI
Step Height in Resist Measured over a 1.0 μm High Polysilicon Step of Varying Widths (HPR 206 Resist)

Dimension (width) (μm)	Step height of polymer (μm)
9	0.50
5	0.40
3	0.35

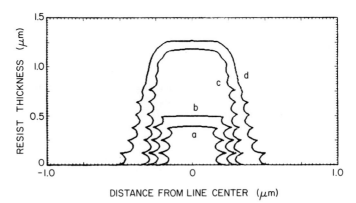

Fig. 4. Simulated resist profile corresponding to the thickness a–d in Fig. 3. (From Neureuther et al. [11].)

with very fine geometries, their usefulness is limited by lack of dimensional control on the wafer. To solve this problem, as it relates to geometries of micrometer and submicron dimensions, is the reason that multilevel resist systems were developed.

Bilevel Resist Systems† B

The resist strategy for bilevel resist systems (portable conformable masks, PCM) is illustrated in Fig. 5.

(1) Composed of two layers: the bottom layer is a planarizing layer; the top layer is a patterning layer and is a resist
(2) Lower level resist (LLR): usually PMMA
(3) Top level resist (TLR): positive resist
(4) Process sequence:
 Spin LLR and cure
 Spin TLR and soft bake
 Pattern TLR; conventional positive resist development
 Expose DUV (210–260 µm) flood exposure
 Develop LLR

The usual developers are

2:1 Acetone/2-propanol,
Chlorobenzene/xylene,
MEK/MIBK.

All of the above mixtures have been used as developers for the LLR. PMMA has been the most popular planarizing film for bilevel resist systems. Since the PMMA is transparent to 436-nm radiation, there is little attenuation of light reflected from the substrate. This reflected light can expose the TLR and cause unacceptable CD variation. The cure appears to be the addition of a dye to the LLR. The improvement in CD control obtained from using a bilevel resist system is illustrated in Table VII.

Some resists have been found to react with the PMMA. Ting *et al.* have found an interlayer film formed between AZ and PMMA [14]. When this film forms, it is a serious impediment to bilevel processing. Others [12] have found no film formation between K809 and PMMA.

Bilevel resists have some applications. This resist strategy fits into a processing line with a minimum of perturbation and is a preferred means of addressing CD control at small dimensions, though it is occasionally inadequate.

† See reference [13].

TABLE VII

Change in Line Width in a 2.0 μm Line Crossing a 1.0 μm Step for Various Step Spacings[a]

	Single level (μm)	Bilevel[b] (μm)
3 × 3 μm line space topology		
Kodak 809 (1.4 μm)		
Top	1.5	2.0
Bottom	2.3	2.1
10 × 10 μm line space topology		
Top	1.5	1.8
Bottom	2.2	2.2

[a] From White [10].
[b] 1 μm PMMA as LLR.

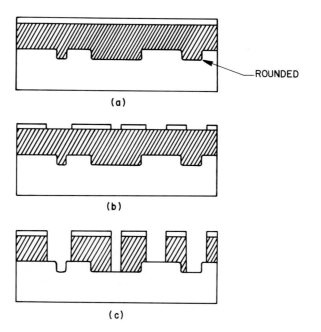

Fig. 5. The bilevel resist procedures for planarizing wafer topology. (a) The first level (LLR) planarizes the wafer features. (b) The TLL is patterned. (c) The lower level is patterned using the TLR as a PCM.

C. Trilevel Resist Processes

There are several reasons for going to a trilevel resist process (optical processing).

(1) PMMA gives poorer topology leveling than do resists. If resists are used, there needs to be a way to separate the TLR from the LLR.
(2) A bake can be used to enhance planarization of a LLR which renders it undevelopable.
(3) The presence of an interlevel film can render CD control on TLR unacceptable. In these and other cases, trilevel resist systems are used.

The most common trilevel resist processes are the following.

(1) Spin LLR: PMMA, resist, polyimide
(2) Cure LLR
(3) Form hard mask (interlevel material): Cr, Al, SiO_2 of 0.3 μm thickness
(4) Spin TLR: positive optical resist
(5) Pattern TLR
(6) Pattern hard mask: plasma CF_4 or wet etch
(7) Pattern LLR: reactive ion etching (RIE) or O_2 plasma

Though these processes have the advantages of processing certainty, they introduce a degree of complexity and expense that are clearly more severe than those encountered in bilevel resists. These disadvantages are justified by the increased processing stability and (hopefully) a lower CD variance.

The use of any of these systems broadens the application of optical lithography and extends the life of the optical patterning equipment. Ultimately, the limit on resolution is the optics. When one wishes to progress beyond these limits, optical patterning must be abandoned.

IV. ELECTRON BEAM PATTERNING

Analogous to the optical resists, the e-beam resists are negative and positive acting. Unlike the optical resists, however, these resists do not have an activator or inhibitor. The electron energy is sufficient to initiate the reaction that results in differential solubility. The negative resists increase their chain length and become less soluble; the positive resist do the opposite. Table VIII lists some of the available e-beam resists. The most commonly used are PBS and COP, which are the original Bell resist materials.

TABLE VIII

Electron Beam Resist Materials

Material	Tone	Sensitivity at 20 kV ($\mu C/cm^2$)	Contrast	Resolution (μm)	Usual thickness (μm)	Comments
PMMA	+	40–80	2–3	0.1	1.0	KTI, not widely used
COP	−	0.2–0.4	0.9–1.2	1.0	0.3	Bell license, widely used, higher defect density than PBS
Waycoat Neg. e-beam	−	0.3	—	1.0	—	Hunt Chemical Corp.
PBS	+	0.8–1.6	1.3–2	0.5	0.4	Mead, Bell license widely used, sensitive to humidity
RE4000-N	−	0.2	1.5	—	—	Hitachi
RE5000-P	+	1.2	1.0	0.5	—	
OEBR 100	−	0.8–1.0	1.6	0.5	0.3	These numbers on OEBR resists are relative numbers and may not be comparable to those for PBS Tokyo Okha
OEBR 1000	+	50	1.6	0.3	0.3	
OEBR 1010	+	60	0.1	0.3	0.3	
OEBR 1030	+	3.0	0.4	0.3	0.3	
CMS Chloromethylated	−	0.3–0.6	1.4	0.5	—	Tokyo Soda excellent dry etch durability;
EXSS	−	0.2	1.2	0.5	—	no post-expose cure required; better resolution than CMS
FBM 110	+	1.0–2.0	5.0	—	—	DAKIN
120	+	0.3–0.5	2.4	—	—	
SEL-N	−	0.3	1.3	0.5	—	Poor adhesion to Cr; Somaru Industries
EBR 1	+	—	—	—	—	TORAY
EBR 9	+	5	—	—	—	TORAY
AZ2415	+	25	0.6–1.8	—	—	Shipley
AZ1450	+	26–45	1.5–3.0	—	—	Shipley

A Negative E-Beam Resists

Negative e-beam resists become less soluble as the electron dose increases. Customarily, the sensitivity of negative resists is defined as the exposure at which one half the initial thickness remains after a development cycle. The insolubility is related to an increase in molecular weight caused by the electron exposure.

19. Resist Technology in VLSI Device Processing

In some resists, such as COP, SEL-N, and OBER-100, this increase in molecular weight is caused by a *chain reaction*. One electron capture leads to the linkage of several polymer molecules. This means that these resists are very sensitive but have poor contrast and low resolution. They also must be vacuum cured to prevent oxygen quenching of the chain-propagation reactions; a rule of thumb is to vacuum cure for a time roughly equal to the writing time. These chain formers are usually epoxy type compounds.

COP: copolymer of glycidyl methacrylate and ethyl acrylate
OBER-100: polyglycidal methacrylate (PGMA)
SEL-N: PGMA hydrolized with maleic acid methyl ester

Other negative resists, which usually contain halide groups, are stepwise polymers in which the electron capture can cause *cross-linking* between two molecules but no chain reaction. The halides tend to quench the free-radical propagation involved in the chain reaction. The newer resists, such as CMS, RE-4000N, and GMC-1, are such polymers. These tend to have characteristics opposite to those of the epoxy resists [15] and are often based on polystyrene.

CMS: chloromethylated polystyrene
RE-4000-N: iodinated polystyrene
GMC-1: copolymer of glycidyl methylacrylate and 3-chlorostyrene

An advantage of these resists is that the vacuum cure is not necessary. This modification is aimed at improving the resolution and contrast of these resists.

Negative resist developers are tabulated below.

Resist	Developer	Rinse I	Rinse II
Epoxy-based			
COP	MEK : ETOH (3 : 1)	MIBK : IPA (1 : 3)	IPA
OBER 100	MEK : ETOH (7 : 1)	MIBK	IPA
SEL-N	MEK : ETOH (1 : 1)	IPA	IPA
Styrene-based			
CMS	iso-amylacetate : cellosolve acetate (2 : 3)	IPA	—
RE-4000-N	P. Dioxane : cellosolve acetate (4 : 3)	IPA	—

Positive E-Beam Resists B

The positive resists are materials that degrade upon electron capture. A well-designed resist fragments to form a distribution of molecular weights that is a completely different from that of the original polymer. The better the separation and the lower the molecular weights of the product, the better the resist contrast and selectivity. Some of these positive resists are based on sulfones, some are modified PMMA materials, and some are Novolac-based materials.

PMMA: poly-methyl methacrylate
EBR-9: poly-2,2,2-trifluoro-ethyl-2 chloro-acrylate
PBS: poly butene-1-sulfone
RE-5000P: poly-methyl pentene-sulfone and Novolac
AZ 2415: Novolac resins

Most of these resists are developed with ketones. The Novolac resists are developed with the alkaline developers used for the positive photoresists.
Positive resist developers (e-beam) are tabulated below.

Resist	Developer	Rinse
PMMA	MEK : MIBK (1 : 1) or chlorobenzene	—
EBR-9	MIBK : IPA (1 : 1)	—
PBS	MIAK : 2 pentanone (1 : 1)	PBS rinse
RE5000P	AZ developer	H_2O
AZ 2415	AZ developer	H_2O

A significant feature of the processing of PBS resist is its sensitivity to the relative humidity in the environment, which affects its development time; the higher the humidity, the more rapid is the development rate. Typically, ±2% relative-humidity control will give good CD control.

C Proximity Effect in E-Beam Lithography

The electrons that expose the resist not only penetrate the resist but also penetrate into the substrate. Some of these electrons are scattered back from the substrate and re-enter the resist film. This scattering means that near an area of intended dosage there is an area of unintended electron dosage. That is, at pattern edges the dosage is different from that within the areas of intended exposure. Both of two geometries that are approaching each other contribute to the dosage of electrons scattered into the space separating them. This contribution causes the geometries to "bloom" toward each other.

This effect is most serious for submicron geometries. Software routines are being written to moderate the exposure in these closely spaced regions but the problem is not simple. Multilevel resist systems tend to alleviate this problem, because the TLR is separated from the substrate by the LLR. Since the LLR is organic, it is a poor electron-scattering material. For good control of e-beam exposures in the submicron region, the proximity exposure problem is significant and needs to be solved in a way that does not severely impact machine productivity.

D Multilevel E-Beam Resists

When masks are being patterned, the lithographic system is simple, almost ideal. The substrate is completely flat and uniform, and variation in resist thickness is minimal.

When wafers are patterned, neither of the above conditions prevail. The wafer topology gives rise to variations in resist thickness and the consequent CD variation. To avoid this problem, multilevel resist systems, analogues to those discussed for optical resists, have been developed. These systems confer several benefits beyond improving CD control. Because a thinner resist can be used, exposure rates and machine productivity can be increased. Also, because the resist no longer rests directly on the substrate but on a thick layer of PMMA, backscattered electrons are not as severe a problem. This situation means edge sharpness and resolution are improved while the proximity effect is weakened.

The multilevel resist systems used in e-beam applications are similar to those employed for optical patterning except that the TLR material must be electron sensitive.

V X-RAY RESISTS

A Resist System

Most of the e-beam resists can be exposed using x-ray radiation. The problem with these resists is that the conventional sources are so weak in x-ray production that long-duration exposures are necessary (10–30 min).

It has been found that the incorporation of elements of atomic number higher than C, O, and N considerably improves the sensitivity of the resist. The most commonly employed elements are the halides. Table IX shows some of the currently used x-ray resists [16]. This table is not intended to be comprehensive. The materials that are candidates for x-ray resists are legion because this field is under active development. The point of the table is that resolution and sensitivity improvements are obtained by incorporating the heavier elements. When there is a near resonance of

TABLE IX

X-ray Resists and their Sensitivities

Resist	Tone	Major absorbing element	Target	Sensitivity (MJ/cm^2)	Resolution (μm)
PBS	+	S	Pd	94	0.5
PMMA	+	O	Al	600–1000	~0.1
Poly (2,2,3,-4,4,4, hexa fluorbutyl Methacrylate (Related to PMMA)	+	F	Mo	52	0.3
COP	–	O	Pd	175	1.0
			Al	15–20	1.0
Poly (2,3) dibromo 1 propyl-acrylate Co-glycidyl acrylate)	–	Br	Rh	15	
Brominated tetrathiafulavalene; functionalized poly(styrene)	–	S, Br	Al	22	0.2

an element's absorption edges with a target emission, an enhanced sensitivity can be obtained.

In addition to the resolution advantages of using these very short x-ray wavelengths, there are several other benefits of x-ray lithography.

(a) A potential for lower wafer defect densities. Most mask contamination is organic and as such is a low x-ray absorber. It is estimated that 90% of all optical mask contamination would not be imaged on the wafer by x-ray sources.

(b) Multilevel resists can be avoided if collimated sources (synchrotron) are used.

(c) There is negligible diffraction, so edge contrast is excellent.

(d) Substrate reflections are absent.

(e) The equipment, which is similar to proximity printers, generally can be of low cost.

The deficits of x-ray lithography are related to sources, machines, and masks.

(a) Conventional (nonsynchrotron) sources are point sources and supply uncollimated radiation. To improve collimation, the sources are located remote from the object. This placement weakens the source strength by $1/r^2$, so all machines have to be a compromise between collimation and source strength. This compromise is the main reason that conventional e-beam resists have had to be modified to improve their x-ray sensitivity.

(b) Because collimation is not good, the mask-to-wafer gap must be extremely well controlled. The newer machine vendors such as Micronix have incorporated six degrees of freedom in the wafer attitude controls [17]. Maintaining this control will be a formidable challenge.

(c) The masks and masking materials are an even more formidable challenge. These substrates must be thin (3–6 μm) and of a material with a low atomic number (for example, BN, polyimide, Si, Ti, and Ta). The masking or absorbing material is usually gold of about 600 nm thickness. These masks are e-beam fabricated. The current state of the art places the defect density at about 1–2/cm^2, which is an order of magnitude greater than the current optical mask state of the art.

B Plasma Developable Resists

A limitation on the negative x-ray resists is the swelling that occurs during development. The swelling can be avoided if a negative resist is dry processed. These systems depend on a differential resistance to O$_2$ plasma attack that is induced by the x-ray exposure. At present, these systems are in the development stage [16].

C Inorganic Resists

Inorganic resists that contain silver have shown good sensitivity for x-ray and other patterning techniques (e-beam and optical). The material Ag$_2$Se–GeSe [18] has been used in a bilevel resist system with RIE patterning of the LLR.

VI CONCLUSIONS

Current VLSI microlithography is in a state of rapid flux and exciting development. Optical patterning techniques currently are those used for the vast majority of all wafers processed; but even in this area, rapid transitions are occurring. The multilevel resist systems give the promise of improved CD control in the region of 1.0 μm geometries, and below, as the DUV and optical resist systems are developed.

The e-beam and x-ray resist systems are also in a state of rapid development. In the larger corporations such as Bell Labs, Texas Instruments, and IBM, both resist systems have been used in fairly broad scale applications. The drive to submicron resolution will test both techniques. X-ray patterning will be hampered by mask quality and mask availability. The e-beam patterning will be hampered by the complexity of the processing and the correction for the proximity effect.

We expect these problems to be exciting in their development and solution for well beyond the decade of the 1980s.

ACKNOWLEDGMENTS

The authors wish to acknowledge the efforts of Mrs. Marion Kull and Mrs. Carol Nadzak in the preparation of this manuscript. One of the authors (RB) wishes to acknowledge the support of his wife Mary in the time this work took away from home. We also wish to acknowledge technical discussions with Dr. Robert Owens (in work on contrast standards in photoresist systems) and Dr. Marlon Shopbell (in work on photoresist systems).

REFERENCES

1. J. M. Roussel, *Proc. SPIE* **275**, 9 (1981).
2. A. Stein, "The Chemistry and Technology of Negative Photoresist: A Waycoat Tutorial." Philip A. Hunt Chemical Corp., 1983.
3. J. Pacansky and J. R. Lyeria, Photochemical decomposition mechanisms for AZ-type photoresists, *IBM J. Res Dev.* **23**, 1, 42 (1979).
4. W. D. Hinsberg and M. L. Gutierrez, Effect of developer composition on photoresist performance, *Kodak Microelectronics Seminar* (October, 1983).
5. A. R. Johnson, Metal-ion free and metal-ion based developer solutions (variations in processing), *Kodak Microelectronics Seminar,* p. 60 (October, 1982).
6. D. J. Kim, W. G. Oldham, and A. R. Neureuther, Characterization of resist development, *Kodak Microelectronics Seminar,* p. 100 (October, 1982).
7. T. Kameko, T. Umegaki, and Y. Kawakami, A practical approach to sub-micron photolithography, *Kodak Microelectronics Seminar,* p. 25 (October, 1980).
8. B. J. Lin, *J. Vac. Sci Tech.* **12**, 1317 (1975).
9. R. F. Leonard and W. F. Cordes, *SPIE Optical Photolithography Seminar* (March, 1983).
10. L. K. White, Planarizing topographical features with spun-on polymer coating, *Kodak Microelectronics Seminar,* p. 72 (October, 1982).
11. A. R. Neureuther, P. K. Jain, and W. G. Oldham, Factors affecting linewidth control including multiple exposure and chromatic aberrations, *Proc. SPIE* **275**, 110 (1981).
12. M. Kaplan, D. Meyerhofer, and L. White, *RCA Review,* **44**, 135 (1983).
13. B. J. Lin and T. H. P. Chang, *J. Vac. Sci. Tech.* **16**, 1669 (1979).
14. C. H. Ting, I. Avigal, and B. C. Lu, Multilevel resist strategy for high resolution lithography, *Kodak Microelectronics Seminar.* p. 139 (1982).
15. S. Fok and G. H. K. Hong, E beam resist evaluation and process development in mask making, *Kodak Microelectronics Seminar* (November, 1983).
16. G. N. Taylor, *Solid State Tech.,* **44**, No. 5,3 (1980).
17. P. S. Burggaaf, X-ray lithography: optical's heir, *Semicond. Internat.,* **6**, No. 9, 60 (1983).
18. K. L. Tai, W. R. Sinclair, R. G. Vadimsky, J. M. Moran, and R. M. Rand, *J. Vac. Sci. Tech.* **16**, 1977 (1979).

Chapter 20

Electron Beam Lithography

R. KENT WATTS

Bell Laboratories
Murray Hill, New Jersey

I.	Introduction	351
II.	Mask Making	352
III.	Direct Writing	354
IV.	Resists	354
	A. Solubility	354
	B. Limits on Resolution	355
	C. Resist Thickness	356
	D. Available Electron Beam Resists	357
V.	Electron Optics	357
	A. Electron Beam Generators	357
	B. Applicable Rules of Light Optics	358
	C. Spot Sizes	358
	D. Aberrations	359
VI.	Raster Scan	360
VII.	Vector Scan	362
	A. Round Writing Beam	362
	B. Other Beam Shapes	362
	References	364

INTRODUCTION I

Electron beam lithography offers higher resolution than optical lithography because of the shorter wavelength of electrons. Resolution is determined in this case by lens aberrations, electron scattering, and resist properties. Electron beam pattern generators are widely used in making photomasks and less widely used in patterning wafers directly. Accurate pattern placement and level-to-level registration are of importance in these applications. Because of the serial nature of the patterning, throughput is much less than with optical projection printers.

II MASK MAKING

The first widespread use of electron beam pattern generation has been in photomask making. Many mask shops now have machines of the EBES type for this purpose. Figure 1 shows the methods of making a photomask. In the older method a 10× reticle was patterned with an *optical pattern generator,* a machine that under computer control exposes pattern elements to form the chip image at 10× scale, making 60 to 100 exposures per minute. The reticle is used in a step-and-repeat camera, which steps the reduced image to fill the desired 1× mask area. Stepping accuracy is interferometrically controlled. The reticle could be used directly in a wafer stepper to expose wafers to a reduced image. If an *e-beam pattern generator* is used to make the reticle, the main advantage for complex chips is speed. A large dense chip can require 20 hours or more of optical pattern generator time; the e-beam pattern generator is one to two orders of magnitude faster. Figure 1(b) illustrates the direct generation of 1× masks by e-beam. The master mask can be used in a projection printer, or copied and the copies used in a contact printer.

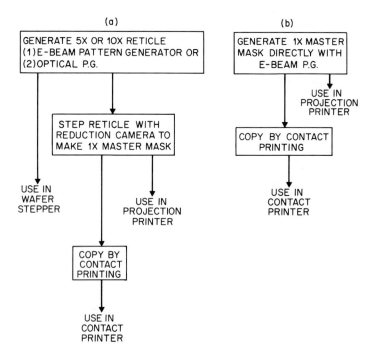

Fig. 1. Mask and reticle generation.

Optically and e-beam generated master masks generally have similar specifications for defect density and critical dimension (CD) tolerances; but stacking errors, the overlay misregistration of masks in a set, are 2 to 4 times smaller for e-beam generated masks. *Pattern placement errors* on a mask are measured by scanning computer-controlled X-Y optical measuring machines or with the E-beam pattern generator itself (Fig. 2). For this mask, maximum placement error is 0.14 μm. Fused quartz mask substrates with their low expansion coefficient can reduce thermal contributions to stacking errors.

Mask defects are either opaque chrome spots in areas that should be transparent or pinholes in the continuous chromium layer. Critical masks are often repaired to reduce the number of defects. Reticles should be defect free. Defect density is measured by an optical instrument that compares one chip with another on the mask, or with the data, and records differences or defects greater than some minimum size (1–2.5 μm).

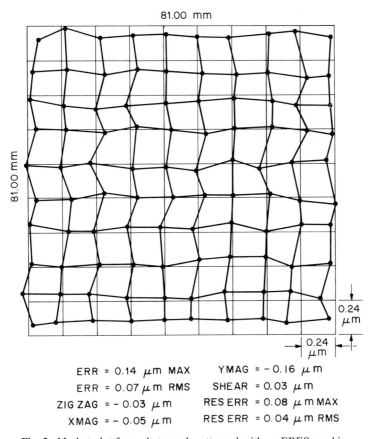

ERR = 0.14 μm MAX YMAG = −0.16 μm
ERR = 0.07 μm RMS SHEAR = 0.03 μm
ZIG ZAG = −0.03 μm RES ERR = 0.08 μm MAX
XMAG = −0.05 μm RES ERR = 0.04 μm RMS

Fig. 2. Market plot for a photomask patterned with an EBES machine.

III DIRECT WRITING

Direct writing of circuit patterns on wafers has been used for experimental devices with small features and for special circuits. In the first application to *low-volume integrated circuits,* custom logic arrays have been direct written at the window and metal levels, the other levels being optically imaged [1]. For *special circuits* that do not need large numbers of wafers there is some number of wafers n_w below which it is less expensive to direct write because of the mask cost associated with replication. If M is the cost of a mask, E the cost of direct writing a wafer, and R the cost of imaging a wafer optically, then $n_w = M/(E - R)$. If we wanted to consider the number of good chips obtained by each method, then the yields of the processes would have to be taken into account. Of course, mask defects play no role in direct writing.

In direct writing, it is necessary to have *alignment marks* on the wafer so that one circuit level can be written over another and aligned properly to it. Marks can be trenches etched into silicon or metal that provides sufficient contrast. Marks are sensed by monitoring the backscattered electrons as in a scanning electron microscope. In a machine with step–repeat stage motion, alignment marks in the corners of the square exposure field can be scanned to set deflection parameters before the pattern in the field is exposed. After exposure, the stage moves to the next field and these operations are repeated. In writing photomasks, an alignment mark on the stage may be scanned periodically during writing. No marks are placed on the mask.

IV RESISTS

The exposure of resists to electrons causes the formation of bonds or crosslinks between polymer chains (for a negative resist) or bond breaking (for a positive resist). The incident electrons have energies far greater than the bond energies in the resist molecules. Thus, there is no resonant absorption of the energy. Whether bond scission or formation predominates determines whether the resist is positive or negative [2].

A Solubility

In a negative resist, electron beam induced crosslinks between molecules make the polymer less soluble in the developer solution. Sensitivity increases with increasing molecular weight. If the molecules are larger, then fewer crosslinks are required per unit volume for insolubility. The polymer molecules in the unexposed resist have a distribution of lengths or molecular weights and thus a distribution of sensitivities to radiation. The narrower the distribution is, the higher the contrast. The exposure dose Q has units of charge deposited by the beam per unit area C/cm^2. As in the case of optical negative resists, after irradiation there may be long-lived active species that produce crosslinking and are quenched on exposure to air. In a positive resist, high molecular weight and narrow distribution are advantageous also.

Limits on Resolution   B

Two factors are of major importance in limiting resist resolution: swelling of the resist in the developer, more important for negative resists, and electron scattering.

Swelling   1

Swelling of negative resists has two deleterious effects. Two adjacent lines of resist may swell enough that they touch. On shrinking in the rinse, they may not completely separate, leaving a bridge here and there. This expansion and contraction weakens adhesion of very small resist features to the substrate and can cause small undulations in narrow (0.5 μm) lines. Both problems become less severe as resist thickness is reduced.

Electron Backscattering   2

When electrons are incident on a resist or other material, they enter the material and lose energy by scattering, producing secondary electrons and x rays. This fundamental process limits the resolution of electron resists to an extent depending on resist thickness, beam energy, and substrate composition. For example, more electrons are scattered back into the resist from a heavy metal substrate than from a silicon substrate. The envelope of the electron cloud in the material is pulled closer to the surface as beam voltage decreases. At higher beam voltage the electrons penetrate farther before being scattered over large lateral distances.

As separation between lines decreases, there is a larger dose of backscattered electrons between the lines where there should be none. This is somewhat similar to the reduced modulation in an optical image at higher spatial frequencies. An exposed pattern element adjacent to another element receives exposure not only from the incident electron beam but also from scattered electrons from the adjacent element. This is called the *proximity effect* and is more pronounced the smaller the gap between elements. For example 0.5 μm lines require 20–30% more exposure when isolated than when separated from each other by 0.5 μm. Thus as pattern density increases, it becomes necessary to adjust exposure for various classes of elements, or in the extreme case, for different parts of the elements.

C Resist Thickness

Resist resolution is better in thinner resist layers. Minimum thickness is set by the need to keep defect density sufficiently low and by resistance to etching in device processing. For photomasks, for which the surface is flat and only a thin layer of chrome must be etched, resist thicknesses in the range 0.2–0.4 μm are used. For device processing in which topographic steps must be covered and dry-etching conditions are more severe, thicknesses of 0.5–2 μm are required. Most electron resists are not as resistant to dry etching as optical resists. One way of alleviating the problems of proximity effect, step coverage, and process worthiness is by using a multilevel resist structure in which the thick bottom layer consists of a process resistant polymer. In one realization of a three-level structure, the uppermost layer of electron resist is used to pattern a thin intermediate layer, such as 1200 Å of SiO_2, which serves as mask for etching the thick polymer below. For electron lithography, a conducting layer can be substituted for the SiO_2 to prevent charge buildup which can lead to beam placement errors. In another two-layer resist structure, both thin upper layer and thick lower layer are positive electron resists, but they are developed in different solvents.

TABLE I

Some Electron Resists

Resist	Type	Sensitivity ($\mu C/cm^2$)	Resolution (μm)
COP (Mead Chem.)	−	0.3 @ 10 kV	1.0
PBS (Mead Chem.)	+	0.8 @ 10 kV	0.5
OEBR-100 (Tokyo Okha)	−	0.5 @ 20 kV	1.0
Sel-N (Somar)	−	0.4 @ 27 kV	0.5
EBR-9 (Toray)	+	3 @ 20 kV	0.5
FBM 120 (Daikin Kogyo)	+	1 @ 20 kV	0.5
PMMA	+	50 @ 15 kV	0.1
Microposit 2400 (Shipley)	+	20 @ 20 kV	0.5

Available Electron Beam Resists D

Table I lists some available electron resists. Polymethyl methacrylate (PMMA) is the highest resolution resist known. Microposit 2400 is a photoresist that is also electron sensitive. Since faster electrons penetrate more deeply, more beam current must be used at higher voltages. A resist is about 2× less sensitive at 20 kV that at 10 kV. The values given for resolution are meant to serve for rough comparison.

ELECTRON OPTICS V

Electron Beam Generators A

E-beam pattern generators are similar to scanning electron microscopes, from which they are derived (Fig. 3). Two or more magnetic

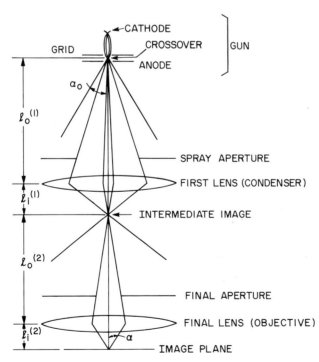

Fig. 3. A simple electron optical system. [From R. K. Watts, "Very Large Scale Integration, Fundamentals and Applications" (D. F. Barbe, ed.), ch. 3. Springer Verlag, New York, 1982.]

lenses form a demagnified image of the source on the wafer image plane. The electrons originate from a cathode, which is usually a tungsten or LaB_6 thermionic emitter. The electrons are accelerated through 10–50 kV in the gun and are focused to a spot of diameter $d_o = 10\text{–}100$ μm, called the crossover. If I_b is the beam current in the crossover, the brightness or current density per unit solid angle is given by

$$\beta = 4I_b/(\pi d_o \alpha_o)^2.$$

A few machines use a thermal field emitter rather than a thermionic cathode. This consists of a tungsten tip of radius 0.5–1 μm that is heated sufficiently to provide annealing of sputtering damage for relaxed vacuum requirements and better stability compared with a cold-field emitter. There is no gun crossover. A very small virtual source is located inside the tip. Thus, the gun and the (magnifying) optics are very different from those we consider. This type of cathode is of interest because it provides higher currents than thermionic cathodes for small image spots.

B Applicable Rules of Light Optics

Some rules of light optics apply to electron optics, such as the thin-lens law relating object distance, image distance, and focal length,

$$1/l_o + 1/l_i = 1/f.$$

The magnification of a lens is $M = l_i/l_o$. An intermediate image of the crossover is formed by the first lens with magnification M_1 and this is further demagnified by the second lens to form a spot in the image plane of diameter

$$d_i = Md_o, \qquad M = M_1 M_2.$$

Typically $M \approx 10^{-3}$ to 10^{-1}. Although the lengths $l_o^{(1)}$ and $l_i^{(2)}$ are fixed, $l_i^{(1)}$ and $l_o^{(2)}$ are variable. If current through the windings of the first lens is increased, $l_i^{(1)}$ and M_1 are reduced, and the beam current passing through the final aperture is reduced. The current density and current in the image plane are

$$J \leq \pi\beta\alpha^2, \qquad I = J(\pi d_i^2/4).$$

C Spot Sizes

Spot sizes of interest are in the range 0.1–2 μm. This is far from the diffraction limit ($\sim 10^{-3}$ μm). Thus resolution is not diffraction limited but is determined by aberrations of the final lens. Figure 4 shows a typical double deflection system arranged so that the beam always passes through the center of the final lens. There will be another set of deflection coils to provide deflection perpendicular to the plane of the figure.

Aberrations D

Aberrations are treated as independent contributions to spot broadening. The actual spot size is then the square root of the sum of squares of the contributions. The aberrations are of two types: aberrations of the undeflected beam and aberrations that are functions of the distance r in the image plane from the axis to the deflected beam. Aberrations of the first type are

spherical aberration $\quad d_s = (C_s/2)\alpha^3,$

chromatic aberration $\quad d_c = C_c(\Delta V/V)\alpha.$

The spread of electron energies is $e\Delta V$, usually a few electron volts. C_s and C_c are constants characterizing the aberrations. *Astigmatism,* a third possibility, can be removed by introducing compensating fields with stigmator coils (Fig. 4). *Deflection aberrations,* such as coma and field

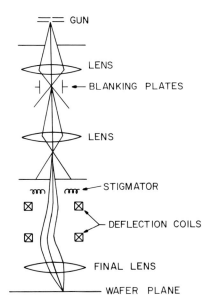

Fig. 4. Electron optical system with scanning. [From R. K. Watts, "Very Large Scale Integration, Fundamentals and Applications" (D. F. Barbe, ed.), ch. 3. Springer Verlag, New York, 1982.]

TABLE II

Aberrations at the Corners of 2 × 2 mm and 4 × 4 mm Exposure Fields for VS1

	2 mm field (μm)	4 mm field (μm)
Spherical, d_s	0.01	0.01
Chormatic, d_c	0.04	0.04
Deflection, d_{df}	0.087	0.35
Total, no dynamic correction	0.095	0.35
Total, dynamically corrected	0.043	0.05

curvature, are proportional to $r^2\alpha$ and $r\alpha^2$; we lump them together and call their contribution to the spot d_{df}. Field distortion and some of the deflection aberrations can be reduced by adding corrections to the deflection fields. Table II shows these aberrations for the IBM VS1 machine [3].

Another source of spot broadening is the mutual Coulomb repulsion of the electrons in the beam. For total column length L this contribution is

$$d_{ee} \approx [LI/(\alpha V^{3/2})] \times 10^8 \; \mu\text{m}.$$

This becomes important only at large currents when small spot sizes are sought.

In the presence of aberrations, the current becomes

$$I = (\pi^2/4)\beta\alpha^2[d^2 - d_s^2 - d_c^2 - d_{df}^2 - \cdots].$$

For a given set of aberrations, an optimum α can be found that gives minimum spot size for a particular current. However, in practice, the final aperture is not changed for each current but optimized for some small spot size. Then the current and beam diameter are related by

$$I \sim (d^2 - \text{constant}).$$

VI RASTER SCAN

In raster scan, the beam is deflected repetitively over the exposure field, as in a television raster. The beam is turned on at various points in the scan to expose the desired pattern. The EBES machine [4], developed by Bell Laboratories and available commercially from Perkin–Elmer ETEC Corp. and Extrion, uses beam deflection in one dimension. The *writing scheme* is shown in Fig. 5. The stage moves continuously in a direction perpendicular to the writing beam. The pattern data are decomposed into a number of stripes, and one stripe is written over all chips of the same type before the next stripe is begun. The stripe is 512 addresses

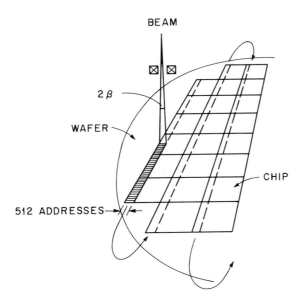

Fig. 5. Patterning with EBES. The curved arrows indicate stage motion. [From R. K. Watts, "Very Large Scale Integration, Fundamentals and Applications" (D. F. Barbe, ed.), ch. 3. Springer Verlag, New York, 1982.]

wide, an address corresponding to 0.5 µm or 0.25 µm. The pattern information comes to the blanking plates from a shift register at a rate of 20 MHz or 40 MHz. Since the total time necessary to write the 512 address scan at 20 MHz is 31.6 µs–50 ns per spot plus 6 µs for flyback, the writing time is approximately 2 cm²/min with the 0.5 µm spot and 0.6 cm²/min with the 0.25 µm spot. The rate is twice this at 40 MHz.

Stage motion is monitored by laser interferometer, and small position corrections are made by beam deflection, larger ones by drive motor speed variation. Beam position errors are checked and corrected periodically by scanning over a fiducial mark on the stage.

The EBES machine is the most widely used electron beam photomask pattern generator. The small beam deflection range $\beta = 9$ mrad and the repetitive nature of the scan provide very accurate pattern placement. For writing on wafers, three or more alignment marks on the wafer are scanned to provide level-to-level registration. In raster scan the whole field is scanned. This is less efficient than vector scan but has the advantage that the polarity of the pattern is easily changed. The exposure is determined by the beam current and the (fixed) 50 ns dwell time. It is not changed during writing. Thus, corrections for proximity effect, which are easily made with vector scan, cannot be made. However, the variable scan speed of vector scan places a severe burden on the deflection system, and eddy current effects must be compensated or eliminated over a wider frequency range of the deflection system.

VII VECTOR SCAN

A Round Writing Beam

In a vector scan machine, the beam is directed in turn to each pattern element to be exposed. No time is wasted scanning over parts of the exposure field where there are no patterns. Figure 6(a) shows how the round writing beam of diameter d fills in a pattern element by scanning it with velocity v and scan spacing s. The minimum time required to expose a fraction η of the field area F^2 with current I and resist sensitivity Q is

$$T_e = \eta F^2 Q / I.$$

Q and I are related to v and s by

$$vs = I/Q.$$

The exposure field should be as large as allowed by distortion and deflection aberrations because stage stepping (and settling) time reduces throughput.

Once the beam current is set, exposure is controlled by varying v. The scan speed can be set to one value for the whole pattern or varied to give different pattern elements different exposures to compensate proximity effects, which become important for pattern dimensions around 1 μm and smaller.

B Other Beam Shapes

Not all machines use round beams. Figure 7 shows the electron column of the JEOL JBX6-A machine, which has a variable rectangular beam. The image of the first square aperture is shifted to cover various portions of the second square aperture, which is the object imaged on mask or

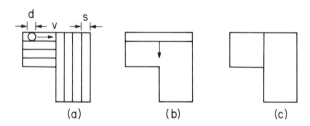

Fig. 6. Exposing a pattern element with (a) a round beam, (b) a line beam, and (c) a rectangular beam.

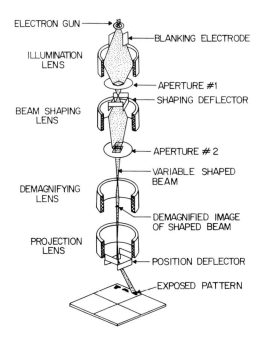

Fig. 7. Electron optical system of the JEOL JBX6-A pattern generator. [From R. K. Watts, "Very Large Scale Integration, Fundamentals and Applications" (D. F. Barbe, ed.), ch. 3. Springer Verlag, New York, 1982.]

wafer. Figure 6c shows how a pattern element can be exposed in only two shots with such a machine, saving exposure time compared with a round beam.

Figure 6b shows how a pattern element is exposed with the variable line beam of the Varian VLS1000 machine. In this case the first aperture is L-shaped and the second is square. Use of a line beam keeps beam current lower than does a rectangular beam; lower beam current causes less edge broadening from electron–electron interactions. Throughput depends on

TABLE III

Some Specifications of Two Vector Scan Pattern Generators

	IBM VS1	Varian VLS1000
Cathode	LaB_6	LaB_6
Image current density (A/cm^2)	300	80
Exposure field (mm^2)	4 × 4	1.6 × 1.6
Beam diameter (μm)	≥0.05	0.2–3.2
Accelerating voltage (kV)	25	10–30
Typical overlay error (μm)	±0.1	±0.15
Stage motion	Step repeat	Continuous

many factors including pattern feature size and pattern density. Throughput of the Varian machine is $\geq 10^4$ in. wafers with 0.75 µm minimum feature size. Table III shows specifications of two vector scan machines.

REFERENCES

1. R. D. Moore, G. A. Caccoma, H. C. Pfeiffer, E. V. Weber, and O. C. Woodard, in "Proc. 16th Symp. on Electron, Ion, and Photon Beam Technology" (D. R. Herriott and J. H. Bruning, eds.), p. 950. A.I.P., New York, 1981.
2. M. J. Bowden and L. F. Thompson, *Solid State Technol.* **22,** 72 (1979).
3. T. H. P. Chang, A. J. Speth, C. H. Ting, R. Viswanathan, M. Parikh, and E. Munro, in "Proc. 8th Int. Conf. on Electron and Ion Beam Science and Technology" (R. Bakish, ed.), p. 377. Electrochemical Society, Princeton, New Jersey, 1976.
4. D. R. Herriott, R. J. Collier, D. S. Alles, and R. W. Stafford, *IEEE Trans. Electron Devices* **ED22,** 385 (1975).

Chapter 21
X-Ray Lithography

R. KENT WATTS

Bell Laboratories
Murray Hill, New Jersey

I.	Introduction	366
II.	X-Ray Proximity Printing	366
	A. Wavelength Region	366
	B. Diffraction Blur	367
	C. Runout Distortion	367
III.	Sources	367
	A. Efficiency	367
	B. Targets	369
	C. Electron Gun	369
	D. Electron Storage Ring	369
	E. Plasma Source	370
IV.	Masks	371
	A. Characteristics	371
	B. Materials	371
	C. Errors	371
V.	Resists	372
	A. X-Ray Absorption	373
	B. Negative Resist Characteristics	374
	C. Positive Resist Characteristics	375
	D. X-Ray Resist Materials	375
VI.	Mask Alignment	376
	A. Contributions to Total Error	376
	B. Alignment Schemes	376
VII.	Exposure Systems	377
	A. The NTT System	378
	B. The Bell Laboratories System	378
VIII.	Applications	378
	References	380

I INTRODUCTION

The x-ray flux from the first x-ray sources used for lithography was very small, and the distances from source to mask and from mask to wafer were kept small, making development of mask alignment apparatus difficult. Higher power sources and more sensitive resists now allow larger spacings without long exposure time, and x-ray systems begin to resemble optical proximity printers in their compact design.

II X-RAY PROXIMITY PRINTING

A Wavelength Region

The wavelength region of interest for x-ray lithography is 4–50 Å. At shorter wavelengths, transmission of mask absorbers and resists is too high; at longer wavelengths, transmission of the "clear" part of the mask is too low. In a practical system the source spectrum, the mask transmission, and the resist absorption must be optimized to minimize exposure time and maximize mask contrast. Most x-ray systems use wavelengths in the 4–10 Å range.

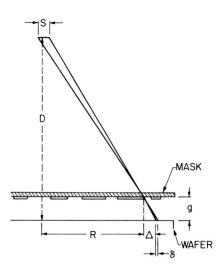

Fig. 1. Geometry of an x-ray lithography proximity printer. The runout error Δ and the blur δ due to the source size S are shown.

Diffraction Blur B

An x-ray proximity printer is similar to an optical proximity printer except that the x-radiation diverges from a point source and is not collimated. Taking $\sqrt{g\lambda}/2$ to be the diffraction blur due to the mask–wafer gap g, we see that with an 800-fold reduction in wavelength λ in the x-ray case and the same gap as in the optical case (10–40 μm), diffraction is usually negligible for x-rays. The blur at a feature edge due to the x-ray source spot size S (Fig. 1) is given by

$$\delta = (g/D)S.$$

For the typical values $g = 40$ μm, $D = 50$ cm, and $S = 3$ mm, the blur is $\delta = 0.24$ μm. Resist contrast is also important in determining resist edge profiles.

Runout Distortion C

Runout distortion or magnification is a feature that is unique to the x-ray point source. Because of the gap g, a feature located at distance R from the center of the mask will be replicated on the wafer at $R + \Delta$, where

$$\Delta = (g/D)R.$$

If g varies from one exposure level to another, the change in Δ represents a misalignment of levels. A vacuum pin chuck is generally used to avoid local changes in gap due to particles on the back of the wafer. The magnification, $1 + g/D$, can be varied by changing g to compensate magnification errors on the mask or expansion of the wafer in processing.

SOURCES III

Efficiency A

In conventional x-ray sources, x-rays are produced by an electron beam striking a metal target. Efficiency is low. Most of the input power is converted into heat, and the x-ray power produced is limited by the ability of the target to dissipate this heat.

The x-ray flux ϕ produced at distance D from a target bombarded by a current I of electrons of energy E_o is given by

$$\phi = \frac{IE_o\eta}{eD^2},$$

where η is the power efficiency (watts/watt steradian). The radiation consists of characteristic lines and a broad continuum (bremsstrahlung), as shown in Fig. 2. The efficiency of generation of the characteristic lines is

$$\eta_c = (K/E_o)(E_o - E_x)^{1.63}f(E_o,\theta),$$

where K is a constant, E_x is the shell ionization energy (K, L, M), and f is a target reabsorption factor that depends on θ, the angle between the flux and the surface normal. Values of η_c for five target materials are listed in Table I. For the continuum radiation, the efficiency of generation in a bandwidth ΔE is

$$\eta_B = K'(1 - E/E_o)f(E,\theta)\,\Delta E.$$

The continuum radiation extends to energy E_o and reduces mask contrast.

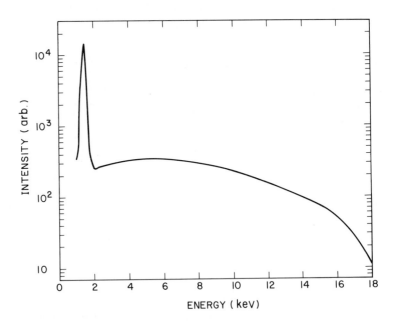

Fig. 2. Spectrum of a high power (18 kW, 20 kV) aluminum source. The characteristic peak is broadened by the detector.

TABLE I

Efficiency of Characteristic X-Ray Generation Normal to Target Surface for Various Anode–Cathode Potentials[a]

	λ(Å)	Target voltage (kV)					
		8	10	15	20	30	40	60
Al Kα	8.34	27	34	46	55	67	73	80
Si Kα	7.13	28	35	49	60	75	83	94
Mo Lα	5.41	26	33	46	53	59	59	54
Rh Lα	4.60	27	37	55	65	79	82	81
Pd Lα	4.37	27	37	55	65	79	82	81

[a] Units are microwatts per watt steradian.

Targets B

Two types of target are in use: rotating and stationary. Rotating anodes are capable of higher power operation [1], but stationary anodes are more reliable [2]. High power x-ray generators with aluminum targets are available from Rigaku Denki in Japan.

Electron Gun C

The electron gun must be capable of high current density at relatively low voltage (<25 kV). Current may be as large as 1 A. The most commonly used guns are the *Pierce gun* and the *ring gun*. In the ring gun, electron paths curve ~180° and strike the target at normal incidence.

Electron Storage Ring D

In an electron storage ring, high-energy electrons are forced into a closed curved path by magnetic fields. An electron has an acceleration directed toward the center of its orbit and emits radiation. The radiation from a circular machine comes from all tangents and has the shape of a disc. Angular divergence of the radiation in the vertical direction is only a few milliradians. The peak of the power spectrum occurs at wavelength λ_p, which is related to the electron energy E_o (in GeV) and magnet bending radius R (in meters) by $\lambda_p = 2.35\, R/E_o^3$. The total power radiated into all angles by a circulating electron is given by

$$P(\lambda) = (7.51 \times 10^{-8}) E_o^7 R^{-3} (\lambda_c/\lambda) \int_{\lambda_c/\lambda}^{\infty} K_{5/3}(x)\,dx, \qquad \lambda_c = 2.38\, \lambda_p.$$

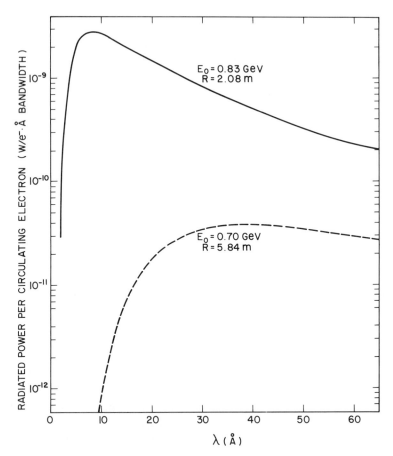

Fig. 3. Spectrum of radiation from a storage ring for two choices of bending radius and electron energy.

This is potted in Fig. 3 for two different parameters. For an exposure station located on a tangent 10 m from the ring, the useful flux can be ~0.7 W/cm² for a circulating current of 500 mA. This compares with typical flux of ~0.1 mW/cm² for a conventional source. The radiation is more nearly collimated. This may be a disadvantage because magnification cannot be adjusted by varying the mask–wafer gap.

E Plasma Source

Plasmas that are sufficiently hot and dense to radiate soft x-rays can be produced by several means. In the most practical method, a small hollow cylinder of gas is caused to collapse when energy from a capacitor bank is discharged through it [3]. Line x-radiation is produced, and the efficiency is much higher than for conventional sources. The radiation spectrum depends on the gas. This technique seems a promising way to obtain a high-brightness source without the high cost of a storage ring. Remaining problems include reducing the amount of debris deposited on the x-ray window and increasing pulse repetition rate while reducing peak power.

IV. MASKS

A. Characteristics

An x-ray mask must meet a number of requirements. The substrate must be sufficiently transparent in the range of high resist sensitivity. The mask absorber must attenuate the radiation. Typically, mask contrast ranges from 6 to 20, where the contrast is defined as the resist dose through the transmissive part of the mask divided by the dose through the absorbing part. The mask must be rugged and dimensionally stable. Its thermal expansion properties should match those of the wafer. The defect density of the mask should be low (less than $1/cm^2$). If the alignment system uses visible light, the mask should be optically transparent.

B. Materials

Gold, which is a strong x-ray absorber, has almost always been used as the *absorber* on x-ray masks (Fig. 4). The *substrate* is a thin membrane on a silicon or glass ring. Membranes that have been used include Si, Ti, SiC, Si_3N_4, Si_3N_4/SiO_2, $Si_3N_4/SiO_2/Si_3N_4$, Al_2O_3, Mylar, polyimide, and BN/polyimide. For the *BN/polyimide* mask substrate used by Bell Laboratories for exposure by Pd radiation, the BN membrane is 5 μm thick, and the polyimide layer thickness is 2 μm. The BN provides dimensional stability, and the polyimide provides resilience. Membranes used at longer wavelengths must be thinner.

C. Errors

Figure 5 is a Market plot of distortion of a BN/polyimide x-ray mask. The RMS *placement error* is 0.082 μm for this 44 × 44 mm field. There is a

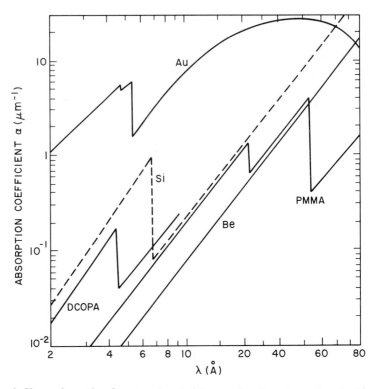

Fig. 4. X-ray absorption for several materials as a function of wavelength. Between edges the absorption is nearly proportional to λ^3.

very small *magnification error* of 0.0012 μm/mm. This would be compensated, if it were larger, by adjustment of the mask–wafer gap. Thus, the significant error is the *residual error* (0.080 μm), which remains after the magnification is subtracted from the data. This plot is made by the EBES pattern generator operating in an SEM mode. Linewidth variation is measured at the same time and can be displayed in a similar plot.

V RESISTS

In positive resists, the areas of the resist that are under the transmissive parts of the mask are removed in the developing step; for negative resists, the irradiated areas remain after developing, and the shadowed areas are removed.

21. X-Ray Lithography

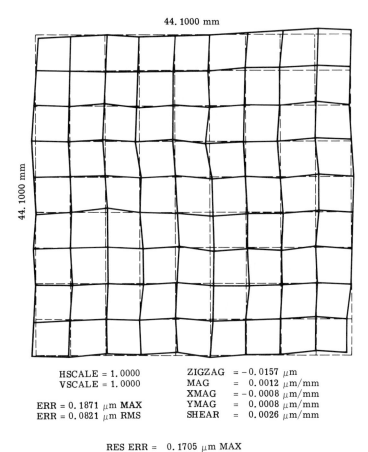

Fig. 5. Market plot of pattern-placement errors on an x-ray mask. Ideal placement is indicated by dashed lines, actual placement by solid lines. Each intersection represents a mark location.

X-Ray Absorption   A

Polymer resists that contain only carbon, hydrogen, and oxygen have x-ray absorption spectra like that of poly(methyl methacrylate), or PMMA, shown in Fig. 4. Since absorption varies over two orders of magnitude in the figure, absorption and sensitivity are strongly dependent on wavelength. Exposure dose Q is generally given in units of incident flux (J/cm^2). For a conventional x-ray source supplying a flux of 0.1 mW/cm^2, resist sensitivity corresponding to a 1-min exposure time would be

$Q = 6$ mJ/cm². Optical exposure tools typically provide a flux three orders of magnitude greater. Therefore, x-ray resists must be much more sensitive than photoresists.

When an x-ray photon is absorbed by a resist molecule, a photoelectron is released from an inner shell. The subsequent reactions—excitation, ionization, and production of free radicals—lead to the breaking of chemical bonds and formation of new bonds. All electron resists are also x-ray resists because similar processes occur in electron bombardment.

B Negative Resist Characteristics

In negative resists, formation of cross-inking bonds between the long chain molecules predominates and is characterized by an *efficiency* G_x, the number of cross-links formed per 100 eV of absorbed energy. At the gel dose Q_{gi}, there is on average one cross-link per molecule. From the equation

$$Q_{gi} = K_1/(G_x \overline{M}_w),$$

we see that *sensitivity* is greater for larger G_x and molecular weight $\overline{M}_w$. The sensitivity of negative resists is often presented as in Fig. 6, in which the ratio of developed thickness t to initial thickness t_o is plotted versus exposure dose. Typical practical exposures lie in the range $t/t_o = 0.5$–0.8. The slope of the steep part of the curve is the *contrast* γ,

$$\gamma = [\log(Q_{1.0}/Q_{gi})]^{-1}.$$

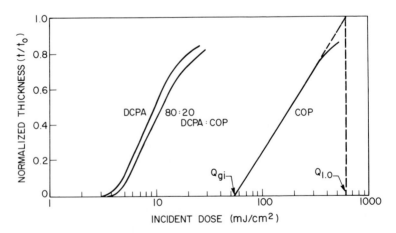

Fig. 6. Fractional thickness remaining after exposure to a Pd radiation dose and subsequent development for three resists.

21. X-Ray Lithography

Contrast is higher the narrower the molecular weight distribution. The *resolution* of negative resists is often limited by the swelling of the cross-linked gel in the developer solvent, as closely spaced lines swell, touch, and stick together.

C Positive Resist Characteristics

In positive resists, chain scission is similarly characterized by an *efficiency* factor G_s. The ratio of dissolution rate in the developer of irradiated resist to unexposed resist is given by

$$S/S_o = (1 + K_2 G_s Q \overline{M}_n)^c, \quad c > 0.$$

Larger molecular weight also leads to greater *sensitivity* for positive resists. M_n is the number average molecular weight.

D X-Ray Resist Materials

PMMA, has the highest resolution among resists. Features smaller than 100 Å have been resolved using synchrotron radiation. PMMA sensitivity is low, as is seen in Table II. Some commercial positive photoresists have x-ray sensitivity similar to that of PMMA.

The negative resist *DCOPA* is a mixture of poly(2,3-dichloro-1-propyl acrylate) (DCPA) and the electron resist COP, or poly(glycidal methacrylate-co-ethyl acrylate). Chlorine has an absorption edge at 4.40 Å (Fig. 4) and increases resist absorption of 4.37 Å Pd characteristic radiation for chlorine-containing resists.

Poly(hexafluorobutyl methacrylate), *FBM*, is a high contrast ($\gamma = 4.5$) positive resist with good resolution. The electron resist poly(butene-1-sulfone) (PBS) is also a positive x-ray resist.

TABLE II

Some X-Ray Resists

Resist	Q (mJ/cm²)	Source	Polarity	Resolution (μm)
PMMA	500	Al	+	<0.1
DCOPA	15	Pd	−	0.8
PDXR	1.5	Pd	−	0.5
FBM	50	Mo	+	0.3
PBS	~15	Al	+	0.3
COP	20	Al	−	0.8

PDXR is a plasma-developed resist consisting of a mixture of poly(2,3-dichloro-1-propyl acrylate) and monomer bis-acryloxybutyltetramethyldisiloxane. Exposure locks in the monomer and reduces the etch rate of the exposed areas in an O_2 plasma. Since there is no developer solution, this negative resist exhibits no swelling.

Because most x-ray resists do not withstand plasma etching as well as commercial photoresists, they are often used in multilevel resist structures. For example, in a tri-level scheme a thin SiO_2 stencil layer is sandwiched between a thick HPR photoresist base layer (~2 μm) and a thin (~0.4 μm) upper x-ray resist layer that is used as a mask for reactive sputter etching of the stencil. The SiO_2 serves as a mask for O_2 sputter etching of the HPR.

VI MASK ALIGNMENT

A Contributions to Total Error

Mask–wafer registration errors have several components ε_i that contribute to the total error ε_t. The contributions are usually assumed to be independent of each other. In this case,

$$\varepsilon_t = \left(\sum_i \varepsilon_i^2\right)^{1/2}.$$

One contribution, e_m, comes from placement errors on the masks used to image the two levels. If both masks have similar errors, equal to that shown in Fig. 5, then $\varepsilon_m = \sqrt{2}\,(0.08\ \mu\text{m}) = 0.11\ \mu\text{m}$. Another contribution, ε_a, comes from the mask alignment system. Since magnification must be controlled as well as in-plane orientation, the precision of setting the mask–wafer gap contributes to ε_a.

B Alignment Schemes

The simplest alignment scheme consists of viewing alignment marks on mask and wafer with a microscope and aligning manually. Precision is at best $\varepsilon_a \approx 0.2\ \mu$m. In one automatic alignment system, rectangular marks are viewed on mask and wafer. Modulation of the alignment signal is obtained by causing the wafer to vibrate. Total registration error [4] is $\varepsilon_t < 0.3\ \mu$m (2σ).

In another alignment system [5], circular zone plate alignment marks on mask and wafer are used as lenses to focus collimated incident light to two

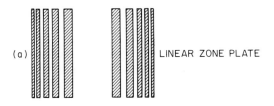

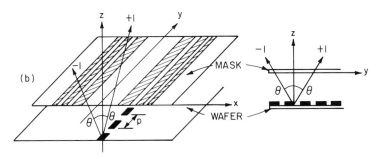

Fig. 7. Alignment scheme for detecting relative displacement in the x direction. (a) Mark on mask is a linear zone plate. (b) Wafer mark is a broken line.

spots above mask and wafer. The system aligns by bringing the two spots into coincidence. If r_1 is the radius of the first zone, normally incident light of wavelength λ is focused at distances along the axis of the zone plate,

$$f_n = r_1^2/n\lambda, \quad n = \pm 1, \pm 3, \pm 5, \cdots.$$

The focal lengths of the zone plates on mask and wafer differ by the mask–wafer gap,

$$f_1^{(w)} = f_1^{(m)} + g.$$

Another method [6] is illustrated in Fig. 7. The mask alignment mark is a linear zone plate, similar to a cylindrical lens, and the wafer mark is a line. The mask mark focuses laser light onto the line. The line is broken to diffract the light away from the zero-order beam. The diffracted beam is detected; it is maximum at alignment. Repeatability of the system is $\varepsilon_a < 0.1 \ \mu\text{m}$.

VII EXPOSURE SYSTEMS

An exposure system consists of an x-ray source, an exposure chamber, a mask and wafer handling apparatus, an alignment subsystem, control electronics, a mask, and a resist.

A The NTT System

A system developed at NTT uses a rotating silicon anode with 20 kW input power [4]. Mask substrate is a $Si_3N_4/SiO_2/Si_3N_4$ sandwich, and gold absorber thickness is 0.8 μm. The alignment system uses the scheme of modulation by vibration. Resists are SEL-N(−) and FBM(+).

B The Bell Laboratories System

In the Bell Laboratories system, a 4-kW stationary Pd anode is used. The resist absorption is matched to the Pd spectrum by incorporation of chlorine. Mask and wafer are manually loaded and aligned. These functions will be automated in a new system, using the circular zone plates for alignment. Values of the proximity printing parameters for this machine are $S = 3$ mm, $D = 50$ cm, and $g = 40$ μm. The wafer backside is flattened by a pin chuck. The wafer is transported to a leveling station. A leveling plate is automatically placed on the stage and contacts the edge of the wafer at three points. These three points place the wafer parallel to the mask and 40 ± 1 μm below it. After this leveling and gap-setting sequence, the mask is placed over the wafer, and the stage is moved under the alignment microscope. Manual alignment is performed by means of a trackball driving four inchworms (x, y, θ, z). Wafer motion occurs in steps of 0.1 μm. The 40-μm mask–wafer gap can next be altered to compensate improper magnification. After alignment, wafer and mask are locked in position, and the stage moves to the exposure chamber. After exposure, the shutter closes, and the stage moves back to the mask storage position where the mask is automatically unloaded and stored for the next exposure. The stage then moves to the unloading position, at which the wafer is manually unloaded and a new wafer is loaded. The complete cycle, including alignment, using DCOPA resist takes about 6 min/wafer. The new automated system and PDXR resist will increase throughput substantially.

VIII APPLICATIONS

X-ray lithography has been applied largely to fabrication of simple MOS devices and circuits. The first reported use is that of Bernacki and Smith [7]. They made diodes, bipolar transistors, and PMOS transistors with minimum linewidth 2.5 μm. C–V measurements on MOS dot capaci-

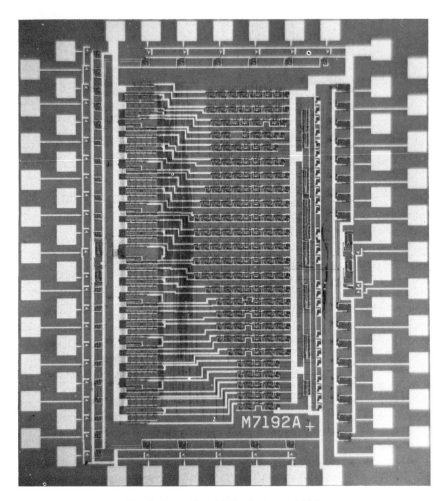

Fig. 8. 16 × 16 multiplier in 1-μm nMOS.

tors showed that an x-ray induced positive charge density of 5×10^{11} cm^{-2} was removed by a 455°C anneal. PMMA and KMNR, a Kodak negative photoresist, were used with a low-power aluminum source.

Stover *et al.* [8] of Hughes Laboratories have made CMOS/SOS circuits, exposing COP resist with an exposure system made by fitting a 10-kW aluminum source to a Cobilt mask aligner. The test chip contained, among other test structures, transistors and ring oscillators with gate lengths 1, 2, 3, 4, 5, and 7.5 μm. A minimum stage delay of 180 ps was obtained for the 1-μm oscillators. The same chip was made with optical lithography, and equivalent performance was obtained for the two lots. (The 1-μm devices were made only with x-ray lithography.) The two lots also showed similar degradation from gamma irradiation and thermal stress. Suzuki *et al.* [4] have also compared MOS transistors and ring oscillators made by x-ray with devices imaged optically.

Over the last six years a program at Bell Laboratories has been producing MOS high-speed logic circuits of various kinds, beginning with simple SSI circuits and extending to LSI [9]. Minimum ring oscillator stage delay obtained is 30 ps. A divide-by-sixteen counter operates at 2.5 GHz. A 750-MHz digital-to-analog converter has been made for the deflection circuit of a new EBES IV electron beam pattern generator. A multiplier chip is shown in Fig. 8.

REFERENCES

1. M. Yoshimatsu and S. Kozaki, *in* "X-Ray Optics, Applications to Solids" (H. J. Queisser, ed.). Springer, New York, 1977.
2. J. R. Maldonado, M. E. Poulsen, T. E. Saunders, F. Vratny, and A. Zacharias, *J. Vac. Sci. Technol.* **16,** 1942 (1979).
3. S. M. Mathews and R. S. Cooper, *Proc. SPIE* **333,** 136 (1982).
4. K. Suzuki, J. Matsui, T. Ono, and Y. Saito, *J. Electrochem. Soc.* **128,** 2434 (1981); S. Yamazaki and T. Hayasaka, *Jpn. J. Appl. Phys.* **19** (Suppl. 19-1), 35 (1979).
5. M. Feldman and A. D. White, U. S. Patent 4037969.
6. B. Fay, J. Trotel, and A. Frichet, *J. Vac. Sci. Technol.* **16,** 1954 (1979).
7. S. E. Bernacki and H. I. Smith, "6th Int. Conf. Electron and Ion Beam Science and Technology" (R. Bakish, ed.), p. 34. Electrochem. Soc., Princeton, New Jersey, 1974.
8. H. L. Stover, F. L. Hause, and D. McGreivy, *J. Vac. Sci. Technol.* **16,** 1635 (1979).
9. M. P. Lepselter, D. S. Alles, H. J. Levinstein, G. E. Smith, and H. A. Watson, *Proc. IEEE* **71,** 640 (1983).

Chapter 22

Oxides for VLSI

DAVID A. BAGLEE

Texas Instruments Incorporated
Houston, Texas

I. Introduction	381
II. Thermal Oxidation	382
A. Growth Rates of Oxides	383
B. Accelerated Oxidations	384
III. Leakage and Breakdown	385
A. Leakage	385
B. Breakdown	386
IV. Oxide Charges	389
A. Use of C-V Curves	389
B. Mobile Charge	391
C. Fixed Oxide Charge	392
D. Interface-Trapped Charge	392
E. Trapped Charge	392
F. Minimizing the Effects of Charges	392
V. Special Considerations	393
A. Field Oxides	393
B. Gate Oxides	396
C. Interlayer (Interlevel) Oxides	397
References	399

INTRODUCTION   I

Silicon dioxide plays many important roles in VLSI circuits. It is used for gate dielectrics in MOSFETs, for junction isolation between active devices, and for isolating the various levels of interconnects. The requirements of the oxide depend on the application. For example, SiO_2 is used as the dielectric medium in the capacitor storage cells of dynamic random access memories (DRAMs).

Increasing levels of integration are driving reduced feature sizes and hence thinner dielectrics. Figure 1 shows how the thickness of this oxide

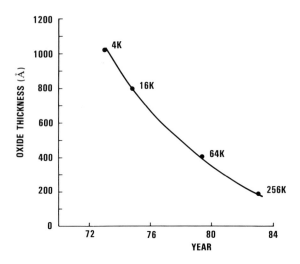

Fig. 1. DRAM capacitor oxide evolution.

has been reduced during the last 12 years. It is projected that a 1-Mbit DRAM will require a capacitor oxide thickness in the region of 100 Å. This will result in large electrical stresses and manufacturing problems.

In this chapter, we shall highlight the methods of growing oxides, summarize the important electrical characteristics, and then highlight special considerations for the various oxide applications in VLSI circuits.

II THERMAL OXIDATION

Before the silicon wafers can be thermally oxidized, they must be cleaned, immediately before insertion into the furnace. Typical preoxidation cleanups could consist of either of the following sequences.

Cleanup A		Cleanup B	
1.	$NH_4OH + H_2O_2$	1.	HF
2.	Deionized water	2.	Deionized water
3.	$HCL + H_2O_2$	3.	Hot HNO_3
4.	Deionized water	4.	Deionized water
5.	Spin dry	5.	Spin dry

SiO_2 is grown in either a wet or a dry oxygen ambient at temperatures between 800 and 1200°C. Small percentages of chlorine (3–8%) are usually added to the oxidizing ambient to minimize mobile-ion contamination during oxidation.

22. Oxides for VLSI

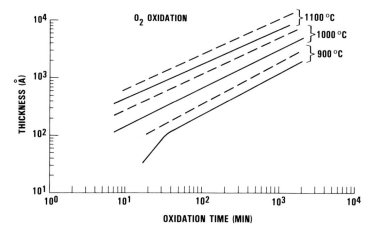

Fig. 2. Growth curves for dry oxidation. ———, no HCl present; ————, 4.5% HCl added.

Growth Rates of Oxides   A

Typical growth curves are given in Figs. 2 and 3. Note that chlorine increases the oxide growth rate.

$$\text{Si} + \text{O}_2 \longrightarrow \text{SiO}_2 \tag{1}$$

$$\text{Si} + 2\text{H}_2\text{O} \longrightarrow \text{SiO}_2 + 2\text{H}_2 \tag{2}$$

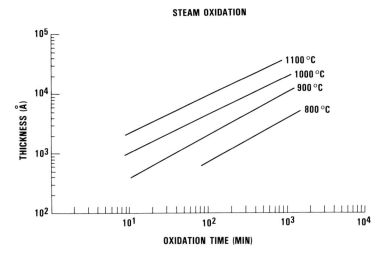

Fig. 3. Growth curves for pyrogenic steam oxidation.

For thin oxides and lower temperatures, the growth rate is a linear function of time because the oxidation takes place at the silicon–silicon dioxide interface; it is the surface reaction rate that is the limiting step. At higher temperatures and for thicker oxides, the oxidation is limited by the rate at which oxygen can diffuse through the silicon dioxide. Then the oxide thickness is proportional to the square root of the time. The expression in Eq. (3) gives the relationship between thickness and time:

$$X_{ox} = \frac{A}{2}\left[\sqrt{1 + \frac{t + \tau}{A^2/4B}} - 1\right]. \tag{3}$$

In the expression, A, B, and τ are constants that depend on oxidation conditions and the surface preparation techniques.

For longer oxidation times, Eq. (3) reduces to

$$X_{ox} = \sqrt{B(t + \tau)} \sim \sqrt{Bt}. \tag{4}$$

For shorter times, we obtain

$$X_{ox} = \frac{B}{A}(t + \tau). \tag{5}$$

B Accelerated Oxidations

Wet oxides grow much faster than dry oxides, which is very useful for growing the thick isolation (or field) oxides in VLSI circuits. The details of the reaction between water vapor and silicon are more complex than in Eq. (2). The water vapor reacts with the bridging oxygen ions:

$$H_2O + Si\text{—}O\text{—}Si \longrightarrow Si\text{—}OH + OH\text{—}Si. \tag{6}$$

These hydroxyl groups then react with the silicon lattice to form

$$\begin{array}{c} Si\text{—}OH \\ + Si\text{—}Si \longrightarrow \\ Si\text{—}OH \end{array} \begin{array}{c} Si\text{—}O\text{—}Si \\ + H_2 \\ Si\text{—}O\text{—}Si \end{array} \tag{7}$$

The hydrogen leaves the oxide layer and may react further with bridging oxygen to form more hydroxyl groups:

$$\tfrac{1}{2}H_2 + O\text{—}Si \longrightarrow OH\text{—}Si \tag{8}$$

The oxidation rate is much higher because of the weaker bonding of the hydroxyl groups compared with the bridging oxygen.

A wet ambient can be achieved by two methods, either by bubbling dry oxygen through water held close to its boiling point or, more commonly, by directly reacting hydrogen and oxygen in the furnance tube.

LEAKAGE AND BREAKDOWN III

Leakage A

The usual *test structure* for oxides is the MOS capacitor with a polysilicon or aluminum top plate. Figure 4 shows the energy band diagram for the Si–SiO$_2$ interface with positive bias on the upper capacitor electrode. Because of the large barrier (3.2 eV) at the interface, conduction is via Fowler–Nordheim tunneling. Electrons tunnel from the silicon conduction band into the oxide conduction band, where they are quite mobile ($\mu = 30$ cm^2/V·s) and are quickly swept out of the oxide. Conduction in SiO$_2$ via holes is negligible due to the higher interface barrier (4.2 eV) and their low mobility ($\mu < 10^{-3}$cm^2/V·s).

Fowler–Nordheim leakage is characterized by

$$J = AE^2 \exp(-B/E), \tag{9}$$

where J is the current density, E is the electric field, and A and B are constants.

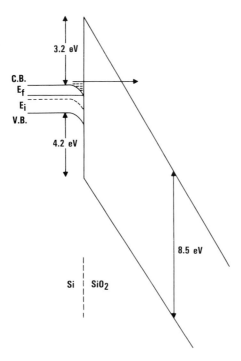

Fig. 4. Energy band diagram for Si–SiO$_2$ interface with positive bias applied to upper electrode. The arrow indicates the direction of electron tunneling.

Figure 5 shows leakage characteristics for various oxide thicknesses. These capacitors had n-doped polysilicon upper electrodes, and the gate oxides were thermally grown in dry O_2 with 4.5% HCl. Measurements were made by using a staircase voltage generator with the current being read one second after each bias was applied. In Fig. 6 we have taken the leakage data for the 200-Å oxide and replotted it to demonstrate the Fowler–Nordheim type behavior in high electric fields.

B   Breakdown

At some applied voltage the oxide will break down. Breakdown usually occurs when the electric field is over 9 MV/cm; it is thickness dependent. Breakdown occurs in localized regions and can be defect or impurity

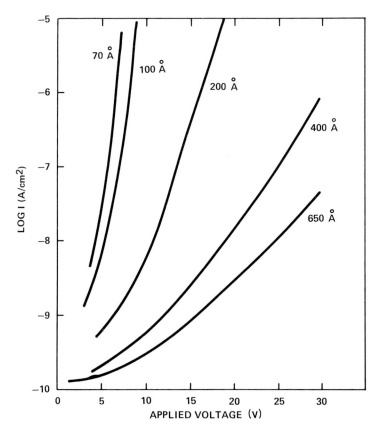

Fig. 5. Leakage current density as a function of applied voltage for various oxide thicknesses.

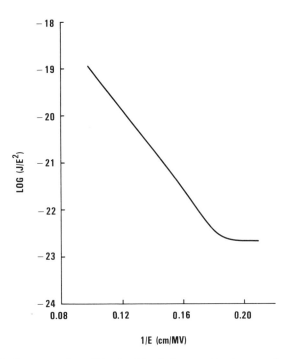

Fig. 6. Demonstration of Fowler–Nordheim tunneling in a 200-Å oxide.

related. For example, ionic contamination results in localized barrier lowering. Electrons injected into the oxide are accelerated under the high electric field, causing impact ionization and breakdown. A typical breakdown distribution for a 200-Å oxide is shown in Fig. 7. In Fig. 8 the dielectric strength of SiO_2 films is shown to increase as the oxide thickness decreases. This phenomenon can result in problems with breakdown measurements. Typically, a current level is chosen; and when the oxide leaks at that level, it is defined as having broken down. However, thin oxides can pass larger currents without breaking down than can thicker oxides; therefore, the definition of dielectric breakdown voltage must be chosen carefully. A simple technique that removes this ambiguity is shown in Fig. 9. After each stress voltage is applied, the capacitor is subjected to a low reference voltage. If the capacitor has broken down, a large current flows when V_{REF} is applied. This eliminates failures that are due merely to Fowler–Nordheim tunneling.

A number often quoted in industry is *defect density*, which can be calculated in a number of ways. Typically,

$$D = -\frac{1}{A} \log(Y), \qquad (10)$$

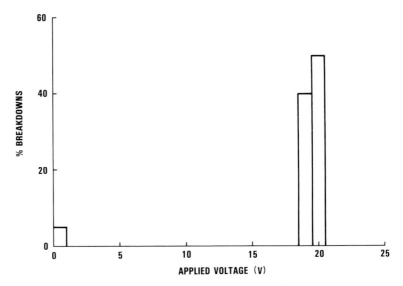

Fig. 7. Breakdown distribution of a 200-Å oxide.

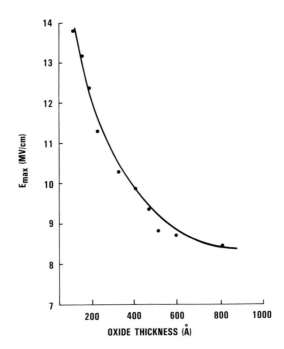

Fig. 8. Maximum dielectric strength as a function of oxide thickness.

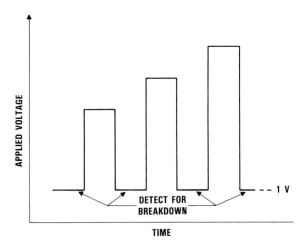

Fig. 9. Technique for obtaining breakdown distribution (in this case $V_{REF} = 1$ V).

where D is defect density, A is area of test device, and Y is yield. A number of different ways have been used to define *yield*. For example, it can be the percentage of good devices at a given voltage or those units breaking down above 80% of the maximum voltage. All techniques have validity, and the choice of technique depends on the application.

IV OXIDE CHARGES

The four types of charges that are usually associated with the silicon–silicon dioxide system are illustrated in Fig. 10, using the standard terminology [1].

A Use of C–V Curves

Important information about the oxide can be ascertained from high frequency C–V curves (Fig. 11). When a capacitor is biased in strong accumulation (negative bias for p-type substrates), the capacitance is that due to the oxide. Therefore,

$$X_{ox} = \varepsilon_o \varepsilon_r A / C_{max}, \tag{11}$$

where X_{ox} is oxide thickness, $\varepsilon_r = 3.9$, $\varepsilon_o = 8.85 \times 10^{-14}$ F/cm, A is area of capacitor, and $C_{max} = C_{ox}$ is capacitance in strong accumulation.

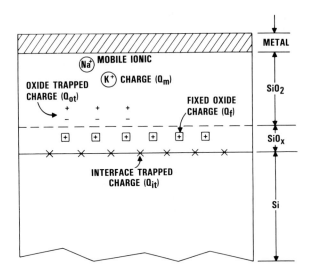

Fig. 10. Standard terminology for oxide charges. (After B. E. Deal, © IEEE.)

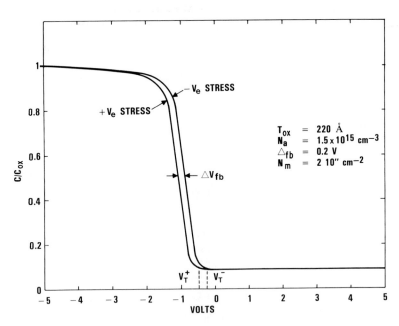

Fig. 11. Typical C–V curves for thin oxide after temperature bias stressing. This oxide was deliberately contaminated to demonstrate flatband voltage shift (typically $N_M \leq 10^{10}$ cm^{-2}).

22. Oxides for VLSI

The *doping density* can be calculated knowing C_{ox} and C_{min} (the minimum capacitance), by iteratively solving:

$$N_a = C \log N_a - B, \tag{12}$$

where

$$C = \frac{6.22 \times 10^5}{A^2} \left(\frac{C_{max} C_{min}}{C_{max} - C_{min}}\right)^2, \quad B = 23.4C,$$

and where A is area in square centimeters and C_{max} and C_{min} are in pico-Farods.

The *flatband voltage* V_{fb} can be measured from the C–V curve, once the doping density is known. At flatband

$$C_{fb} = \frac{C_{ox} C'}{C_{ox} + C'}, \quad C' = \frac{\varepsilon_s \varepsilon_o}{L_D}, \tag{13}$$

where L_D is extrinsic Debye length:

$$L_D = \sqrt{\frac{\varepsilon_o \varepsilon_s kT}{q^2 N_a}}, \quad \varepsilon_s = 11.8.$$

Thus, the flatband voltage is the point on the C–V curve at which $C = C_{fb}$. Monitoring flatband voltage gives a direct indication of process stability. The *threshold voltage* V_t can also be calculated:

$$V_t = V_{fb} + 2\phi_{fp} + \frac{\sqrt{2\varepsilon_o \varepsilon_s q N_a (2\phi_{fp})}}{C_{ox}}, \tag{14}$$

and

$$\phi_{fp} = \frac{kT}{q} \ln\left(\frac{N_a}{n_i}\right).$$

Threshold voltage is the onset of strong inversion in the device.

Mobile Charge   B

Mobile charge Q_m is due to ionic impurities in the oxide. These are usually sodium ions but can be potassium and lithium. When an external bias is applied, these ions drift through the oxide, because of their small size, causing shifts in the flatband voltage of the device. Mobile charge is easily measured by alternately biasing a capacitor negative and positive at a high temperature and obtaining a high frequency C–V curve after each stress condition.

The amount of mobile charge can be calculated from the shift in the C–V curve:

$$Q_m = C_{ox} \Delta V \tag{15}$$

where $Q_m = qN_m$ is mobile charge, C_{ox} oxide capacitance per unit area, and ΔV shift in C–V curve between negative and positive stress. A typical example is given in Fig. 11. Mobile charge is minimized by the addition of a small amount of chlorine during the oxidation process.

C Fixed Oxide Charge

Fixed oxide charge Q_f results from the stressed Si—O bonds near the silicon–oxide interface. This positive charge is located within 30 Å of the interface. It can be reduced by lowering the oxidation temperature and slowing the wafer cooling rate. It is difficult to separate Q_f from the metal–semiconductor work function difference (ϕ_{ms}). However, the two can be calculated by measuring V_{fb} for different thickness oxides. Since

$$V_{fb} = \phi_{ms} - qQ_F X_{ox}/\varepsilon_r\varepsilon_o, \tag{16}$$

plotting V_{fb} as a function of X_{ox} will yield ϕ_{ms} where $X_{ox} = 0$, and the slope will be proportional to Q_f.

D Interface-Trapped Charge

At the oxide–silicon interface there will be structural imperfections resulting in interface-trapped charge Q_{it}. These states can be filled or emptied depending on the bias conditions and cause a "smearing out" of the C–V curve. Interface states can be minimized by a hydrogen anneal at 450°C.

E Trapped Charge

The fourth type of charge, Q_{ot} is due to holes and electrons trapped in the bulk of the oxide. Trapped charge is usually a result of ionizing radiation, hot electron injection, or some similar process.

F Minimizing the Effects of Charges

It is important in VLSI processing to understand and be able to minimize the effects that each of these charges have on device operation. One of the advantages of continued scaling in oxide thickness is that the C_{ox} term is increasing and thereby reducing the effects that small amounts of oxide charge have on device operation. For example, if we assume $Q_m = 10^{11}$ cm^{-2} and $T_{ox} = 1000$ Å, then $\Delta V_{fb} = 0.46$; but if $Q_m = 10^{11}$ cm^{-2} and $T_{ox} = 100$ Å, then $\Delta V_{fb} = 0.046$.

SPECIAL CONSIDERATIONS V

Field Oxides A

Because the purpose of the field oxide is to electrically isolate various active regions, it is typically several thousand angstroms thick. A steam process called LOCOS (local oxidation of silicon) is usually used to make oxidation times reasonable (Fig. 12). The channel-stop implant (boron for n-channel devices) improves the isolation by further raising the field threshold voltage of the channel under the thick oxide. Oxide is grown only where the nitride was removed because Si$_3$N$_4$ acts as a diffusion block to oxygen. It has recently been found that high pressure oxidation

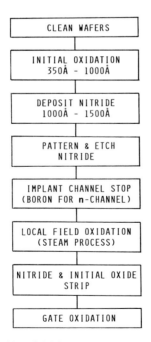

Fig. 12. LOCOS process flow diagram.

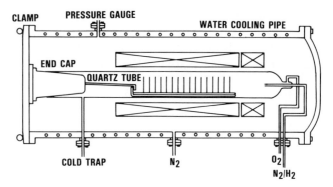

Fig. 13. Schematic diagram of high pressure oxidation (HIPOX) system. (After Hirayama [2], © IEEE.)

(HIPOX) is advantageous for growing field oxides. Pressures used are between 8 and 25 atmospheres, and oxidation time is significally reduced. A typical system is shown in Fig. 13. HIPOX reduces stacking faults and increases the field threshold voltage. Results of an experiment comparing HIPOX with atmospheric steam oxidation (APOX) are shown in Fig. 14. Field oxide thickness was approximately 6000 Å and V_{fon} is defined as the gate voltage required for 1 pA leakage with $V_{ds} = 8$ V and $V_{bb} = 0$ V.

The formation of birds' beaks and the diffusion of the channel stop during LOCOS processing is forcing technologists to look for alternatives. One possible approach is to use a *buried oxide* (Box) technique.

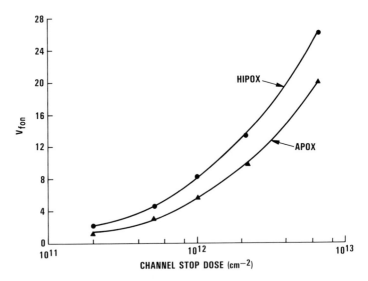

Fig. 14. Demonstration of improvement in field transistor turn-on voltage by use of HIPOX, for various boron channel stop doses.

This is shown schematically in Fig. 15a. It can be seen that the bird's beak has been eliminated.

A second alternative is *trench isolation*. This may be particularly beneficial in scaled CMOS circuits in which latchup is a problem. Figure 15b shows a cross section through a typical structure. The trench is etched by RIE and then refilled with either CVD oxide or undoped polysilicon. The problem with this technique is leakage along the trench sidewalls due to oxide charges; at this time, further development is needed to demonstrate a manufacturable processes.

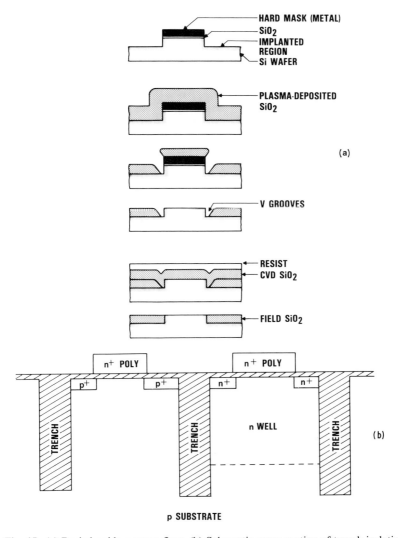

Fig. 15. (a) Buried-oxide process flow. (b) Schematic cross section of trench isolation.

B Gate Oxides

The cross section shown in Fig. 16 shows a typical DRAM cell, which features a transistor that transfers charge into and out of the storage capacitor. To achieve high capacitance from a small area, the capacitor oxide must be as thin as possible. Dynamic RAMs have large areas of thin capacitor oxide.

The *defect density* in the oxide must be very low to give reasonable yields of fully functional devices. It is also a requirement that the oxides be reliable. In Fig. 17 wear-out data are presented for a 100-Å oxide and are shown as a function of applied voltage. These data are typical of a log–normal distribution

$$F(t) = \frac{1}{\sqrt{2\pi}\,\sigma t} \exp\left[-\frac{1}{2}\frac{\ln(t) - \mu^2}{\sigma}\right], \qquad (17)$$

where μ is the mean and σ is the variance.

Increasing the *electric field* in the oxide increases the failure rate by a factor

$$A_{\text{ef}} = \exp\left[\frac{(E_\text{o} - E_\text{s})}{E_{\text{ef}}}\right], \qquad (18)$$

where E_{ef} is an empirical constant, E_s the stress field, and E_o the desired operational field. In Fig. 17 $A_{\text{ef}} \approx 70\text{–}100/\text{MV}\cdot\text{cm}$, which, for 100-Å oxides, means that increasing the *voltage stress* by 1 V accelerates failures by a factor of 70–100.

Similarly, increasing the temperature accelerates the rate of failure. The *temperature acceleration* A_t can be derived from an Arrhenius plot of time to reach a given failure percentage versus $1/T$

$$A_\text{t} = \exp\left[\frac{E_\text{a}}{k}\left(\frac{1}{T_\text{o}} - \frac{1}{T_\text{s}}\right)\right], \qquad (19)$$

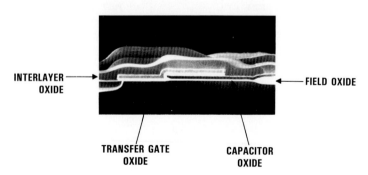

Fig. 16. Cross section through a typical DRAM storage cell.

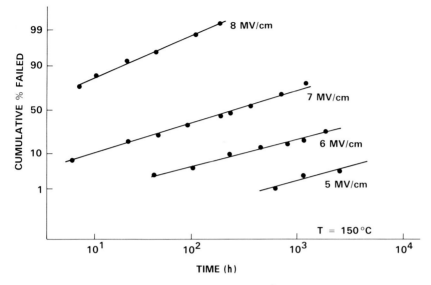

Fig. 17. Field accelerated wear-out in 100-Å oxides. $T = 150°C$.

where T_s is the stress temperature, T_o the operating temperature, k Boltzmann's constant, and E_a the activation energy.

Activation energies in the range 0.3–1.0 eV have been reported for silicon dioxide. In addition to oxide wear-out, gate oxides are also susceptible to *hot electron injection and trapping*. This is usually a problem with short channel MOS transistors (<1.5 μm). Carriers in the pinch-off region of the transistor channel are accelerated under the strong electric field. Collisions with the lattice make it possible for them to generate electron–hole pairs; and a percentage of the carriers will have sufficient energy to be injected into the oxide. In addition to contributing to gate current, some carriers become trapped, modifying the electric field near the drain of the transistor. This leads to a decrease in current drive capabilities and threshold voltage shifts in the MOSFETs. Hot carrier effects can be reduced by employing various graded drain structures to lower the peak channel electric fields.

Interlayer (Interlevel) Oxides C

Interlayer oxides play an important role in separating various levels of interconnect from each other and from the active devices. These oxides are typically quite thick (1 μm) and cannot be thermally grown as in the case of field oxides or gate oxides. The oxides must, however, have certain characteristics. They must be economical to deposit (i.e., have high deposition rates), they must have low particulate generation to reduce pinhole density, and they must have a low relative permittivity to reduce capacitative coupling. One final factor is that the films should have low tensile stress to reduce cracking in the films. Techniques for interlayer-oxide deposition have been evolving from atmospheric CVD oxides to LPCVD oxides to plasma-enhanced LPCVD oxides. All techniques rely on the reaction of silane with oxygen in the temperature range of 400–700°C.

$$SiH_4 + O_2 \longrightarrow SiO_2 + 2H_2 \tag{20}$$

These deposited oxides are usually doped with phosphorous up to approximately 8 wt. %. This allows the oxide film to reflow when heated to about 950°C, reducing the abrupt steps at contacts and improving subsequent metal step coverage. Phosphorous also acts as a gettering agent for any ionic impurities. Increasing the phosphorous concentration increases the amount of reflow. However, above 8% phosphorous can lead to reliability problems on finished devices.

Moisture in the package can react with the phosphorous to give phosphoric acid, which attacks aluminum metallization, resulting in failures after limited field operation. LPCVD deposition techniques have been found to reduce particulate contamination, but this is at the expense of slower deposition rates (100 Å/min for LPCVD vs 1000 Å/min for atmospheric CVD oxides). Plasma-enhanced LPCVD oxides offer deposition rates of around 500 Å/min, and the stress is compressive, reducing the likelihood of cracking in the film. The penalty that is paid is in higher capacitative coupling due to a larger relative dielectric constant, as shown below.

Oxide type	Relative dielectric constant
Atmospheric CVD oxide	3.8–4.1
Low-pressure CVD oxide	3.8–4.1
Plasma-enhanced LPCVD oxide	4.6
Thermal SiO_2	3.9

This chapter was intended to provide some insight into the demands placed on oxides for VLSI circuits, how these films are formed, and what electrical properties can be expected. The references [3–7] are intended to provide the reader with an initial starting point for obtaining further information.

REFERENCES

1. B. E. Deal, *IEEE Trans. Electron Devices* **27,** 607 (1980).
2. M. Hirayama, H. Miyoshi, N. Tsubouchi, H. Abe, *IEEE J. Solid State Circuits* **17,** 133 (1982).
3. S. K. Lai, *in* "Semiconductor Silicon 1981" (H. R. Huff and R. J. Kriegler, eds.), p. 416. Electrochemical Society, Princeton, New Jersey, 1981.
4. E. H. Nicollian and J. R. Brews, "MOS Physics and Technology." Wiley (Interscience), New York, 1982.
5. S. M. Sze, "Physics of Semiconductor Devices." Wiley (Interscience), New York, 1981.
6. D. J. Dimaria, *in* "The Physics of SiO_2 and Its Interfaces" (S. T. Pantalides, ed.), p. 160. Pergammon, Oxford, 1978.
7. D. A. Baglee, P. L. Shah, Ultra-thin gate dielectric processes for VLSI applications, *in* "VLSI Electronics: Microstructure Science," vol. 7, ch. 4. Academic Press, New York, 1984.
8. J. W. McPherson, D. A. Baglee, *in* "Proceedings of 23rd International Reliability Physics Symposium," (1985).

Chapter 23
Nitrides for VLSI

MICHAEL P. DUANE
DAVID A. BAGLEE
Texas Instruments Incorporated
Houston, Texas

I. Introduction	401
II. Film Formation	402
A. Thermal Nidridation of Silicon	402
B. Chemical Vapor Deposition of Nitrides	404
C. Thermal Nitridation of SiO_2	405
D. Implantation	407
III. Electrical Properties	407
A. Conduction	407
B. Capacitance–Voltage Characteristics	409
IV. Applications of Nitrides	411
A. Circuit Isolation	411
B. Nonvolatile Memories	413
References	414

INTRODUCTION I

Silicon nitride, a counterpart to SiO_2, is increasingly used in the microelectronics industry. However, the properties of silicon nitride are not as well understood, partially because of the many ways in which the films can be prepared and the difficulties of achieving pure stoichiometric Si_3N_4. Depending on the method of preparation, the films can be silicon rich, can incorporate oxygen as in nitridated oxides (oxynitrides), or can incorporate large atomic percentages of hydrogen. The characteristics of the nitride are dependent on the final composition of the film.

II FILM FORMATION

A Thermal Nitridation of Silicon

The direct growth of silicon nitride is usually accomplished by heating the silicon substrates to a high temperature in nitrogen or ammonia ambients. In either case, the growth is *self-limiting* because the high density of the resulting nitride retards the diffusion of the reactants through the film. (This is in contrast to silicon dioxide, through which oxygen can readily diffuse.) The maximum thickness of nitrides grown in this manner is typically less than 70 Å. This feature can be exploited to grow nitrides of reproducible thicknesses.

1 Nitridating Ambients

If the nitridating ambient is *nitrogen,* then temperatures greater than 1200°C are required to dissociate the nitrogen molecule, but the reaction with *ammonia* can occur at temperatures of less than 1000°C because of the weaker bonding of the ammonia molecule. One reason why ammonia is more commonly used is that VLSI processes are moving towards lower-temperature steps to minimize induced stresses in large diameter silicon wafers.

2 Growth Rate in Ammonia

Figure 1 shows the growth rate of silicon nitride in ammonia for various temperatures on (100) silicon substrates [1]. Unlike the oxidation of silicon, little effect is seen by using (111) substrates or increasing the ambient pressure. Grown nitride thickness has been found empirically to fit the equation

$$X = at^n,$$

where x is film thickness, t is time, and a and n are constants.

3 Effects of Moisture and Oxygen

Uniform films can be grown only if the oxygen and moisture concentrations are below 1 ppm, because these contaminants react with the silicon more readily than the nitrogen will. Care must also be taken to minimize the native oxide thickness. At 1200°C, this oxide can be removed by H_2 or HCl etching. For lower-temperature work, the silicon should remain in an

23. Nitrides for VLSI

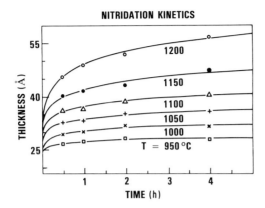

Fig. 1. Growth curves for silicon nitride in an ammonia ambient. $n = 2.0$. (After Moslehi et al. [1], originally presented at the Spring 1983 Meeting of the Electrochemical Society, Inc., held in San Francisco, California.)

inert atmosphere after removal from an HF dip. Silicon carbide coated susceptors and cold wall reactors also reduce oxygen contamination.

Growth by Plasma Excitation 4

If thicker thermal nitrides are desired, films up to 200 Å thick can easily be grown by plasma excitation of the ammonia gas. However, these films tend to incorporate more oxygen than do non-plasma films. A typical system is illustrated in Fig. 2.

Reaction Kinetics 5

The reaction kinetics for nitride growth are not well understood. Nitridation probably occurs throughout the film and not just at the insulator–silicon interface as in the case for oxides. The use of a plasma increases the nitridation rate by increasing the atomic nitrogen concentration.

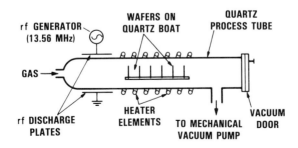

Fig. 2. Schematic diagram of plasma-enhanced nitridation reactor.

B Chemical Vapor Deposition of Nitrides

Chemical vapor deposition (CVD) is the most common method for realizing silicon nitride films in integrated circuits. There is a wide variation in the processing conditions for these films. A typical reaction is the combination of silane (SiH_4) and ammonia (NH_3) in an inert atmosphere, usually N_2, Ar, or H_2. The flow ratio of NH_3/SiH_4 can vary from 1 to 1000, with the films being silicon-rich for the lower ratios (<300). Less oxygen is incorporated in the final film if the reaction takes place at low pressure (several hundred millitorr). Dichlorosilane (SiH_2Cl_2) is used instead of silane in low-pressure CVD (LPCVD) systems for uniformity. The NH_3/SiH_2Cl_2 ratio is typically between 1 and 100. Temperatures for both methods range between 600 and 900°C, with 750 to 800°C being typical.

1 Plasma and Photoenhancement

Silicon nitride can be deposited at much lower temperatures with plasma enhancement (PECVD) or photoenhancement (laser or UV light). PECVD nitrides are commonly used as the final passivation layer in integrated circuits, but the plasma results in radiative damage that often is not annealed out. These films have moderate compressive stress, as opposed to thermal CVD films which are tensile. Figure 3 shows the variation of stress in Si_3N_4 versus temperature [2]. The 1-μm thick film was deposited

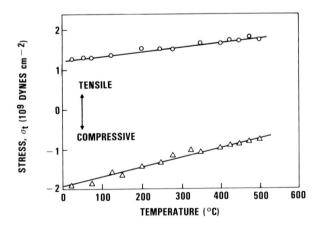

Fig. 3. Stress of PECVD films deposted at 280°C on silicon. SiN is tensile, Si_3N_4 is compressive. (After Sinha *et al.* [2].)

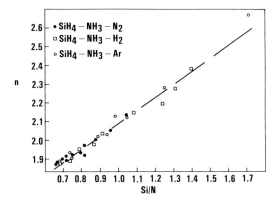

Fig. 4. Refractive index of PECVD films versus Si/N ratio. (After Claasen et al. [3], originally presented at the Spring 1983 Meeting of the Electrochemical Society, Inc., held in San Francisco, California.)

at 280°C and 350 W directly on silicon. A SiN film deposited at 250 W was slightly tensile.

Plasma nitrides tend to be silicon-rich, with the more stoichiometric layers being grown at high NH_3/SiH_4 ratios. The Si/N ratio is usually determined by Auger electron spectroscopy (AES) or Rutherford backscattering (RBS), but there is a good correlation between this ratio and the refractive index or the etch rate in buffered HF. As Si/N approaches 0.75, the etch rate increases and the refractive index decreases. This latter relationship is illustrated in Fig. 4. PECVD nitrides contain large atomic percentages of hydrogen, often as high as 20 or 30%. The hydrogen content can be reduced by depositing at higher temperatures or using N_2 instead of NH_3. However, NH_3 films have better thickness and refractive index uniformity.

Plasma Excitation Frequency 2

Another parameter in the deposition process is the plasma excitation frequency. At lower frequencies, the silane dissociates better, resulting in films that are less silicon-rich.

Thermal Nitridation of SiO_2 C

Silicon nitrides are generally more conductive and have higher surface state densities than SiO_2, but they also have higher breakdown fields and are better diffusion barriers. Nitridation of a SiO_2 layer retains the interface properties of the oxide and strengthens the film. The resulting film composition depends on the deposition conditions and the ammonia purity and is often called a nitridated oxide. Of the two reactions,

$$2SiO_2 + 2NH_3 \longrightarrow Si_2N_2O + 3H_2O,$$

$$3SiO_2 + 4NH_3 \longrightarrow Si_3N_4 + 6H_2O,$$

the first is more thermodynamically likely, but it is best to think of a nitridated oxide as a film whose nitrogen fraction varies with depth.

1 Nitrogen Profiles in Films

The nitrogen profile depends on the pressure, temperature, time, and ammonia purity. For long times, the nitrogen concentration becomes fairly uniform [4], as shown in Fig. 5, where some hydrocarbon contamination is present on the film surface. For shorter times, the nitrogen concentration peaks at the surface and decreases with depth. A concentration increase, often seen at the silicon interface, is thought to be due to a process similar to the *Kooi effect*. This peak is not seen when high-purity ammonia is used. The final film can be thicker than the starting SiO_2 film by tens of angstroms.

2 Advantages and Disadvantages of Nitridation

Nitridated oxides can have a breakdown field of 12 MV/cm. They are less susceptible to radiation damage from later processing than non-nitridated oxides and can withstand high-temperature hydrogen annealing.

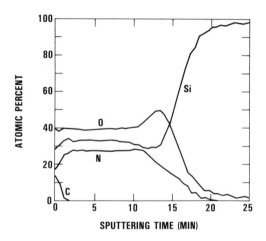

Fig. 5. Auger depth profile of an oxide nitridated in ammonia at 1150°C for 6 h. (After Chang *et al.* [4].)

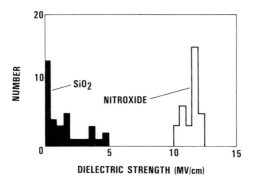

Fig. 6. Breakdown field comparison between oxide and nitridated oxide after 1100°C forming-gas annealing. (After Ito *et al.* [5].)

Figure 6 shows the improved resistance of a nitridated oxide to 5% H_2 forming-gas annealing [5]. The original breakdown strength of the oxide was near 10 MV/cm. After 30 min of 1100°C forming-gas annealing, the non-nitridated oxides have greatly degraded whereas the nitridated oxides have not. On the negative side, the electron-trap density in nitridated oxides tends to be higher, but this property is sometimes used to create memory devices.

Implantation D

Si_3N_4 can be formed by implanting very high doses of nitrogen into Si and then annealing. The implantation is typically done at energies less than 20 keV with doses of $10^{17}/cm^2$. This does not result in a high quality film, but it illustrates one more technique for nitride formation. Such a method requires long implant times and is therefore not as economical as, for instance, a CVD nitride.

ELECTRICAL PROPERTIES III

Conduction A

A primary technique for characterizing nitride films is the determination of *I–V* characteristics. The conduction mechanism appears to be strongly dependent on the manner in which the films are deposited or grown.

Early work in silicon nitride films demonstrated that conduction in MNS (metal–nitride–silicon) devices was dominated by a Frenkel–Poole mechanism in which carriers are emitted from traps in the band gap into the silicon nitride conduction band as a result of field-enhanced thermal excitation. The Frenkel–Poole effect can be characterized by

$$J = KE \exp\left\{-\frac{q}{KT}\left[\phi_b - \left(\frac{qE}{\pi\varepsilon_0\varepsilon_d}\right)^{1/2}\right]\right\},$$

where J is current density, K a constant, E the applied electric field, ϕ_b the trap depth (electron volts), and ε_d the dynamic dielectric constant. In Fig. 7, we have plotted the results of various workers. The data of Mar (curve 1) and Nagy (curve 2) are from plasma deposited nitride films; the data of Sze (curve 3) and Brown (curve 4) are from CVD nitrides. Taguchi's data (curve 5) were obtained from films grown using a plasma enhanced thermal nitridation process at 1000°C, and those of Moslehi

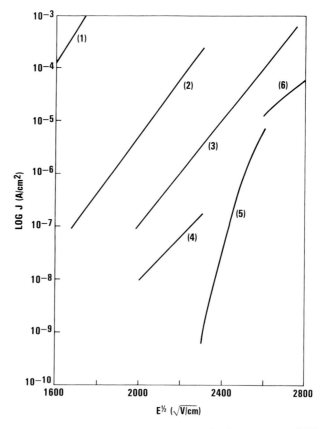

Fig. 7. Leakage data of various workers: (1) Mar *et al.* [6]; (2) Nagy *et al.* [7]; (3) Sze [8]; (4) Brown *et al.* [9]; (5) Taguchi *et al.* [10]; (6) Moslehi *et al.* [1].

(curve 6) from thermal nitridation in a cold wall rf-heated reactor. It is evident from the data that the current density can vary by several orders of magnitude at any given electric field. Notice, however, that the slopes of the first four curves are very similar and in good agreement with a Frenkel–Poole conduction mechanism.

It is possible to estimate the depth of the Frenkel–Poole traps. The dynamic dielectric constant ε_d must first be calculated from the slope of the log J versus $E^{1/2}$ curves in Fig. 7. A thermal activation energy must then be determined from an Arrhenius plot of conduction current (at a given electric field) versus reciprocal temperature.

Having calculated these parameters, we determine the trap depth from

$$\phi_b = E_a/q + (qE/\pi\varepsilon_o\varepsilon_d)^{1/2}.$$

Nagy et al. [7] determined $\varepsilon_d = 6.02$, $E_a = 0.303$ eV, and trap depth = 0.84 eV.

The data of Taguchi and Moslehi do not give a good fit to a Frenkel–Poole mechanism but do fit a Fowler–Nordheim tunneling process similar to that seen in thermally grown oxides. Fowler–Nordheim tunneling is characterized by

$$J = AE^2 \exp(-B/E).$$

In Fig. 8, the data of Taguchi are replotted to demonstrate the fit to this equation. It can be seen that the plot of $\log(J/E^2)$ versus $1/E$ yields a straight line over several orders of magnitude.

Capacitance–Voltage Characteristics   B

Nitride versus Oxide   1

Theory similar to that discussed in Chapter 22 on oxides can be applied to nitrides, with several qualifications: The higher relative dielectric constant of $\varepsilon_r \sim 7$ for Si_3N_4 versus 3.9 for oxides; the insensitivity to mobile contaminants, such as sodium, which are unable to diffuse through the nitride; and the ability of silicon nitride to trap large amounts of charge (positive or negative).

Effects of Electric Fields   2

Figure 9 shows the shifts in C–V curves caused by the application of positive or negative electric fields. Note that the C–V curve is shifted in the same direction as the polarity of the applied stress field. A positive electric field causes the flatband voltage to become more positive, indicating that the dielectric is trapping negative charge; a negative field causes

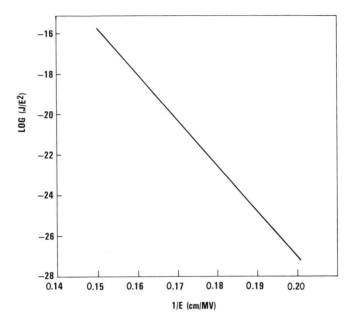

Fig. 8. Data of Taguchi *et al.* [10] replotted to illustrate Fowler–Nordheim tunneling.

positive charge to be trapped. Long-term measurements show that the flatband voltage decays with time due to the detrapping of carriers in the nitride.

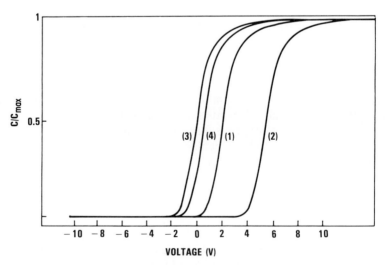

Fig. 9. $C-V$ curves for 400-Å MNS structure, illustrating charge trapping: (1) Initial. (2) After +20 V for 1 s. (3) After −20 V for 1 s. (4) After 1000-min delay. (After Nagy *et al.* [7], originally presented at the Spring 1983 Meeting of the Electrochemical Society, Inc., held in San Francisco, California.)

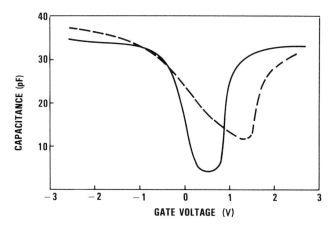

Fig. 10. Effect of boron diffusion in 100-Å oxides on low-frequency C–V curves: ——— initial C–V; ---- after 950°C 1 h anneal. Note shift in flatband voltage. (After Wong et al. [11].) Reprinted by permission of the publisher, The Electrochemical Society, Inc.

3 Blockage of Boron Diffusion

With increasing interest in CMOS processing, it is becoming desirable to use p-type polysilicon as opposed to the more conventional n-type polysilicon. (The work function for p-type is approximately 1.1 V higher.) Unfortunately, the boron used to dope the gate electrode is very mobile in the subsequent high-temperature process steps. It readily drifts through thin gate oxides and modifies the transistor channel doping profile, causing a change in threshold voltage. In Fig. 10, the effects of boron drifting through the gate oxide can be seen in the change in the low-frequency C–V curves [11]. Nitridated oxide films have been found to be very effective in blocking boron diffusion, and no shift in C–V curves is seen from initial characteristics after high-temperature treatments. This is probably due to the thin nitride surface layer that is usually formed in oxynitrides. Such a layer, as stated earlier, is an effective diffusion barrier.

IV APPLICATIONS OF NITRIDES

A Circuit Isolation

1 The LOCOS Process

Silicon nitride is commonly used in the industry standard *local oxidation of silicon* (LOCOS) process for growing thick oxide regions to isolate

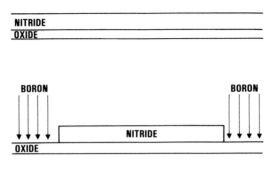

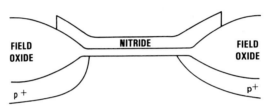

Fig. 11. Basic steps in LOCOS isolation.

active devices. A flow diagram for a typical LOCOS process is given in Fig. 12 of Chapter 22 in this book. In Fig. 11, we illustrate the basic steps.

2 Problems with the LOCOS Process

In Chapter 22, we also alluded to some of the problems with a LOCOS process. The first is the Kooi or white ribbon effect, which is a localized thinning of the gate oxide at the edge of isolation oxides, resulting in unexpectedly low breakdown voltages. The Kooi mechanism is illustrated in Fig. 12, along with a second problem: Particulate contamination or voids created in the nitride during deposition cause pinholes or locally thinned regions in the nitride. During the steam oxidation, some nitridation of the silicon surface occurs, resulting in a locally thin gate oxide with low breakdown voltages. The growth and strip of a sacrificial gate oxide has been shown to eliminate the Kooi effect [12].

Elimination of the second problem can be achieved by ensuring that the pad oxide is free from contaminants and that the number of particles generated is minimized. High-pressure scrubbing of slices after nitride deposition is not recommended: It has been shown to increase the number of defects in nitride films.

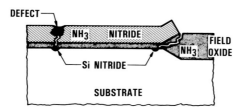

Fig. 12. Potential defects induced during LOCOS isolation formation. (After Goodwin et al. [12].) Reprinted by permission of the publisher, The Electrochemical Society, Inc.

Nonvolatile Memories   B

The use of nitrides in nonvolatile memories falls into two categories: (a) as part of a dual dielectric in MNOS (metal–Si_3N_4–SiO_2–Si) device and (b) in EEPROMs (electrically erasable programmable read-only memories).

Use in MNOS Devices   1

A MNOS device is similar in many ways to a conventional MOSFET except that the gate dielectric is composed of a thin oxide layer and a thicker, deposited silicon nitride layer. The threshold voltage is controlled by the presence (or absence) of charge trapped at the oxide–nitride interface. Writing is achieved by grounding the source and drain, and raising the gate to a high potential; this is illustrated in Fig. 13a. The current in the oxide is due to the expected Fowler–Nordheim tunneling, whereas that in the nitride is Frenkel–Poole emission. Due to the large number of traps at the oxide–nitride interface, some electrons become trapped. Trapping of negative charge raises the threshold voltage of the MNOS transistor and represents a logical-1 data-bit.

Erasing is achieved by grounding the gate and raising the substrate potential (Fig. 13b). Electrons are then emitted from the interface traps, resulting in a negative threshold voltage. This erased condition represents a stored 0 data-bit. Charge trapped in MNOS devices is much more stable than in MNS devices, such as those described in Section IIIB.

Use in EEPROMs   2

The second application of nitrides in nonvolatile memories involves the use of oxynitrides in EEPROMs. In this memory device, data are stored by charging or discharging a floating polysilicon gate through a very thin insulator (typically 150 Å). Initially, this insulator was silicon dioxide but

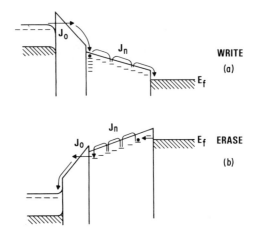

Fig. 13. Write and erase mechanisms in a MNOS memory cell.

severe degradation in performance was observed, with the voltage window between a '1' and a '0' closing up because of charge trapped in the oxide. Nitridated oxides, such as those described in Section IIC, offer significant improvements, including a higher breakdown field and more efficient data writing due to a lower tunnel barrier. A more efficient write cycle results in less charge becoming trapped on the insulator. It has been found that an EEPROM with a nitridated oxide is usable for more than 10^7 write/erase cycles, approximately 100 times as many as with pure SiO_2.

REFERENCES

1. M. M. Moslehi and K. C. Saraswat, *in* "Proc. Symp. on Silicon Nitride Thin Insulating Films," p. 324. Electrochemical Society, Princeton, New Jersey, 1983.
2. A. K. Sinha, H. J. Levinstein, and T. E. Smith, *J. Appl. Phys.* **49** (4), 2423 (1978).
3. W. A. Claassen, W. G. Valkenberg, F. H. Habraken, and Y. Tamminga, *J. Electrochem. Soc.* **130** (12), 2420 (1983).
4. S. T. Chang, N. M. Johnson, and S. A. Lyon, *Appl. Phys. Lett.* **44** (3), 316 (1984).
5. T. Ito, T. Nakamura, and H. Ishikawa, *IEEE J. Solid State Circuits* **17** (7), 2951 (1982).
6. K. M. Mar and G. M. Samuelson, *Solid State Technol.* **23** (4), 137 (1980).
7. T. E. Nagy, C. D. Fung, and W. H. Ko, *in* "Proc. Symp. on Silicon Nitride Thin Insulating Films," p. 167. Electrochemical Society, Princeton, New Jersey, 1983.
8. S. M. Sze, *J. Appl. Phys.* **38** (7), 2951 (1967).
9. G. A. Brown, W. C. Robinette, and H. G. Carlson, *J. Electrochem. Soc.* **115** (9), 948 (1968).
10. M. Taguchi, T. Ito, T. Fukano, T. Nakamura, and H. Ishikawa, *IEDM Tech. Dig.* p. 400 (1981).
11. S. S. Wong, C. G. Sodini, T. W. Ekstedt, H. R. Grinolds, K. H. Jackson, and S. H. Kwan, *J. Electrochem. Soc.* **130** (5), 1139 (1983).
12. C. A. Goodwin and J. W. Brossan, *J. Electrochem. Soc.* **129** (5), 1066 (1982).

Chapter 24

Silicides

MARC-A. NICOLET

California Institute of Technology
Pasadena, California

S. S. LAU

University of California at San Diego
La Jolla, California

Introduction to Tables	415
References	433

INTRODUCTION TO TABLES

The application of transition-metal silicides in VLSI circuits is twofold:

(1) Silicides are used as part of the contact that joins an interconnecting line to the silicon substrate in a contact window; there, the electric current flows mainly perpendicularly to the silicide layer and across the silicide–silicon interface. The electrical characteristics of the contact between silicide and silicon are the prime interest (Table V).

(2) Silicides are used in interconnection lines in which the electric current flows mainly along the plane of the silicide film. A main parameter of interest is the resistivity of the silicide (Table IV).

There is also a general concern for the formation and the behavior of silicides under high-temperature annealing in subsequent processing steps. The tables in this chapter provide data on pertinent parameters concerning these subjects. References to further compilations are also given [1–9].

TABLE I

Crystallographic Data of Transition Metals and Their Silicides[a]

(1) Element	(2) Silicide	(3) Crystal system	(4) Name of structure	(5) Strukturberichte classification
[22]Ti	Ti_5Si_3	Hexag.	Mn_5Si_3	$D8_8$
	TiSi	Orthor.	—	C_{2v}^1 or D_{2h}^1
	$TiSi_2$	Orthor.	—	C54
			$ZrSi_2$	C49
[23]V	VSi_2	Hexag.	$CrSi_2$	C40
[24]Cr	$CrSi_2$	Hexag.	—	C40
[25]Mn	MnSi	Cubic	FeSi	B20
	$MnSi_2$	Tetrag.	—	—
[26]Fe	Fe_3Si	Cubic	Cu_2MnAl	—
	α'-Fe_3Si	Cubic	BiF_3	$D0_3$
	FeSi	Cubic	FeSi	B28
				B20
	α-$FeSi_2$ ($>960°C$)	Tetrag.	—	—
	β-$FeSi_2$ ($<960°C$)	Orthor.	Deformed Cl, represented by $CoSi_2$	—
[27]Co	Co_2Si	Orthor.	$PbCl_2$	C23
	CoSi	Cubic	FeSi	B20
			FeSi	B28
	$CoSi_2$	Cubic	CaF_2	C1
[28]Ni	δ-Ni_2Si (LT)	Orthor.	$PbCl_2$	C23
	θ-Ni_2Si (HT)	Hexag.	—	—
	η-NiSi	Orthor.	MnP	B31
	NiSi	Cubic	FeSi	B28
	$NiSi_2$	Cubic	—	C1
	ζ-$NiSi_2$	Rhomb.	—	—
[39]Y	$YSi_{\sim 1.7}$ (Si rich)	Hexag.	—	—
	YSi_2		AlB_2	—
		Orthor.	$GdSi_2$	—
		Tetrag.	$ThSi_2$	—
[40]Zr	$ZrSi_2$	Orthor.	—	C49
[41]Nb	$NbSi_2$	Hexag.	$CrSi_2$	—
			$CrSi_2$	—
[42]Mo	$MoSi_2$	Tetrag.	—	C11
	[$T \leq 600°C$]	Hexag.	[$CrSi_2$]	C40
	[$T \geq 700–800°C$]	Tetrag.*	$MoSi_2$	$C11_b$

24. Silicides

(6–8) Lattice constants of unit cell (Å)			(9) No. of formula units per unit cell	(10) No. density of formula units ($10^{22}/cm^3$)	(11) Density (g/cm³)	(12)
a	b	c			D_x	D_m
7.429	—	5.1392	2	[0.8140]	4.376	—
3.618	6.492	4.973	4	[3.440]	4.34	4.21
8.236	4.773	8.523	[8]*	[2.388]	[4.126]*	4.02
3.62	13.76	3.605	4	[2.278]	3.85	3.9
4.571	—	6.372	3	[2.602]	[4.628]	—
4.428	—	6.363	3	[2.777]	[4.988]	4.91
4.558	—	—	4	[4.226]	5.826	—
5.513 ± 3	—	17.422 ± 5	16	[2.997]	5.53	5.3
2.824 ± 1	—	—	—	—	—	—
5.6554	—	—	4	[2.211]	[7.182]	—
4.487	—	—	4	[4.428]	[6.171]	—
—	—	—	—	—	—	—
2.6901	—	5.134	1.99 Si+	[2.523]	[4.260]	4.54
2.684	—	5.128	0.82 Fe	[2.537]	[4.283]	—
2.684	—	5.128	1	[2.706]	5.034	4.74
9.863	7.791	7.833	16	[2.649]	4.93	4.94
4.918	3.737	7.109	4	[3.062]	7.42	—
4.4426	—	—	4	[4.562]	[6.592]	6.506
4.447 ± 2	—	—	4	[4.548]	6.571	—
5.365	—	—	4	[2.590]	4.950	4.94
4.99	7.03	3.72	4	[3.065]	7.405	7.23
7.40	9.92	7.05	—	—	—	—
3.805	—	4.890	2	[3.262]	7.880	6.85
5.18	5.62	3.34	4	[4.114]	5.982	5.86
4.446	—	—	4	[4.552]	6.559	—
5.395 ± 3	—	—	4	[2.547]	4.859	—
8.881	—	α = 90°23′	—	—	—	—
3.836 ± 1	—	4.139 ± 1	—	—	—	—
3.85	—	4.14	—	—	—	—
4.052	3.954	13.360	—	—	—	—
4.04	—	13.42	—	—	—	—
3.72	14.61	3.67	4	[1.993]	4.878	4.88
4.78	—	6.56	3	[2.311]	5.720	—
4.7971	—	6.592	3	[2.283]	5.652	—
3.200 ± 5	—	7.861 ± 5	2	[2.470]	6.24	~6.0
4.605	—	6.559	[3]	[2.491]	[6.29]	—
3.206 or 3.203	—	7.877 or 7.887	2	[2.852] or [2.854]	[7.209] or [7.208]	—

(Continued)

TABLE I (*Continued*)

(1) Element	(2) Silicide	(3) Crystal system	(4) Name of structure	(5) Strukturberichte classification
^{44}Ru	Ru$_2$Si$_3$	Tetrag.	—	—
		Orthor.	—	—
^{45}Rh	Rh$_2$Si	Orthor.	PbCl$_2$	C23
	RhSi	Orthor.	MnP	B31
		Cubic	FeSi	B2
			—	B20
	Rh$_4$Si$_5$	Monocl.	—	—
	Rh$_3$Si$_4$	Orthor.	—	—
	RhSi$_2$	Cubic	CsCl (with trace Al)	B-2
^{46}Pd	Pd$_2$Si	Hexag.	Fe$_2$P	C22
	PdSi	Orthor.	MnP	B31
^{57}La	LaSi$_2$	Tetrag.	α-Si$_2$Th	C$_c$
^{72}Hf	HfSi	Orthor.	FeB	B27
		Hexag.	—	—
	HfSi$_2$	Orthor.	ZrSi$_2$	C49 (12)
^{73}Ta	TaSi$_2$	Hexag.	CrSi$_2$	C40
^{74}W	WSi$_2$			
	[$T \geq 600°C$]	Tetrag.	MoSi$_2$	C11$_b$ (12)
	[$T < 550°C$]	Hexag.	CrSi$_2$	C40
^{75}Re	ReSi	Cubic	FeSi	B20
	ReSi$_2$	Tetrag.	MoSi$_2$	C11$_b$
^{76}Os	OsSi	Cubic	FeSi	B20
		Cubic	CsCl	—
	Os$_2$Si$_3$	Tetrag.	Ru$_2$Si$_3$	—
		Orthor.	RuSi$_{1.5}$	—
		Pseudohexag. or tetrag.	—	—
		Orthor.	—	—
	OsSi$_{1.8}$ (?)	—	—	—
	OsSi$_2$	Orthor.	β-FeSi$_2$	—
		Monocl.	OsGe$_2$	—
			As$_2$Nb (?)	—
^{77}Ir	IrSi	Orthor.	MnP (18)	—
	IrSi$_{1.75}$	Monocl.	—	—
	IrSi$_2$	—	—	—
	IrSi$_3$	Hexag.	AsNa$_3$ (?)	D0$_{18}$
^{78}Pt	Pt$_2$Si	Tetrag.	—	Deformed C1
		Cubic	CaF$_2$	—
	α-Pt$_2$Si (LT)	Tetrag.	ZrH$_2$	—

24. Silicides   419

(6–8) Lattice constants of unit cell (Å)			(9) No. of formula units per unit cell	(10) No. density of formula units (10^{22}/cm^3)	(11) Density (g/cm^3)	(12)
a	b	c			D_x	D_m
11.075 ± 1	—	8.954 ± 1	16	[1.457]	6.93	6.83
11.057	8.934	5.533	—	—	6.96	—
5.408	7.383	3.930	4	[2.549]	9.899	—
5.531	6.362	3.063	4	[3.711]	[8.071]	—
2.9630 ± 5	—	—	1	[3.862]	8.4	—
4.675 ± 1	—	—	4	[3.908]	8.5	8.3
12.335	3.508	5.924	2	[0.792]	[7.26]	—
	$\beta = 100°\ 30'$					
18.819	3.614	5.813	4	[1.012]	[7.075]	—
2.963	—	—	—	—	—	—
6.49	—	3.43	3	[2.397]	9.589	—
13.055	—	27.490	96	[2.366]	9.462	—
5.599	6.133	3.381	4	[3.445]	7.693	—
4.281	—	13.75	4	[1.587]	5.14	5.05
6.855	3.753	5.191	4	[2.995]	—	—
6.86	—	12.60	—	—	—	—
3.677	14.550	3.690	4	[1.945]	7.642	7.2
4.7821	—	6.5695	3	[2.304]	9.072	—
3.212 ± 5	—	7.880 ± 5	2	[2.446]	9.75	~9.4
4.614	—	6.414	3	[2.537]	10.12	—
4.774	—	—	4	[3.682]	13.1	13.4
3.131	—	7.676	2	[2.659]	10.7	—
4.729	—	—	4	[3.782]	[13.708]	—
2.963	—	—	—	—	—	—
11.158	—	8.962	8	[0.7169]	5.531	—
—	—	—	—	—	—	—
11.158	—	8.962	—	—	—	—
5.58	—	4.48	—	—	—	—
11.124	8.932	5.570	—	—	11.15	—
—	—	—	—	—	—	—
10.150	8.117	8.223	—	—	—	—
8.77	3.00	7.38	~4	[~2.337]	[~9.560)	—
	$\beta = 118°\ 30'$		—	—	—	—
5.558	6.273	3.211	4	[3.572]	13.067	—
5.542	14.166	12.426	—	—	—	—
	$\beta = 120.61°$					
—	—	—	—	—	—	—
4.350	—	6.630	2	[1.847]	8.48	8.64
5.554	—	5.923	—	—	—	—
5.63	—	—	—	—	—	—
3.933	—	5.910	2	[2.188]	[15.196]	—
	(at 34 at.% Si)					

(Continued)

TABLE I (*Continued*)

(1) Element	(2) Silicide	(3) Crystal system	(4) Name of structure	(5) Strukturberichte classification
	β-Pt₂Si (HT > 695°C)	Hexag.	Fe₂P	C22
	PtSi	Orthor.	MnP	B31
Lanthanides				
⁶⁴Gd	GdSi₂	Orthor.	LaSi₂	
	GdSi₂ (460°C)		ThSi₂	
⁶⁵Tb	TbSi₁.₇	Hexag.	Defect AlB₂	
	TbSi₂	Orthor.		
⁶⁶Dy	Dy₃Si₅ (DySi₁.₇)	Orthor.	LaSi₂	
	DySi₂ (585°C)		ThSi₂	
⁶⁷Ho	Ho₂Si₃	Hexag.	AlB₂	
	HoSi₂	Orthor.	LaSi₂	
⁶⁸Er	ErSi₂ (ErSi₁.₇)	Hexag.	AlB₂	

^{a} Adapted from Nicolet and Lau [1]. Elements are listed in order of increasing atomic number from ^{21}Sc to ^{78}Pt, excluding ^{43}Tc and the lanthanides (^{58}Ce to ^{71}Lu). The transition elements ^{30}Zn, ^{47}Ag, ^{48}Cd, ^{79}Au, and ^{80}Hg are also omitted because they do not form compounds with Si. For each element, we list the main polymorphous forms of the element first and then all compounds reported in the sources surveyed, in order of increasing atomic fraction of Si. Parentheses ( ) after a chemical formula indicate that this compound is stabilized by the element in the parentheses as an interstitial impurity or by the constraint indicated in the parentheses.

Column (3) gives the crystal system. Columns (4) and (5) give the name of the structure (by a representative equivalent structure) and the structure type by the Strukturberichte classification. Columns (6)–(8) give the lattice constants of a unit cell. Column (9) gives the number of formula units (of "molecules") contained in that unit cell for the silicides and the number of atoms per unit cell for the elements. Column (10) gives the density of formula units (of "molecules") or of atoms per cubic centimeter. Columns (11) and (12) give the density calculated from the x-ray data (D_x) and as measured (D_m). The value given for D_x is that quoted by the source and may not necessarily agree exactly with that computed from the lattice constants of the unit cell and the number of formula units contained in the unit cell, columns (5)–(8). When no value for D_x was quoted by the source, the value given is that computed from columns (6)–(9). In all cases, the number density of formula units given in column (10) has been determined from D_x and the number of formula units per unit cell given in column (9).

Brackets indicate that the entry has been provided by the present authors in lieu of missing information. An asterisk indicates that our entry has been substituted in lieu of obviously erroneous information.

Entries for a few lanthanides are listed separately at the end of the table.

(6–8) Lattice constants of unit cell (Å)			(9) No. of formula units per unit cell	(10) No. density of formula units (10^{22}/cm^3)	(11) Density (g/cm^3)	(12)
a	b	c			D_x	D_m
6.436	—	3.569	3	[2.343]	16.271	—
5.595	5.932	3.603	4	[3.342]	12.394	—
4.09	4.01	13.43				
4.10	—	13.61				
3.847	—	4.146				
4.05	3.96	13.38				
4.03	3.94	13.33				
4.03	—	13.388				
3.816	—	4.107				
4.03	3.92	13.29				
3.799	—	4.090				

TABLE II

Characteristics of Transition-Metal Silicides Formed by Thermal Annealing[a]

Element	Silicide	Formation temperature (°C)	Activation energy (eV)	Growth rate
[22]Ti	TiSi	500	—	—
	$TiSi_2$	600	—	—
[23]V	VSi_2	600	2.9, 1.8	$t, t^{1/2}$
[24]Cr	$CrSi_2$	450	1.7	t
[25]Mn	MnSi	400–500	—	—
	$MnSi_2$	800	—	—
[26]Fe	FeSi	450–550	1.7	$t^{1/2}$
	$FeSi_2$	550	—	—
[27]Co	Co_2Si	350–500	1.5	$t^{1/2}$
	CoSi	375–500	1.9	$t^{1/2}$
	$CoSi_2$	550	—	—
[28]Ni	Ni_2Si	200–350	1.5	$t^{1/2}$
	NiSi	350–750	1.4	$t^{1/2}, t$
	$NiSi_2$	≥750	—	—
[40]Zr	$ZrSi_2$	700	—	—
[41]Nb	$NbSi_2$	650	—	—
[42]Mo	$MoSi_2$	525	3.2	t
[45]Rh	RhSi	350–425	1.95	$t^{1/2}$
	Rh_2Si	400	—	—
	Rh_4Si_5	825–850	—	—
	Rh_3Si_4	925	—	—
[46]Pd	Pd_2Si	100–300	1.5	$t^{1/2}$
	PdSi	850	—	—
[72]Hf	HfSi	550–700	2.5	$t^{1/2}$
	$HfSi_2$	750	—	—
[73]Ta	$TaSi_2$	650	3.7	t
[74]W	WSi_2	650	3.0	$t, t^{1/2}$
[77]Ir	IrSi	400–500	1.9	$t^{1/2}$
	$IrSi_{1.7}$	500–1000	—	—
	$IrSi_3$	1000	—	—
[78]Pt	Pt_2Si	200–500	1.5	$t^{1/2}$
	PtSi	300	1.6	$t^{1/2}$

[a] Adapted from Nicolet and Lau [1]. The silicides listed are those observed upon thermal annealing of a thin metal film on a silicon substrate. Cases characterized by a laterally uniform growth with well-defined kinetics are identified by their mode of growth [linear (t) or parabolic ($t^{1/2}$) in time] and the activation energy of the process. Typical formation temperatures are also given.

TABLE III

Experimentally Observed Sequence of Silicide Phases by Thermal Annealing of a Metal Film on a Silicon Substrate[a]

	Ti	V	Cr	Mn	Fe	Co	Ni
	$Ti_5Si_3(?)$	$V\underline{Si}_2$	$Cr\underline{Si}_2$	$MnSi$	Fe_3Si	Co_2Si	$\underline{Ni}_2Si$
	$TiSi(?)$			$Mn\underline{Si}_2$	$FeSi$	$\overline{CoSi}$	$\overline{NiSi}$
	$Ti\underline{Si}_2$				$Fe\underline{Si}_2$	$\ddagger Co\underline{Si}_2$	$\ddagger Ni\underline{Si}_2$
$\ddagger Y\underline{Si}_{1.7}$		$Nb\underline{Si}_2$	$Mo\underline{Si}_2$		Ru_2Si_3	Rh_2Si	Pd_2Si
						$RhSi$	$\overline{PdSi}$
						$\ddagger R\overline{h}_4Si_5$	
						$Rh_3\overline{Si}_4$	
$\ddagger GdSi_{1.7}$	$HfSi$	$Ta\underline{Si}_2$	$W\underline{Si}_2$		Os_2Si_3	$IrSi$	Pt_2Si
$\ddagger TbSi_{1.7}$	$\ddagger Hf\underline{Si}_2$				$OsSi_2$	$Ir\overline{Si}_{1.75}$	$\overline{PtSi}$
$\ddagger DySi_{1.7}$						$\ddagger IrSi_3$	
$\ddagger HoSi_{1.7}$							
$\ddagger ErSi_{1.7}$							

[a] Adapted from Nicolet and Lau [1]. The silicides observed upon thermal annealing of a thin metal film on a Si substrate are listed in order of their appearance, with the uppermost forming first. A double dagger ($\ddagger$) indicates that the compound grows in a laterally nonuniform fashion. Underlining indicates the diffusing species.

TABLE IV

Electrical Resistivity, Thermal Conductivity, Coefficient α of Thermal Linear Expansion, and Specific Heat c_p of Transition-Metal Silicides Near Room Temperature[a]

Element	Silicide	Resistivity (10^{-6} Ω cm)	Thermal conductivity (cal/cm Ks)	α (10^{-6} K^{-1})	c_p (10^{-1} cal/gK)
^{22}Ti	Ti$_5$Si$_3$	55 ± 4	0.0363	11.0	9.4
	TiSi	63 ± 6	0.041	8.8	7.4
	TiSi$_2$	10–25	0.11	10.5	1.2
^{23}V	VSi$_2$	50–55	0.040	11.2	2.0
^{24}Cr	CrSi$_2$	>250–1420	0.019	12–13	1.3
^{25}Mn	MnSi	200–260	0.022	16.3	—
	MnSi$_2$	6500–13000	0.017	—	—
^{26}Fe	Fe$_3$Si	130 ± 19	0.041	14.4	1.2
	FeSi	260–290	0.024	13.0	1.4
	FeSi$_2$	—	0.028	6.7	1.4
	α-FeSi$_2$	455	—	—	—
	β-FeSi$_2$	6.67×10^5	—	—	—
^{27}Co	Co$_2$Si	60–130	—	—	—
	CoSi	90–170	0.03	10.0	1.2
	CoSi$_2$	18–65	0.036	9.4	1.9
^{28}Ni	Ni$_2$Si	20–25	0.06	16.5	—
	NiSi	14–50	—	—	—
	NiSi$_2$	34–60	0.02	—	—
^{39}Y	YSi$_2$	69	—	—	—
^{40}Zr	ZrSi$_2$	35–40	0.03	8.6	1.1
^{41}Nb	NbSi$_2$	6.3–100	0.04–0.1	8.4	—
^{42}Mo	MoSi$_2$	21–200	0.12	7.8–9.2	1.0
^{46}Pd	Pd$_2$Si	25–35	—	—	—
^{57}La	LaSi$_2$	350–2800	0.03	7.8	—
^{72}Hf	HfSi$_2$	—	0.043	8.6	—
^{73}Ta	TaSi$_2$	8.5–55	0.052	8.8	7.2
^{74}W	WSi$_2$	50–200	0.11	6.3–8.8	7.4
^{75}Re	ReSi	~730	0.057	—	—
	ReSi$_2$	6000–8000	0.045	6.6	—
^{77}Ir	IrSi$_3$	460–580	—	—	—
^{78}Pt	PtSi	28–40	—	14.0	—

[a] Adapted from Nicolet and Lau [1].

TABLE V

Typical Schottky Barrier Height ϕ_B and Work Function ϕ of Transition-Metal Silicides[a]

Element	Silicide	ϕ_B^n (eV)[b]	ϕ_B^p (eV)[c]	ϕ (eV)
[22]Ti	Ti	0.5	0.61	
	Ti$_5$Si$_3$			3.61–3.73
	TiSi			3.68–3.94
	TiSi$_2$	0.59–0.61		3.94–4.18
[23]V	V			
	VSi$_2$			3.2
[24]Cr	Cr	0.61	0.50	
	CrSi$_2$	0.57–0.58		3.7
[25]Mn	Mn			
	MnSi	0.76		
[26]Fe	Fe	0.63		
[27]Co	Co	0.61		
	Co$_2$Si			
	CoSi	0.66–0.68		
	CoSi$_2$	0.64–0.67		
[28]Ni	Ni	0.61–0.67	0.51	
	Ni$_2$Si	0.66		
	NiSi	0.65–0.67		
	NiSi$_2$	0.66–0.7		
[39]Y	Y		0.68–0.69	
	YSi$_2$	0.39	0.74–0.75	3.0
[40]Zr	Zr			
	ZrSi$_2$	0.55–0.56		3.95, 4.57
[41]Nb	Nb			
	NbSi$_2$	0.63		4.2
[42]Mo	Mo	0.68	0.42	4.73, 5–6
	MoSi$_2$	0.55–0.57		
[45]Rh	Rh			
	RhSi	0.68–0.69		
[46]Pd	Pd	0.71		
	Pd$_2$Si	0.73–0.75		
	PdSi			
[64]Gd	Gd			
	GdSi$_2$	0.37	0.71	
[65]Tb	Tb		0.63–0.65	
[66]Dy	Dy			
	DySi$_2$	0.37	0.73	
[67]Ho	Ho			
	HoSi$_2$	0.37		
[68]Er	Er		0.68	
	ErSi$_2$	0.39	0.76–0.78	
[70]Yb	Yb		0.64–0.67	
	YbSi$_2$			
[72]Hf	Hf	0.58		
	HfSi$_2$	0.53–0.6		4.51
[73]Ta	Ta			
	TaSi$_2$	0.59–0.60		4.71
[74]W	W	0.67	0.45	4.62, 5–6
	WSi$_2$	0.64–0.66		
[77]Ir	Ir			
	IrSi	0.91–0.93		
[78]Pt	Pt	0.9		
	Pt$_2$Si	0.78–0.79		
	PtSi	0.84–0.88		
[79]Au	Au	0.79–0.81	0.34	

[a] Adapted from Nicolet and Lau [1].
[b] ϕ_B^n = barrier height on n-type Si
[c] ϕ_B^p = barrier height on p-type Si

TABLE VI

Standard Heats of Formation ΔH_f° of Transition-Metal Silicides and Oxides at 25°C[a]

Element	Compound	Silicide ΔH_f°	Compound	Oxide ΔH_f°
[14]Si			SiO_2 (am)	−68.5
[21]Sc			Sc_2O_3	−82.2
[22]Ti	Ti_5Si_3	−17.3	TiO	−67.5
	TiSi	−15.5	Ti_2O_3	−75.0
	$TiSi_2$	−10.7	Ti_3O_5	−73.4
			TiO_2	−73.0
[23]V	V_3Si	−6.8	VO	−53.0
	V_2Si (?)	−12.3	V_2O_3	−60.0
	V_5Si_3	−12.0	VO_2	−58.7
	VSi_2	−24.6	V_2O_5	−54.7
[24]Cr	Cr_3Si	−8.3	Cr_2O_3	−53.9
	Cr_5Si_3	−9.8	CrO_2	−46.3
	CrSi	−9.5	CrO_3	−34.2
	$CrSi_2$	−9.6		
[25]Mn	Mn_3Si	−6.8, −8.2	MnO	−46.0
	Mn_5Si_3	−6.0, −6.9	Mn_3O_4	−47.3
	MnSi	−13.3, −11.6	MnO_3	−46.4
	$MnSi_2$	−2.6	MnO_2	−41.5
			Mn_2O_7	−19.3
[26]Fe	Fe_3Si	−5.6	FeO	−31.8
	FeSi	−8.8	Fe_3O_4	−38.1
	Fe_5Si_3	−4.6, −7.3	Fe_2O_3	−39.3
	FeSi	−9.6		
	$FeSi_2$	−5.6, −6.5		
[27]Co	Co_2Si	−9.2	CoO	−28.6
	CoSi	−12.0	Co_3O_4	−30.0
	$CoSi_2$	8.2		
[28]Ni	Ni_3Si	−8.9	NiO	−29.2
	Ni_5Si_2	−10.3	Ni_2O_3	−23.4
	Ni_2Si	−10.5, −11.8		
	Ni_3Si_2	−10.7		
	NiSi	−10.2		
	$NiSi_2$	−6.9		
[29]Cu			Cu_2O	−13.6
			CuO	−18.8
[39]Y	YSi	−16.1	Y_2O_3	−91.1
[40]Zr	Zr_4Si	−10.4	ZrO	−87.3
	Zr_2Si	−13.0, −16.7	ZrO_2	−87.2
	Zr_5Si_3	−17.2		
	Zr_3Si_2	−18.4		
	Zr_6Si_5	−18.6		
	ZrSi	−18.5		
	$ZrSi_2$	−12.7		

24. Silicides

TABLE VI (*Continued*)

Element	Compound	Silicide ΔH_f°	Compound	Oxide ΔH_f°
^{41}Nb	Nb$_3$Si	-7.9	NbO	-58.0
	Nb$_5$Si$_3$	$-10.9, -14.5$	NbO$_2$	-63.7
	NbSi$_2$	$-7.3, -11$	Nb$_2$O$_5$	-66.1
^{42}Mo	Mo$_3$Si	-5.8	MoO$_2$	-43.3
	Mo$_5$Si$_3$	-8.5	MoO$_3$	-45.1
	MoSi$_2$	-9.3		
^{44}Ru	RuSi	$-8.0, -10.1$	RuO$_2$	-17.7
			RuO$_4$	-11.5
^{45}Rh	RhSi	-8.1	Rh$_2$O	-7.6
			RhO	-10.8
			Rh$_2$O$_3$	-13.7
^{46}Pd	Pd$_3$Si	-15.1	PdO	-10.5
	Pd$_2$Si	$-6.9, -19.1$		
	PdSi	$-6.9, -17$		
^{57}La	LaSi	-15.0	La$_2$O$_3$	-85.7
	LaSi$_2$	-14.6		
^{72}Hf	Hf$_2$Si	-15	HfO	$+6.0$
	Hf$_5$Si$_3$	$-27.4, -16.7$	HfO$_2$	-89.3
	HfSi	-17		
	HfSi$_2$	-18		
^{73}Ta	Ta$_5$Si	-5.4	Ta$_2$O$_5$	-71.4
	Ta$_{4.5}$Si	-5.9		
	Ta$_3$Si	-9.2		
	Ta$_2$Si	-10.1		
	Ta$_5$Si$_3$	$-9.0, -10.0$		
	TaSi$_2$	$-8.0, -9.5$		
^{74}W	W$_5$Si$_3$	$-5.0, -5.8$	WO$_2$	-45.4
	W$_3$Si$_2$	-6.0	W$_2$O$_5$	-48.3
	WSi$_2$	-7.5	WO$_3$	-50.2
^{75}Re	Re$_3$Si	-3.1	ReO$_2$	-33.7
	Re$_5$Si$_3$	-4.8	ReO$_3$	-36.8
	ReSi	$-5.1, -6.3$	Re$_2$O$_7$	-33.3
	ReSi$_2$	$-5.5, -7.2$	Re$_2$O$_8$	-30.8
^{76}Os	OsSi	$-7.8, -11.0$	OsO$_2$	-20.5
			OsO$_3$	-11.4
			OsO$_4$	-16.0
^{77}Ir	IrSi	$-8.0, -16.0$	IrO$_2$	-15.3
	IrSi$_2$	-6.1		
	IrSi$_3$	-4.6		
^{78}Pt	Pt$_3$Si	-12.6	PtO	-8.5
	Pt$_2$Si	$-6.9, -17.0$	Pt$_3$O$_4$	-5.6
	PtSi	$-7.9, -20.1$	PtO$_2$	-10.7

[a] Adapted from Nicolet and Lau [1] and Murarka [2]. This table gives values in kilocalories per gram-atom, i.e., the heat of formation of one mole of the compound divided by the number of atoms in the chemical formula. No distinction between polymorphs of a particular compound is made.

TABLE VII

Kinetics of SiO_2 Growth by Thermal Annealing of Silicide Films on Various Substrates in Oxidizing Ambient.[a] (Adapted from Ref. 3.)

(1)	(2)	(3)		(4)	(5)–(6)		(7)	(8)–(9)		(10)	(11)	(12)
		Kinetics[c]			Rate constants[c]			Activation energy (eV)			Oxide thickness [Å] 1 h @	
Silicide film	Substrate	$x =$ thickness $t =$ time		Condition	B/A (cm/s)	B (cm^2/s)	Temperature (°C)	B/A	B	Temperature range (°C)	900°C	Reference[b]
Si	Si (?)	$x^2 + Ax = B(t + \tau)$		Wet	3.53×10^{-8}	7.97×10^{-13}	1000	1.96	0.71	920–1200	2000	10
	Si (?)	$x^2 + Ax = B(t + \tau)$		Dry	1.97×10^{-9}	3.25×10^{-14}	1000	1.99	1.24	650–1200	190	10
TiSi$_2$	Si(poly)	$x^2 + Ax = B(t + \tau)$		Wet	5.56×10^{-8}	6.39×10^{-13}	1000	2.0	1.39	900–1050	2200	11
TiSi$_2$	Si(poly)	$x^2 + Ax = B(t + \tau)$		Wet	6.58×10^{-8}	6.39×10^{-13}	1000	2.06	1.51	700–1100	2200	12
TiSi$_2$	Si(poly)	No oxidation		Dry	—	—	—	—	—	—	—	13
CoSi$_2$	Si(111)	$x^2 = Bt$		Wet	—	5.08×10^{-13}	1000	—	1.05	700–1000	2700	14
	(100)	$x^2 = Bt$		Dry	—	3.09×10^{-14}	1000	—	1.39	650–1100	600	14
NiSi$_2$	Si(111)	$x^2 = Bt$		Wet	—	2.22×10^{-13}	900	—	1.05	700–900	2700	15
	Si(100) Si(poly) SiO$_2$	$x^2 = Bt$		Dry	—	1.11×10^{-14}	900	—	1.5	700–900	600	15

Silicide	Substrate	Kinetics	Condition	A	B	E_A	E_B	Temp range	Thickness	Ref
$MoSi_2$	SiO_2	$x^2 + Ax = B(t + \tau)$	Dry	2.8×10^{-9}	1.9×10^{-14}	1.9	1.6	900–1100	200	16
$MoSi_2$	Si ⟨?⟩	$x^2 + Ax = Bt$	Wet	1.2×10^{-7}	1.0×10^{-12}	0.84	1.1	800–1000	—	17[d]
$RhSi$	Si ⟨?⟩	$x^2 + Ax = Bt$	Wet	1.1×10^{-7}	1.0×10^{-12}	0.92	1.1	800–1000	—	17
Rh_3Si_4	Si ⟨?⟩	$x^2 + Ax = Bt$	Wet	1.0×10^{-7}	1.0×10^{-12}	0.99	1.1	800–1000	—	17
$TaSi_2$	Si(poly)	$x^2 = Bt$	Wet	—	1.4×10^{-13}	—	1.4	900–1050	1200	18
$TaSi_2$	Si(poly)	$x^2 = Bt$	Dry	—	2.8×10^{-14}	—	1.2	800–1200	700	19
$TaSi_2$	Si ⟨?⟩	$x^2 + Ax = Bt$	Wet	8.2×10^{-8}	1.0×10^{-12}	0.93	1.1	800–1000	—	17
$TaSi_2$	Si(poly)	$x^2 + Ax = Bt$	Wet	9.7×10^{-8}	8.3×10^{-13}	2.04	1.17	800–1000	2700	20
WSi_2	Si(poly)	$x^{1.82} = B^*t$	Wet	—	$B^* = 4.43 \times 10^{-12}$	—	1.0	1000–1200	2900[e]	21
	SiO_2	$x^{1.8} = B^*t$		—	$B^* = 6.99 \times 10^{-12}$	—	0.35	1000–1200	—	21
WSi_2	Si ⟨?⟩	$x^2 + Ax = Bt$	Wet	1.3×10^{-7}	6.8×10^{-13}	1.0	1.3	800–1000	2890	17
$IrSi_{1.75}$	Si ⟨?⟩	$x^2 + Ax = Bt$	Wet	5.8×10^{-8}	8.5×10^{-13}	1.3	1.2	800–1000	—	17

[a] Column (2) specifies the substrate on which the silicide film rests. Column (3) gives the kinetics of the oxide growth. Column (4) specifies the oxidizing condition. Columns (5) and (6) give the rate constants of the kinetics at the temperature given in column (7). Columns (8) and (9) give the activation energies of the rate constants; the applicable temperature range is in column (10). Column (11) gives the oxide layer thickness after annealing at 900°C for 1 h. Column (12) gives the literature reference listed in footnote b. For comparison, the values valid for bulk Si single crystal are listed first.

[b] References [10–21] are found in the end-of-chapter reference list.

[c] Units of B^* are centimeters and seconds with appropriate powers.

[d] Data from reference [17] are extracted from Figs. 1 and 2 in the temperature range of 800–900°C.

[e] Extrapolated error is about 15%.

TABLE VIII

Volumetric Changes in Silicide Formation[a]

Amount of metal reacting and its thickness t_M, normalized to 1		Amount of Si reacting and its thickness t_{Si} in units of t_M		Amount of resulting silicide t_{sil} in units of t_M	$\dfrac{t_M}{t_{sil}}$	$\dfrac{t_{Si}}{t_{sil}}$	$\dfrac{t_M + t_{Si}}{t_{sil}}$
[22]Ti 1.00	+	Si 1.12	→	TiSi 1.65	0.60	0.68	1.28
Ti 1.00	+	2Si 2.24	→	TiSi$_2$ 2.50	0.40	0.90	1.30
[23]V 1.00	+	2Si 2.86	→	VSi$_2$ 2.74	0.36	1.04*	1.40
[24]Cr 1.00	+	2Si 3.35	→	CrSi$_2$ 3.00	0.33	1.12*	1.45
[25]Mn 1.00	+	Si 1.65	→	MnSi 1.94	0.52	0.85	1.37
Mn 1.00	+	2Si 3.29	→	MnSi$_2$ 2.73	0.37	1.21*	1.57
[26]Fe 1.00	+	Si 1.70	→	FeSi 1.92	0.52	0.89	1.41
Fe 1.00	+	2Si 3.41	→	FeSi$_2$ 3.36	0.30	1.01*	1.31
[27]2Co 1.00	+	Si 0.91	→	Co$_2$Si 1.47	0.68	0.62	1.30
Co 1.00	+	Si 1.81	→	CoSi (FeSi, B20) 1.98	0.51	0.91	1.42
Co 1.00	+	2Si 3.63	→	CoSi$_2$ 3.49	0.28	1.04*	1.33
[28]2Ni 1.00	+	Si 0.92	→	Ni$_2$Si (δ, PbCl$_2$, C23) 1.49	0.67	0.61	1.29
Ni 1.00	+	Si 1.84	→	NiSi (η, MnP, B31(P)) 2.22	0.45	0.82	1.28
Ni 1.00	+	2Si 3.67	→	NiSi$_2$ 3.59	0.28	1.02*	1.30
[40]Zr 1.00	+	2Si 1.74	→	ZrSi$_2$ 2.17	0.46	0.80	1.26

TABLE VIII (Continued)

Amount of metal reacting and its thickness t_M, normalized to 1	Amount of Si reacting and its thickness t_{Si} in units of t_M	Amount of resulting silicide t_{sil} in units of t_M	$\dfrac{t_M}{t_{sil}}$	$\dfrac{t_{Si}}{t_{sil}}$	$\dfrac{t_M + t_{Si}}{t_{sil}}$
⁴¹Nb 1.00	+ 2Si 2.22	→ NbSi₂ 2.39	0.41	0.93	1.35
⁴²Mo 1.00	+ 2Si 2.58	→ MoSi₂ (C11) 2.60	0.38	0.98	1.37
⁴⁵Rh 1.00	+ Si 1.45	→ RhSi (FeSi, B20) 1.85	0.54	0.78	1.32
3Rh 1.00	+ 4Si 1.94	→ Rh₃Si₄ 2.38	0.42	0.82	1.24
⁴⁶2Pd 1.00	+ Si 0.68	→ Pd₂Si 1.42	0.70	0.48	1.18
Pd 1.00	+ Si 1.36	→ PdSi 1.97	0.51	0.69	1.20
⁷²Hf 1.00	+ Si 0.90	→ HfSi (FeB, B27) 1.50	0.66	0.60	1.27
Hf 1.00	+ 2Si 1.80	→ HfSi₂ 2.30	0.43	0.78	1.21
⁷³Ta 1.00	+ 2Si 2.22	→ TaSi₂ 2.40	0.42	0.92	1.34
⁷⁴W 1.00	+ 2Si 2.53	→ WSi₂ 2.58	0.38	0.98	1.37
⁷⁵Re 1.00	+ Si 1.37	→ ReSi 1.85	0.54	0.74	1.28
Re 1.00	+ 2Si 2.73	→ ReSi₂ 2.55	0.39	1.07*	1.46
⁷⁷Ir 1.00	+ Si 1.43	→ IrSi 1.99	0.50	0.72	1.22
Ir 1.00	+ 3Si 4.28	→ IrSi₃ 3.84	0.26	1.11*	1.38
⁷⁸2Pt 1.00	+ Si 0.66	→ Pt₂Si (α, ZnH₂) 1.51	0.66	0.43	1.09
Pt 1.00	+ Si 1.32	→ PtSi 1.98	0.51	0.67	1.18

[a] Adapted from Nicolet and Lau [1]. This table lists the changes in the volume of individual elements when a silicide is formed from two elemental films. It is assumed that the volume changes are accommodated fully by adjusting only the thickness. The first column lists the amount of metal involved in the reaction and its thickness, normalized to 1. The second column lists the amount of Si involved in the reaction and its thickness, normalized to that of the metal. The third column lists the amount of the resulting silicide and its thickness, normalized to that of the metal. The last three columns express the thicknesses of the metal and Si layers that are involved in the reaction, and their sum in terms of the thickness of the resulting silicide. The silicides are listed in order of increasing atomic number. Asterisks identify cases with decreasing Si thickness ($t_{Si} > t_{sil}$).

TABLE IX

Comparison of Properties Pertinent to Microelectronic Application of Silicides of Ti, Ta, Mo, and W[a]

Property	TiSi$_2$	TaSi$_2$	MoSi$_2$	WSi$_2$
Resistivity[b] ($\mu\Omega$ cm)	~25	~50	~90–100	~70
Etch rate in[c]				
10:1 BHF (Å/min)	≥2000	100–300	Low	Low
Dry etching in				
Plasma	Yes	Yes	Yes	Yes
Reactive sputter-etching	—	Yes	—	—
Stability in				
Dry O$_2$	Yes	No, yes[d]	Yes	Yes
Wet O$_2$	Yes	Yes	Yes	Yes
Stability of the metal oxide	Good	Good	Poor, vaporizes	Poor, vaporizes
Oxidation stability on silicide on				
Silicon	Good	Good	Good	Good
Oxide	Good	Fair	Poor	Poor
Al–Silicide interaction temperature	>500°C	>500°C	>500°C	>500°C
Recommended for use				
On oxide	Yes	Yes[e]	No	No
On Si	Yes	Yes	Yes	Yes
Barrier heights on n-Si (eV)	~0.6	~0.6	~0.55	~0.65
Formation temperature (°C) on				
crystalline Si substrate	600	650	525	650
Dominant diffusing species	Si	Si	Si	Si
Adhesion on SiO$_2$	Good	Good	Not as good	Not as good

[a] Adapted from Murarka [2].

[b] Codeposited (sputtering or evaporation), sputtered from sintered target followed by annealing.

[c] See Reference 2 for additional information.

[d] TaSi$_2$ can be oxidized in dry O$_2$ according to K. C. Saraswat, Electronic Materials Conference, Santa Barbara, California (June 24–26, 1981), Abstract T-3.

[e] Not to be exposed to oxidizing ambience at high temperatures.

24. Silicides

REFERENCES

1. M-A. Nicolet and S. S. Lau, Formation and characterization of transition-metal silicides, *in* "VLSI Electronics: Microstructure Science," vol. 6 (N. G. Einspruch and G. B. Larrabee, eds.). Academic Press, New York, 1983.
2. S. P. Murarka, "Silicides for VLSI Applications." Academic Press, New York, 1983.
3. M. Bartur and M-A. Nicolet, Thermal oxidation of transition-metal silicides on Si: Summary, *J. Electrochem. Soc.* **131,** 371 (1984).
4. R. Kieffer and F. Benesovsky, "Hartstoffe" Springer-Verlag, Berlin and New York, 1963.
5. H. Nowotny, Alloy chemistry of transition element borides, carbides, nitrides, aluminides, and silicides, *in* "Electronic Structure and Alloy Chemistry of the Transition Elements" (P. A. Beck, ed.), p. 179. Wiley (Interscience), New York, 1963.
6. P. T. B. Shaffer, "Handbooks of High-Temperature Materials," vols. 1 (Materials Index) and 2 (Properties Index) (G. V. Samsonov, ed.). Plenum Press, New York, 1964.
7. B. Aronsson, T. Lundström, and S. Rundqvist, "Borides, Silicides and Phosphides. Methuen, London, 1965.
8. H. J. Goldschmidt, "Interstitial Alloys," Ch. 7. Plenum Press, New York, 1967.
9. G. V. Samsonov and I. M. Vinitskii, "Handbook of Refractory Compounds." IFI/Plenum Press, New York, 1980.
10. B. E. Deal and A. S. Grove, *J. Appl. Phys.* **36,** 3770 (1965).
11. J. R. Chen, M. P. Houng, S. K. Hsiung, and Y. C. Liu, *Appl. Phys. Lett.* **37,** 824 (1980).
12. J. R. Chen, Y. C. Liu, and S. D. Chiu, *J. Elect. Mater.* **11,** 355 (1982).
13. J. R. Chen, Y. C. Liu, and S. D. Chiu, Abstract T-4 from *Technical Program of the Electronic Materials Conference, Santa Barbara, California* (June 24–26, 1981).
14. M. Bartur and M-A. Nicolet, *Appl. Phys. A* **29,** 69 (1982).
15. M. Bartur and M-A. Nicolet, *Appl. Phys. Lett.* **40,** 175 (1982).
16. T. Mochizûki and M. Kashiwagi, *J. Electrochem. Soc.* **127,** 1128 (1980).
17. J. E. E. Baglin, F. M. d'Heurle, and C. S. Petersson, *J. Appl. Phys.* **54,** 1949 (1983).
18. S. P. Murarka, D. B. Fraser, W. S. Lindenberger, and A. K. Sinha, *J. Appl. Phys.* **51,** 3241 (1980).
19. R. R. Razouk, M. E. Thomas, and S. L. Pressacco, *J. Appl. Phys.* **53,** 5342 (1982).
20. J. M. DeBasi, R. R. Razouk, and M. E. Thomas, *J. Electrochem. Soc.* **130,** 2478 (1983).
21. F. Mohammadi, K. C. Saraswat, and J. D. Meindl, *Appl. Phys. Lett.* **35,** 529 (1979).

Chapter 25

Metallization for VLSI

W. H. CLASS
Allied Corp., Semi-Alloys Div.
Mt. Vernon, New York

J. F. SMITH
Materials Research Corporation
Orangeburg, New York

I. Introduction	435
II. Aluminum and Its Alloys	436
A. Reasons for Choosing Aluminum	436
B. Silicon Diffusion and Junction Spiking	436
C. Electromigration	441
D. Hillocks	443
E. Step Coverage	444
F. Film Deposition Methods for Pure and Alloy Aluminum	446
III. Barrier Layers for Metallization	447
A. Reason for Using Barrier Layers	447
B. Use in Bipolar and MOS Technologies	447
C. Requirements for an Effective Barrier Layer	447
D. Interdiffusion of Metallization Layers	448
E. Practical Barrier Layers	449
References	452

INTRODUCTION   I

In this chapter we deal with aluminum and its alloys and with barrier materials, which, together with refractory metal silicides, are expected to be the metallization systems most commonly used for VLSI interconnection. Optimum results, in terms of manufacturing yields and circuit performance, are achieved when the proper combination of materials is selected. The principles underlying this choice are the main content of this chapter.

II ALUMINUM AND ITS ALLOYS

A Reasons for Choosing Aluminum

The reasons for choosing an aluminum-based interconnection system are well documented [1,2,3,4] and include

(1) excellent adherence to silicon and SiO_2,
(2) ease of forming both ohmic and Schottky contacts with silicon,
(3) compatibility with CVD oxide or nitride depositions,
(4) excellent bondability, and
(5) high conductivity.

Very important is the fact that aluminum can frequently be used as a single metal system, which simplifies manufacturing processes. VLSI devices require that the aluminum metallization exhibit the following characteristics:

(1) reliable contact over shallow junctions <0.5 μm in depth,
(2) reliable contact in small contact areas approaching 1 × 1 μm,
(3) reliable contact when 1 × 1-μm contact windows are etched in SiO_2 layers that exceed 1 μm in thickness,
(4) smooth, hillock-free coatings, and
(5) acceptable electromigration resistance of metallization linewidths that approach 0.5 μm.

Central to the achievement of these characteristics is the choice of alloy and the methods employed for its deposition.

B Silicon Diffusion and Junction Spiking

The limitations of pure aluminum metallization stem principally from the ability of aluminum to dissolve a small but significant amount of silicon and from the rapidity with which diffusion processes take place in aluminum at modest ambient temperatures. Reasonably rapid diffusion is expected at temperatures that exceed one-half the melting temperature (in kelvins). For aluminum, this corresponds to an ambient of ~200°C. Thus, silicon migration, hillock growth, electromigration, grain growth, and

Solubility of Silicon in Aluminum 1

The solubility of silicon in aluminum is apparent in the Al–Si binary phase diagram (Fig. 1) [5]. The solubility curve (A–B) indicates a retrograde silicon solubility, decreasing from 1.59 wt. % at 577°C, to 0.8 wt. % at 500°C, to 0.5 wt. % at 450°C, and to 0.25 wt. % at 400°C. Crossing the solubility curve into the (Al) + Si region of the diagram as the temperature is decreased indicates that silicon precipitation will occur. Similarly, the (Al) region is entered as the temperature is increased, and silicon dissolution occurs. The silicon that precipitates from Al–Si is saturated with aluminum. Since aluminum is an acceptor, these precipitates are p-type.

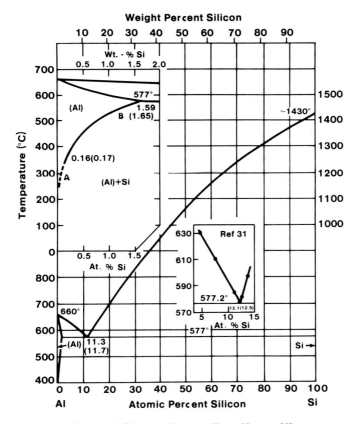

Fig. 1. Al–Si phase diagram. (From Hansen [5]).

2 Phase Conditions

The phase conditions depicted in Fig. 1 are predicted on the condition that sufficient time is allowed for thermodynamic equilibrium to be achieved. In aluminum thin films, these conditions are readily achieved, despite short high-temperature dwell times, because of the rapid diffusion of silicon in thin-film aluminum.

3 Silicon Diffusivities

This rapid diffusion may be inferred from data that compare silicon diffusivities for bulk and thin-film aluminum [1,6]. The data of McCaldin and Sankur, Van Gurp, and Nanda all pertain to vapor deposited films in Fig. 2 [6]. Data for magnetron sputtered films are shown in Table I [7].

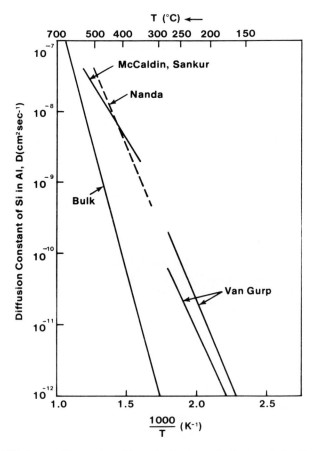

Fig. 2. Diffusion coefficients for silicon in aluminum for bulk and thin-film aluminum. (From Pramanik et al. [1]).

TABLE I

Silicon Diffusivities D in Magnetron Sputtered Films of Pure and Alloy Aluminum[a]

Alloy	Temperature (°C)	Diffusivity (cm^2/s)
Pure Al	400	4.93×10^{-9}
	450	1.07×10^{-8}
	500	1.89×10^{-8}
	560	4.08×10^{-8}
Al + 0.013 at.% In	400	1.69×10^{-9}
	450	4.12×10^{-9}
	500	9.23×10^{-9}
	560	2.30×10^{-8}
Al + 0.014 at.% Sn	400	2.67×10^{-8}
	450	7.90×10^{-9}
	500	1.60×10^{-8}
	560	2.85×10^{-8}

[a] From Garg [7].

The data for pure aluminum are in substantial agreement with the McCaldin and Sankur data. The diffusivity reductions produced by the indium and tin additions indicate that these alloys could be valuable in suppressing silicon migration.

Junction Spiking 4

The most important result of silicon dissolution and migration in aluminum is junction spiking, which is the process whereby silicon dissolves into the aluminum metal and migrates along the metallization line, to be replaced by aluminum that penetrates the junction. The driving force for this process is the solubility of silicon in aluminum. The degree of junction spiking is determined by time and the diffusivity of silicon. The latter is determined by the mean silicon migration distance, given by

$$l \simeq \sqrt{Dt}, \tag{1}$$

where t is the time and D the silicon diffusivity at the process ambient temperature.

Junction spiking takes place when the silicon wafer is heated after metal deposition to effect contact between the silicon and aluminum. Here, the aluminum chemically reduces the native SiO_2 film to form an intimate Al–Si interface. The SiO_2 is penetrated along microflaws, from which silicon dissolution and migration take place. This process is depicted in Fig. 3, which is roughly to scale for a contact sintering temperature of 450°C.

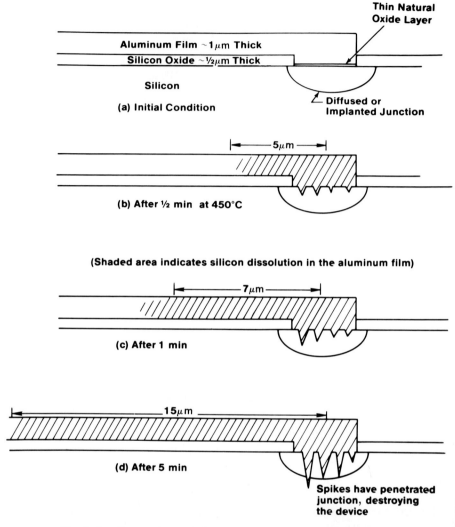

Fig. 3. Junction spiking and silicon migration during contact sintering.

5 Suppression of Junction Spiking

Junction spiking is suppressed by adding silicon to the aluminum, thereby saturating it to eliminate the driving force for dissolution. Other processes, however, ensue. These include [1,2]

(1) silicon precipitation during film growth,
(2) silicon precipitation on oxide step edges,

(3) formation of epitaxial silicon layers at the junction, and
(4) preferential dissolution of polysilicon and its subsequent reprecipitation elsewhere on the device.

Central to all the preceding processes is the rapid transport of silicon at processing temperatures. The driving forces for this may be other than chemical (concentration gradient), including

(1) surface area reduction (precipitate coarsening and step decoration),
(2) grain boundary area reduction (preferential poly Si attack),
(3) internal energy reduction (transport of amorphous silicon to yield epilayer growth), and
(4) stress reduction.

The metallization engineer should never lose sight of the fact that many silicon transport processes can occur during device fabrication, as well as during device operation.

Electromigration C

Self-diffusion of aluminum is also important in wafer metallization and device performance. As with silicon, diffusivities much greater than bulk may be inferred from measured activation energies and the speed with which the phenomena occur. These high diffusivities are the consequence of preferred diffusion along structural flaws such as grain boundaries, surfaces, and dislocations. Thus, many of the resulting phenomena are structure dependent.

One such structure dependent property is *electromigration*. Electromigration is a perturbation of the normal random walk of the self-diffusing aluminum atoms: A preferred drift in the direction of an applied driving force occurs. The drift velocity v is given by the Einstein relation

$$v = DF/kT, \qquad (2)$$

where D is the aluminum self diffusivity, k Boltzmann's constant, T temperature (K), and F the applied driving force.

Electron Wind Effect 1

For electromigration, this applied force results from a momentum transfer between current carrying electrons and the diffusing atoms; hence the term *electron wind effect* (d'Heurle and Rosenberg [8] provide an excellent review).

For aluminum, at temperatures where electromigration is of concern (i.e., ambient device-operating temperatures), it is due predominantly to the effect of the electron wind on grain boundary diffusion. The amount of electromigration correlates with the grain structure of the film, being dependent on both grain size and grain orientation. It is also affected by the structure of the grain boundary and can, therefore, be influenced by alloy additions that preferentially segregate at the grain boundary.

2 Divergences

The *drift velocity* that results from the electron wind refers to atoms migrating along grain boundaries in the film. These boundaries behave as "rivers" for mass transport. The resulting flux of migrating atoms can further be subject to divergences according to variations in diffusive mobility along the network of interconnecting grain boundaries. A *negative divergence* implies an accumulation of mass in a region, which causes the growth of hillocks which can lead to electrical short circuits between adjacent conductors. A *positive divergence* implies loss of matter from a film region with ensuing void formation. Both phenomena result in conductor failures. These failures follow the general expression [9]

$$MTF = Aj^{-n} \exp(Q/T), \tag{3}$$

where MTF is the median time to failure; A a structural parameter that is a function of film geometry, film structure, substrate, and film overcoating; J the current density (A/cm^2); n an exponent that usually varies between 1 and 2; and Q an activation energy that, for aluminum, lies in the range of 0.6 ± 0.2 eV.

3 Median Time to Failure

Figure 4 shows a compilation of some typical median times to failure, which vary between 10 and 5000 h at a test temperature of 250°C and a current density of 1×10^6 A/cm^2. The effects of copper additions on prolonging film life are well documented [8]. Copper does segregate at film grain boundaries, causing a reduction in aluminum diffusivity along these boundaries. Optimum performance is achieved with copper concentration in the 3–5 wt. % range [8].

Other correlations between film structure and MTF have been observed. These include the following [2,10].

(1) Films with a preferred (111) orientation exhibit improved MTF.
(2) MTF increases with increasing grain size.
(3) MTF decreases with increasing variation in grain size.

25. Metallization for VLSI

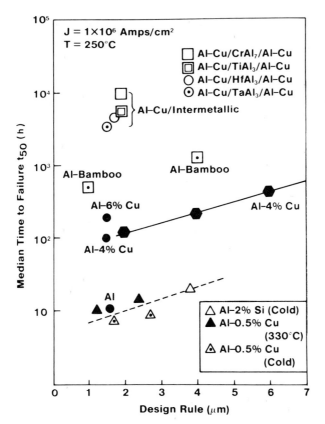

Fig. 4. Median time to failure (MTF) for various metallizations. (From Pramanik et al. [1]).

These three effects were combined by Vaidya et al. [10] into a single parameter η:

$$\eta = (S/\sigma^2) \log(I_{111}/I_{200})^3,$$

where S is the mean film grain size, σ the variation in grain size, I_{111} the intensity of the film (111) x-ray reflection, and I_{200} the intensity of the film (200) x-ray reflection. Here, large values of η imply a long MTF; and the relative importance of S, σ, and preferred (111) orientation are quantitively expressed.

Hillocks D

Hillock formation is also a diffusion dependent phenomenon. Hillocks are mounds that protrude from the film surface; they cause the aluminum film to assume a milky, diffuse appearance, whereas hillock-free films are usually specular. Hillocks result from compressive stress in the film. Any rise in wafer temperature after the initiation of deposition results in a compressive stress in the aluminum already present, since the thermal expansion of the film is greater by a factor of ten than that of the silicon substrate. This stress induces grain boundary sliding and rotation, which extrudes hillocks from the film surface, thereby causing the relaxation of the compressive stress [11]. The process is depicted in Fig. 5a, which shows in cross section a typical columnar film structure. The arrows directed along grain boundaries depict the grain boundary shear stresses that result from the compressive stress. Figure 5b depicts the extruding grain and the lateral motion of neighboring grains that results in stress relaxation.

Such grain boundary sliding is dependent on grain boundary diffusivity. Grains tend to lock along steps and triple points, and unlocking requires diffusive mass transport which is the rate-limiting process. Thus, resistance to hillock growth is induced by the addition of diffusion-inhibiting agents. Copper, tin, and indium are excellent for this purpose.

E Step Coverage

Step coverage is another film property that is sensitive to diffusion. Problems with step coverage result from the fact that substrate steps tend to shadow nearby regions, thereby causing nonconformal coating. The phenomenon is depicted in Fig. 6. It is usually observed around oxide steps such as those that define a contact window. Poor step coverage is characterized by a sidewall film thickness that may be one-fourth the film thickness at the top of the step. A *microcrack* is also frequently observed at the base of the step; this can result in poor contact or enhanced lateral etching during pattern definition, resulting in etching yield losses. The thin sidewall metallization results in premature electromigration failure because of the higher current density in this region.

The application of heat during metallization results in healing of the microcrack [3] and also causes thickening of the sidewall metallization. Further, the application of sufficient ion bombardment during film growth can cause the same beneficial effect, by resputtering the film at the base and top edge of the step [12] and by enhancing the surface mobility of the film atoms [13,14]. This latter phenomenon causes the surface migration of aluminum atoms up the sidewall. Thus, modern deposition methods make use of both heat and ion bombardment during film growth to achieve enhanced step coverage.

25. Metallization for VLSI 445

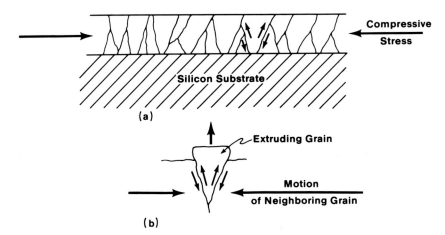

Fig. 5. Grain-boundary-sliding model of hillock growth.

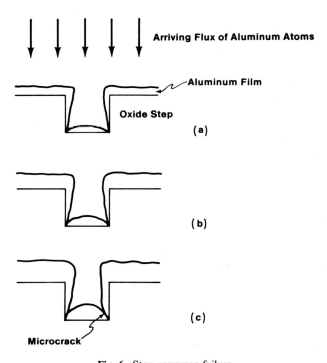

Fig. 6. Step coverage failure.

F Film Deposition Methods for Pure and Alloy Aluminum

Three deposition methods are currently in use for silicon wafer metallization: *evaporation, sputtering,* and *chemical vapor deposition*. At the present time, sputtering is the preferred method. Evaporation methods are declining in importance, and chemical vapor deposition is currently a laboratory process, although it holds some promise for future metallization requirements [15].

Effective physical vapor deposition of aluminum requires an environment that

(1) is very pure with regard to residual reactive gases,
(2) allows for the application of heat before and during film growth,
(3) permits ion bombardment before and during film deposition,
(4) permits compositional control during film deposition, and
(5) minimizes particulate contamination.

Conditions (3) and (4) favor the use of a sputtering method, and condition (5) favors vertical wafer orientation in the sputtering system. Minimal residual gas contamination requires that such sputtering systems be vacuum load-locked if practical equipment productivity is to be achieved. Residual gas contaminants can

(1) increase hillock formation,
(2) increase film resistivity,
(3) create etching problems, and
(4) decrease electromigration MTF.

Chapman [16] and Vossen and Kern [17] provide an excellent review of the sputtering process and VLSI deposition methods.

VLSI sputter deposition methods for aluminum alloy metallization make use of *magnetron techniques*. This approach utilizes a magnetic field that confines the plasma close to the sputtering cathode, which is the source of the deposited material. This confining field results in very efficient ionization of the argon gas maintained at a pressure of ~ 10 mTorr within the sputtering chamber. Thus, a high degree of argon ion bombardment and consequent rapid sputtering of the cathode occur. Substrates placed nearby are typically coated at a rate of 1 μm/min [18].

Ion bombardment during film growth is achieved by applying a negative bias to the substrate. Thus, positive ions from the nearby plasma are attracted to the substrate to bombard the growing film. This method is called *bias sputtering* and makes use of both dc and rf methods. Ion bombardment is also used prior to film deposition to remove any adsorbed atmospheric gases from the substrate surface. For both these applications, typical voltages are between -25 and -150 V [16].

BARRIER LAYERS FOR METALLIZATION   III

Reason for Using Barrier Layers   A

Essentially, the motivation for employing barrier layers in semiconductor metallization schemes is the prevention of the interdiffusion and reaction of materials that exhibit an affinity for each other. Such phenomena, of course, are more pronounced at elevated temperatures; hence the main role of the barrier is usually to inhibit the mixing of materials at high processing temperatures. The barrier is interposed between materials that would ideally be juxtaposed.

Use in Bipolar and MOS Technologies   B

Barrier layers can be utilized in both bipolar and MOS technologies. In the former, ohmic contacts are usually established by first forming a layer of PtSi or Pd_2Si on the silicon surface. Aluminum is then connected to this via a barrier layer because the aluminum–silicide interface is unstable. In MOS devices, aluminum is deposited directly on the silicon surface. The aluminum usually contains silicon, which precipitates out and forms an epitaxial, p-doped layer in the contact area. This can lead to contact resistance problems which will have to be alleviated eventually by the use of intermediate materials.

Requirements for an Effective Barrier Layer   C

Diffusion occurs in a heterogeneous system to eliminate chemical potential gradients. Such gradients between two materials are unaffected by interposing a third material between them, so such a material can serve only to influence the kinetics of diffusion.

The requirements for an effective barrier layer have been discussed in the literature [19,20]:

(1) The barrier should inhibit interdiffusion of the adjoining materials. In the case of Al–Si intermixing, silicon diffuses far more readily into aluminum than vice versa, and hence the minimization of silicon migration is of prime importance.

(2) The barrier should not react with or diffuse into the adjoining layers. In practice it is well understood that an interfacial transition region of varying composition and structure is inevitable.

(3) The barrier should interact with the adjacent layers only sufficiently to promote good adhesion and low contact resistance to them.

(4) The barrier layer should have a coefficient of expansion not markedly different from that of the rest of the thin-film structure to minimize thermal stresses due to temperature cycling.

(5) The barrier material should be amenable to dry etching.

(6) The barrier material should adhere to SiO_2.

D Interdiffusion of Metallization Layers

1 Boundary versus Bulk Diffusion

Diffusion, being a thermally activated process, is described by an Arhennius-type expression. However, in thin films, because the defect concentration is relatively high, diffusion along grain boundaries, stacking faults, and dislocations can predominate over volume diffusion [21]. Since the activation energy for diffusion along internal surfaces is reduced, the boundary diffusion coefficient is less sensitive to temperature than is the bulk diffusion coefficient; therefore, at a certain temperature (generally about one-half of the melting point of the material) bulk diffusion becomes more significant. In absolute terms, the amount of material transported is determined by, for example, the total boundary area and boundary orientation; it is desirable, therefore, to optimize barrier layer deposition processes for maximum grain size and, perhaps, a random grain orientation which would reduce the number of conductive paths intersecting the layers. Boundary diffusion, in fact, depends not only on the activation energy; species with the same activation energy but different boundary affinities may diffuse at different rates because boundary adsorption may occur, which reduces the effective boundary width.

2 Intermetallic Compound Formation

When the interdiffusion of metal layers occurs, phases appear in accordance with the equilibrium phase diagram. For example, in the simple system Al–Ti–Si, in which the intermediate titanium layer is present to prevent the Al–Si interaction, it has been found that the titanium layer is gradually consumed at 400°C, with the formation of the intermetallic compound $TiAl_3$ [22]. In this case, the aluminum diffuses through the titanium, which is quite impervious to silicon. A volume decrease usually takes place on the formation of intermediate phases, resulting in the development of tensile stress. Also, intermetallic compounds tend to be relatively brittle, which can lead to bonding problems and can increase the overall resistivity.

3. Porous Layer Formation

Porosity can develop in a thin-film couple if the diffusivities of the intermixing species are different, which is nearly always the case. This is manifested by the appearance of compressive stress on the side of the couple, to which a net mass flow occurs, and a tensile stress on the other side. Chromium is unstable as a barrier in contact with aluminum because of such phenomena; intermixing at the Cr–Al interface leads to stress-related cracking [23]. Although vacancy creation and annihilation mechanisms that can preclude this do occur under some circumstances, prolonged exposure to high temperatures can result in void coalescence and the development of a porous layer that can seriously detract from both the electrical and mechanical properties of the film structure.

E. Practical Barrier Layers

We shall now discuss barrier layers that are currently utilized or have recently been investigated for applications with aluminum metallization.

1. Titanium

The behavior of pure titanium as a barrier layer depends strongly on whether it is in contact with silicon, $PdSi_2$, or PtSi.

(1) *With silicon.* When interposed between aluminum and silicon, titanium inhibits the diffusion of silicon, while the compound $TiAl_3$ begins to form at 400°C [24]. Rapid silicon diffusion begins when all of the titanium has been converted, typically after 30 min at 450°C [22]. Titanium is not a barrier to aluminum diffusion, but this is irrelevant here because of the limited solubility of aluminum in silicon. Consumable barriers of this type have been termed *sacrificial barriers* and have the advantage that their effective lifetime can often be calculated for given processing conditions [19].

(2) *With $PdSi_2$.* When interposed between aluminum and $PdSi_2$, titanium begins to react with the silicide at 200°C, ultimately converting it to $PdAl_3$. This reaction provides the driving force for substantial aluminum migration through the film. Typically, the electrical properties of Pd_2Si–Si contacts degrade after 30 min at 400°C.

(3) *With PtSi.* In the case of PtSi contacts, a 1000-Å titanium layer is stable for up to 30 min at 450°C, which are common annealing conditions [23]. Figure 7 shows the variation in Schottky barrier height after annealing 1000-Å titanium films for 30 min at different temperatures. The barrier height decreases abruptly upon conversion of PtSi to $PtAl_2$.

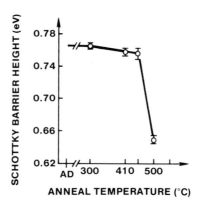

Fig. 7. Schottky barrier height versus anneal temperature for a Si–PtSi–Ti–Al structure. [From Merchant and Amano [23].)

2 Tungsten

It has been shown [25] that tungsten, by virtue of a low silicon diffusivity, is considerably more effective in reducing aluminum spiking than is the traditional remedy of adding silicon to the aluminum. A 1200–1500 Å film was deposited by CVD techniques, contact being established by the formation of 200–400 Å of WSi_2. The barrier degradation mechanism in this case is the formation of the compound WAl_{12}, which commences at 500°C. One of the main problems with tungsten, as opposed to titanium, for instance, is that it will not dissolve the native oxide on the silicon surface; contact areas must, therefore, be cleaned immediately prior to tungsten deposition, by sputter etching for example [26].

3 Titanium–Tungsten

Alloys of titanium and tungsten are now widely used and have been discussed extensively in the literature [27,28,29]. Compositions of around 10% Ti–90% W are most typical. It has, in fact, been shown that the aluminum barrier properties of Ti–W can be improved three orders of magnitude by the incorporation of up to 10^{-3} wt. % of nitrogen in the growing film [30]. This is probably due to the decoration of high diffusivity paths by nitrogen, that is, by grain boundary *stuffing*.

It was reported that 1000-Å films of 10% Ti–90% W interposed between PtSi and aluminum were stable for up to 30 min at 550°C [23]. During annealing, a decrease in contact resistance was observed as native oxide dissolution occurred. The stability of the interface is demonstrated by the Schottky barrier data shown in Fig. 8. Similarly, in the case of the

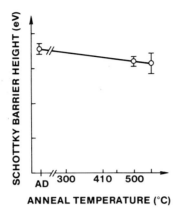

Fig. 8. Schottky barrier height versus anneal temperature for a Si–PtSi–(Ti–W)–Al structure [23].

Al/Ti–W system, it was shown by Auger depth profiling [27] (Fig. 9) that after 30 min at 500°C intermixing was evident at the Al/Ti–W interface but was absent from the Ti–W film and the Ti–W/Si interface.

The Ti–W films are sputter deposited from hot pressed targets. Since gas scattering of sputtered material invariably occurs, with light atoms in general being deflected through larger angles than heavy atoms, the titanium content of Ti–W films is generally less than that of the target material. Also, residual gases which can be adsorbed during sputtering can strongly influence Ti–W barrier properties.

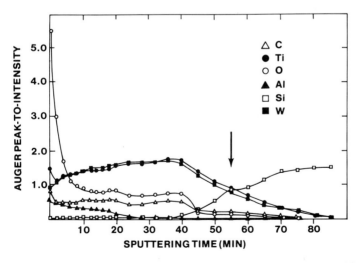

Fig. 9. Auger sputter profiles of Al–(Ti–W)–Si structure after 30 min at 500°C. (From Ghate *et al.* [27].)

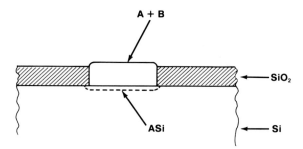

Fig. 10. Noble metal–refractory metal contact after annealing.

4 Pd–W and Pt–W Alloys

The use of noble and near-noble metals along with refractory metals is of particular interest for shallow contacts because of the inherent limitation on silicon consumption during silicide formation. The alloy must be a solid solution. Basically, on annealing, the noble or near-noble metal reacts to form a silicide, leaving behind a refractory metal barrier layer (usually still containing some of the other constituent) as shown in Fig. 10. Thus, in this approach, two distinct layers are produced in only one deposition step. Sputtering is the preferred method for depositing high melting point alloys because the film composition is determined only by the target composition.

It has been shown that post-deposition annealing to form the silicide can be eliminated by the use of appropriate substrate temperatures during deposition [31]. For a Pd_2Si contact from an alloy of 20% Pd–80% W this should be 100°C; for a PtSi contact from 20% Pt–80% W it should be 300°C.

5 TiN

Considerable interest is now being shown in the use of reactively sputtered TiN barrier layers. This barrier was found to be effective in a PtSi–TiN–Al structure after 30 min at 550°C [32], but it decomposed through the formation of AlN at 600°C. It appears that nitrogen-rich films exhibit significantly better barrier properties than do titanium-rich films [33].

REFERENCES

1. D. Pramanik and A. N. Saxena, *Solid State Technol.* **26**(1), 127–333 (1983).
2. D. Pramanik and A. N. Saxena, *Solid State Technol.* **26**(3), 131–137 (1983).
3. A. J. Learn, *J. Electrochem. Soc.* **122**, 1127 (1975).
4. W. H. Class and J. F. Smith, *32nd MRC Sputtering School Transaction* **St-111**, 1–39 (1983).

5. M. Hansen, "Constitution of Binary Alloys," McGraw-Hill, New York, 1958.
6. G. J. Van Gurp, *J. Appl. Phys.* **44,** 1040–2050 (1973).
7. N. Garg, "Diffusion of Silicon in Aluminum and Dilute Aluminum Alloy Films," Ph.D. dissertation. Polytechnic Inst., New York, 1981.
8. F. M. d'Heurle and R. Rosenberg, "Physics of Thin Films" Vol. 7 pp. 257 (G. Hass, M. H. Francombe, R. W. Hoffman, eds.). Academic Press, New York, 1973.
9. P. B. Ghate, *Solid State Technol.* **26**(2), 113 (1983).
10. S. Vaidya, D. B. Fraser, and A. K. Sinha, *IEEE Proc. 18th Annu. Conf. Reliability Physics,* pp. 165–170 (1980).
11. A. K. Sinha and T. T. Sheng, *Thin Solid Films* **48,** 117–126 (1978).
12. J. L. Vossen, *J. Vac. Sci. Technol.* **8,** 512 (1971).
13. J. F. Smith, *Solid State Technol.* **27**(1), 135–138 (1984).
14. J. E. Greene, *CRC Critical Reviews* **11,** 47 (1983).
15. M. L. Green and R. A. Levy, *Proc. 1984 Int. Conf. on Metallic Coatings* (to be published in *Thin Solid Films*).
16. B. N. Chapman, "Glow Discharge Processes." Wiley, New York, 1980.
17. J. L. Vossen and W. Kern (eds.), "Thin Film Processes." Academic Press, New York, 1978.
18. W. H. Class, *Solid State Technol.* **26**(6), 103–106 (1983).
19. M.-A. Nicolet and M. Bartur, *J. Vac. Sci. Technol.* **19,** 786 (1981).
20. C. Y. Ting and M. Wittmer. *Proc. Int. Metallic. Conf.* p. 327, San Diego, California, April 5–8 (1982).
21. D. Gupta and P. S. Ho, *Thin Solid Films* **72,** 399 (1980).
22. C. Y. Ting and B. L. Crowder, *J. Electrochem. Soc.* **129,** 2590 (1982).
23. P. Merchant and J. Amano, *J. Vac. Sci. Technol.* **1,** 459 (1983).
24. R. W. Bower, *Appl. Phys. Lett.* **23,** 99 (1973).
25. N. E. Miller and I. Beinglass, *Solid State Technol.* **25,** 85, December, 85 (1982).
26. P. A. Gargini, *Ind. Res. Dev.* **25,** 141, March, 141 (1983).
27. P. B. Ghate, J. C. Blair, C. R. Fuller, and G. E. McGuire, *Thin Solid Films* **53,** 117 (1978).
28. M. Hill, *Solid State Technol.* **23,**53, January, 53 (1980).
29. V. Hoffman, *Solid State Technol.* June, 119 (1983).
30. R. S. Nowicki, J. M. Harris, M.-A. Nicolet, and I. V. Mitchel, *Thin Solid Films* **53,** 195 (1978).
31. M. Eizenberg, G. Ottaviani, and K. N. Tu, *Appl. Phys. Lett.* **37,** 87 (1980).
32. M. Wittmer, *J. Appl. Phys.* **53,** 1008 (1982).
33. A. I. Pan, A. E. Morgan, and K. Ritz, *Proc. Int. Metallic Conf.,* San Diego, California, April 9–13 (1984).

Chapter 26

Application of Ion Implantation in VLSI

K. L. WANG

University of California
Los Angeles, California

I.	Introduction	456
II.	Aspects of Ion Implantation	457
	A. Range Distribution	457
	B. Anomalous Distributions and Annealing Behavior	457
	C. Small Geometry Effects	460
	D. Masking	463
	E. Ion Implantors	465
III.	Doping Applications in MOS Technology	466
	A. Junction Formation	466
	B. Threshold Voltage Control	466
	C. Buried Channel Formation	467
	D. CMOS	468
	E. Punchthrough Stopper	468
	F. Graded Sources and Drains	471
	G. Application Example	471
IV.	Doping Applications in Bipolar Technology	472
	A. Predeposition	472
	B. Buried Collector or Subcollector Implantation	473
	C. Implanted Resistors	474
	D. Process Simplification	474
V.	Recent Advances and Other Applications	476
	A. Damage Gettering	476
	B. Enhanced Etching	476
	C. Local Oxidation	478
	D. Buried Insulator Layer Formation	480
	E. Silicon-on-Insulator Using Oxygen Implantation	480
	F. Radiation Enhanced Diffusion	481
	G. Silicidation of Metal–Si Reaction	482
	H. Beam Writing and Ion Beam Lithography	483
	References	484

I INTRODUCTION

Ion implantation is a process by which ionized impurity atoms are accelerated to high energy and subsequently impinged into a substrate to alter its electric properties. Most applications of ion implantation in microelectronics are to provide a powerful, direct alternative to diffusion as a means for doping semiconductors in device and circuit fabrication. This technique has many unique advantages over the conventional diffusion process and has become an essential, indispensable process technique in VLSI. The advantages of ion implantation are the following.

(a) Mass separation is provided by a mass spectrometer, resulting in an extremely pure dopant beam; a single machine can be used for a variety of impurities without cross contamination of different impurity species.

(b) High accuracy of doping is offered by direct measurement of the ion current, which gives accurate doping concentration to within $\pm 1\%$ over fluences ranging from 10^{10} to 10^{17} cm^{-2}.

(c) Ion implantation is a nonequilibrium kinetic process rather than a chemical process. The impurity distribution is controlled by the ion energy and mass rather than by the diffusivity of the species used; thus, a precise control of the doping distribution is possible.

(d) A variety of doping profiles can be obtained by superposition of many profiles by using multiple implantation.

(e) Because of the near-Gaussian shape of the distribution, an abrupt junction can be obtained.

(f) With scanning of the beam or the sample, a uniform doping of better than $\pm 5\%$ over the entire wafer (5-in. diameter) can be easily achieved.

(g) The lateral spread of the doping can be reduced because the transverse scattering is limited.

(h) Since ion implantation is usually performed at room temperature, low temperature material can be used to mask the selected areas for doping.

With all these attributes, the applications of ion implantation have gone beyond simply providing doping in facilitating VLSI fabrication.

Because ion implantation also results in damage to the crystalline semiconductor, annealing at elevated temperatures (between 800 and 1100°C for Si) is necessary to repair some or all of this damage.

ASPECTS OF ION IMPLANTATION II

Range Distribution A

When an energetic ion with energy E impacts a substrate, the ion encounters the statistical processes of collisions and finally come to rest. The total distance traveled by the ion before coming to rest is defined as the range R. The range projected to the direction of the ion beam is defined as the projected range R_p, and the projected straggling, denoted by ΔR_p, is used to characterize the projected spread of the ion distribution. The implanted atoms for an amorphous target can be approximately described by a one-dimensional Gaussian distribution [1],

$$N(x) = \frac{Q_0}{\sqrt{2\pi}\,\Delta R_p} \left[-\tfrac{1}{2}\left(\frac{x-R_p}{\Delta R_p}\right)^2 \right].$$

The parameters R_p and ΔR_p depend on the projectile and target masses as well as the ion energy and can be calculated from the two principal energy-loss mechanisms: electronic and nuclear stoppings. The projectile range and the straggling have been calculated using the LSS theory [1] (Linhard, Scharff, and Schiott) and are tabulated in the literature [2]. Figures 1 and 2 show R_p and ΔR_p, respectively, for various dopants in Si. Note that these distributions were obtained using a first-order approximation. Higher-order corrections and other effects are known to cause deviation from the Gaussian function, as discussed in the following section. The transverse straggling, which characterizes the lateral spread of ions, is also given in Fig. 3 for several important dopants in Si [3]. The effect of the transverse straggling on VLSI fabrication will be discussed in Section II.C.

Anomalous Distributions and Annealing Behavior B

Ion Distribution 1

When one takes into consideration the second and higher orders of the range distribution theory and other effects such as ionization enhanced diffusion and channeling along the major crystalline axes or planes, the calculated ion distribution deviates from the Gaussian shape, particularly near the tail of the distribution. These deviations can have significant

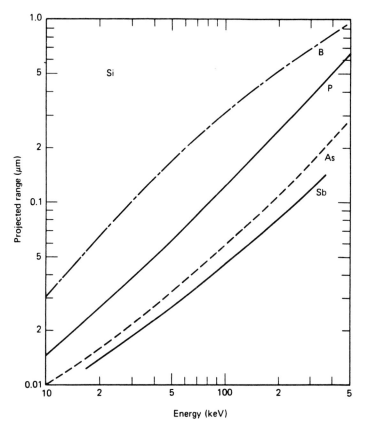

Fig. 1. Projected range of boron, phosphorus, arsenic, and antimony in silicon. (After Gibbons *et al.* [2].)

effects on the reproducibility of the formation of the extremely shallow junctions needed for VLSI circuits. On the other hand, the above effects can be used to carry out special deep implantation without having to resort to the use of expensive high-energy implantors. The distributions caused by the various effects are illustrated in Fig. 4.

2 Annealing

The annealing behavior of implanted silicon is relatively well understood. The 550–600°C annealing of the implanted Si results in solid-phase epitaxial regrowth of the implanted layer and an activation of 50–90% of the implanted ions, but very little recovery of the carrier lifetime. The solid-phase epitaxial regrowth is accomplished by the formation of dislocation loops, and considerable damage remains in the implanted layer.

26. Application of Ion Implantation in VLSI

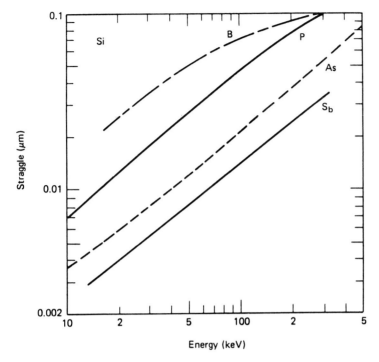

Fig. 2. Straggle for boron, phosphorus, arsenic, and antimony in silicon. (Adapted from Gibbons *et al.* [2].)

The carrier lifetime is severely degraded by the presence of the residual damage. Dislocation loops grow with increasing annealing temperatures, and the recovery of the lifetime takes place at annealing temperatures beyond 950°C, with full recovery by 1000–1100°C.

Impurity Redistribution 3

At such high annealing temperatures, impurities are redistributed (sometimes anomalously), and this redistribution presents problems in controlling the base doping of bipolar transistors, in which recovery of most of the lifetime is essential.

Annealing Research 4

Recent research in annealing focuses on *rapid* or *flash annealing* by laser or other heat sources to anneal the implanted damage in a relatively short time (from a few seconds to nanoseconds), while the distribution

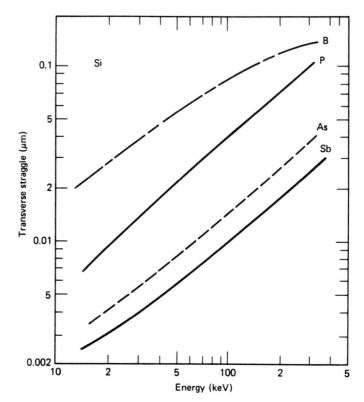

Fig. 3. Transverse straggle for boron, phosphorus, arsenic, and antimony in silicon. (After Furakawa et al. [3].)

profile is kept intact (see, for example, reference [5]). This technique is potentially useful in application to extremely shallow junctions for VLSI.

C Small Geometry Effects

As device dimensions continue to decrease in the VLSI or ULSI era, lateral distributions and geometry effects can significantly change the peak concentration and the shape of the distribution in the doped layer. For small VLSI devices, extremely narrow mask cuts are required. From the standpoint of designing and fabricating small devices, detailed knowledge about the lateral impurity distribution and the range distributions (in the beam direction) is becoming extremely important. Accurate ion distributions and lateral spreadings should take the transverse straggling into account.

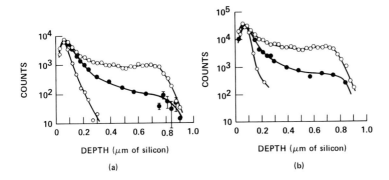

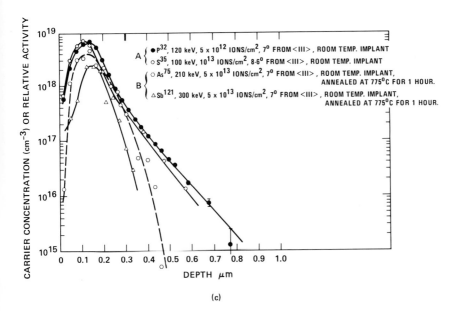

Fig. 4. (a) The dependence of the range distribution of P^{32} ions implanted into silicon on the orientation of the crystal with respect to the $\langle 110 \rangle$ axis at 40 keV: ○, aligned; ●, 2° misalignment; and ○, 8° misalignment (after Dearnaley et al. [4]). (b) The dose dependence of the range distribution of 40-keV P^{32} ions implanted along the $\langle 110 \rangle$ direction into silicon: ○, $<10^{13}$ ions/cm^2; ●, 9×10^{13} ions/cm^2; ○, 7×10^{14} ions/cm^2. (c) Distribution of 120-keV P^{32} and 100-keV S^{35} implanted into misoriented crystals, showing a similar exponential tail due to enhanced diffusion. Also shown are the donor distributions for 210-keV As^{75} and 300-keV Sb^{121} implantations (after Dearnaley et al. [4]).

1 Distribution Function

For an amorphous target, the spatial distribution of ions implanted through a narrow mask cut is given approximately by

$$N(x,y,z) \cong \frac{Q_0}{(2\pi)^{1/2} \Delta R_p} \left\{ \left[-\frac{1}{2} \left(\frac{x - R_p}{\Delta R_p} \right)^2 \right] \left[\frac{1}{\sqrt{\pi}} \operatorname{erfc} \left(\frac{y - a}{\sqrt{2} \, \Delta R_t} \right) \right] \right\},$$

where the parameter a is the distance measured from the center of the mask cut to the edge and the transverse straggling R_t has been obtained as shown in Fig. 3. This distribution function and lateral spreadings have been incorporated into a computer simulation program known as SUPREM [6]. For more accurate data, Takeda and Yoshi have solved the two-dimensional Boltzmann transport equation using a Monte Carlo technique; the results obtained show substantial improvement in the accuracy of the range distribution compared with those obtained from the LSS theory, especially for heavy ions [7]. Most significantly, the small geometry effects on the distribution function due to the transverse straggling are clearly demonstrated.

2 Peak Concentration

Figure 5 shows that the peak concentration is constant for an implanted region for a diameter A_i greater than 1000 Å, as expected; however, as the

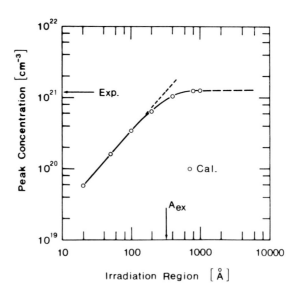

Fig. 5. Relations between peak concentration and irradiation region A_i for 50-keV, 5 × 10^{15} cm^{-2} As implants in the calculation. (After Takeda et al. [7].)

implanted region becomes smaller, the peak concentration proportionally decreases [7]. This can be understood from the relative lateral broadening of the implanted region, caused by ion straggling, as shown in Fig. 6; the relative dimension of the effective implanted region significantly increases for small values of A_i. It can be seen from Fig. 6 that as the irradiation region A_i is further reduced, the position of the peak concentration moves toward the outside edge of A_i. Thus, for device dimensions below 1000 Å, the two-dimensional effects of lateral broadening must be taken into account in designing complex and fine structures for VLSI.

Masking D

For an implanted Gaussian profile, a mask layer of thickness $R_p + 3.72 \Delta R_p$ is required to achieve a masking effectiveness of 99.99% (or allowing just 10^{-2}% of the total implanted ions to penetrate into the substrate), or $R_p + 4.27 \Delta R_p$ for 99.999%. Figures 7a, 7b, and 7c show the minimum thickness required for 99.99% effective masking in using photoresist, SiO_2, and Si_3N_4, respectively [8]. Deposited polysilicon and metal films can also be used for implantation masking, and the required thickness can be calculated from the R_p and ΔR_p of Si and metal (obtainable from Refs. 1 and 2).

In LSI processing, polysilicon gate and insulator thicknesses used in

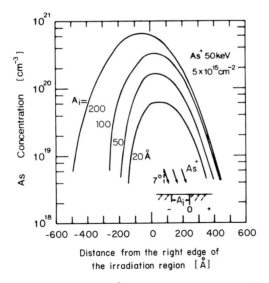

Fig. 6. As lateral distributions at 360 Å in depth for implants with 50 keV, 5×10^{15} cm^{-2}, 7° off in the right direction, and small irradiation region A_i values. (After Takeda et al. [7].)

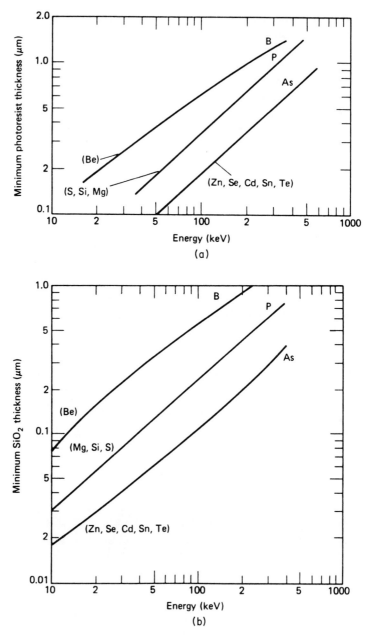

Fig. 7. Minimum film thickness for a masking effectiveness of 99.99% for (a) photoresist, (b) SiO_2, and (c) Si_3N_4. (After Ghanhdi [8].)

26. Application of Ion Implantation in VLSI

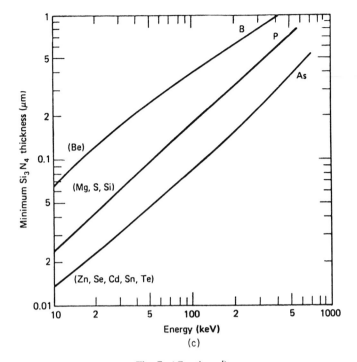

Fig. 7. (*Continued*)

the processing are generally sufficient to block ion penetration into the underlying layer. In VLSI, the gate and insulator thicknesses are much reduced according to the scaling rules; some ions, particularly those with small masses, can penetrate into the oxide insulator and the silicon substrate beneath the masking films, causing undesired effects. One common solution to this unreliable implant masking is to retain the pattern-defining photoresist of these gates and insulators. However, the ion implantation step is restricted to the process step immediately following the patterning, and usually an additional photolithographic step is needed to pattern the photoresist layer.

E. Ion Implantors

The principle of operation for various kinds of commercial ion implantors can be found in the existing literature [9]. Recent advances in the manufacture of ion implantors have achieved high ion current at extremely low or high implant energies (<1 keV or >800 keV). At high energy, power dissipation on the wafer and the uniformity of the dose (by scanning the wafer) are the ultimate concerns of the manufacturers of ion implantors, because excessive heating and uneven heat sinking can cause the redistribution of the implanted impurities and, thus, nonuniformity as well as contamination. It is necessary to carry out periodic testing of dose uniformity as well as dose error calibration of the ion implantor to insure the quality of the circuits and devices manufactured with these machines. At low energy, beam quality and intensity are of primary concern. The methods of testing the electrically active doping density can be found in various texts [9].

III DOPING APPLICATIONS IN MOS TECHNOLOGY

A Junction Formation

For Si, pn junctions are usually formed by ion implantation with dosages above 10^{14} cm^{-2}, followed by annealing at temperatures higher than 900°C to activate the dopants electrically. The pn junctions formed by ion implantation have a curvature at the window edge; this curvature results in an electric field that is stronger than that of an ideal, parallel-plane structure and causes premature breakdown. This problem is especially severe in the VLSI junctions, which are shallow and have small radii of curvature.

VLSI-type devices require source- and drain-junction depths of less than 0.5 μm. Ion implantation with its control of impurity type, concentration, vertical profile, and lateral extent is the preferred technology. Owing to the different diffusivities of the species, the anomalous distribution tail can be reduced by the proper choice of the species, such as As over P (as discussed in Section II.B).

B Threshold Voltage Control

One of the most important applications of ion implantation in MOS technology is for controlling the threshold voltage of MOS structures. This is done by implantation through the gate oxide at relatively low energy; and if the deposited dopants are placed in the channel region, the threshold gate voltage changes by $\Delta V_T = -\Delta Q_s/C_{ox}$, as illustrated in Fig. 8, where ΔQ_s is the change of the sheet ionized dopant charge in the channel, and C_{ox} is the gate oxide capacitance per unit area. A more detailed formulation that takes into account the change of the quasi-fermi potential, as well as the depletion region, gives a more accurate estimate [10]. The change of sheet charge can be calculated with SUPREM. The

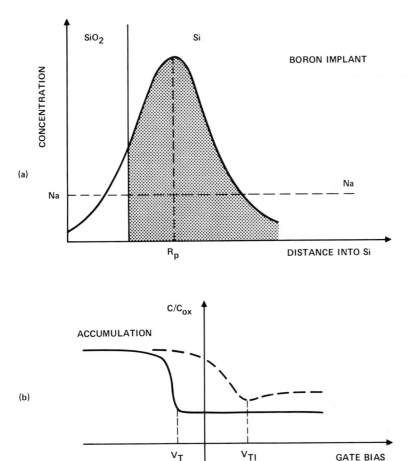

Fig. 8. (a) Implanted charge represented by the shaded area (*n*-type dopants into a *p*-type substrate). (b) The C–V curve indicating the threshold voltage shift from V_T to V_{TI}.

use of boron atoms results in a positive shift of V_T (and a phosphorus implant results in a negative shift of V_T). Similar applications can be used for the field oxide to prevent surface inversion. By the proper use of threshold control, enhancement and depletion mode devices for both *n*- and *p*-channel devices (CMOS) can be fabricated on the same substrate.

C. Buried Channel Formation

A buried channel device [11] is one in which the signal charge is either stored or transported in the bulk of the semiconductor beneath the SiO_2–

Si interface. The effect due to interface state trapping is markedly reduced in buried channel devices, and high transfer-efficiency charge-coupled devices have been fabricated. For buried channel devices, ion implantation is usually performed using an n-type dopant to form a junction with an underlying p-type substrate resulting in the structure shown in Fig. 9a. The potential distribution for this structure under a positive V_G bias is illustrated in Fig. 9b for the case of zero channel charge, and the channel potential maximum (energy minimum for electrons) occurs between the surface and the junction. The channel depth decreases with the gate bias, and the maximum channel charge, beyond which the location of the potential maximum reaches the surface, depends upon the gate oxide thickness, the implanted fluence, and the junction depth.

Buried channel transistors have also been demonstrated a lower $1/f$ noise than that of surface channel transistors, as illustrated in Fig. 10a. The residual defects caused by ion implantation and annealing for surface channel transistors increase the $1/f$ noise, as indicated in Fig. 10b.

D CMOS

An extremely important use of ion implantation is in the formation of the p well in n-type substrates for CMOS. The p region is needed for the fabrication of n-channel transistors. In this application, ion implantation is used to control threshold voltage for both n- and p-channel transistors; so a single power supply is adequate. Ion implantation also replaces the conventional diffusion for source and drain doping with improved and reproducible results.

E Punchthrough Stopper

Although the speed of MOS devices improves substantially from downscaling structural dimensions in VLSI, various undesirable short-channel effects arise. One of the most significant degradation effects is *reduced drain-breakdown voltage* BV_{DS}. At high drain bias, the depletion boundary of the drain can expand to reach the boundary of the source for VLSI devices, and *source-to-drain breakdown* occurs. This condition is called *punchthrough*. Punchthrough is usually prevented by a deep boron implant under the gate in n-channel MOSFETs to suppress the expansion of the depletion region from the drain; the results of the device characteristics show a small threshold variation in addition to prevention of premature punchthrough. Normally, the deep boron implant is done at high energy (e.g., 1×10^{12} cm^{-2} at energies greater than 100 keV).

26. Application of Ion Implantation in VLSI

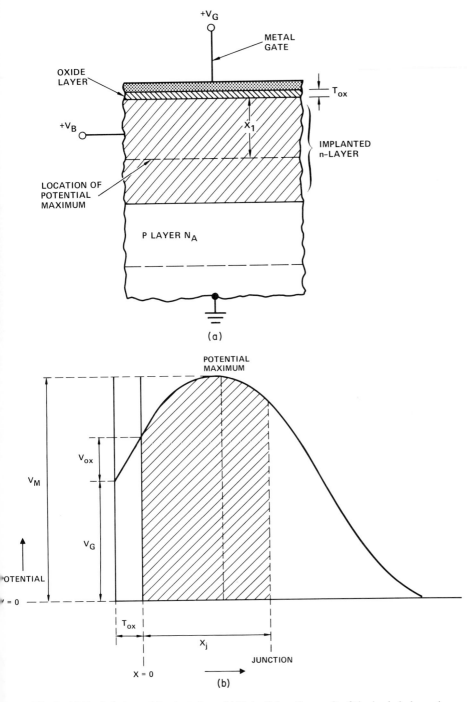

Fig. 9. (a) Buried channel implantation. (b) Potential as the result of the buried channel implantation.

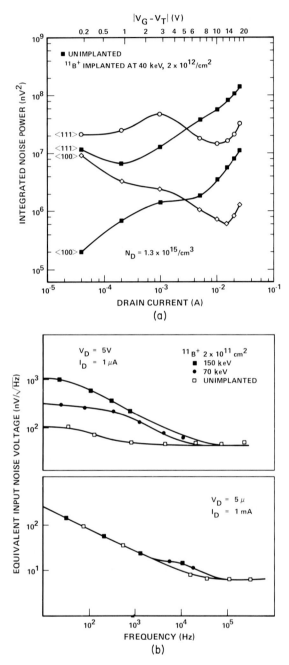

Fig. 10. (a) Noise power versus drain current for surface *p*-channel FETs in both the $\langle 100 \rangle$ and $\langle 111 \rangle$ orientations. Two noise power curves for ^{11}B$^+$-implanted samples are also illustrated to show the improvement for buried channel FETs when operated in the high drain current range (after Wang [12]). (b) $1/f$ noise of B^{11+} implanted *p*-channel transistors annealed at 1000°C for 10 min. (the implantation fluence is sufficiently low to avoid buried channel) (after Nakamura *et al.* [12]).

Graded Sources and Drains F

For *n*-channel FETs, a low-dose, shallow *n*-implant is performed, in addition to a high-dose As source-and-drain implant. Figure 11 shows one of many ways to achieve a graded source and drain with the poly-Si gate as a mask. By this means, a tapered distribution of graded junctions is formed, which provides the control of the depletion width of the drain and source junctions necessary for preventing punchthrough. The graded junction implant can be done with 10^{14} cm^{-2} followed by the As source-and-drain implant.

Application Example G

With combinations of buried channel, graded drain, and punchthrough-stopped implants, as illustrated in Fig. 12, the drain breakdown voltage

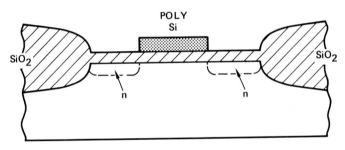

(a) Low-Energy Implant

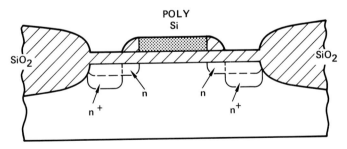

(b) High-Energy Implant

Fig. 11. (a) Double implant performed by using first a low-energy boron implant. (b) A high-energy P or As implant to form the n^+ source and drain after oxidation to form the graded source and drain.

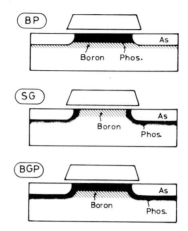

Fig. 12. Device structures experimentally investigated. Symbols S, B, G, and P denote surface channel, buried-channel, graded-drain, and punchthrough stopper. (After Sunami *et al.* [13].)

(discussed in Sections III.E and F) can be further increased. Figure 12 illustrates three such possible structures: (a) buried channel with punchthrough stopper, BP; (b) graded drain, SG; and (c) buried channel and graded drain along with the punchthrough stopper, BGP. The deep boron implant is used to form a punchthrough stopper. The resulting improvement in BV_{DS} for all these cases is compared with values for surface channel devices, S, as shown in Fig. 13a. Along with this result, the substrate current decreases as shown in Fig. 13b, similar to the improved trend of BV_{DS}. Also shown in the figure, the subthreshold current increases substantially for BGP and BP devices, whereas that of SG structures decreases. Two drawbacks—increased junction capacitance and an increased back gate bias effect—are noticeable for BGP and BP.

IV DOPING APPLICATIONS IN BIPOLAR TECHNOLOGY

A Predeposition

Ion implantation can be used instead of conventional diffusion for predeposition prior to the base, the emitter, and the collector drive-in's. Because of its precise and reproducible control of the Gummel number in bipolar transistors, the all-implanted process is preferred for advanced VLSI circuit fabrication.

Ion implantation is especially useful in forming extremely small and

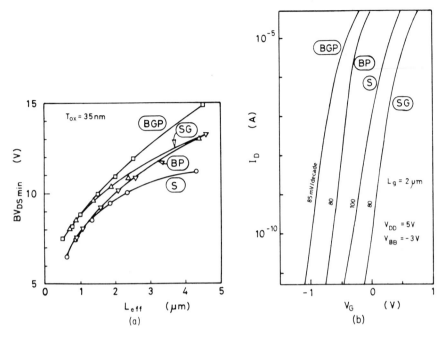

Fig. 13. (a) Channel-length dependence on minimum source-to-drain breakdown voltage (after Sunami et al. [13]). (b) Subthreshold current characteristics for 2-μm gate devices (after Sunami et al. [13]).

shallow emitter junctions for VLSI circuits. In this application, As ions are the preferred species, because the resulting doping profile after heat treatment is relatively abrupt and free from deep tailing effects. Precautions and applications are discussed in Section III.A. Fully implanted bipolar VLSIs have been demonstrated in engineering and development environments for many years. However, problems remain in the volume production of ion-implanted bases. The difficulties involve controlling the distribution tail while simultaneously attempting to anneal the residual defects. To achieve maximum control over the final impurity profiles, ion implantation—coupled with low temperature oxidation, annealing, and plasma oxide—should be used extensively except in collector annealing. The use of the flash anneal as discussed in Section II.B.4. can result in a major advance in the bipolar technology.

B Buried Collector or Subcollector Implantation

Buried layers were previously formed with diffusion followed by epitaxial growth of an *n*-collector layer, because the required collector depths were too large to be accomplished by direct ion implantation. However, for VLSI, direct high-energy ion implantation (100 keV to 1 MeV) can be used to provide a heavily doped subcollector with a doping density greater than 2×10^{17} cm^{-3}, because the collector layer may be only 200 nm thick. Such thin buried layers could be produced selectively and with a higher yield than by CVD epitaxy. With a fully ion-implanted process, the processing complexity can also be much reduced (see Section IV.D).

C Implanted Resistors

High value resistors are often made with an extremely shallow ion implant step. With ion implantation, reproducible sheet resistances up to 20 kΩ/□ (p-ion implant in the n-epilayer) are possible. The higher the sheet resistance, the less accurate and reproducible the resistance is because of variation in the doping of the epilayer, spreading of the ion distribution, and the presence of oxide charge in the oxide layer on top of the resistor (this can cause a depletion or enhancement layer at the oxide interface). The highest resistor values are limited by the fact that the p-type implant must overdope (compensate) the n-epilayer by at least a factor of two or so.

D Process Simplification

Processing in bipolar VLSI can become so complex that the number of the masking steps can exceed the point of diminishing returns for die yield. Thus, the major thrust is to use the attributes of ion implantation, along with the self-aligned feature, to simplify processing. A self-aligned and full ion-implanted process, using photoresist for implant masks, is illustrated in Fig. 14.

(a) Figure 14a shows the structure after field implant and isoplanar field oxidation, followed by self-aligned transistor masking (SAT) to define various components of the transistor by selectively removing the nitride layer. The exposed thin oxide area is oxidized again to form SAT oxide.

(b) The nitride layer is subsequently removed, as in Fig. 14b. The self-aligned feature is based on the differential oxide thickness of the thin oxide on the epilayer, SAT oxide, and the thick field oxide.

(c) A masked phosphorus sink implantation provides the collector contact (Fig. 14c).

26. Application of Ion Implantation in VLSI 475

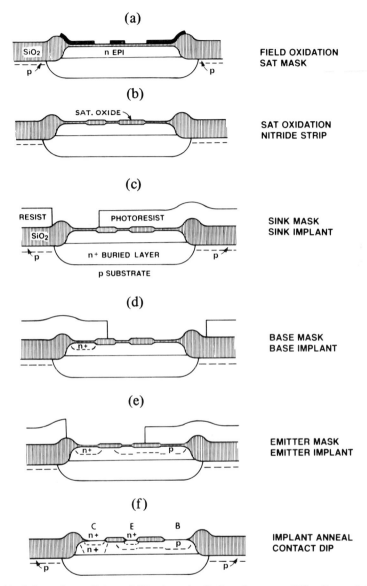

Fig. 14. Schematics outlining a full implanted self-aligned process. (After Ko et al. [14].)

(d) Next comes a masked boron implant for the base (Fig. 14d). Two-energy boron implantations are used: The shallow (low energy) implant is provided for good ohmic contacts to the metal, and the deep (high energy) implant gives the desired Gummel number control for the n–p–n transistor.

(e) Figure 14e illustrates masking of the shallow As implantation for the emitter and the collector contact.

(f) Figure 14f shows the completed structure after 1000°C annealing.

All the implantation steps shown in Figs. 14b–f use photoresist films as the implantation masks. The self-aligned scheme employed results in the structures ready for metallization; no additional contact window mask is needed.

V RECENT ADVANCES AND OTHER APPLICATIONS

A Damage Gettering

One of the major consequences of ion implantation is damage to the crystalline substrate. However, the damage can be utilized to getter unwanted impurities that are introduced during processing [15]. When the substrate is heated above 800°C, unwanted impurities diffuse to the damaged area, for instance, to the back of the wafer or to a selected location on the front, produced by implantation of heavy species, such as BF_2, or inert species, such as Ar or Xe. (Recently, frontside gettering has been found to be more reproducible than backside gettering.) The unwanted impurities then precipitate in the implanted region, and the circuit yield can be increased markedly. The gettering implant is always performed prior to a subsequent heat treatment, 850–950°C for 30 min. The gettering effectiveness decreases as the postannealing temperature increases beyond 1000°C due to the fact that the implanted defects are annealed, and the substrate recovers its crystallinity. As a result, the damaged gettering sites are removed. The gettering is not effective for heat treatment below 800°C since most of the unwanted impurities cannot diffuse to the damaged site at such low temperatures. The desired lower processing temperature (800–1000°C) called for in VLSI will make the ion-implantation gettering technique more attractive. As illustrated in Fig. 15, the improved carrier lifetime made possible by ion implantation gettering requires that effective gettering be performed in the last high-temperature step (850–1000°C) in the processing sequence, because additional high-temperature treatments degrade gettering effectiveness and, thus, carrier lifetime.

B Enhanced Etching

Implantation-induced damage can alter the etching rate of thin films, and it usually increases the etching rate of SiO_2 and Si_3N_4 and hardens the

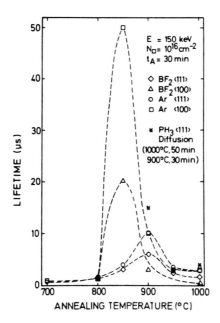

Fig. 15. Lifetime versus gettering temperature for argon and BF_2 in (100) and (111) silicon. (After Ryssel et al. [15].)

photoresist to UV exposure. For example, Ar implantation at a moderate dose (5×10^{13} cm^{-2}) can result in a sixfold increase of the etching rate of SiO_2 in BHF. Thus, the damage implant can be performed through a resist window to facilitate fine-line patterning without sidewall over-etching, as illustrated in Fig. 16.

On the other hand, sharp corners on the top of a window can result in breaks in metal lines or other overlayers. A very shallow implant can be

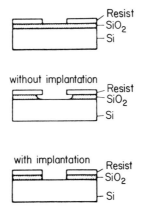

Fig. 16. Etching of narrow lines using implantation damage. (After Ryssel et al. [16].)

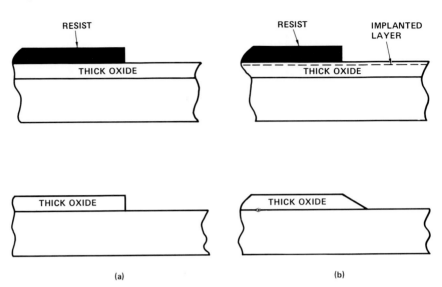

Fig. 17. (a) A sharp corner resulting from patterning on a thick oxide film. (b) A very shallow implant is performed prior to the patterning, resulting in a tapered window after etching.

done first to increase the etching rate on the surface as illustrated in Fig. 17a; enhanced etching of the surface layer results in the tapered step shown in Fig. 17b. The tapered slope angle of the etched windows with various etching techniques begins to change with a dose greater than 10^{13} cm^{-2} and approaches about 30° for a dose greater than 10^{14} cm^{-2}, as shown in Figs. 18a and b for Ar and As implants, respectively [16].

C Local Oxidation

In conventional local oxidation, CVD Si_3N_4, which inhibits thermal oxidation of Si, can be deposited and followed by patterning to define the oxidation window. As an alternative to CVD deposition, nitrogen can be implanted in a selected region of Si to form a planar Si_3N_4. The implanted nitrogen fluence required for providing effective masking need not exceed the value that is sufficiently high to form stoichiometric Si_3N_4. This oxidation masking technique provides a new LOCOS technique and avoids several shortcomings of the conventional CVD technique: stress in the CVD-deposited Si_3N_4 interface, lateral oxidations, an anomalous oxidation at the edge of the deposited nitride layer, and pinholes in the CVD-deposited Si_3N_4 film. A retardation factor of 30 in the oxidation rate has been achieved for a 2.4×10^{17} N cm^{-2} implant, as illustrated in Fig. 19

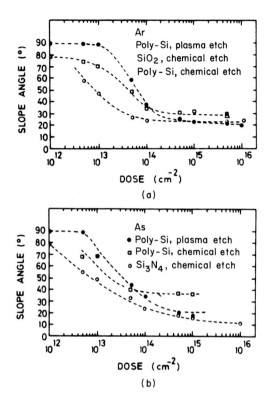

Fig. 18. Slopes of tapered windows as a function of implantation dose of (a) argon and (b) arsenic. (After Ryssel et al. [16].)

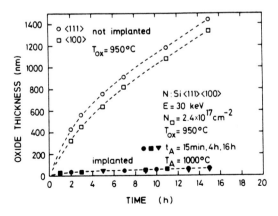

Fig. 19. Retarded oxidation for a 2.4×10^{17} cm^{-2} nitrogen implant for (111) and (100) silicon. (After Ryssel et al. [16].)

(almost as effective as a factor of 36 if a CVD-deposited Si_3N_4 mask is used). The attributes of implantation masking provide many advantages for fabricating fine-line circuits and for increasing the VLSI yield.

D Buried Insulator Layer Formation

The heavily-doped buried layer for use as a subcollector in bipolar transistors is described in Section IV.B. In addition to these applications, high-energy implantations using nitrogen or oxygen have been used to produce a buried SiO_2 or Si_3N_4 layer having a breakdown field greater than 5×10^6 V cm^{-1}. With ion implantation, it is possible to fabricate isolated islands on which VLSI devices and circuits can be built by standard processes. One possible process procedure for making completely isolated islands is illustrated in Fig. 20. A masking and patterning step was done first (Fig. 20a) before a high-energy N implantation (e.g., 3×10^{17} cm^{-2} at 1.2 MeV) to produce a deep Si_3N_4 layer after annealing. Trenches were fabricated using standard lithography with a CVD-deposited Si_3N_4 film (Fig. 20b). With the remaining Si_3N_4 used as an oxidation mask, the sidewalls of the silicon islands are oxidized, which produces the isolated Si islands (Fig. 20c). Similarly, an oxygen implant can be used to replace the nitrogen implant to produce a buried oxygen layer, as discussed in Section V.E.

E Silicon-on-Insulator Using Oxygen Implantation

There is growing interest in the formation of silicon-on-insulator (SOI) substrates as a means of achieving vertical isolation and three-dimensional integration in VLSI device technology.

Silicon on a buried SiO_2 layer that is formed by high-dose oxygen implantation (with a fluence greater than 1×10^{18} O$^+$ cm^{-2}), followed by annealing, is one of the most promising technologies currently available for the fabrication of VLSI circuits on SOI substrates.

The buried oxide layer formed using this technique is shown to be stoichiometric silicon dioxide and has abrupt Si–oxide interfaces on both sides of the oxide film resulting from internal oxidation. During annealing, the implanted oxygen diffuses into the Si–SiO_2 boundaries where it reacts with the remaining Si. An epitaxial layer can be subsequently grown on the annealed substrate. However, for the best result (i.e., the lowest yield in RBS channeling data), the epitaxial film grown on a preannealed substrate (for 2 h at 1150°C) should be followed by a postepitaxial growth anneal of 4 h at 1150°C in Ar. (The use of N_2 as an ambient gas can cause

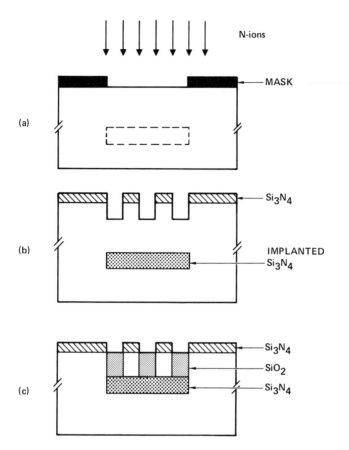

Fig. 20. (a) N-ion implantation performed to form a buried Si_3N_4. (b) Etching to remove Si. (c) Masked oxidation to form isolated islands. (After Ryssel et al. [16].)

reaction of Si + N_2 and result in pitted Si surfaces after annealing.) To obtain a low leakage current in the oxide, an extremely high fluence of oxygen is required. It was demonstrated that the leakage current of the buried SiO_2 decreases to 10^{-8} A cm^{-2} at ± 10 V for samples implanted with a fluence greater than 1.5×10^{18} cm^{-2} (at 200 keV) followed by annealing at 1150°C for 4 h [17].

Radiation Enhanced Diffusion F

Radiation enhanced diffusion is the increase of the impurity diffusion rate in Si due to the presence of the abundant vacancies that are created by irradiation with high-energy light-mass ions at relatively low tempera-

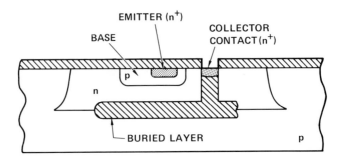

Fig. 21. Fabrication of a deep collector contact to the buried layer using ionization enhanced diffusion. Without the use of the ionization enhanced diffusion, the contact would have to be performed by a high-energy implant or a deep drive-in, which causes the lateral broadening.

tures. The vacancies provide abundant additional sites for dopant atoms to move and increases the rate of diffusion. Significant enhancement is possible with hydrogen implantation; other inert species such as He and Ar have also been tried, but these species tend to produce bubbles and other damage that cannot be annealed [16]. One application of ionization enhanced diffusion is illustrated in Fig. 21. Here, the collector contact to the buried layer can be achieved by enhanced diffusion of the n^+ layer vertically by proton implantation, without suffering the isotropic broadening of the conventional diffusion. This technique circumvents the need for an expensive high-energy implantor and can reduce the contact size and save Si real estate in VLSI design. However, an additional masking is needed in carrying out the implantation for achieving the desired ionization enhanced vertical diffusion for the selected region.

G Silicidation of Metal–Si Reaction

High-dose implantation can be used to promote ion beam mixing of metal and silicon in producing metal silicide [18]. This ion mixing technique provides excellent reproducibility in silicidation and results in silicide films of superior uniformity and integrity, because the interface native oxide layer, which often causes the problems, is "destroyed" by the energetic ions. In addition, this technique gives improved surface morphology over these silicides formed by thermal reaction. Heavy ions are found to be more effective than light ions for the ion mixing. (For example, As ions are more effective than P ions.) This process is particularly promising for VLSI source-and-drain doping and contacting, in which extremely stable junctions with junction depths less than 0.1 μm are desired. As illustrated in Fig. 22, an As dose ranging from 5×10^{14} cm^{-2} to

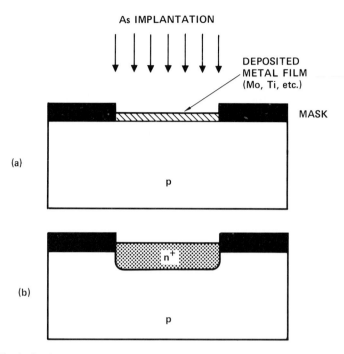

Fig. 22. As implantation through a thin Mo film to form a silicided n^+/p. (a) Prior to implantation. (b) The resulting n^+/p junction.

10^{16} cm^{-2} is used for an n^+/p case. The implantation is performed through a thin metal film (e.g., Mo or Ti) that was previously deposited onto a Si substrate, and a silicided n^+/p junction is formed after 1000°C annealing for 20 min. In this structure, a contact resistance comparable to the conventional n^+/p junction was obtained, and the leakage current of the pn junction was comparable to that of conventional n^+/p junctions [18]. A Mo-silicided contact formed by ion implantation has been shown to be a stable diffusion barrier for Al overlayers up to 500°C. Because of the low mass of boron atoms, silicided p^+ layers need to be formed by a two-step implantation: a high-dose implantation with electrically inert ions such as Si and Ar for ion beam mixing, followed by a boron implantation to provide the necessary p^+ layer. The implantation process in the reverse order is also feasible.

Beam Writing and Ion Beam Lithography   H

Other applications of ion implantations may include fine-line direct doping using a focused beam as well as the improvement of lithograph resolution using an ion beam to expose the resist.

Focused ion beam doping will require an ion beam with an extremely high current density (on the order of 1 A/cm^2 or higher) for practical writing speed. Such high current density, which is difficult to obtain in a microbeam often generates sufficient heat to cause the Si to melt. To date, this direct micro-ion beam doping is not feasible. A more feasible application is ion beam lithography in which a focused ion beam is used to replace the electron beam presently used commercially for writing on the resist to produce fine-line patterns. The polymer resist currently used for electron lithography is already a factor of approximately 100 more sensitive to ions than to electrons. With ion beams, the primary ions expose the resist with little lateral scattering, whereas in electron beam lithography, the scattered secondary electrons do the exposure. Since the lateral scattering of ion beams is very restricted, ion beam lithography ultimately provides a much higher resolution than electron beam lithography. With the advancement of highly bright ion sources (e.g., using liquid metal field ionization ion sources), ion beam lithography is projected to have a bright future in applications to VLSI or ULSI.

ACKNOWLEDGMENT

The author would like to acknowledge partial support of the Semiconductor Research Corporation for this work.

REFERENCES

1. J. Linhard, M. Scharff, and H. E. Schiott, *Mat. Fys. Medd. Dan. Vict. Selsk.* **33** (14) (1963).
2. J. F. Gibbons, W. S. Johnson, and S. W. Mylorie, "Projected Range Statistics." Dowden, Hutchinson, and Ross, Inc., Stroudsburg, Pennsylvania, 1975.
3. S. Furukawa, H. Matsumura, and H. Ishiwara, Theoretical considerations on lateral spread of implanted ions, *Jpn. J. Appl. Phys.* **11,** 134 (1972). Also quoted by S. K. Ganhdi, "VLSI Fabrication Principles." Wiley, New York, 1983.
4. G. Dearnaley, J. H. Freeman, G. A. Card, and M. A. Wilkins, *Can. J. Phys.* **46,** 587 (1968). Also in G. Dearnaley, J. H. Freeman, R. S. Nelson, and J. Stephen, "Ion Implantation." North Holland, Amsterdam, 1973.
5. J. F. Gibbons, L. D. Hess, and T. W. Sigmon (eds.), "Laser and Electron Beam Solid Interactions and Material Processing. North Holland, Amsterdam, 1981.
6. Stanford University Processing Engineering Model. See D. A. Antoniadis and R. W. Dutton, *IEEE J. Solid State Circuits* **SC-14,** 412 (1979).
7. T. Takeda and A. Yoshii, *IEEE Electron Device Lett.* **EDL-4,** 430 (1983).
8. S. K. Ghanhdi, "VLSI Fabrication Principles," p. 350. Wiley, New York, 1983.
9. G. Dearnaley, J. H. Freeman, R. S. Nelson, and J. Stephen, "Ion Implantation," p. 255 *ff*. North Holland, Amsterdam, 1973.
10. T. W. Sigmon and R. Swanson, *Solid-State Electron.* **16,** 1217 (1973).
11. A. Mohsen and M. F. Tompsett, *IEEE Trans. Electron Devices* **ED-21,** 701 (1974).

26. Application of Ion Implantation in VLSI

12. K. L. Wang, *IEEE Trans. Electron Devices* **ED-25,** 478 (1979). K. Nakamura, O. Kudoh, M. Kamoshida, and Y. Haneta, "Ion Implantation in Semiconductors (S. Namba, ed.). Plenum Press, New York, 1975.
13. H. Sunami, K. Shimohigashi, and N. Hashimoto, *IEEE Trans. Electron Devices* **ED-29,** 607 (1982).
14. W. C. Ko, T. C. Gwo, P. H. Yeung, and S. Radigan, *IEEE Trans. Electron Devices* **ED-30,** 236 (1983).
15. H. Ryssel, H. Kranz, P. Bayerl, and B. Schmiedt, *Int. Conf. on Ion Beam Modification of Materials, Budapest, 4–8 September 1978*. T. E. Seidel, R. L. Meek, and A. G. Gullis, *Inst. Phys. Conf. Ser.* **23,** 494 (1975). H. J. Geipel and W. K. Tree, *IBM J. Res. Develop.* **24,** 310, (1980).
16. H. Ryssel and I. Ruge, *4th European Conf. Electrotechnics, 24–28 March 1980,* p. 63 (1980).
17. R. F. Pinizotto, B. L. Vaandrager, S. Matteson, H. W. Lam, S. D. S. Malhi, A. H. Hamdi, and F. D. McDaniel, *IEEE Trans. Nuclear Sci.* **NS-30,** 1718 (1983).
18. K. L. Wang, F. Bacon, and R. F. Reihl, *J. Vac. Sci. Technol.* **16,** 1909 (1979). E. Nagasawa, M. Morimoto, and H. Okabayashi, *1982 Sym. on VLSI Technology Digest, Osio, Japan,* p. 26 (1982).

Chapter 27
Plasma Processing for VLSI

BARBARA A. HEATH
LEE KAMMERDINER

INMOS Corporation
Colorado Springs, Colorado

I. Introduction 487
II. Etching 488
   A. Anisotropy 488
   B. Selectivity 488
   C. Etch Profiles 488
   D. Equipment 490
   E. Parameters Affecting Anisotropy and Selectivity 491
   F. Mechanisms for Producing Anisotropy and Selectivity 491
   G. Summary 494
III. Sputtering 494
   A. Description 494
   B. Equipment 494
   C. Typical Applications 497
IV. Plasma Enhanced Chemical Vapor Deposition 499
   A. Equipment Configurations 499
   B. Typical Process 500
   C. Applications 501
   References 502

INTRODUCTION I

The introduction of plasma processing to the manufacture of integrated circuits over the last 15 years has contributed significantly to the advent of VLSI. Plasma-assisted etching techniques have replaced chemical etching, particularly where strict dimensional control is required. Sputtering has replaced evaporation for the deposition of metals, and plasma enhanced chemical vapor deposition (PECVD) is providing an attractive alternative to conventional atmospheric and low pressure CVD methods.

II ETCHING

In this section the basic principles of plasma-assisted etching are introduced. Melliar-Smith and Mogab [1] have reviewed the subject, and the mechanisms of these processes have been discussed in detail by Flamm et al. [2]. Reviews of specific techniques such as reactive ion etching [3] and ion milling [4] are also available.

A Anisotropy

The greatest advantage of dry or plasma-assisted etching processes over wet chemical etching techniques for VLSI circuit fabrication is the superior dimensional control. Fundamental to dimensional control, the accurate replication of a photoresist feature into the underlying film, is the concept of anisotropy. In an anisotropic etch, the rate of etching perpendicular to the wafer surface is much greater than that lateral to the surface. In an isotropic etch, the etching rates are equal in both directions, which causes undercutting of the masking material and a change in dimension of the etched pattern by at least twice the film thickness.

B Selectivity

In addition to a degree of anisotropy, a VLSI etching technique requires selectivity of the film to be etched over both the masking resist and the often thin underlying substrate. Further, the best results have been obtained when the etch yields volatile products which can be removed by the plasma reactor vacuum system. The requirements for selectivity and volatile etch products have led to the use of gases containing halogens (F, Cl, and Br) as plasma etchants. In the plasma environment, compounds made up of the halogens as well as carbon, sulfur, nitrogen, silicon, or hydrogen dissociate to form reactive atoms, radicals, and ions which participate in the etching reactions. These plasmas have been conveniently grouped into those that primarily generate free F atoms, unsaturated fluorocarbons, and those generating Cl and Br species [2].

C Etch Profiles

Four types of etch profiles resulting from plasma-assisted etching processes are shown in Fig. 1.

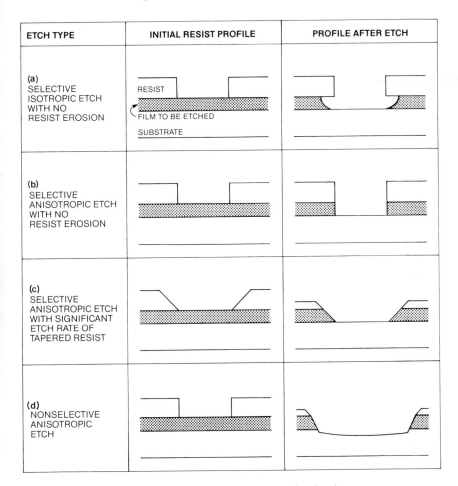

Fig. 1. Etch profiles resulting from plasma-assisted etch processes.

(a) A selective isotropic etch with little or no etching of the initial resist pattern yields an undercut profile (Fig. 1a). Dimensional control is difficult with this type of etch, because after the film has been cleared, lateral etching often occurs at an increased rate due to the increased availability of reactant.

(b) A selective anisotropic etch results in excellent pattern transfer from resist to underlying film (Fig. 1b).

(c) The application of a selective anisotropic etch prepares a controlled slope on the edge of the etched film (Fig. 1c). Plasma parameters are adjusted so that the resist has a significant anisotropic etch rate, resulting in the transfer of the original resist profile into the underlying material. This technique can be useful when the steep step prepared by method (b) would complicate patterning of subsequently deposited films.

(d) The results of a nonselective anisotropic etch are shown (Fig. 1d). Due to the sputtering nature of this process, significant erosion of the resist and underlying film has occurred, as well as a change in pattern dimension.

D Equipment

Four basis types of equipment used for plasma-assisted etching are shown in Fig. 2.

(a) In a *barrel etcher* (Fig. 2a), the plasma serves only to produce reactive etchant species [2]. The plasma is often isolated from the wafers by an etch "tunnel," leaving the reactant species to reach the wafers by

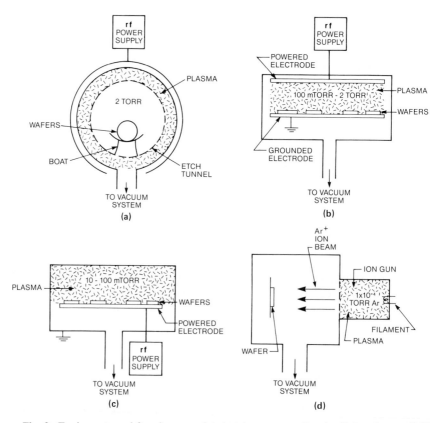

Fig. 2. Equipment used for plasma-assisted etch processes (for simplicity, the manifold used to introduce the etch gases into the system is not shown). (a) Barrel etcher. (b) Planar plasma (or diode) reactor. (c) Reactive ion etcher. (d) Ion milling apparatus.

diffusion. The design of this reactor and the high pressures usually used result in isotropic etch profiles. Si_3N_4 and polycrystalline Si can be etched with this type of equipment, but its most common use today is in photoresist stripping.

(b) In a *planar plasma* (or *diode*) *reactor* (Fig. 2b), the wafers are loaded on the grounded electrode.

(c) In a *reactive ion etcher* (RIE) (Fig. 2c), the wafers are loaded on the powered electrode. The difference in size of the grounded electrode (the whole chamber) and the powered electrode, coupled with operation at low pressure (<100 mTorr) cause a negative dc bias to appear on the powered electrode of the RIE system. This causes significant positive ion bombardment of the electrode surface, resulting in an ion-assisted anisotropic etch. Though the planar plasma etcher shown in Fig. 2b operates at a higher pressure (100 mTorr–several Torr), it can be made to produce significant ion bombardment and hence anisotropy by using a reduced frequency (<1 MHz). For a discussion of this point, see Ref. 2. Both the planar etcher and the reactive ion etcher schematically described in Figs. 2b and c are popular for patterning films where dimensional control and selectivity are required, such as Si (polycrystalline and single crystal), SiO_2, and Al alloys.

(d) In an *ion milling apparatus* (Fig. 2d), the ion producing plasma is isolated from the sample to be etched. An ion beam is produced and directed onto the sample, resulting in a purely physical etching process and an etch profile such as that shown in Fig. 1d.†

Parameters Affecting Anisotropy and Selectivity E

Many parameters affect the anisotropy and selectivity of a plasma-assisted etch. Most important are the chemical nature of the reactant gas or gases, the reactor type, gas pressure, and applied rf power and frequency. The gas flow, electrode materials and spacing, wafer temperature, and number of wafers etched at one time (loading) can also have significant effects.‡

Mechanisms for Producing Anisotropy and Selectivity F

The mechanisms active in producing anisotropy and selectivity in plasma-assisted etching processes have been described in detail in Ref. 2.

---

† For a review of this type of ion etching, see reference [4].
‡ For detailed discussions of these parameters, see references [1–4].

Briefly, anisotropy as produced by positive ion bombardment can be achieved by three mechanisms.

(a) Ions can sputter adsorbed reaction products from the surface.

(b) Ions can produce lattice damage, creating more active sites for reaction.

(c) Ion bombardment can remove an adsorbed layer that inhibits etching (by reacting with the etchant or simply by inhibiting access to the surface by the etchant).

Since the ions are directed perpendicular to the surface, all three mechanisms result in anisotropic etching as described in Fig. 1b. Selectivity is more difficult to engineer into an etch process. One method is to adjust the plasma parameters and etch-gas chemistry so that one material is essentially only physically sputtered whereas the other is etched at an enhanced rate by chemical reaction. Alternatively, when chemical reaction limits the selectivity, the etch-gas composition can be adjusted to favor the chemically enhanced etching of one material over the other. Selectivity is most often achieved empirically.

TABLE I

Summary of Plasma Etching with F-Atom Source Plasmas

Materials	Gases	Anisotropy	Selectivity
Si	F_2, CF_4-O_2, $C_2F_6-O_2$, $C_3F_8-O_2$, SF_6-O_2, SiF_4/O_2, NF_3, ClF_3	Isotropic	High over SiO_2, Si_xN_y, metals, silicides
SiO_2		Isotropic at low ion energy; anisotropic at high ion energies	Very high over III–V compounds
Si_xN_y		Similar to SiO_2	Selective over SiO_2 in isotropic range
$TiSi_2$, $TaSi_2$, $MoSi_2$, WSi_2		Partially anisotropic at high ion energies	High over SiO_2, Si_xN_y; ~1:1 over Si
Ti, Ta, Mo, W, Nb		Anisotropic at high ion energies	High over SiO_2 at high powers; selective over TaO_2
Ta_2N		Isotropic at low ion energy	Selective over TaO_2, SiO_2, Al_2O_3

TABLE II
Summary of Etching with Unsaturated Fluorocarbon Plasmas

Materials	Gases	Anisotropy	Selectivity
Si	CF_4, C_2F_6	Partially anisotropic at low pressure, high ion energy	—
SiO_2, Si_xN_y	CF_4, C_2F_6, C_3F_8, CHF_3, CF_4 or C_2F_6 with H_2, CH_4, C_2H_2, or C_2H_4 additions	Anisotropic at high ion energies	High over Si and III–V compounds
TiO_2			High over Ti
V_2O_5			High over V

TABLE III
Summary of Plasma Etching in Chlorine- and Bromine-Containing Plasmas

Materials	Gases	Anisotropy	Selectivity
Si	Cl_2, CCl_4, CF_2Cl_2, CF_3Cl, Cl_2/C_2F_6, Cl_2/CCl_4, Br_2, CF_3Br	Anisotropic at high and low ion energies; isotropic for doped Si under high pressure conditions with pure Cl_2	High over SiO_2
III–V Compounds	Cl_2, CCl_4, CF_2Cl_2, CCl_4/O_2, Cl_2/O_2, Br_2	Anisotropic at high and low ion energies (for chlorocarbon and BCl_3 mixtures); isotropic for Br_2 and Cl_2 under some conditions	High over SiO_2, Al_2O_3, Cr, MgO
Al	Cl_2, CCl_4, $SiCl_4$, BCl_3, Cl_2/CCl_4, Cl_2/BCl_3, Cl_2/CH_3Cl, $SiCl_4/Cl_2$	Anisotropic at high and low (for chlorocarbon mixtures and BCl_3) ion energies; isotropic for Cl_2 without surface inhibitor	High over Al_2O_3 and some photoresists
Ti	Br_2		
Cr	$Cl_2/O_2/Ar$ $CCl_4/O_2/Ar$	Anisotropic at low ion energies	$1:1$ vs CrO_2 at $>20\% \, O_2$
CrO_2	Cl_2/Ar, CCl_4/Ar		High vs Cr
Au	$C_2Cl_2F_4$, Cl_2		

G Summary

Tables I, II, and III summarize plasma-assisted etching with F-atom source plasmas, unsaturated fluorocarbon plasmas, and Cl and Br plasmas [2]. The materials most commonly etched in each type of plasma are listed, along with etch gases and comments on the selectivity and anisotropy of each process. Many papers have been written on plasma-assisted etching of specific materials [1–4].

III SPUTTERING

A Description

Sputtering is the physical deposition of a thin film by ion bombardment of a suitable target material. The process is performed in a high vacuum system equipped with a target assembly and platens for holding wafers. An inert gas, usually argon, is introduced into the system in the 1–10 mTorr range to provide the proper operating pressure. Either an rf voltage or a negative dc voltage is applied to the target, ignites a plasma (a plasma igniter is provided if needed) and creates a negative potential in the 1–10 keV range on the target with respect to the plasma. This voltage causes heavy ion bombardment which, through momentum transfer, results in the ejection of atoms or molecules from the target.

The fundamental parameter of interest is the *sputtering yield,* defined as the number of atoms (or molecules) ejected per incident ion. This depends primarily on the target voltage, target material, and bombarding ion. Table IV shows sputtering yields for most of the metals of interest in VLSI processing. The *deposition rate,* defined as thickness per unit time, depends on the type of sputtering system and the sticking coefficient of the depositing material. These yields can be used as guides, but deposition rates must be determined empirically.†

B Equipment

Equipment for sputtering has evolved significantly over the past 20 years. Early machines were not suitable for semiconductor processing,

---

† Reviews on the details of sputtering and sputtering equipment can be found in references [5–10]. Examples of sputtering applications in VLSI processing can be found in references [11–13].

TABLE IV

Sputtering Yield of Elements at 500 eV[a]

Element	He	Ne	Ar	Kr	Xe
Be	0.24	0.42	0.51	0.48	0.35
C	0.07	—	0.12	0.13	0.17
Al	0.16	0.73	1.05	0.96	0.82
Si	0.13	0.48	0.50	0.50	0.42
Ti	0.07	0.43	0.51	0.48	0.43
V	0.06	0.48	0.65	0.62	0.63
Cr	0.17	0.99	1.18	1.39	1.55
Mn	—	—	—	1.39	1.43
Fe	0.15	0.88	1.10	1.07	1.00
Co	0.13	0.90	1.22	1.08	1.08
Ni	0.16	1.10	1.45	1.30	1.22
Cu	0.24	1.80	2.35	2.35	2.05
Ge	0.08	0.68	1.1	1.12	1.04
Zr	0.02	0.38	0.65	0.51	0.58
Nb	0.03	0.33	0.60	0.55	0.53
Mo	0.03	0.48	0.80	0.87	0.87
Ru	—	0.57	1.15	1.27	1.20
Rh	0.06	0.70	1.30	1.43	1.38
Pd	0.13	1.15	2.08	2.22	2.23
Ag	0.20	1.77	3.12	3.27	3.32
Hf	0.01	0.32	0.70	0.80	—
Ta	0.01	0.28	0.57	0.87	0.88
W	0.01	0.28	0.57	0.91	1.01
Re	0.01	0.37	0.87	1.25	—
Os	0.01	0.37	0.87	1.27	1.33
Ir	0.01	0.43	1.01	1.35	1.56
Pt	0.03	0.63	1.40	1.82	1.93
Au	0.07	1.08	2.40	3.06	3.01

[a] Data from G. K. Wehner, Rep. No. 2309, General Mills, Minneapolis, 1962.

mainly due to the low deposition rates. The breakthrough occurred with the invention of the *Magnetron,* which solved this problem. Although there are many different configurations, the central feature of the Magnetron involves a set of magnets, located behind the target surface, that provides a magnetic field that concentrates the plasma in the immediate vicinity of the target. This leads to very high ion bombardment currents which produce deposition rates an order of magnitude higher than those of systems without magnetic fields. In addition, the magnetic field keeps secondary electrons, which are emitted during the sputtering process, from reaching the wafers and causing excess heating which can limit the sputtering rate.

1 Basic Configurations

Although all production-oriented sputtering machines use Magnetrons, the actual machine configuration can vary considerably. Figure 3 shows three basic configurations.

(a) A machine suitable for sputtering alloys from separate targets is shown in Fig. 3a, but it has the disadvantage that it is not load locked and does not produce as clean a film as the close-spaced systems.

(b) The machine in Fig. 3b has a load lock and a fairly simple loading system.

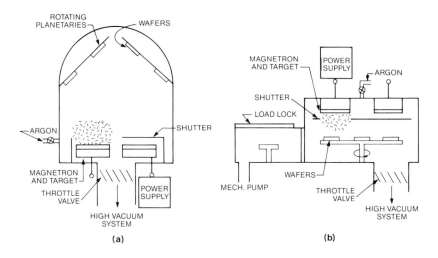

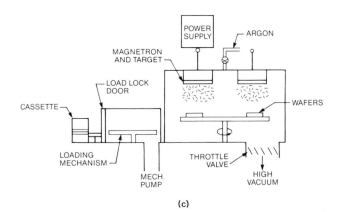

Fig. 3. Equipment used for sputtering. Location of the plasma is shown with and without shutters in place. (a) Long-spaced, planetary, manual load. (b) Close-spaced, load-locked, manual load. (c) Close-spaced, load-locked, automatic load.

27. Plasma Processing for VLSI

TABLE V

Sputtering Machine Capabilities

Vacuum	<10^{-7} Torr
Substrate heat	to 400°C
Substrate bias	dc or rf to 300 V
Sputter etch	rf to 2 kW
dc Power supply for metals	
rf Power supply for insulators	
RGA for vacuum monitor	

(c) When designed properly, the system in Fig. 3c, which has closely spaced electrodes, load lock, and automated wafer handling, is the most desirable.

2 Machine Capabilities

Table V is a list of desired capabilities for production sputtering equipment. Though all these may not be required for every application, they are recommended. Typical sputter parameters are best shown for a specific case. Table VI shows parameters common to a close-spaced deposition, such as that used for Al. The rate shown is the instantaneous rate and does not give the actual cycle time which is dependent on the substrate load and motion. A qualitative comparison of sputtering with evaporation is shown in Table VII. As can be seen, the overwhelming advantage of sputtering is the capability of sputtering alloys. Most other deposition results also favor sputtering for many applications.

C Typical Applications

Typical applications of sputtering to VLSI processing are shown in Table VIII.

TABLE VI

Typical Aluminum Deposition Parameters

Target power	8 kW
Target voltage	600 V
Ion current	13.3 A
Argon pressure	6–12 m Torr
Substrate heat	~300°C
Substrate bias	100–200 V
Deposition rate	~10,000 Å/min

TABLE VII

Qualitative Comparison of Sputtering and Evaporation

Deposition properties	Sputtering	Evaporation
Alloy deposition capability	Excellent	Poor
Insulator deposition capability	Fair	Fair–poor
Grain size	Small–medium	Medium
Step coverage	Fair	Fair–poor
Thickness uniformity	Excellent	Good
Adhesion	Excellent	Good

Sputtering is especially well suited to *aluminum alloy depositions* and has the important advantage that the composition is easily controlled simply by purchasing readily available targets. Deposition rates are very high, making this a process easily adapted to production.

TABLE VIII

Typical Applications

Material	Target	Preferred system type (Fig. 3)	Deposition rates
Interconnect aluminum alloys			
Al/Si, Al/Si/CU	Composite	B or C	Very high
Low resistance contacts and Schottky contacts			
Pd	Pure metal	B or C	High
Pt			High
Ti			High
Barrier metals			
Ti, W	Pure metal	B or C	High
Ti/W	Composite	B or C	Medium
Interconnect silicides			
$CoSi_2$	Composite	B or C	Medium
$MoSi_2$	Cosputter	A	Low
$TiSi_2$	Layering	B or C	Medium
$TaSi_2$			
WSi_2			
Dielectrics			
SiO_2	Composite	B or C	Low
SiN_3	Silicon with reactive gas	B or C	Medium
Backside metals			
Cr + Au	Pure metals	B or C	High
Ti + Ti/Ni + Ni	Pure metals	B or C	High

27. Plasma Processing for VLSI

Low resistance contacts, *Schottky contacts*, and *barrier metals* are also well suited to sputtering. This application usually takes the form of a three-layer film having a thin silicide contact, a middle diffusion barrier, and a thick aluminum alloy layer for maximum conductivity.

Another common application of sputtering is the formation of the *gate interconnect*. In this case, both silicon and a refractory metal are deposited on top of a standard doped 2000–4000 Å polysilicon film. (The silicide is then formed from the sputtered material in a subsequent high temperature annealing step.) The materials can be sputtered from either composite targets or the pure elements. Though the former is preferable for production, it may be difficult to obtain the desired composition. Both composite and silicon targets are brittle; this limits the power that can be applied to them, which results in lower deposition rates.

Dielectrics can also be sputtered by using rf instead of dc power supplies. These can produce high integrity films, though the deposition rates are lower than for metals.

Finally, *backside metallizations* are also well suited to sputtering, especially for the refractory metals. Gold can be sputtered at very high rates, though the economics are questionable when this method is compared with resistive filament evaporation in a barrel type system.

PLASMA ENHANCED CHEMICAL VAPOR DEPOSITION IV

Plasma enhanced chemical vapor deposition (PECVD) is a chemical deposition technique used in VLSI processing to fabricate both insulating and conducting films. This method is similar to low pressure chemical vapor deposition (LPCVD) except that, in addition to thermal energy, plasma excitation is also provided. Typically, this results in lower-temperature depositions without sacrificing the integrity of the film. In particular, even at lower temperatures, PECVD films are more dense and have more desirable step coverage properties than LPCVD.†

Equipment Configurations A

Figure 4 schematically represents the two most common PECVD equipment configurations suitable for VLSI requirements.

(a) The first type (Fig. 4a) is similar in configuration to both close-space sputtering systems and planar plasma etchers and can be automated for cassette-to-cassette operation.

† See references [14,15] and additional references therein for greater detail.

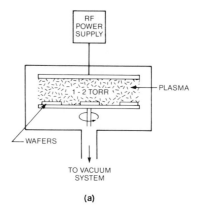

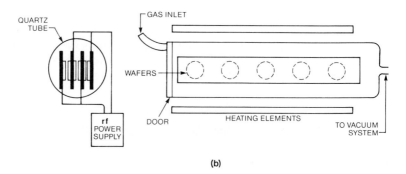

Fig. 4. Equipment used for plasma enhanced chemical vapor deposition. (a) Horizontal deposition system. (b) Longitudinal deposition system.

(b) The second system (Fig. 4b) is built in an LPCVD tube except that the wafer boat consists of powered electrodes, as shown, for providing the plasma excitation. This configuration can also be automated using robotics, though it is more difficult than with the horizontal reactor.

B Typical Process

A typical process is best illustrated by example. Table IX lists typical parameters for a silicon nitride and silicon dioxide deposition in a longitudinal reactor. These are similar to plasma etch conditions except that the reaction products are nonvolatile and the power densities are significantly lower.

TABLE IX

Typical Si_3N_4 and SiO_2 Deposition Parameters

	Si_3N_4	SiO_2
Gases	SiH_4, 200 sccm	SiH_4, 125 sccm
	NH_3, 1500 sccm	N_2O, 3000 sccm
		O_2, 35 sccm
Pressure	2 Torr	1.2 Torr
Temperature	360°C	380°C
Power	80–100 W	30–40 W
Rf frequency	450 kHz	450 kHz
Boat load	70 wafers	70 wafers
Deposition rate	300–350 Å/min	400 Å/min
Refractive index	2.0	1.5

C. Applications

Applications of PECVD to VLSI processing along with suitable reactant gases are shown in Table X. The most common application of PECVD is the use of Si_3N_4 for the *top layer passivation*. Nitride is especially desirable as a barrier to Na^+ and water vapor on devices in plastic packages. When a lower dielectric constant is needed, oxides or oxynitrides can be used in place of the nitride, though they do not provide as good a barrier to contaminants.

Other possible applications are not yet as well accepted as the passivation process. These include

(a) doped glasses for the interlevel dielectric, which can provide superior step coverage and, in the case of borophosphosilicate glass, can be reflowed at temperatures below 1000°C, and

TABLE X

Applications of PECVD to VLSI

Applications	Layer	Reactive gases
Undoped insulating layers	SiO_2	SiH_4, N_2O
	Si_3N_4	SiH_4, NH_3
	SiO_xN_y	SiH_4, NH_3, N_2O
Doped insulating layers	PSG	SiH_4, N_2O, PH_3
	BSG	SiH_4, N_2O, B_2H_6
	BPSG	SiH_4, N_2Om, PH_3,
	BPSG	SiH_4, N_2O, PH_3, B_2H_6
Conducting layers	Si	$Si_2Cl_2H_2$, Ar
	$TiSi_2$	SiH_4, Ar, $TiCl_4$

(b) the deposition of silicides (e.g., $TiSi_2$) for the gate interconnect on top of a standard polysilicon, which gives better step coverage than sputtering.

REFERENCES

1. C. M. Melliar-Smith and C. J. Mogab, *in* "Thin Film Processes" (J. L. Vossen and W. Kern, eds.), p. 497. Academic Press, New York, 1978.
2. D. L. Flamm, V. M. Donnelly, and D. E. Ibbotson, *in* "VLSI Electronics: Microstructure Science," vol. 8 (N. Einspruch and D. M. Brown, eds.). Academic Press, New York, 1984.
3. B. Gorowitz and R. J. Saia, *in* "VLSI Electronics: Microstructure Science," vol. 8 (N. Einspruch and D. M. Brown, eds.). Academic Press, New York, 1984.
4. R. E. Lee, *in* "VLSI Electronics: Microstructure Science," vol. 8 (N. Einspruch and D. M. Brown, eds.). Academic Press, New York, 1984.
5. G. K. Wehner and G. S. Anderson, *in* "Handbook of Thin Film Technology" (L. I. Maissel and R. Glang, eds.), ch. 3. McGraw-Hill, New York, 1970.
6. L. Maissel, *in* "Handbook of Thin Film Technology (L. I. Maissel and R. Glang, eds.), ch. 4. McGraw-Hill, New York, 1970.
7. J. L. Vossen and J. J. Cuorno, *in* "Thin Film Processes" (J. L. Vossen and W. Kern, eds.), p. 12. Academic Press, New York, 1978.
8. D. B. Fraser, *in* "Thin Film Processes" (J. L. Vossen and W. Kern, eds.), p. 115. Academic Press, New York, 1978.
9. R. K. Waits, *in* "Thin Film Processes" (J. L. Vossen and W. Kern, eds.), p. 131. Academic Press, New York, 1978.
10. B. Chapman, "Glow Discharge Process," p. 177. Wiley, New York, 1980.
11. R. S. Nowicki, *in* "VLSI Electronics: Microstructure Science," vol. 8 (N. Einspruch and D. M. Brown, eds.). Academic Press, New York, 1984.
12. A. C. Adams, *in* "VLSI Technology" (S. M. Sze, ed.), p. 124. McGraw-Hill, New York, 1983.
13. D. B. Fraser, *in* "VLSI Technology" (S. M. Sze, ed.), p. 347. McGraw-Hill, New York, 1983.
14. J. R. Hollahan and R. S. Rosler, *in* "Thin Film Processes (J. L. Vossen and W. Kern, eds.), p. 335. Academic Press, New York, 1978.
15. T. B. Gorczyca and B. Gorowitz, *in* "VLSI Electronics: Microstructure Science," vol. 8 (N. Einspruch and D. M. Brown, eds.). Academic Press, New York, 1984.

Chapter 28

Silicon-on-Insulator for VLSI Applications

H. W. LAM
Texas Instruments Inc.
Dallas, Texas

R. F. PINIZZOTTO
Ultrastructure Inc.
Richardson, Texas

A. F. TASCH, JR.
Motorola Inc.
Austin, Texas

I. Introduction	503
II. Heteroepitaxy	505
III. SOI by Thin Film Recrystallization	506
IV. Formation of Buried Insulating Layers by Ion Implantation	508
V. Full Isolation by Porous Oxidized Silicon (FIPOS)	509
VI. Epitaxial Lateral Overgrowth	510
VII. LPCVD Polysilicon SOI Thin Film Transistors (TFT)	511
VIII. Grain Boundary Passivation	512
IX. Three-Dimensional Integrated Circuits	512
X. Summary	514
References	514

INTRODUCTION   I

Complementary metal oxide semiconductor (CMOS) is the preferred device technology for silicon-on-insulator (SOI) circuit implementation, whereas a horizontal bipolar or a merged CMOS–bipolar technology can also significantly improve overall circuit performance with the incorpora-

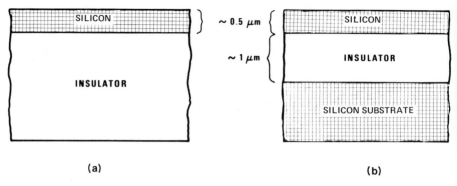

Fig. 1. Cross section of the two basic SOI structures. (a) Insulator serves as entire substrate. (b) Insulating film lies on a silicon substrate.

TABLE I

Advantages of an SOI Technology

Advantages	Primary reasons	Application
Improved packing density and improved lateral isolation	Minimum device separation due to dielectric isolation and circumvention of need for relatively deep (and thus lateral) diffusions	VLSI
Reduced capacitance	Reduced junction and line capacitances due to optimum choice of dielectric thickness and use of SiO_2	High-speed electronics
Immunity to latchup	Dielectric isolation	VLSI CMOS
Immunity to transient upset	Small active volume for electron–hole generation	Soft-error free memories; defense electronics
High voltage isolation	Dielectric isolation	High voltage electronics; fault tolerant circuits; nonvolatile memories
Resistance to gamma radiation	Ability to conveniently apply negative bias at back interface	Space and defense electronics

TABLE II

Advantages of CMOS for VLSI

Advantages	Primary reasons
Low speed–power product	Low active power
Low stand-by power	Static logic design
High noise margin	Ratioless logic increases device parametric margin
Better analog compatibility	Ease of design and larger margin in device parametric variations

28. Silicon-on-Insulator for VLSI Applications   505

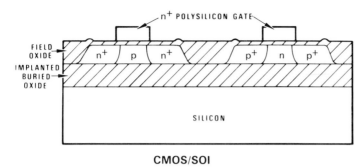

Fig. 2. Cross section of CMOS devices on SOI.

tion of SOI. Furthermore, high voltage and rad-hard requirements can benefit from SOI technologies.

Figure 1 shows two generic SOI structures:

(a) a structure with a thick dielectric substrate; a typical example is silicon-on-sapphire (SOS);

(b) a structure with a thin dielectric layer sandwiched between a silicon substrate and a thin, device quality silicon layer.

Table I shows the advantages of an SOI technology and the potential areas of application.

Figure 2 shows a cross-sectional view of a CMOS/SOI structure. The advantages of using CMOS technology in VLSI circuits are summarized in Table II. Combining these advantages with those intrinsic to SOI, we expect that a CMOS/SOI circuit will be ideal for VLSI applications, exploiting aggressively scaled (to submicron limits) MOS devices.

HETEROEPITAXY II

When a proper substrate is chosen, a silicon epitaxial layer can be successfully grown. The most common example is silicon grown on sapphire. Sapphire crystals, ($1\bar{1}02$)-oriented, are generally used for growing (100) silicon; (0001)-oriented substrates are used for (111) silicon epitaxy layers. Though other materials such as spinel, cubic zirconia, calcium, and barium fluoride have been used as the substrate, their usage has so far been of limited interest. Generally, heteroepitaxial SOI possesses all the advantages of SOI but has significant disadvantages. The disadvantages of heteroepitaxy, and particularly of SOS, are listed in Table III.

Several approaches are being pursued to solve the aforementioned problems. They are listed in Table IV.

TABLE III
Disadvantages of Heteroepitaxy SOI Systems

Disadvantages	Principle causes
High and uncontrollable impurity concentration at heteroepitaxy interface	Autodoping resulting from reaction of silane with host crystal; in SOS, sapphire is reduced to form aluminum, which diffuses into the silicon layer
High defect density	Mismatch between lattice constants of silicon and host crystal
Compressive–tensile stress in silicon film; stress in SOS is compressive	Difference in coefficient of linear thermal expansion between silicon and host crystal
Difficult to grow device-quality thin epitaxy layer	Defect due to lattice constant mismatch is most dense at interface

The last approach in Table IV, in which silicon is epitaxially grown on a thin crystalline dielectric layer that is in turn epitaxially grown on a bulk silicon wafer, holds promise as an approach for three-dimensional integrated circuits, because theoretically there is no limit to the number of silicon and dielectric layers one can grow. One example of such a structure has been demonstrated with silicon on spinel on silicon [1].

III SOI BY THIN FILM RECRYSTALLIZATION

Many SOI fabrication techniques are based on recrystallization of thin polycrystalline or amorphous silicon films. The general principles and a

TABLE IV
Approaches Being Pursued to Overcome Difficulties of Heteroepitaxy SOI Systems

Approach	Why it may work
Silicon self-implantation and solid phase epitaxy (SPE)	Implantation amorphizes defective layer at epitaxial interface; subsequent SPE with less defective top silicon layer as seed improves overall defect density
Double silicon self-implant and SPE	Repeat above process; amorphize top layer in the second step
Thin silicon epitaxy on thin dielectric layer epitaxially grown on bulk silicon wafer	Thermal expansion of bulk silicon wafer dominates the structure; since the epitaxial silicon has same coefficient as the bulk, stress is minimized

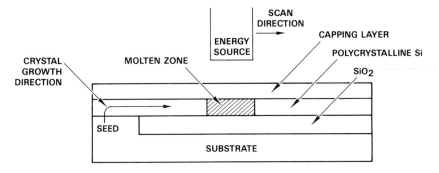

Fig. 3. SOI by recrystallization of polycrystalline silicon.

sample configuration are illustrated in Fig. 3. Procedures for preparing the sample are as follows:

(a) A standard silicon wafer is thermally oxidized to form an SiO_2 insulating layer, typically between 0.5 and 1.0 μm thick. Recessed oxides, as shown in the illustration, yield single crystal material with smaller defect densities than structures with large oxide steps.

(b) Openings are etched through the oxide to reveal part of the original substrate. These areas are the seeds for subsequent crystal growth. Seeds are required for orientation control; nonseeded techniques yield material of inferior quality.

(c) A layer of either polycrystalline or amorphous silicon (about 0.5 μm thick) is deposited over the entire sample, covering both oxide and seed regions.

(d) A capping layer composed of SiO_2 and/or Si_3N_4 is deposited. Each independent research group has a particular type of cap that works best for them. The reasons for the differences from group to group are not known.

(e) Recrystallization of the top silicon layer is usually achieved by *melting,* if the samples are polycrystalline, or by *solid phase recrystallization* of amorphous silicon.

When melting is used in step (e), a temperature gradient is exploited to control the movement of the liquid–solid interface. This can be done by spatially defining the energy input with a narrow beam and then scanning the beam across the sample. Energy sources include lasers, *e*-beams, graphite and tungsten wire strip heaters, and incoherent light. As the beam is scanned, a molten zone follows the energy source, and the liquid–solid interface traverses the sample. If seeds are used, the recrystallized layer has the same orientation as the substrate. Alternatively, unfocused energy sources may be used. In this case, the thermal gradients are controlled by the substrate configuration. The heat conductivity of silicon is far greater than that of silicon dioxide; hence, during cooling crystal growth proceeds from the open silicon seed windows and progresses over the silicon-on-oxide regions. The process with a stationary light source and a relatively thick oxide substrate is called *lateral epitaxial overgrowth over oxide* (LEGO) by the research group who is developing this process. The direction of crystal growth and the temperature gradients are the same for both focused and unfocused beams. It is also possible to form similar temperature gradients using thickness variations in either the underlying oxide or in the capping layer. Another variation is to embed material with different heat absorbing power in the capping layer itself.

The solid phase recrystallization of amorphous silicon is based on a different principle. The amorphous silicon recrystallizes because the crystalline form is the stable equilibrium phase. The crystalline–amorphous interface moves through the amorphous layer at moderate temperatures (approximately 600°C). The temperature is chosen to minimize the intrinsic crystalline nucleation rate while still maintaining enough atomic mobility to permit the amorphous to crystalline phase conversion. The crystalline growth direction relative to the seed is the same as for the cases described previously, but no temperature gradients are needed. The driving force is the free energy of the phase transformation. In the absence of other nuclei, growth occurs only at the crystalline–amorphous interface.

IV FORMATION OF BURIED INSULATING LAYERS BY ION IMPLANTATION

Figure 4 illustrates the basic principles of the formation of SOI by ion implantation.

(a) High-energy implanted oxygen or nitrogen ions come to rest with a Gaussian distribution at a fixed depth below the sample surface (Fig. 4a). The peak of the crystal damage distribution profile is slightly closer to the sample surface than the peak of the impurity distribution profile. During implantation, it is mandatory that the damage at the front surface remain less than the critical value needed for the formation of amorphous silicon. This is usually achieved through sample heating caused by the ion beam itself.

(b) As the dose is increased, the maximum concentration reaches the stoichiometric limit for the insulating material, usually SiO_2 or Si_3N_4 (Fig. 4b). If the implanted ion is oxygen, the stoichiometric ratio is never exceeded.

(c) Implantation of larger doses leads to the formation of a thicker buried SiO_2 layer (Fig. 4c). Polycrystalline silicon and damage zones are formed on both sides of the buried layer.

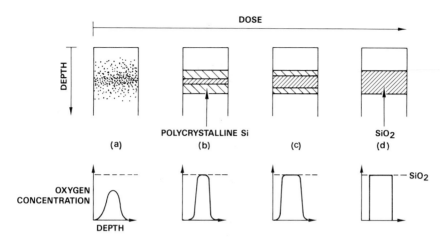

Fig. 4. SOI by oxygen ion implantation.

(d) Eventually, the dose is large enough so that the thickness of the buried layer is larger than the width of the Gaussian implant distribution. When this occurs, sharp interfaces are formed on both sides of the buried oxide layer. The structure is shown in Fig. 4d.

Implantation of nitrogen ions above stoichiometric level does not yield sharp interfaces because the nitrogen concentration is not limited at the stoichiometry of Si_3N_4. Presumably, a layer of Si_3N_4 with N_2 inclusions is formed. After implantation, the material is annealed at high temperature to remove implantation-induced defects and recrystallize the top silicon layer. Finally, an epitaxial layer can be deposited to increase the SOI layer thickness for device fabrication. Rutherford backscattering spectroscopy measurements have shown that the buried-oxide materials system is superior in crystal quality to silicon-on-sapphire.

FULL ISOLATION BY POROUS OXIDIZED SILICON (FIPOS) V

Silicon forms a porous structure when anodized under the proper conditions in HF solutions. The pore size can vary from a few nanometers to several tens of nanometers in diameter. The pores meander from the surface downwards into the sample and can form several branches and subbranches. The detailed microstructure is not fully known. *P*-type silicon anodizes much more rapidly than *n*-type because holes are needed to break the silicon–silicon bonds and allow silicon–fluorine bonds to form. The soluble species is thought to be SiF_3^+. It is possible to form isolated *n*-

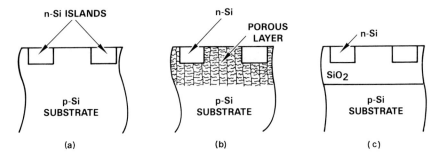

Fig. 5. Process flow of SOI formation by FIPOS.

type islands on porous *p*-type silicon substrates. The porous layer can be converted to SiO_2 at low temperatures in short times because of the large silicon surface area. Oxygen diffuses in the gas phase throughout the porous layer. The interpore thickness is only several tens of nanometers, which is easily oxidized at low temperatures. It is possible to adjust the porosity of the layer to compensate for the increased volume of the SiO_2.

The basics of the process are shown in Fig. 5.

(a) *n*-Type silicon islands are formed by ion implantation into a *p*-type substrate (Fig. 5a). Alternatively, *n*-type islands may be formed in n^+ material. This latter method has been shown to yield sharper island-to-porous layer interfaces.

(b) After anodization, the *p*-type silicon is porous, but the *n*-type islands remain unaffected (Fig. 5b).

(c) The assembly is then oxidized, leaving the *n*-type islands completely surrounded by SiO_2. (Fig. 5c).

A variation of this approach uses a uniformly anodized substrate with no predefined islands. After anodization, molecular beam epitaxy is used to form a continuous top silicon layer. Low temperature MBE prevents the collapse of the porous material, which occurs if it is annealed at the high temperatures needed for conventional epitaxy. Islands are formed in the top layer using conventional photolithography and etching. Low temperature oxidation of the porous layer completes the process flow.

VI EPITAXIAL LATERAL OVERGROWTH

It is possible to form an SOI structure using only a slight modification of conventional epitaxial silicon deposition (Fig. 6). A substrate is oxidized to form a continuous silicon dioxide layer. Holes are etched through this

28. Silicon-on-Insulator for VLSI Applications

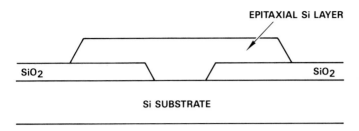

Fig. 6. Silicon epitaxial overgrowth to achieve SOI.

layer to reveal the substrate in selected areas. These areas are used as seeds for subsequent epitaxial growth. The growth conditions are chosen so that deposition takes place in the seeds but not on the oxide layer. As the deposition proceeds, the epitaxial layer becomes thicker than the oxide, and lateral growth can then occur.

The technique depends critically on two important phenomena.

(a) The lateral growth rate should be larger than the vertical growth rate. This is required for thin epitaxial films of large dimensions.

(b) The nucleation rate of polysilicon on SiO_2 must be much lower than the growth rate of silicon in the silicon seed regions.

These two phenomena can be controlled by appropriate choice of the deposition conditions and regular etchback of the deposited material. The etchback removes any polysilicon nuclei that have formed on the oxide during the growth of the silicon along the seeds. At the present time, the lateral extent is limited because the ratio of the vertical to the lateral growth rate is limited to 3–4. Additionally, defects are formed where growth fronts from different seeds on the substrate intersect.

VII. LPCVD POLYSILICON SOI THIN FILM TRANSISTORS (TFT)

Polysilicon has been used in semiconductor processing for passive circuit elements such as resistors and interconnect materials. There has been continued interest in using polysilicon as an active element, but usage has not been widespread because significant applications have not been found in which a polysilicon active device can replace its counterpart and result in improved performance or reduced cost.

The development of three-dimensional IC concepts and grain-boundary passivation methods have changed the outlook. Stacked CMOS [2] allows the stacking of a polysilicon SOI device on top of a bulk MOS device to improve packing density. The use of a stacked polysilicon SOI *p*MOS transistor in place of a polysilicon resistor in a cross-coupled latch sRAM structure results in improved performance, especially if the grain boundaries in the polysilicon transistor are passivated.

VIII GRAIN BOUNDARY PASSIVATION

Grain boundaries are surfaces that are characterized by surface states caused by dangling bonds. Passivation of grain boundaries involves a reduction of these states by reducing the number of dangling bonds. A number of atomic species, most notably hydrogen, can diffuse rapidly along grain boundaries and form Si–H bonds.

A typical passivation process exposes the polysilicon material in an ambient of atomic hydrogen, usually generated with rf excitation, at an elevated temperature (about 300–400°C). Alternatively, hydrogen atoms can be implanted into the polysilicon and activated by a subsequent low temperature anneal. Passivation can also be carried out by annealing in a molecular hydrogen ambient, but this is not very effective.

SOI transistors that are fabricated with passivated polysilicon show an increase in drive current of several orders of magnitude and a decrease in leakage current of several orders of magnitude over nonpassivated transistors. Most passivation effects are reversible if the temperature is raised above 450°C, because of dissociation of the Si–H bonds. Hence, the passivation treatment can be applied only after all high temperature processing of the device, which imposes some limitations on the fabrication process.

IX THREE-DIMENSIONAL INTEGRATED CIRCUITS

The use of SOI transistors in integrated circuits frees the designer from the limitations of device fabrication in the planar top surface of a bulk wafer, making feasible three-dimensional circuits in which active device elements are stacked in many layers. This approach can increase the device packing density beyond VLSI. Three approaches for 3-D ICs are presented in Fig. 7:

(a) In *stacked CMOS*, a pMOS device is stacked on top of and shares the same gate with a bulk nMOS device [2].

(b) In *staggered CMOS* [3], a layer of SOI devices is arranged to form a compact cross-coupled latch structure with the underlying bulk devices.

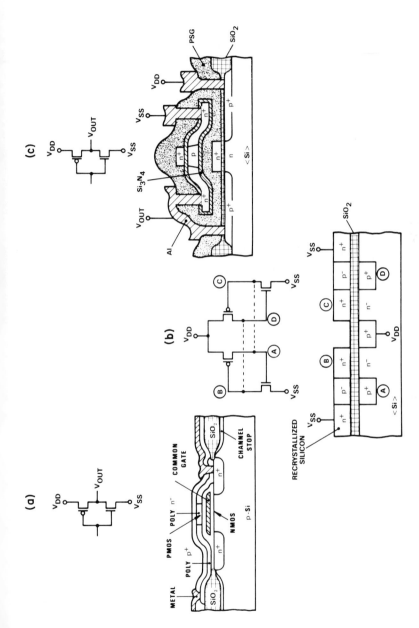

Fig. 7. Three approaches to three-dimensional integrated circuits. (a) Stacked CMOS; (b) staggered CMOS; (c) multilayer CMOS.

TABLE V

Choice of SOI Technology Based on Requirement of the Application

Application	SOI technologies most likely to be used
VLSI electronics (MOS)	SOS, implanted buried oxide, FIPOS
VLSI high-density MOS memory	Stacked CMOS, passivated polysilicon TFT
VLSI electronics (bipolar)	FIPOS, epitaxial overgrowth, implanted buried dielectric
High-voltage electronics	SOS, implanted buried oxide, LEGO, FIPOS
Radiation-hardened electronics	SOS, implanted buried oxide
Three-dimensional IC	Recrystallized SOI, SPE, heteroepitaxy (such as silicon on spinel on silicon)
Display	Recrystallized SOI, passivated polysilicon TFT

(c) In the *multilayered structure* shown in Fig. 7c [4] layers of *p*- or *n*-channel devices are added on top of each other. If this process is cleverly designed, it can be relatively simple, because all the ion implantation can be done without photoresist masking.

X SUMMARY

We have discussed the advantages of SOI for VLSI and briefly surveyed the various SOI technologies. Table 5 looks at SOI technologies and the applications in which they will most likely be used.

Readers are referred to the two review articles [5,6] for more extensive reading. Most of the topics discussed in this chapter can be found in these two references.

REFERENCES

1. M. Mikami, Y. Hokari, K. Egami, H. Tsuya, and M. Kanamori, *Extended Abstracts of the 15th Conference on Solid State Devices and Materials,* Japan Society of Applied Physics, Tokyo, August, (1983).
2. J. F. Gibbons and K. F. Lee, *IEEE Electron Devices Letts.* **EDL-1,** 117 (1980).
3. E. W. Maby and D. A. Antoniadis, in "Comparison of Thin Film Transistor and SOI Technologies" (H. W. Lam and M. J. Thompson, eds.). North Holland, New York, 1984.
4. S. Kawamura, N. Sasaki, T. Iwai, R. Mukai, M. Nakano, and M. Takagi, *Technical Digest IEE International Electron Device Meeting,* p. 365, Washington, D.C., December, 1983.
5. G. W. Cullen (ed.), Single crystal silicon on non-single-crystal insulators, *J. Crystal Growth* **63,** (3), (1983).
6. H. W. Lam, A. F. Tasch, Jr., and R. F. Pinizzotto, Silicon-on-Insulator for VLSI and VHSIC, *in* "VLSI Electronics: Microstructural Science," vol. 4 (N. Einspruch, ed.). Academic Press, New York, 1983.

Chapter 29

Testing of VLSI Parametrics

CHARLES A. BECKER
General Electric Company
Schenectady, New York

I. Purpose	515
A. Process Control	516
B. Process and Device Development	517
C. Circuit Modeling	517
II. Implications of VLSI	517
A. Density and Scaling	517
B. Tradeoffs	518
C. Tighter Control	519
D. Improved Analysis Tools	519
III. Test Types	519
IV. Test Structures	520
V. Instrumentation	521
A. Typical Parametric Testers	521
B. Alternative Tester Architecture	522
C. Electrical Requirements	523
D. Computer System Requirements	524
VI. Data Analysis–Information Retrieval	525
References	526

I. PURPOSE

Measurement of the electrical characteristics of individual devices on special test chips on fabricated wafers is a well established tool in the modern manufacture of integrated circuits [1–3]. Parametric testing, especially at the wafer level, addresses the need for rapid, accurate characterization of the individual devices and components of which circuits are made. It serves as a crucial source of information for process control, for engineering development of new processes and devices, and of the device characteristics required for circuit modeling and design.

The demands of high-speed data acquisition and analysis, as part of a highly automated VLSI manufacturing process, have guided the evolution of parametric testing from a few measurements with bench top instruments and curve tracers to thousands of measurements per wafer on specialized, computer-driven test systems.

Each phase of the IC development and manufacturing sequence places its own special requirements on parametric evaluation. For each of these applications, however, the primary result of the test effort must be information for engineering decisions and not merely the raw data produced by the tests. For this reason, flexible and sophisticated analysis of the data and rapid dissemination of the results of these analyses are key elements of VLSI parametric testing.

A Process Control

1 In-Process Measurements

In-process measurements, such as four-point-probe sheet resistivity, ellipsometric thickness measurements, and optical or SEM linewidth studies, provide immediate evaluation and feedback to one or a small group of closely related manufacturing steps. However, the cumulative effects of all such steps can be determined only on the completed wafers.

2 Circuit Performance

Though testing actual circuits is a most useful measure, and one that is close to the desired end product, its interpretation in terms of process conditions and variations is extremely difficult; thus, it is of limited use as a process control feedback tool, especially for processes of the complexity needed to produce VLSI circuitry.

3 Parametric Tests

Parametric testing of completed devices provides a final evaluation of the myriad interactions between materials and processing that occur during the IC manufacturing sequence. Properly designed parametric tests and analyses can effectively isolate yield- or performance-limiting effects to close the process control loop. The primary requirements of parametric testing as a process control tool are *speed, reliability,* and *compatibility* with the (computerized) process control and data base systems. The ability to supply easily understood and accurate information to the process line in a timely fashion is the single most important parametric testing requirement in this application.

Process and Device Development   B

The engineering-development stage of new IC processes and devices places the greatest burden on the precision, sensitivity, and flexibility of parametric testing. In the absence of reliable models for accurately predicting the outcome of the more than 100 processing steps in a modern fabrication sequence, detailed and rapid analysis of the finished components is a necessity. The cost and elapsed time of fabrication for each experimental wafer lot demand that complete characterization be performed as rapidly and accurately as possible to feed the results back to wafer lots that are still in the processing and planning stages.

Because of its inherently dynamic and exploratory nature, process and device development requires flexibility and high sensitivity to the study of subtle and often unexpected effects. Raw data must be analyzed in terms of the best available physical models, and occasionally, the results of parametric tests are useful in extending and improving such models.

Circuit Modeling   C

The same test structures and equipment that provide feedback for process control can also be used to gather the parameters needed for computer simulations of circuit performance. This usually requires more time-consuming measurements and detailed analyses of device characteristics than are needed for the process control task. The ever-shrinking device dimensions of VLSI, however, require far more sophisticated models and thus will place an increasing emphasis on automated parametric testing to supply the data needed for estimates of best-case, worst-case, and typical performance, as well as expected yield of circuits.

IMPLICATIONS OF VLSI   II

Density and Scaling   A

The most salient feature of VLSI circuitry is its density, which approaches or exceeds one million transistors per chip. The scaling of tran-

TABLE I

MOS VLSI Trends: Impact on Parametric Testing

Trend	Test implications
Shorter channel length	Higher precision in I and V measurements Greater emphasis on leakage measurements Greater sophistication in device parametrization for circuit modeling More complex test sequences
Thinner gate oxides	Improved transient protection Very low current measurement capabilities High precision I and V limits on forcing supplies
Complex doping profiles	Improved CV and IV measurements, structures, and analyses
Multilevel structures, complex interconnects	Increased emphasis on random fault tests and yield prediction
Latch-up susceptibility (CMOS)	Pulse capability for V and I sources Transient analysis capability Design to avoid latch-up in FET and other test structures
Rapid process evolution	Shorter turnaround time required for test masks and software More sophisticated, adaptable test algorithms
Larger wafer size	More test die per wafer Greater need for spatial analysis (wafer maps)
Larger chip size	Need for intrachip variation studies
Use of wafer steppers	Ability to change test chip locations and distribution rapidly
Greater process and device complexity	Large, flexible data base and computerized analysis tools

sistors to the necessary dimensions [4] for this packing density causes several of the trends for MOS ICs noted in Table I, with their attendant ramifications for parametric testing. Similar lists can be generated for VLSI implemented in bipolar and other technologies, each with its own special needs. In MOS, short channel IGFET transistors, with their lower threshold and punchthrough voltages, thinner gate dielectrics, and complex doping profiles, require very high performance test systems and test and analysis algorithms (see Sections III and IV).

B Tradeoffs

For other than memory applications, VLSI is likely to be applied primarily in specially designed very high performance circuits requiring fine-tuning of the manufacturing process for specific tradeoffs in terms of performance, cost (yield and complexity), and reliability. Parametric testing will become even more vital as a supplier of the large quantities of information needed to make these tradeoff decisions. Flexibility and rapid turnaround in test chip design, testing algorithms, and data analysis are required to meet the challenges of the rapid evolution in process technology.

C. Tighter Control

The sheer complexity and cost of VLSI fabrication processes will force tighter control of the individual manufacturing steps and earlier detection and correction of drifts from optimal conditions. This factor, combined with an increase in the local intelligence of test equipment made possible by microprocessor control, will force an increase in the application of sophisticated, special-purpose parametric tests and analyses, such as electrical alignment and linewidth measurements, as in-process monitors and feedback systems.

D. Improved Analysis Tools

Other trends noted in Table I are characteristic of the general maturation of IC processing. The performance of parametric test and analysis systems must evolve rapidly to serve the needs of this dynamic industry. In general, there is a need for much more sophisticated application of computer data base and analysis tools to handle the complexity presented in VLSI manufacture.

III. TEST TYPES

Traditionally, parametric testing has been primarily the measurement and display of the capacitance and dc current and voltage characteristics of capacitors, resistors, diodes, and transistors on special test chips that are fabricated on wafers at the same time as actual circuit chips. A recent review discusses both the structures and their associated measurements in some detail [1]. Some of the typical current and voltage measurements made on MOS transistors are listed in Table II. Other device characteristics and model parameters, such as carrier mobility, can be computed from these measurements.

TABLE II

Typical MOS Transistor Parametric Measurements

Threshold voltage
Conduction factor
Channel resistance
Effective channel length
Effective channel width
Transconductance
Saturation current
Punchthrough voltage
Channel leakage

It is crucial that all users of parametric test data understand clearly the test procedures applied, including their limitations and bases in physical models. Threshold voltage for MOS transistors, for instance, can be measured in a variety of ways, each useful for some devices and measurement purposes and inappropriate for others.

In general, as higher device performance is required, simplifying assumptions often used in the analysis of device behavior can no longer be applied. Test and analysis algorithms must reflect the level of sophistication required at any given time to explain observed device behavior [5]. Often much effort is applied to the correlation of the very simple, high-speed measurements used for process control feedback with the more detailed (and time-consuming) measurements that are used in process development. Such a strategy is of doubtful long-term use, because the correlations often apply over only a limited range and tend to obscure subtleties that can be sensitive indicators of process drifts. In addition, as the rate of process evolution continues to increase, the correlation process itself becomes a significant resource burden. More sophisticated measurements, applying more sophisticated instrumentation, if necessary, will be required to provide the detail needed to control the complexities of VLSI manufacture.

IV TEST STRUCTURES

Recent work by the National Bureau of Standards and others has pointed out the advantages of a *modular approach* to test structure layout, particularly one in which the probing pads are arranged in a 2-by-*N* array [1]. The benefits include:

29. Testing of VLSI Parametrics

(a) standardization of debugged test structures,

(b) rapid assembly of new test chip designs from libraries of standard modules, and

(c) a tendency to reduce the amount of multiplexing or pad-sharing among separate test structures

The last point deserves special mention because it is frequently overlooked in test mask design, which leads to structures that can be tested only if all devices work as expected. All too often, unexpected processing or design difficulties render an entire interconnected set of devices useless because of a single failure.

Another advantage of modular test mask design, not generally noted in the literature, is the ability to create complementary standard software testing modules, optimized for each of the entries in the test cell library. Since the development and debugging of test software remains one of the largest cost elements in parametric testing, the ability to produce a test program for a new chip by assembling existing routines, each known to work for a given structure, represents a significant gain in overall efficiency.

V. INSTRUMENTATION

Selection of test instrumentation and control computer systems must be approached carefully, with a clear understanding of present and probable future test needs for the application at hand. In general, flexibility, accuracy, and ease of software generation should be the primary considerations for VLSI applications. Actual overall system efficiency, including personnel training, software generation, testing to the required precision, and data analysis, is far more important than the raw speed of a single measurement or the instruction cycle time of the control computer. A real-time demonstration of the process of writing, debugging, executing, and analyzing both an unusual parametric test and a large number of tests in a single pass is a useful means of determining the relative throughput capabilities of various systems.

A. Typical Parametric Testers

Figure 1 shows a generalized schematic of commercially available wafer parametric testers. The crosspoint matrix is used to connect the appropriate forcing and measuring devices to the probes contacting the device

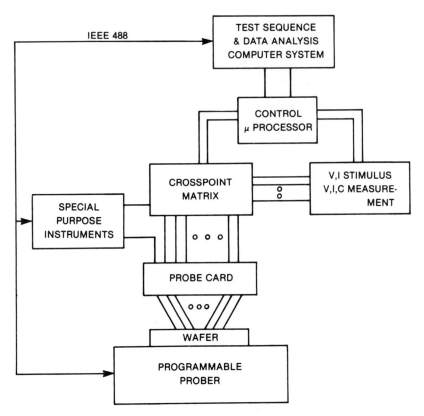

Fig. 1. Schematic of typical automated parametric test system. Details of digital interfaces and equipment partitioning can vary.

under test. Since it is involved in nearly all measurements, the quality and reliability of this matrix subsystem is a crucial element in overall tester performance. In some designs, the matrix is located adjacent to the probe card, thus minimizing parasitic noise, leakage, and capacitance effects. This design also usually permits all measurements to be made with equal quality at any device pin, thus permitting greater freedom in test chip design and greater flexibility in the diagnosis of unusual or unexpected phenomena.

B Alternative Tester Architecture

Alternative architectures, particularly ones in which each pin is equipped with dedicated electronics, have been suggested and imple-

TABLE III

VLSI Parametric Test System: Desirable Electrical Capabilities

Function	Range	Number of units
Voltage force	1 mV–100 V	≥4
Voltage measure	1 μV–100 V	≥4
Current force	1 pA–100 mA	≥4
Current measure	0.1 pA–100 mA	≥4
Capacitance measure	1 ff–2 nf	1
Capacitance dc bias	0–100 V	1
Frequency counter	dc–200 MHz	1
Pulse generator	dc–40 MHz	1

mented. The difficulty and cost of providing high accuracy and precision, uniform calibration, and timing coordination for each set of pin electronics has thus far prevented the use of this approach in commercially available test systems.

C. Electrical Requirements

Table III lists a set of electrical force and measurement capabilities desirable for VLSI parametric testing. In general, it is useful to provide several current and voltage units, as listed in the third column of the table, to allow complete characterization of a four-terminal device with a minimum of reconnections. Also useful in this regard are multipurpose units, each capable of acting as either a current source with voltage measurement or as a voltage source with current measurement. When a number of similar force and/or measurement units are available, it is generally sufficient for only one to satisfy the highest resolution requirements.

Table III includes two devices not generally available as part of parametric test systems: a frequency counter and a pulse or waveform generator. These devices are useful in the evaluation of such test structures as ring oscillators, delay chains, and on-chip measurement circuits, which are becoming more prevalent as a means of increasing effective sensitivity.

Unfortunately, no single commercially available test system satisfies all of the specifications given in Table III. It is thus necessary either to compromise on some or all capabilities or to build or enhance a system in-house. Very careful consideration of present and future needs must be given before sacrificing these basic functions for short-term expedience.

D Computer System Requirements

The list of requirements given in Table IV for the computer system is aggressive, but necessary for the full realization of the power of parametric testing. In most installations, the functions of archival data storage and analysis are handled by a larger host computer that can serve several test systems. Though this reduces some of the demands on the test system computer, such as those for graphics and large archival data stores, many of the items in Table IV still apply.

Most present-day testers are based on a computer with an outdated architecture having limited memory-addressing abilities. Though such machines can in some cases be equipped with up to one Mbyte of memory, this memory is highly segmented; thus, a task can usually access less than 64 Kbytes of program space. The result is that for any realistically sized test program, overlaying or memory-switching techniques are required, which can seriously increase program execution time and greatly complicate the software development task.

A truly interactive mode or an interactive programming language is essential for debugging purposes. As noted, this must operate in the same environment as the full-speed test routines to be fully useful.

TABLE IV

VLSI Parametric Test System: Desirable Computer Characteristics

Function	Specification
RAM memory	>256 Kbyte, unpartitioned, available to programs and data
Graphics	High resolution (>300 × 300) screen and hardcopy (color?)
	Interactive, flexible software
Secondary storage	≥10 Mbytes for programs
	≥20 Mbytes for data store
	Removable medium (floppy or tape) for interchange and update
Reliable, convenient backup	Single medium (disk or tape) for full system backup
Host computer communication	Local area network
	>10 Kbit/s
High-level language	Structured BASIC and FORTRAN
	PASCAL
	(ATLAS?)
Interactive mode	Same sequence and test-condition environment as full-speed test routines
Data base	Modern architecture and keyed access
	Uniform, accessible format
	Flexible analysis of retrieved data

A particular weakness of all available systems is their inability to handle flexibly and efficiently the very large quantities of data generated by the testing function. In process development, it is not uncommon to generate several kilobytes of data on a single test wafer. State-of-the-art hardware and data-base software are required in this application. For this reason, in most instances data analysis is done on a more powerful host machine, necessitating an efficient communication capability between the tester and the host.

VI DATA ANALYSIS–INFORMATION RETRIEVAL

Because the generation of useful engineering information is the primary function of parametric testing, a number of numerical and graphical analysis tools have been developed, some of which are shown in Table V. Without such easily understood synopses of the collected data, it is impossible to comprehend fully the meaning of the measurements because of the sheer volume of data.

Trend charts, histograms, statistical summaries, and wafer maps are of particular value in the process control task, because their results are usually easily interpretable in terms of process conditions. Wafer maps are a sufficiently powerful tool that they should be available in a variety of forms; numerical, gray scale (or color), and contour mapping are the most prevalent. Each type is useful for particular applications and user preferences.

Vertical yield maps, applied to both parametric (i.e., within specification) and device function yield, can provide extremely useful feedback on the existence of defects and yield-reducing process artifacts [6].

Correlation or *scatter plots* are useful in a variety of analyses. The study of parameter-to-parameter interactions, sensitivity of circuit perfor-

TABLE V

Parametric Test Data Analysis Displays

Analysis	Selection by		
	Lot	Wafer	Arbitrary group
Trend or control charts	X	—	X
Summary statistics	X	X	X
Histograms	X	X	X
Wafer maps	—	X	—
Vertical yield maps	X	—	—
Correlation (scatter plot)	—	X	—

mance and yield as a function of parameter distribution, and parameter values as a function of process conditions can all be aided by the visual feedback inherent in such displays.

To exploit fully the information contained in parametric test data, it is necessary to apply the enormous power of modern computer data-base and analysis systems. The full exploitation of such tools will be the dominant challenge of VLSI parametric testing for several years to come.

REFERENCES

1. M. G. Buehler, in *"VLSI Electronics: Microstructure Science"* (N. G. Einspruch and G. B. Larrabee, eds.), p. 529. Academic Press, New York, 1983.
2. G. P. Carver, L. W. Linholm, and T. J. Russell, *Solid State Technol.* **23,** 85 (September 1980).
3. P. H. Singer, *Semicond. Int.* **6,** 84 (September 1983).
4. J. L. Prince, in *"Very Large Scale Integration (VLSI)"* (D. F. Barbe, ed.), p. 4. Springer-Verlag, Berlin Heidelberg, 1980.
5. S. M. Sze, "Physics of Semiconductor Devices," 2nd ed. Wiley, New York, 1981.
6. C. L. Mallory, D. S. Perloff, T. F. Hason, R. M. Stanley, *Solid State Technol.* **26,** 121 (November 1983).

Chapter 30

VLSI Testing from Design through Production

CHARLES J. McMINN

Megatest Corporation
San Jose, California

I. Semiconductor Testing — 528
   A. Definition and Scope — 528
   B. Location in the Manufacturing Process — 528
   C. Typical Semiconductor Tests — 530
II. Testing with Automatic Test Equipment — 531
   A. Types of Test Equipment — 532
   B. Writing a Test Program — 534
III. Test Descriptions — 538
   A. Shorts Test — 538
   B. Opens Test — 538
   C. Functional Test — 539
   D. Maximum Current Test — 539
   E. Leakage Test — 539
   F. Output Driver Test — 539
   G. Breakdown Voltage Test — 539
IV. Logic Testing — 539
   A. Synchronization at Reset — 540
   B. Speed Binning — 540
   C. Output Relative Timings — 540
V. Memory Testing — 541
   A. Algorithmic Pattern Generation — 542
   B. Memory Failure Modes — 542
   C. Common Memory Test Patterns — 543
   D. Topological Scrambling — 546
VI. Testing Throughput — 546
VII. Quality Assurance and Sample Testing — 547
   A. Sampling Plans — 547
   B. Tester Accuracy and Testing Guardbands — 548
References — 550

I SEMICONDUCTOR TESTING

In most industries, testing is used to screen out a small fraction of the total production volume at various stages in the manufacturing process. Defective products are usually repaired and then returned to the manufacturing flow.

A Definition and Scope

The situation is very different in the semiconductor industry. Typically, over half of all semiconductors produced are thrown away because of defects in their fabrication that render them useless. As a result, testing takes on a new, more important role in the semiconductor industry. Exhaustive functional and electrical testing must be performed on every semiconductor device that a manufacturer produces, and testing must be repeated at every step in a semiconductor manufacturing process to weed out defective devices as early as possible. Further, testing is not used just to exercise the devices and prove their functionality; it is also used to determine their electrical and operating limits. To decrease the amount of semiconductor product that is rejected, manufacturers usually support multiple speed and performance classifications of semiconductor devices. One of the major roles of testing is to classify working circuits into several operating ranges. This places a premium on complete and accurate functional and electrical testing of semiconductor devices on the one hand, and high throughput, low-cost testing on the other.

B Location in the Manufacturing Process

Figure 1 shows the various places in the semiconductor manufacturing process where testing is done [1]. Although some testing is done during device fabrication (chiefly on the process itself), most device testing is performed after the semiconductor wafers have been fabricated. The first such test, known as *wafer sort* or *probe,* is used to differentiate potentially good semiconductor circuits from those that are defective. Defective circuits are usually inked at this point. The wafer is scribed and cut, and the potentially good devices are collected and packaged.

For new semiconductor designs, the next step is engineering characterization. Here semiconductor circuits are exhaustively stressed and tested to find their operating and electrical limits. This results in a verification of the design target specifications and the establishment of production speci-

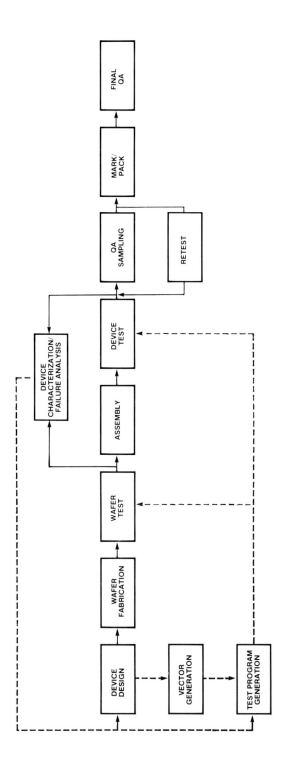

Fig. 1. Typical VLSI semiconductor manufacturing and test flows.

fications for the timing and dc parametrics of the device. On all future runs, production testing is then used to perform go/no-go testing that makes sure that each semiconductor circuit meets these specifications. Production testing is done to a specification that is somewhat tighter than the actual published user specification to guarantee or guardband the user specification.

As a check on the production testing process, a quality assurance test is performed on a sampling of the semiconductor devices. This test, performed to the user data-sheet specifications, is used to decide whether a production test run has been properly performed. Various quality assurance sampling plans are used to determine whether testing on a particular lot is adequate or needs to be redone.

Finally, the purchasers of the semiconductor circuit are also likely to test these devices. Typically, they will perform the quality assurance sampling tests on a fraction of their purchased volume as a part of the incoming inspection procedure for a part.

C Typical Semiconductor Tests

1 dc Parametric Tests

In all of the above stages, various types of testing are performed on the semiconductor circuit. Dc parametric tests are used to verify that the voltage and current operating specifications of the device are met, such as maximum operating current, high and low voltage thresholds for device inputs, drive current for device outputs, and leakage current specifications for each pin input.

2 ac Parametric Tests

Ac parametric tests include tests of minimum and maximum operating frequency and clock duty cycle. These tests are performed over the complete power supply range of the device.

3 Functional Tests

Ac and dc parametric tests are usually performed on a device using a minimal functional subset of the total device instruction set. Complete functional testing is performed only on parametrically good devices. These tests may also be performed at multiple speed and voltage combinations to stress the device fully and categorize it into one of several operate ranges.

Nonelectrical Tests 4

Some nonelectrical tests are also performed on a semiconductor device. These include hermeticity tests to prove that the semiconductor die is properly sealed in the package, burn-in tests to catch any infant mortality problems, and accelerated life tests, usually performed on new devices, to guarantee that the semiconductor's operating life is not too low.

These same four areas of testing are performed on all semiconductor devices, whether they are VLSI logic devices such as microprocessors and peripheral products, VLSI memory devices such as dynamic RAMs, EPROMs, or ROMs, analog devices such as codecs or filters, or mixed signal parts that may contain some analog and some logic functions such as digital signal processors.

TESTING WITH AUTOMATIC TEST EQUIPMENT II

Semiconductor device testing is done on test equipment designed to simulate the operating environment that a semiconductor circuit would see in normal use. A typical VLSI tester such as that shown in Fig. 2 is capable of operating at 50–100 MHz. Automatic test equipment can test semiconductor devices with pin-counts as high as 256. The tester can contain as many as 50,000–75,000 integrated circuits to test a single semiconductor device properly. The accuracy achievable with such equipment can be 500 ps.

Fig. 2. VLSI ATE combine a computer, a timing system (mainframe), and a positionable test head.

A   Types of Test Equipment

Figure 3 shows a typical block diagram of a VLSI tester. A VLSI tester starts with a source of patterns used to stimulate and to check the response of a semiconductor device. This pattern describes a logical sequence or truth table of device states, much like the information captured by a logic analyzer. The logical information describes the state of each pin in each clock cycle of the test program. Semiconductor testers use two methods for storing this pattern:

(a) *Stored-response VLSI testers* store in memory within the tester the logical information used to stimulate the device. They have the advantages of being able to store and manipulate the test pattern much like any other computer data and of not requiring a known good device to serve as the reference.

(b) *Reference device testers* actually run a device similar to the one being tested to generate the proper patterns at each step in the test. This tester can be manufactured at a lower cost and tests parts more economically in the production environment.

In addition to the functional pattern, a VLSI tester also provides timing and wave formatting of the logical information into waveforms of the appropriate shape and timing for the device under test. In the newest generation of testers, these timing and wave formatting resources are provided on a per-pin basis (Fig. 4a) so that each pin can be independently programmed and manipulated. In past generation testers (Fig. 4b), these formatting and wave formatting functions were shared among groups of pins. The decisions about which pins to share timing generators and wave formatting functions were left to the user and resulted in more complex and more restricted test programming.

Once a waveform has been generated within a VLSI tester, it is conditioned in voltage and current by a pin electronics function that actually drives the semiconductor device inputs and compares semiconductor outputs against expected values. Newer testers also provide programmable current load circuits that will actively load semiconductor device outputs so that proper operation can be tested at the limits of the device's rated loading.

New semiconductor testers also provide a parametric test unit and a parametric measurement capability. The test unit is used to check the dc parametric specifications of a device in production. The parametric measurement unit can be used much like a high accuracy voltage and current meter to force and sense voltage and/or current from any designated device pin.

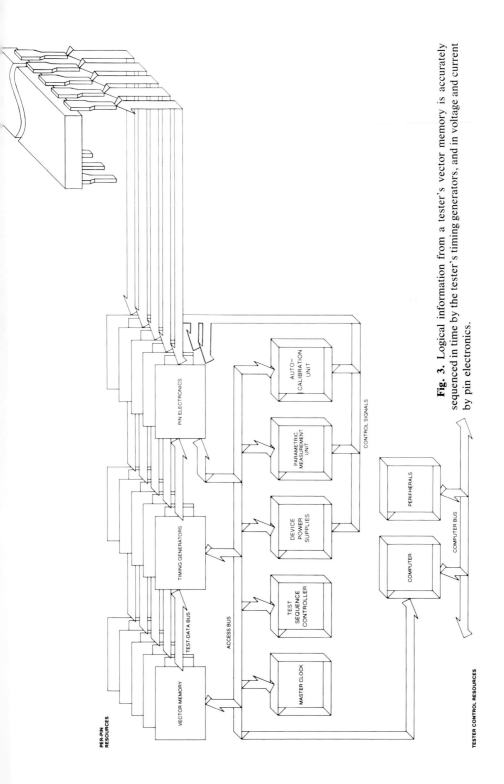

Fig. 3. Logical information from a tester's vector memory is accurately sequenced in time by the tester's timing generators, and in voltage and current by pin electronics.

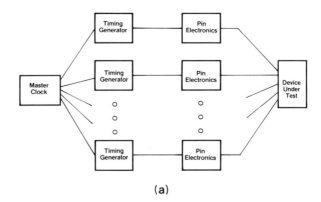

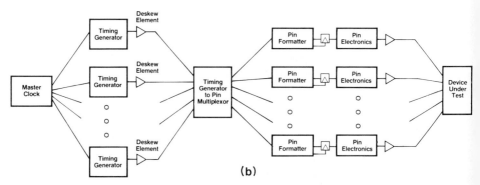

Fig. 4. (a) Newer VLSI testers provide timing and pin electronics resources on a per-pin basis. (b) Older generation testers shared a limited number of these resources.

Most new VLSI test systems also include some form of system autocalibration to maintain system accuracy.

B Writing a Test Program

Writing a program for a VLSI device is a matter of describing the sequence of tests to be performed and the proper limits to test for in each step.

1 Generating the Test Vectors

Writing a test program starts with the creation of the pattern information that will be used to sequence the device through its logical states.

30. VLSI Testing from Design through Production

This pattern or vector file is a clock-by-clock description of the state of each pin of the device. In new VLSI devices it can be as large as 100,000 to 200,000 vectors. Because of its size, it is typically generated not by hand but as part of the chip design process, as an output of a computerized device simulation. Patterns can also can be captured from a device itself, much as a logic analyzer captures device states. In either case, the designer is required to specify a description of the desired sequence of test steps for the device.

Describing the Control Flow 2

The remainder of the test program describes the proper electrical and parametric conditions under which the patterns should be presented and checked by the tester. Figure 5 shows the sequence of testing that is typically performed on a VLSI circuit. Table I presents an example device, a two-input NAND gate, and Table II illustrates the steps required to perform each of the desired tests. The following is a brief description of each of these tests.

TABLE I

Simplified Test Specifications for an Example Device

Device Pinout				Device Truthtable			
A	1	6	V_{cc}	A	B	Q1	Q2
B	2	5	Q_2	L	L	H	L
GND	3	4	Q_1	L	H	H	L
				H	L	H	L
				H	H	L	H

Symbol	Parameter	Min	Max	Units	Test conditions
dc Characteristics[a]					
V_{IL}	Input low voltage	—	0.8	V	—
V_{IH}	Input high voltage	2.0	—	V	—
V_{OL}	Output low voltage	—	0.45	V	$I_{OL} = 2.0$ mA
V_{OH}	Output high voltage	2.4	—	V	$I_{OH} = 240$ μA
I_{IL}	Input leakage	−10	10	μA	$V_{in} = V_{CC}$ to 0 V
I_{CC}	V_{CC} supply current	—	20	mA	—
ac Characteristics[a]					
t_D	Output delay	10	—	ns	80-pF load

[a] ($T_A = 0$ to 70°C, $V_{CC} = 5$ V ± 10%)

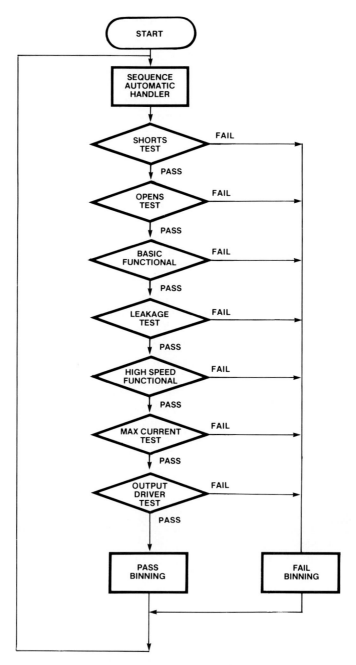

Fig. 5. Flowchart of a typical VLSI test sequence.

TABLE II
Example Device Test Summary

Test	Test for	Setup[a]	Pattern
Shorts	Pin to pin, pin to ground, or pin to V_{CC} shorts	$V_{CC} = 0.0$ V $V_{IL} = 0.0$ V $V_{OH} = 0.3$ V $I_{OL} = 1.0$ mA $V_{CM} = 0.5$ V	H L H L L H L H
Opens	Open Circuits on pins	$V_{CC} = 0.0$ V $V_{OH} = -1.9$ V $I_{OH} = -1.0$ mA $V_{CM} = 2.0$ V	H H H H
Basic functional	Proper operation at loose timings	$V_{CC} = 5.0$ V $V_{IL} = 0.0$ V $V_{IH} = 3.0$ V $V_{OL} = 1.0$ V $V_{OH} = 2.0$ V $I_{OL} = 0.0$ mA $I_{OH} = 0.0$ mA $V_{CM} = 1.5$ V $V_{TH} = 2.0$ V $V_{TL} = 0.8$ V	L L H L L H H L H L H L H H L H
Leakage	Input current leakage of device pins	$V_{CC} = 5.5$ V $V_{IL} = 0.0$ V $V_{IH} = 3.0$ V $V_{OL} = 1.0$ V $V_{OH} = 2.0$ V $V_{TH} = 2.0$ V $V_{TL} = 0.8$ V $I_{IL} = \pm 10.0$ μA	None
High speed functional	Proper operation at full ac timing	$V_{CC} = 4.5$ to 5.5 V $V_{IL} = 0.45$ V $V_{IH} = 2.4$ V $V_{OL} = 0.8$ V $V_{OH} = 2.0$ V $V_{TH} = 2.0$ V $V_{TL} = 0.8$ V	L L H L L H H L H L H L H H L H
Maximum current test	Check for I_{CC} within specification	$V_{CC} = 5.5$ V $V_{IL} = 0.0$ V $V_{IH} = 3.0$ V $V_{OL} = 1.0$ V $V_{OH} = 2.0$ V $I_{OL} = 2.0$ mA $I_{OH} = 240$ μA $V_{CM} = 1.5$ V $V_{TH} = 2.0$ V $V_{TL} = 0.8$ V	L L H L L H H L H L H L H H L H

(Continued)

TABLE II (*Continued*)

Test	Test for	Setup[a]	Pattern
Output driver test	Sufficient output drive current	$V_{CC} = 4.5$ V $V_{IL} = 0.0$ V $V_{IH} = 3.0$ V $V_{OL} = 0.45$ V $V_{OH} = 2.4$ V $I_{OL} = 2.0$ mA $I_{OH} = 240$ μA $V_{CM} = 1.5$ V $V_{TH} = 2.0$ V $V_{TL} = 0.8$ V	L L H L L H H L H L H L H H L H

[a] V_{CC} = power supply voltage
V_{IL} = input low drive voltage
V_{IH} = input high drive voltage
V_{OL} = output low sense voltage
V_{OH} = output high sense voltage
I_{OL} = output low current sinking
I_{OH} = output high current sourcing
V_{CM} = current source/sink crossover voltage
V_{TH} = high time reference voltage
V_{TL} = low time reference voltage
I_{IL} = input current leakage limit

III TEST DESCRIPTIONS

A Shorts Test

A shorts test is usually performed first on a semiconductor device to verify that each pin is electrically isolated from every other pin. A typical failure mode checked for is shorting of adjacent bond wires.

B Opens Test

An opens test checks that there is a current path to every device pin and, therefore, a complete path from every bond pad to every package pin of the semiconductor circuit. Typical failure modes checked for are missing or poorly connected bond wires and electrostatic damage to the device.

30. VLSI Testing from Design through Production 539

Functional Test C

A functional test verifies the complete logical operation of the semiconductor device. This test is usually performed many times at different settings of the voltage and at different clock frequencies to prove proper functionality of a part and to classify it into one of several operating ranges.

Maximum Current Test D

A maximum current test verifies that the maximum current drawn by the device during functional operation is less than the rate specification.

Leakage Test E

A leakage test verifies that input pins draw or source less than the specified amount of current.

Output Driver Test F

An output driver test verifies that the test device outputs operate properly under worst case conditions (their maximum rated current drive).

Breakdown Voltage Test G

The breakdown voltage test is performed on a sampling of devices to verify that the semiconductor circuit continues to operate after being subjected to a higher than normal voltage environment.

LOGIC TESTING IV

Functional testing is by far the most complex step in the testing of semiconductor circuits. It is also the most varied. Functional testing consists of several substeps that may or may not apply to a particular device. All functional testing begins with a basic functional test that quickly differentiates properly functioning from nonfunctioning devices. The goal here and in all testing is to weed out bad devices as quickly as possible so as to spend a minimum amount of time and cost on them.

A Synchronization at Reset

For many logic devices, testing requires synchronization of the device. This may involve the repeated sequencing of the tester through a pattern until the semiconductor device responds appropriately, indicating that the tester and the device are in synchrony.

B Speed Binning

Speed binning classifies the circuit into one of several operating ranges. This involves varying the clock rate/clock duty cycle of the semiconductor circuit and the operating voltage of the circuit. During engineering characterization, this variation is done in a systematic way by a process called *device schmooing* in which a particular operating parameter of the device is examined at every point in a matrix composed of speed and voltage. This information is plotted on a diagram known as a *schmoo plot* (for its resemblance to the Lil' Abner character of that name). Schmoo plots quickly show whether the device will operate correctly over all combinations of voltage and speed. In general, schmooing and schmoo plots can be used to compare graphically the performance of a particular semiconductor circuit over combinations of any two or more variations in operating parameters. Figure 6 shows an example of a schmoo plot.

C Output Relative Timings

In many semiconductor circuits, timings of various pins or modes of the device can depend on other timings, specifically output timings of the circuit. Since the performance of a semiconductor circuit is not known until it is tested, the test may have to be modified to guarantee that the appropriate output relative timing is maintained. This involves determining the time at which a particular signal occurs on a particular device and using that time to determine when other signals should be presented or tested. Output relative timings are determined by the VLSI tester in a process known as *binary* or *linear searching,* in which the tester sequences through multiple passes of a test, moving the strobe window of

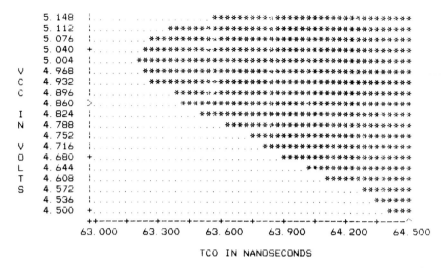

Fig. 6. Semiconductor performance as a function of two testing variables can be quantitatively presented using a schmoo plot.

the output time signal until the go/no-go point of the test is found. This information is then used to scale the other timings in the test appropriately.

MEMORY TESTING V

In the past, memory testing has differed from logic testing in the source and number of patterns used. For LSI, device circuits do not interact extensively with each other, and thus the determination of whether a device is functional or nonfunctional is fairly straightforward. In memory testing, however, because of the large arrays of closely packed elements, various data and addressing-dependent problems cannot be detected without exhaustive pattern testing. As a result, the number of patterns required to test a memory circuit has been much greater than that required to test an LSI logic circuit, and the focus on testing subtle data and address-dependent faults in the device has been higher.

VLSI is changing the situation in logic testing. The increasing logic circuit densities used in VLSI circuits are causing these circuits to exhibit some of the typical memory faults discussed below. The trend of incorporating logic and memory on the same device, such as in microcontrollers, also increases the need to check for memory problems on VLSI logic circuits [2].

A Algorithmic Pattern Generation

The size of the pattern in memory testing is so large that it is not economical to store it in vector memory as logic vectors are stored. Fortunately, the pattern sequence is regular and can be generated "on-the-fly" as the device is being tested in a tester subsystem known as an *algorithmic pattern generator*. Various algorithmic patterns are used to test for different failure mechanisms of memory devices [3,4].

B Memory Failure Modes

The major fault mechanisms in today's semiconductor memories fall into several classes, which are summarized in the following paragraphs.

1 Cell Faults

Cell faults include *stuck-at-zero* and *stuck-at-one* faults caused by poor mask alignment or metallization as well as gross device functional problems. Also included here is *adjacent cell coupling (multiple writes)*. This is the easiest class of faults to detect.

2 Address Decode Faults

Address decode faults occur when address decoders fail completely (nonaddressability of cells) or when they select multiple cells (address nonuniqueness). These problems can occur during read or write accesses. This class of faults is also relatively straightforward to detect.

3 Address Sequencing Faults

Address sequencing faults are the next level of addressing problems in memory circuits. Only specific sequences of addresses result in a chip failure. Addressing sensitivities can be column-to-column, row-to-row, diagonal-to-diagonal, or cell-to-cell (the most difficult case to detect). This class of problems can be caused by noisy address decoders, excessive address propagation delays, and capacitive loading on the device output drivers.

4 Data Sensing Faults

Even if device addressing is not at fault, the memory data can be incorrect if there are data sensing faults present. This class of faults

30. VLSI Testing from Design through Production

includes data coupling in multibit memory devices, slow sense amplifier recovery time (incorrect sensing after an extended period of sensing the opposite state), slow write recovery time, and surrounding cell coupling (not "hard" coupling that would be detected when examining for cell faults). This class of faults can be caused by capacitive coupling between cells, saturated or coupled sense amplifiers, or skewed address decoders.

Data Retention Faults 5

Data retention problems can occur in static memory devices as loss of data over time either in a clocked or unclocked mode. They can also occur in dynamic memory devices as loss of data in the refresh interval and in nonvolatile memory devices such as EPROMs, EEPROMs, and NVRAMs. This last class of devices can also exhibit programming problems as well as retention problems.

Common Memory Test Patterns C

Some of the patterns used to test for these classes of memory faults are shown in Table III. Each pattern is designed to check for one or more of the fault classes described. Different patterns also trade off speed of execution for probability of fault detection. Test execution time is governed by the length of the pattern. As shown in Table III, pattern size ranges from order N to order N^2, where N is the number of elements in the memory array. For a 64K DRAM this means the difference between a subsecond test and a 20-min test for each pattern. Clearly every effort must be made in design and engineering characterization to determine the cell failure mechanisms and develop N and $N^{3/2}$ patterns to detect these faults.

Basic Patterns 1

Basic patterns are order N patterns that are used on writeable memory devices (RAMs, NVRAMs, and EEPROMs) to test for gross device functionality as well as some cell faults. The *straight data pattern* writes data sequentially to each address and then reads each address sequentially. The pattern is then repeated with complement data. The *checkerboard pattern* sequences through all addresses in the same way as the straight data pattern but uses a pattern of alternating ones and zeros. The *Multiple address test* (also known as the address complement test) starts by writing a checkerboard pattern to memory and then reads out the array in a first cell, last cell, second cell, next to last cell, ... sequence followed by a sequential sequence.

TABLE III

Common Memory Test Patterns

Test type	Test name	Test time	Cell fault	Address decode	Address sequence	Data sense	Data retention
Basic	Straight data	N	X	—	—	—	—
	Multiple address	$10N$	—	X	—	—	—
	Checkerboard	$6N$	X	—	—	—	XX
Marching	Data	$10N$	XX	X	—	X	—
	Diagonal	$4N$	X	X	—	X	—
Walking	Data	$2N^2$	XX	XX	—	X	—
	Diagonal	$2N^{3/2}$	XX	X	—	X	—
	Column/row	$2N^{3/2}$	XX	X	—	X	—
	Dual column/row	$N^{3/2}$	XX	X	—	X	—
Galloping	Data	$4N^2$	XX	XX	XX	X	—
	Diagonal	$4N^{3/2}$	X	X	X	X	—
	Column/row	$4N^{3/2}$	X	X	X	X	X
	Butterfly (column and row)	$2N^{3/2}$	X	X	X	X	—
	Write recovery	$8N^2$	XX	XX	XX	XX	—
	Block ping pong	$4N^2$	XX	XX	XX	XX	—
	Moving inversion	$2N^{3/2}$	XX	XX	X	X	X
Surround disturb	Data	$\sim N \log_2 N$	X	X	—	X	—
	Write recovery	$\sim N \log_2 N$	X	X	—	XX	—
	Column/row	$\sim N \log_2 N$	X	X	—	X	X

[a] X = moderate, XX = good.

2 Marching Patterns

Marching patterns are also common order N patterns. The *marching data pattern* writes the array to zero, sequentially reads each cell, and then writes its complement. When the end of the array is reached, the process is reversed; each cell is read out and recomplemented in reverse order. The *marching diagonal pattern* is similar except that an entire diagonal undergoes this read, complement, reverse read, recomplement sequence.

3 Walking Patterns

Walking patterns are characterized by their common addressing sequence. In this N^2 pattern, the array is first filled with a background value. Then a test cell (or cells) is read and complemented, and the entire remainder of the array is reverified for the background value. The test cell (or cells) is cycled through all the positions in the array, and at each step the entire array is read out. The *walking diagonal, column, row, dual column,* and *dual row* patterns are shortened versions of the walking data pattern; these test for specific subclasses of errors.

Galloping Patterns 4

Galloping patterns are among the most complete classes of test patterns but also take the longest time to execute. Galloping patterns start by writing a background value to the memory array and then reading and complementing a test cell. The test cell is then alternately read out with every other cell in the array. Each cell in the array is selected in sequence to be the test cell, and the readout process is repeated. The *galloping column pattern* uses the same readout sequence as *galloping data* but reads out only array elements in the same column as the test cell. *Galloping row, diagonal,* and the *butterfly pattern* are similar modifications of the galloping data pattern.

Galloping write recovery and *block ping pong* extend the gallop data address sequencing to detect some device write faults as well. In *galloping write recovery* the test cell is alternately read out between *writes* to every other cell in the array. In *block ping pong,* the initial background pattern is alternating blocks of ones and zeros where the block length is varied according to test experience with a particular device.

The last galloping pattern, *moving inversion,* is a shortened version of the full galloping pattern test. First, a background value is written into the array. Next, alternate rows of the array are complemented, all rows are read, and then all rows are recomplemented. This process is repeated using alternate columns. Then, the whole sequence is repeated using every second row and column, every fourth row and column, and so on.

Surround Disturb Patterns 5

The last major pattern class is *surround disturb*. The surround disturb data pattern starts by writing a background value and then writing a complement into the selected test cell. The eight surround neighbors to the test cell are then read (sometimes several times), the test cell is reverified, and then it is restored to the background value. Each array element in turn becomes the test cell. *Surround disturb write recovery* uses the same addressing sequence but writes to the surrounding eight neighbors rather than reading them.

Column/row surround disturb uses the entire column or row as the test cell and disturbs by repeatedly reading (or writing) to the first and last elements of the row or column. Each row or column is cycled through in this manner.

D Topological Scrambling

One other complexity must be considered when testing memory devices. Many of the test patterns just described depend on exercising physically adjacent cells of the memory array. This requires, at a minimum, understanding the row and column layout of the device. The problem is complicated by the fact that to optimize cell layout and yield, a memory manufacturer designs the memory device so that contiguous addresses are not necessarily adjacent in the array. The addressing structure of the array is topologically scrambled. New column and row redundancy techniques in VLSI memories further complicate the problem by allowing this scrambling to be different from device to device.

To compensate for this scrambling, most memory testers include a topological scrambling element that scrambles the desired memory pattern before it is presented to the test device and precompensates for the device layout.

VI TESTING THROUGHPUT

As mentioned earlier, testing completeness and accuracy is just one part of the testing problem. The other part of the testing problem is to perform these tests as quickly as possible to increase the throughput and decrease the cost per test. Testing throughput depends on a number of factors, starting with the yield of the semiconductor circuit. *Yield* is the number of good devices out for the number of total devices in. Where multiple tests occur at different temperatures, yields of each test socketing multiply. As a result, four to ten test socketings may be performed for every good device out.

Additional factors that affect throughput are the setup time of the tester, the amount of time it takes to configure the tester to test a particular class of devices; the actual test time including all parametric and functional tests; the production efficiency, or the amount of time actually devoted to semiconductor production testing, as opposed to other uses of the tester such as engineering characterization, maintenance, or quality

30. VLSI Testing from Design through Production

assurance; and the number of test heads available on the tester. Additional test heads, to a total of two or three, are added by semiconductor test equipment manufacturers to increase test throughput [5]. The following equation presents a quantitative description of the interaction of many of these throughput factors [6].

$$T = \frac{(K_1)(\text{units})}{(K_2)(\text{units})(\text{test time}) + (\text{QA time}) + (\text{setup time})} \left[\frac{\text{production}}{\text{utilization}}\right] \times \left[\begin{array}{c}\text{effective number} \\ \text{of test heads}\end{array}\right],$$

where $K_1 = (\text{yield}_{hot})(\text{yield}_{cold})$ and $K_2 = (1 + \text{yield}_{hot})$.

The test program itself also has a significant impact on the throughput of semiconductor testing. In general, test flow should be chosen to minimize the amount of time spent testing a bad semiconductor device. This means putting tests for the most probable failure modes of a device early in the test flow and putting less probable failure modes and long tests near the end of the test flow so that they are performed only on devices that have passed previous tests.

In many testers the collection of test data also has an impact on the total test time. In general, testing to a predetermined limit is faster than acquiring the actual value of a parameter. Testing to limits, of course, must be traded off against collecting information that can be used to feed back into the fabrication process and improve overall yields.

Many of today's testers also support parallel testing of parametrics on multiple pins, whereas past generation testers required serial testing of parametrics. This also can have a dramatic impact on throughput, because serial parametric tests require the parametric measurement unit to be sequentially connected from one pin to another, a process that can take longer than the entire rest of the test.

VII QUALITY ASSURANCE AND SAMPLE TESTING

A Sampling Plans

Sample testing is done both by the quality assurance group in a semiconductor manufacturer and by the incoming inspection group in the semiconductor user's facility. Sample testing is performed on a lot-by-lot basis and is used on a subset of the devices in that lot to determine

TABLE IV

Lot size	Sample size	Acceptable quality level												
		0.010		0.015		0.025		0.040		0.065		0.10		0.15	
		Ac	Re	Ac	Re	Ac	Re	Ac	Re	Ac	Re	Ac	Re	Ac	Re
2–8	2	0	1	0	1	0	1	0	1	0	1	0	1	0	1
9–15	3	0	1	0	1	0	1	0	1	0	1	0	1	0	1
16–25	5	0	1	0	1	0	1	0	1	0	1	0	1	0	1
26–50	8	0	1	0	1	0	1	0	1	0	1	0	1	0	1
51–90	13	0	1	0	1	0	1	0	1	0	1	0	1	0	1
91–150	20	0	1	0	1	0	1	0	1	0	1	0	1	0	1
151–280	32	0	1	0	1	0	1	0	1	0	1	0	1	0	1
281–500	50	0	1	0	1	0	1	0	1	0	1	0	1	0	1
501–1200	80	0	1	0	1	0	1	0	1	0	1	0	1	0	1
1201–3200	125	0	1	0	1	0	1	0	1	0	1	0	1	0	1
3201–10 k	200	0	1	0	1	0	1	0	1	0	1	0	1	1	2
10 k–35 k	315	0	1	0	1	0	1	0	1	0	1	1	2	1	2
35 k–150 k	500	0	1	0	1	0	1	0	1	1	2	1	2	2	3
150 k–500 k	800	0	1	0	1	0	1	1	2	1	2	2	3	3	4
over 500 k	1250	0	1	0	1	1	2	1	2	2	3	3	4	5	6

whether the entire lot should be accepted or rejected [7]. This accept/reject decision is based on the acceptable quality level desired, the lot size, the sample size, and total number of failing devices detected in the sample, as shown in Table IV for the most widely used sampling program, the Mil Standard 105D program [8].

B Tester Accuracy and Testing Guardbands

Production testing typically is performed to a tighter specification than the customer specification to guarantee or guardband the customer specifications and allow for any inaccuracy in the tester or testing process. The necessity for guardbanding customer specifications is illustrated in Fig. 7, in which production testing must be performed to specification limits that are tighter than the customer data sheet to guarantee under worst case inaccuracies that parts are tested correctly.

The accuracy of today's VLSI testers is affected by a number of factors and can range from 500 ps to several ns. Table V shows the sources of tester error that can contribute to the guardbands required to test semiconductor devices properly.

30. VLSI Testing from Design through Production

Mil-Std-105D Sampling Plan

(normal sampling)

0.25		0.40		0.65		1.0		1.5		2.5		4.0		6.5		10	
Ac	Re	Ac	Re	Ac	Re	Ac	Re	Ac	Re	Ac	Re	Ac	Re	Ac	Re	Ac	Re
0	1	0	1	0	1	0	1	0	1	0	1	0	1	0	1	1	2
0	1	0	1	0	1	0	1	0	1	0	1	0	1	0	1	1	2
0	1	0	1	0	1	0	1	0	1	0	1	0	1	1	2	1	2
0	1	0	1	0	1	0	1	0	1	0	1	1	2	1	2	2	3
0	1	0	1	0	1	0	1	0	1	1	2	1	2	2	3	3	4
0	1	0	1	0	1	0	1	1	2	1	2	2	3	3	4	5	6
0	1	0	1	0	1	1	2	1	2	2	3	3	4	5	6	7	8
0	1	0	1	1	2	1	2	2	3	3	4	5	6	7	8	10	11
0	1	1	2	1	2	2	3	3	4	5	6	7	8	10	11	14	15
1	2	1	2	2	3	3	4	5	6	7	8	10	11	14	15	21	22
1	2	2	3	3	4	5	6	7	8	10	11	14	15	21	22	21	22
2	3	3	4	5	6	7	8	10	11	14	15	21	22	21	22	21	22
3	4	5	6	7	8	10	11	14	15	21	22	21	22	21	22	21	22
5	6	7	8	10	11	14	15	21	22	21	22	21	22	21	22	21	22
7	8	10	11	14	15	21	22	21	22	21	22	21	22	21	22	21	22

TABLE V

Common Sources of Tester Inaccuracy

Error source	Description
Pin skew	In shared resource testers, the error caused by differing electrical path lengths between timing generators and multiple connected pins
Format skew	Error caused by the format logic in the high-speed signal path that is used to transform a programmed pulse from a timing generator into a user defined waveform
Fixturing skew	Error caused by differing electrical path lengths and loading of the tester to device connections
Rise/fall time skew	Error caused by differing rise and fall times of tester drive signals across pins
Timing generation error	Errors due to the inherent resolution in timing generation circuit itself
Pin interaction	Errors due to the interaction of one pin's signals on another's

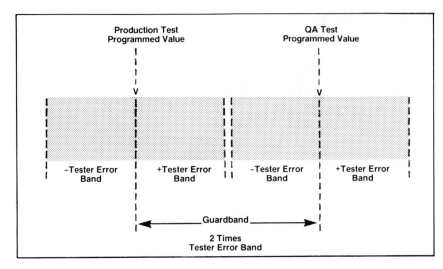

Fig. 7. Production test guardbands of twice the tester accuracy are required to guarantee the published specification under worst-case conditions.

REFERENCES

1. J. T. Healy, "Automatic Testing and Evaluation of Digital Integrated Circuits." Reston Publishing Company, Reston, Virginia, 1981.
2. C. Mead and L. Conway, "Introduction to VLSI Systems." Addison Wesley, Reading, Massachusetts, 1980.
3. R. O. Bower and R. G. Thomas, "Selecting Test Patterns for 64K Rams." Application Note TS300, Eaton Corporation, Woodland Hills, California, 1983.
4. T. Byrd, "Semiconductor RAM Testing." Application Note 17, Xincom Systems Division, Fairchild Camera and Instrument Corporation, Chatsworth, California, 1979.
5. D. Brendan, "The Economics of Automatic Testing." McGraw-Hill, New York, 1982.
6. C. J. McMinn, The impact of a VLSI test system on the test throughput equation, *IEEE Int. Test Conf. Proc.*, pp. 354–361. Philadelphia, Pennsylvania (1983).
7. J. M. Juran, "Quality Control Handbook." McGraw-Hill, New York, 1979.
8. U.S. Department of Defense "Sampling Procedures and Tables for Inspection by Attributes." Mil-Std-105D, U.S. Government Printing Office, Washington, D.C., 1964.

Chapter 31

VLSI Failure Analysis

LAWRENCE C. WAGNER

Texas Instruments, Inc.
Dallas, Texas

I. Introduction	551
II. Initial Nondestructive Procedures	552
A. Electrical Characterization	553
B. Other Nondestructive Tests	553
C. Types of Electrical Failures	553
III. Input–Output Failures	553
A. The Bake	554
B. Decapsulation and Delid	554
C. Nondestructive Tests	554
D. Second Destructive Testing	556
IV. Single Node Failures	556
V. Nonfunctional Defects	561
VI. Conclusion	563

I. INTRODUCTION

Failure analysis is a systematic approach to the isolation of device failures and the identification of failure mechanisms. The identification of failure mechanisms permits the failure analyst to pinpoint the process step(s) that caused or contributed to the failures. The failure analysis of semiconductor devices plays an important role in improving the manufacturing yields, quality, and reliability of those devices. The data acquired through failure analysis are used to identify statistically the most significant problems affecting a single device type or a family of devices. Failure analysis can be used to identify the source of specific problems such as a high failure rate after burn-in on a particular device type. In such cases, small sample sizes may be adequate to identify the most significant cause of failure. Failure analysis can also be used to attack more general problems such as incoming quality levels seen by customers. In such cases,

TABLE I

Key Destructive and Nondestructive Techniques for VLSI Failure Analysis

Destructive techniques	Nondestructive techniques
Initial evaluation: none	Electrical characterization: ATE, curve tracer, bench test, visual inspection of package, x-ray, PIND, leak test
Bake	Electrical retest
Decapsulation and delid	Electrical retest
	Microscopic/SEM inspection
Protective overcoat removal	Electrical retest
	Probe and isolation
	Microscopic/SEM inspection
Delayering	Microscopic/SEM inspection after each layer removed

the failure mechanisms are likely to be numerous and the statistical identification of the most significant mechanisms will require a much larger sample to be analyzed. Such data are frequently acquired over a period of months. Failure analysis can have an impact on all aspects of semiconductor manufacturing: design, silicon fabrication, packaging, and testing.

The basic tools and techniques for the failure analysis of VLSI circuits do not differ in many respects from those used in the analysis of smaller integrated circuits. Those tools and techniques uniquely required for VLSI failure analysis are in general applicable to, but not necessary for, the analysis of smaller circuits. They address the electronic complexity and the manufacturing sophistication of VLSI circuits, in particular the number of silicon fabrication process steps and the very small geometries.

Two key perspectives dominate the approaches to failure analysis of any type of circuit. First, failure analysis uses a combination of destructive and nondestructive techniques (see Table I). Since destructive techniques can destroy or obscure information about the failure mechanism, it is imperative to perform all nondestructive techniques first. Second, it is essential, particularly for VLSI circuits, for efficient failure analysis to constantly reduce the scope of the analysis. Since it is not practicable to test each component or node on a VLSI circuit, the analyst must exclude large areas of the circuit from failure analysis. This is achieved predominantly through electrical circuit analysis.

II INITIAL NONDESTRUCTIVE PROCEDURES

The two best opportunities for the use of nondestructive techniques are at the start of analysis and immediately after delid or decapsulation. The initial evaluation is frequently the most critical.

Electrical Characterization   A

The most important of the nondestructive tests is the electrical characterization of the failure. This includes automatic test equipment (ATE) data as well as bench-top characterization. Though for small-scale integrated circuits the bench testing of devices for failure analysis is useful, bench testing of VLSI circuits is normally limited to curve tracer analysis and functional operation of the failing nodes.

Other Nondestructive Tests   B

Other nondestructive tests employed include x-ray analysis for die attach or bonding problems (die attach quality becomes a significant factor on very large bars because poor die attach can increase the already severe stresses induced by plastic encapsulation), PIND test, leak test of hermetic packages, and visual inspection at low magnifications for package defects.

Types of Electrical Failures   C

The most common type of electrical failure is an input–output failure, which can be detected by curve tracer analysis of the failing pin. Another common type of failure can be isolated to a limited number of nodes by ATE characterization. A third category, which is the most difficult to analyze, is nonfunctional devices in which all or a large portion of the circuit is totally nonfunctional. These three categories include most failures, and they are usually analyzed in different ways. The failure analysis procedures for each category of failure are presented in the following sections.

INPUT–OUTPUT FAILURES   III

Input–output failures are defined as those that can be detected by pin-to-pin curve tracer analysis. They are primarily *continuity, leakage,* and *parametric* failures. These are the most common because most mechanical, chemical, and electrical stresses impact the input–output circuitry more severely than the rest of the circuit. Overstress voltages and currents predominantly affect these parts of the VLSI circuit. Mechanical stresses are most severe at the bar periphery, which is the normal location for bond pads and the associated circuitry. VLSI mechanical stresses are particularly severe because of the size of the bars. Temperature cycle and thermal shock failure mechanisms that result from thermal expansion coefficient mismatches become more significant with distance from the center of the package. Chemical contamination problems from the packaging environment are also most severe on the areas not protected by a passivation layer, predominantly the bond pads. The curve tracer analysis usually limits the search to a few components. Since this circuitry is usually the largest on the bar, analysis is not very different from the analysis of small-scale integrated circuits.

A The Bake

For many analyses, the first destructive analysis step is a bake or vacuum bake followed by an electrical retest. Such a bake can relieve certain failure mechanisms or improve the device operation. They are predominantly failure mechanisms that involve component instability, usually related to the accumulation of charges in the dielectrics. Some mechanical stress failure mechanisms and leakage that is associated with moisture on the chip surfaces can also be relieved by baking.

B Decapsulation and Delid

The second destructive procedure is the decapsulation or delid of the device. Hermetic package lids are normally removed mechanically. The methods for decapsulation of plastic encapsulated devices are numerous, and the choice of method depends on factors such as the maximum temperature to which the device is exposed and whether a wet chemical environment is desirable. Standard wet chemical procedures that use fuming nitric or fuming sulfuric acid remain the classical techniques, although jet etch decapsulation with sulfuric acid, now commercially available, is quicker. Alternatives that may be useful in specific cases are plasma etch and dry decapsulation. These techniques are slower than wet chemical techniques and result in the loss of continuity to the external pins. Because of these limitations, these techniques are used only in special situations, such as when metallization corrosion is suspected and the use of wet chemicals is not desirable.

C Nondestructive Tests

After decapsulation, the second opportunity for extensive use of nondestructive techniques occurs.

Electrical Retesting   1

The most important technique is to retest the part electrically to confirm that the circuit is still failing in the same manner. This may consist of ATE data or bench testing of the failed modes of operation. This procedure of rechecking the electrical behavior of the device after each potentially destructive procedure is important for the efficient failure analysis of any type of circuit.

Hot Spot Detection   2

Various other types of electrical characterization can be performed at this time to help in isolating the defective area. The most useful for input–output failures are hot spot detection schemes, which usually depend on thermal identification of the area where excessive heat is being dissipated, that is, a short or shunt in the circuit. Excess heat can be detected by a number of techniques, including application of liquid crystals to the surface of the bar, application of freon which boils at hot spots, and infrared scanning.

Microscopic Inspection   3

EBIC is a SEM technique that can be helpful in identifying a shunted or leaky site. It should be noted that though SEM is basically a nondestructive technique, the possibility of trapping charges associated with the electron beam is real and can alter device operation. Inspection of the circuit at high magnification is usually limited to areas of VLSI circuit that could be associated with the failure, with only cursory inspection of the rest of the circuit.

Other Techniques   4

Other inspection tools such as infrared and acoustic microscopes may be useful in identifying defects in the bulk silicon. SEM and other E-beam tools such as Auger and x-ray analysis are essential for VLSI failure analysis. As geometries become smaller and the number of layers in the VLSI circuit increases, E-beam tools become more significant because of the limitations on resolution and depth of field of optical microscopy. A broad range of nondestructive procedures are available after decapsulation; thus, one of the keys to efficient failure analysis is the selection of the proper procedures for each analysis.

D Second Destructive Testing

The next destructive step is the removal or thinning of the protective overcoat to facilitate probe and isolation. Again the device is retested electrically. Though simpler, low magnification (approximately 250×) probe stations are adequate for analyzing most input–output failures, a mechanically stable, high magnification (1000× range) probe station is essential for general VLSI failure analysis. *Isolation techniques* for severing conductors are numerous. They range from simple techniques, such as capacitive discharge, mechanical abrasion with a rigid probe tip, and ultrasonic cutting, to more elaborate schemes such as laser cutting, photomask and etch, and techniques using metal migration. Once isolated, *delayering techniques,* described later, are employed to remove the various layers of the VLSI device until the failure mechanism is identified. Once the uppermost conducting layer is removed, only limited electrical characterization by probing is possible. Analysis becomes largely limited to microscopic and/or SEM inspection after the removal of each layer.

As previously suggested, the input–output failures in VLSI are not usually more difficult to analyze than those in small-scale integrated circuits, because the scope of analysis is reduced to a very small bar area with relatively large geometries. However, the general approach of using destructive and nondestructive techniques is carried on into the analysis of single node and functional failures.

IV SINGLE NODE FAILURES

The next most frequently analyzed type of failure is the single (or at worst, a few) node failure. These are different from input–output failures in two significant ways:

(1) The failure is manifested only in the functional operation of the VLSI circuit.

(2) The failure site is away from the input–output circuitry and frequently in the minimum spacing geometry of the device.

A Types of VLSI Circuits

Here it is convenient to distinguish two types of VLSI circuits. There are very *highly structured devices,* predominantly memories, and *less structured devices* such as microprocessors. The highly structured devices give readily interpretable information from ATE. Single-bit, row, or column failures are accurately identified. When combined with a descrambling or decoding table, these ATE data isolate the failure to a relatively small area of the circuit. Ease of isolation makes the analysis of such circuits staightforward.

B. Delayering

Usually, limited probe and isolation are required, and delayering of the device becomes the key for identification of the failure mechanism. Since the number of layers that may have to be removed is directly related to the number of process steps in the silicon fabrication, the delayering of a VLSI circuit can be quite challenging. An example of delayering in a memory cell is illustrated in Fig. 1. The chemical etches used to remove some of the more common materials are listed in Table 2.

C. Failure Analysis Example

The less structured VLSI devices pose a much more demanding electrical analysis problem. It is critical for efficient failure analysis to maximize the utilization of ATE data in narrowing the location of the failure. It is necessary at least to establish the interrelation of failing output and input signals as illustrated in the following example. The pin 20 output of the microprocessor locked up during operation. After delid, the device was programmed to function in an infinite loop that would exercise the failing output. This output circuit has three inputs as shown in Fig. 2. The proper signals at the three inputs operating in the infinite loop were established on a good device. Probing identified input 1 as nonfunctional. In this case,

TABLE II

Chemical Agents Used in Failure Analysis Delayering[a]

Material to be etched	Etchant(s)
Silicon nitride	CF_4, O_2 plasma
Aluminum	Strong acids or bases
Gold	Aqua regia
Silicon oxide	HF (10%); HF (49%); SO etch: acetic acid (4), H_2O (3), 40% NH_4F (1); common oxide etch
Polysilicon	10-to-1 etch: HNO_3 (10), Acetic acid (7), HF (1)
Silicon etches	Wright–Jenkins etch; SECO etch

[a] A wide variety of etches is used; this listing is typical but far from complete.

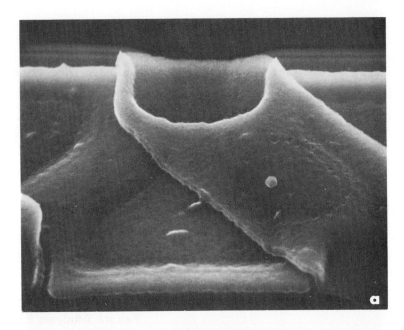

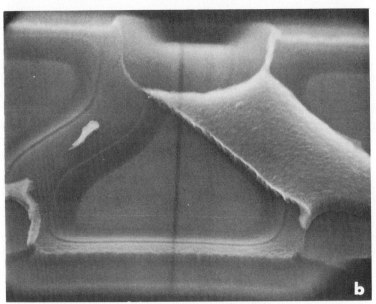

Fig. 1. Single cell DRAM failure is shown after successive delayering procedures. SEM views show failing cell after (a) passivation, metal, multilevel oxide, and polysilicon 2 reoxidation are removed; (b) polysilicon 2 is also removed; (c) interlevel oxide and polysilicon 1 are removed; (d) oxide is removed and stained with Wright–Jenkins etch. The failure was due to epi layer defects observed in the failed cell and surrounding cells. The defects are well defined by the silicon etch in SEM micrograph (d).

31. VLSI Failure Analysis

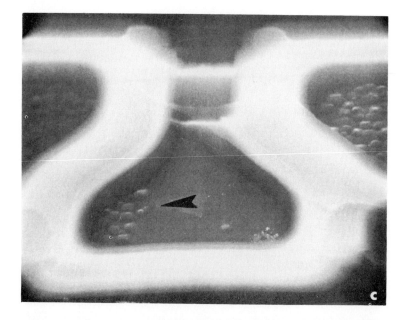

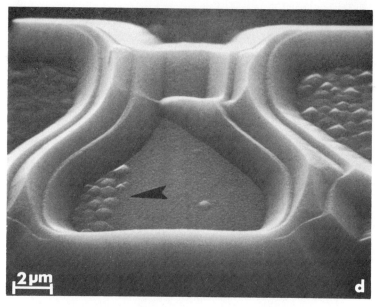

Fig. 1. (*Continued*)

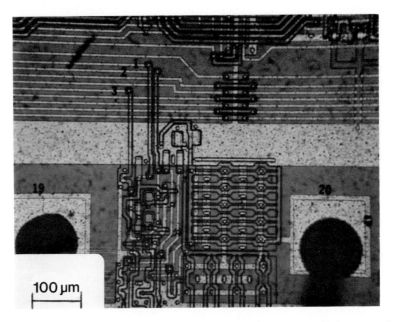

Fig. 2. The pin 19 and 20 area of failing microprocessor is shown with inputs to pin 20 circuitry labeled 1 through 3.

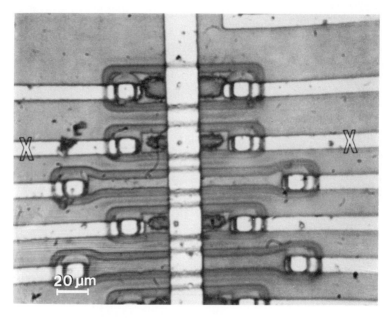

Fig. 3. Microprocessor failure was isolated in an open circuit between the Xs and later identified as a bad aluminum-to-polysilicon contact.

the nonfunctional circuit was traced back to a poor contact at a polysilicon bridge as shown in Fig. 3.

Though this example is simple, the logical procedure of tracing circuit operation is a key element for VLSI failure analysis. The failure must be analyzed in blocks and only when the failing block of the circuit has been isolated, can component level analysis be started.

The above example also illustrates the need for properly functioning devices to verify testing procedures and establish the proper functionality of internal circuit nodes. A number of noncontacting techniques for monitoring circuit operation have been developed that are particularly useful in this type of analysis. The best developed of these is *voltage contrast,* in which the device is operated in a SEM. Both dc and ac techniques are routinely used. This is also an excellent tool for monitoring circuit operation when the circuit is not well understood. *Infrared microscopy* can be used to detect component turn-on and is particularly useful in evaluations of circuit latch-up. *Laser scanning techniques* are also used to monitor circuit operation. Single node failures are usually point defects, often latent defects from the silicon fabrication process.

V NONFUNCTIONAL DEFECTS

This category of failures covers those devices that show no functionality either in the whole circuit or a large area of the circuit. This can be the most difficult type of analysis because the electrical data provide little or no insight into the location of the problem. These failures are normally due either to a silicon fabrication process that is abnormal or to a point defect that affects a critical portion of the circuit. Supply voltage and operation frequency variation can often restore full or partial operation, which may provide some indication of the cause or location of the problem. The general approach employed on nonfunctional failures is to test the operation of the key blocks of the circuit. The following analysis of a full array DRAM failure is illustrative.

Probing (voltage contrast would also be a suitable approach) indicated that most of the critical circuit blocks including the input, output, and write clocks were functioning normally. However, the sense amplifiers were not functioning. With the scope of the analysis thus reduced, the failure could be isolated to an open circuit in a metallization stripe in the sense amplifier, as shown in Fig. 4. The failure mechanism was identified by SEM as metal migration. The key to the failure analysis of nonfunctional failures is the reduction of the scope of analysis.

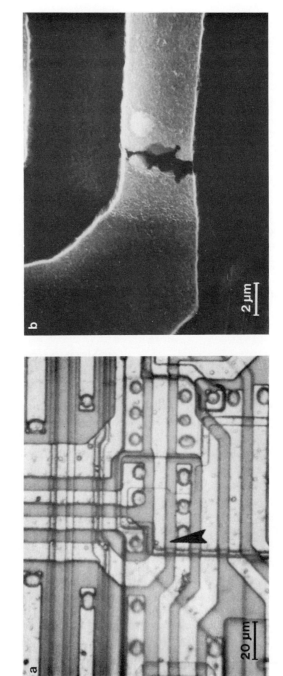

Fig. 4. (a) Optical micrograph indicates that location of the sense amplifier open circuit. (b) SEM micrograph shows a void caused by metal migration.

However, when silicon fabrication abnormalities occur, they usually affect most of the circuit with marginal or nonfunctional behavior observed throughout the VLSI device. These failures tend to show a strong dependence on supply voltages and frequency because they result from unstable resistances or component parameters. Since these failures are not localized, the failures are not truly isolated but rather are deduced from observations of the process abnormality in a few locations.

VI CONCLUSION

Though the classical skills of failure analysis remain important, the emphasis for VLSI is toward the use of electrical data, particularly ATE data, to reduce the scope of the analysis. Without such reduction in the scope of analysis, the problems of VLSI failure analysis become insurmountable. The failure analyst must still have all of the materials analysis skills previously associated with failure analysis in interpreting optical and SEM views of defects and in performing skillful delayering of devices. But these skills must be combined with skills in the interpretation of ATE data, setting up complex bench tests, and the correlation of electrical results to the VLSI bar layout. Thus, the failure analyst must work closely with test programmers and bar designers to work efficiently.

The future emphasis in VLSI failure analysis is likely to be in the area of E-beam analysis. As geometries become smaller, the limitations of optical microscopy in magnification and depth of field become more evident. E-beam techniques, particularly SEM, will become more important not only for routine inspection but also for electrical isolation via voltage contrast and EBIC.

Although the obvious and immediate goal of failure analysis is to locate the causes of failures, the more significant underlying purpose is to provide opportunities for yield, quality, and reliability improvements. As VLSI develops with smaller geometries, differences in the distribution of failure mechanisms should occur. Enhanced susceptibility to metal migration due to reduced metal stripes and to dielectric breakdowns due to thinner gate and interlevel dielectrics pose obvious reliability hazards. As VLSI electronic complexity increases, the bars will become larger and have more interconnects. New packages to accommodate the higher pin count will introduce new failure mechanisms. The larger bars will probably lead to more mechanical stress related mechanisms. As the key factors affecting yield and quality and reliability evolve with VLSI technology, failure analysis must play a key role in identifying them.

ACKNOWLEDGMENT

I would like to thank the staff at the Failure Analysis Laboratories of Texas Instruments for their assistance in the preparation of this chapter. In particular, I would like to thank P. B. Ghate and Tom Hausken for their comments and Olen Adams and Mike Peters whose failure analysis work was used as examples.

Chapter 32

Radiation Effects and Radiation Hardening of VLSI Circuits

BOBBY L. BUCHANAN

Radiation Hardening Technology Branch
Solid State Sciences Division
Hanscom Air Force Base, Massachusetts

I. Introduction	565
A. Scope	566
B. Radiation Requirements and Radiation Hardness	566
II. Radiation-Induced Effects in IC Materials	567
A. Crystalline Silicon	567
B. Silicon Dioxide–Silicon Interface	568
III. Principal Radiation Effects in Devices and ICs	569
A. Permanent (γ) Ionizing Effects	569
B. Transient ($\dot{\gamma}$) Ionizing Effects	572
C. Permanent (ϕ) Displacement Effects	573
IV. Radiation Hardening of Semiconductor Devices and ICs	573
A. Dielectric Isolation Processes	573
B. Techniques Other Than Dielectric Isolation	575
V. Single Event Upset	575
VI. Radiation Hardness of GaAs ICs	577
A. Total Ionizing Dose	577
B. Ionizing Dose Rate	578
C. Permanent ϕ Displacement Effects	578
D. SEU Effects	578
VII. Hardness Trends of Silicon Devices	578
References	579

INTRODUCTION   I

The recent expansion of the area of radiation hardened electronics has been for the most part in response to the need for VLSI in military systems and to the hardening challenges imposed by VLSI. The purposes of this chapter are to provide sufficient information to define these challenges and to give enough perspective to enable the VLSI designer to avoid major pitfalls and be aware of the first-order design compromises for satisfying low- to medium-level hardening requirements.

A Scope

Semiconductor devices are sensitive to radiation from nuclear weapons, nuclear reactors, space, and even from alpha particles emitted from packaging materials. The four principal categories of radiation damage to semiconductors are *displacement, ionization, single event upset,* and *thermomechanical*. The thermomechanical effects are associated with very high-level dose rates or EMP and will not be considered here, because they are usually handled at the systems level. The single event upset (SEU) effects are single particle induced random soft errors in small device structures and will be addressed separately. It is fortunate that most of the first-order ionization and displacement effects can be characterized in terms of three constituent parameters γ, $\dot{\gamma}$, and ϕ, as defined in Table I. It is assumed that, during irradiation, γ, $\dot{\gamma}$, and ϕ are uniformly distributed throughout the semiconductor active device region.

B Radiation Requirements and Radiation Hardness

The radiation requirement of a particular system is given to the electronics designer as an assignment of hardness values for each of the γ, $\dot{\gamma}$,

TABLE I

Radiation Hardening Requirement Levels (Approximate Range)

	Ionizing dose[a] γ (rads)	Dose rate upset[b] $\dot{\gamma}$ (rads/s)	Neutron fluence[c] ϕ (neutrons/cm^2)
Low	$<10^4$	$<10^7$	$<10^{12}$
Medium	10^4-10^5	10^7-10^8	$10^{12}-10^{13}$
High	10^5-10^6	10^8-10^9	$10^{13}-10^{14}$
Very high	$>10^6$	$>10^9$	$>10^{14}$

[a] γ is the total absorbed ionizing dose measured in rads(Si). (In addition to gamma radiation, this includes electrons, the ionizing component of neutrons, etc.). 1 rad = absorption of 100 ergs/g.

[b] $\dot{\gamma}$ is the absorbed ionizing dose rate measured in rads/s. (Pulse length and shape should also be specified for the incident ionizing radiation.)

[c] ϕ is the number of neutrons incident per square centimeter with 1 MeV equivalent energy. (Displacement damage caused by other radiation is included.)

and ϕ constituents. The meaning of the hardness values making up the system radiation requirement is that the electronics must operate without failure before, during, and after exposure to radiation levels of the required values for each constituent. To ascertain which constituent limits the hardening requirement level that an IC technology can satisfy, the approximate range of hardening requirement levels is given in Table I. Many systems specify only a γ constituent. Virtually all systems with a ϕ requirement have γ and $\dot{\gamma}$ requirements. The γ constituent will be emphasized here because it is specified in virtually all systems and because it causes the most subtle and limiting problems for hardening LSI/VLSI up to and including the high-level requirements.

RADIATION-INDUCED EFFECTS IN IC MATERIALS II

Even though semiconductors and insulators other than silicon (Si) and silicon dioxide (SiO_2) have similar radiation-induced effects, this section is limited to effects in Si and SiO_2.

Crystalline Silicon A

The ϕ induced displacement damage permanently reduces the initial majority carrier lifetime τ_o and reduces the initial majority carrier concentration N_0. A second-order permanent effect is mobility reduction. (Most so called permanent effects are only more or less permanent because they anneal with time and temperature). The functional dependence of τ and N on ϕ is given in Fig. 1. The expressions in Fig. 1 are approximations. Since the initial carrier removal rate varies from approximately 6 to 10 carriers removed per cm^3 per neutron/cm^2, a worst case rule of thumb for estimating the onset of carrier removal effects is that when $\phi = N_0/100$, then 10% or less of the carriers have been removed. A rule of thumb for estimating the onset of minority carrier lifetime degradation is that when $\phi = 10^4/\tau_o$, then the minority carrier lifetime is approximately 90% of its initial value.

The γ and $\dot{\gamma}$ have very little first-order permanent effect on bulk silicon. The basic transient effect is the generation of electron–hole pairs. The generation of one electron–hole pair requires an average energy of 3.6 eV thus, by definition, the generation rate is $g(t) = g_o\dot{\gamma}(t)$, where $g_o = 4 \times 10^{13}$ hole–electron pairs/cm^3 rad(Si) and $\dot{\gamma}(t)$ = time-dependent absorbed ionizing radiation, rads (Si)/s.

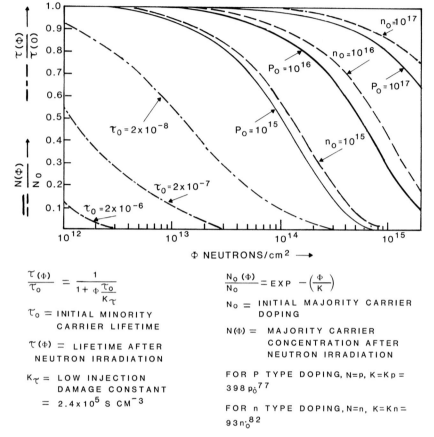

Fig. 1. Minority carrier lifetime reduction and majority carrier removal as function of neutron fluence, initial lifetime, and initial carrier concentration.

B Silicon Dioxide–Silicon Interface

The γ-induced trapping of positive charge at the Si–SiO$_2$ interface occurs at low levels and is the most important permanent radiation effect relative to VLSI. When ionizing radiation passes through SiO$_2$, free electrons and holes are created. Electrons are more mobile and are swept out of the oxide. The less mobile holes tend to be trapped near the silicon interface, and the trapped positive charge is near the silicon interface. The trapped positive charge induces a negative charge in the Si at the SiO$_2$ interface, which can cause changes in the electrical characteristics of component MOS devices and can create inversion layers in p-type silicon. If the inversion layer connects two n-type regions of different potential, a *parasitic MOS* can be turned on by the ionizing radiation, creating a permanent leakage path.

The average energy for generating an electron-hole pair in SiO_2 is approximately 18 eV [1]. Therefore, one rad (Si) generates approximately 8×10^{12} pairs/cm³ in SiO_2. The positive trapped charge density, Q_{sr} (cm^{-2}), at the interface, generated by γ rads is $Q_{sr} = 8 \times 10^8 (\gamma)(T_{ox}) P_T$ (cm^{-2}), where T_{ox} is the oxide thickness in microns and P_T is the probability that a generated hole is trapped, at the interface.

Another important γ-induced effect is the creation of interface states at the Si–SiO_2 interface. In a p-type Si–SiO_2 interface, used for *n*-channel transistors, the interface states are charged neutral or negative and therefore, to some extent, can compensate the trapped positive charge. In an *n*-type Si–SiO_2 interface, the interface states show very little effect.

PRINCIPAL RADIATION EFFECTS IN DEVICES AND ICs III

The big difference between radiation effects in discrete devices and radiation effects in integrated circuits is caused by the isolation methods used to separate one device from another on the monolithic chip. The isolation methods create parasitic MOS and parasitic bipolar devices that can be turned on by ionizing radiation, causing circuit failure. The increased seriousness of the problems in integrated circuits, indicated in Table II, is caused by the parasitic devices.

Permanent (γ) Ionizing Effects A

The total dose ionizing effects in semiconductor devices can be explained in terms of the effects in MOS devices. Any oxide–silicon interface is treated as if it were a MOS structure even if it does not have a metal (or polysilicon) gate electrode. MOS structures are divided into two types, *circuit component* and *parasitic*.

The effects of ionizing radiation on an *n*-channel component MOS transistor and its transfer characteristic are illustrated in Fig. 2. The square root of drain current is plotted as a function of gate voltage for a fixed source-to-drain voltage. The initial unirradiated characteristics curve 0 is a straight line with a slope that is proportional to the gain and an intercept on the voltage axis that is the threshold voltage. After a medium-level γ irradiation, because of the positive-charge trapping in the gate oxide, the transfer characteristic curve shows a parallel shift toward smaller threshold voltage levels and may shift into depletion mode as depicted by curve

TABLE II

Impact of Principal Radiation Damage Mechanisms on Devices and ICs[a]

Nonhardened device/IC	Displacement		Ionization	
	Carrier removal	Lifetime degradation	Trapped charge	Transient photocurrents
Discrete device				
Resistor	X	—	—	X
Diode	X	X	X	XX
Bipolar	X	XX	X	XX
MOS	—	—	XXX	XX
JFET	X	—	—	XX
MESFET	X	—	—	XX
Integrated circuits				
Bipolar	X	XX	XXX	XXX
MOS	—	—	XXX	XXX
JFET	X	—	X	XX
MESFET	X	—	X	XX

[a] X: could be problem; XX: serious problem; XXX: very serious problem.

1 in the figure. At higher radiation levels, interface state effects may become more prominent, and cause the curve to shift back toward its initial position with a corresponding decrease in slope as depicted by curve 2. The charge buildup mechanism usually dominates threshold shifts, except at high γ levels, and is much better understood than the interface states. Regions of trapped charge and interface state domination are indicated in Fig. 3. The magnitude of the radiation induced threshold shift is strongly dependent on oxide thickness, the processing of the MOS devices, and the voltage bias applied to the gate insulator during irradiation. Many early commercial MOS transistors showed threshold shifts of

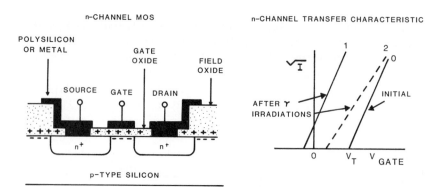

Fig. 2. Cross section of NMOS transistor illustrating trapped positive charge effects and transfer characteristics after two ionizing radiation levels.

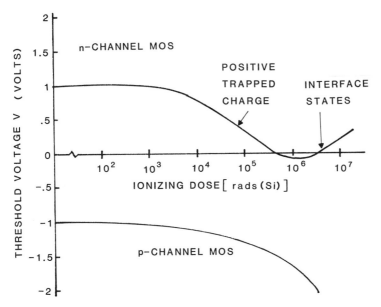

Fig. 3. Threshold voltage of *n*- and *p*-channel MOS transistors as a function of ionizing radiation illustrating the typical interface state induced turn around in the *n*-channel threshold voltage at high radiation levels.

40 V or more when irradiated at 10^6 rad(Si) while under a 10-V bias. If the only mechanism for threshold shift is trapped positive charge, then $\Delta V_T = -(.038)(T_{ox})^2(P_T)\gamma$. It can also be shown [2] that the magnitude of the negative threshold shift $\Delta V_T < 0.038(T_{ox})^2\gamma$ for *n*-channel devices if all threshold shift mechanisms are considered.

The effects of ionizing radiation on parasitic MOS structures are shown in Figs. 4, 5, and 6. The back-channel leakage [3] illustrated in Fig. 4

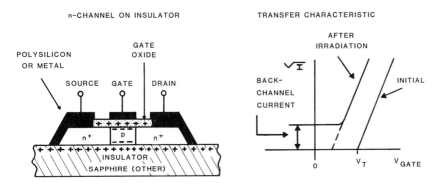

Fig. 4. Cross section of NMOS transistor on insulating substrate illustrating trapped positive charge induced parasitic back-channel and effect on transfer characteristics.

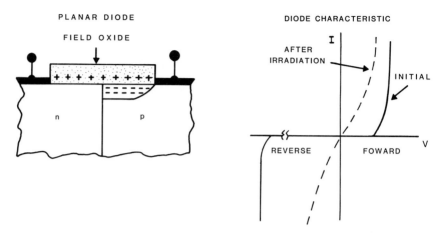

Fig. 5. Illustration of effect of ionizing radiation on cross section of diode and diode characteristics.

occurs for most unhardened SOS structures and other SOI structures at medium γ levels. The diode leakage effects in Fig. 5 can also be important as gain degradation mechanisms in bipolar structure. One of the problems limiting the total dose hardness of advanced bipolar ICs, utilizing side wall isolation, is illustrated in Fig. 6. Failure levels of $<10^4$ rads have been reported [4] for this type structure, even though the γ hardness of the component bipolar is usually $\gamma > 10^5$.

B Transient ($\dot{\gamma}$) Ionizing Effects

The $\dot{\gamma}$ effects are very circuit dependent. All diodes have $\dot{\gamma}$-generated primary photocurrent. In bipolar circuits [5] the emitter and collector

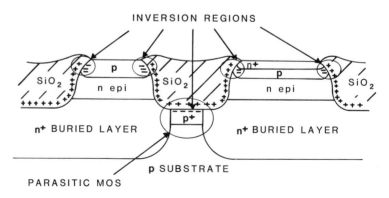

Fig. 6. Cross section of typical recessed field oxide bipolar structure utilizing walled emitters showing potential ionizing radiation induced parasitic inversion regions.

diode photocurrents are called primary photocurrents and the enhanced response caused by the transistor gain is called secondary photocurrent. Most of the effects are transient and cause temporary logic upsets that cannot be adequately hardened against by the IC designer at the systems level. A very important $\dot{\gamma}$ effect that can be addressed at the IC designer level is *latch-up*. Latch-up occurs as the result of a parasitic SCR being turned on by the $\dot{\gamma}$ radiation. All possible *p-n-p-n* four-layer paths can potentially cause latch-up in both junction isolated bipolar ICs and bulk CMOS. Latch-up is not possible in dielectrically isolated ICs. (See reference [5] for an excellent review of transient ionizing effects.)

Permanent (ϕ) Displacement Effects   C

Displacement effects mainly affect bipolar devices. However, design trends for high-speed digital bipolar ICs during the past ten years have (with a few I²L exceptions) tended toward designs that were naturally resistant to ϕ-induced effects through the high-level requirements range. Early bipolar transistors with wide basewidths were very susceptible to ϕ-induced effects. The gain β was controlled by the minority carrier lifetime in the base, so that the β degradation was given by $\beta(\phi)/\beta_0 \simeq \tau(\phi)/\tau_o$. Possible ϕ-induced degradation in the resistance R of resistors and the transconductance G_m of JFETs and MESFETs that might occur because of ϕ-induced carrier removal can be obtained from Fig. 1 by the relations $G_m(\phi)/G_{mo} \simeq N(\phi)/N_o \simeq R_o/R(\phi)$. The relations in Fig. 1 can also be used to determine the (ϕ) degradation in diodes by substituting $N(\phi)$ and $\tau(\phi)$ for τ and N in the functional relationships describing diode properties.

RADIATION HARDENING OF SEMICONDUCTOR DEVICES AND ICs   IV

Hardening processes and techniques as applied at different levels of complexity are shown in Table III. The system level hardening (except for shielding) is applied to satisfy dose-rate requirements that cannot be satisfied at the IC level.

Dielectric Isolation Processes   A

Historically, the isolation method used for bipolar ICs was reverse biased *pn* junctions. A junction isolated integrated circuit is referred to as a JIIC by the "hardening community." The isolation junctions are large in

TABLE III

Application of Semiconductor Hardening Processes and Techniques

Component device	Integrated circuits	Systems
Minimize pn junctions	Dielectric isolation	Shielding
High interface doping	Highly doped substrate	Current limiting resistors
High majority carrier concentration	Thin film resistors	Circumvention
Kill minority carrier lifetime	Block parasitic NMOS channels	Power strobing
Hardened insulators	Circuit design	
Thin insulators	High noise margins	
Low-temperature processing	Photocurrent Compensation	
	All device hardening processes	

comparison with component device junctions and result in large photocurrent generation and potential latch-up paths when subjected to $\dot{\gamma}$. Because of the significant $\dot{\gamma}$ problems caused by JIICs, junction isolation has been replaced by so-called dielectric isolation (DI) in hardened SSI–MSI bipolar integrated circuits. The bipolar dielectrically isolated integrated circuits are referred to as DIICs. (Recently commercial bipolar ICs with only side wall dielectric isolation have emerged for bipolar LSI/VLSI).

For many years DI had a special meaning pertaining only to bipolar technology. The present writer views DI as a general way to isolate component IC devices spanning all potential VLSI technologies including the sapphire in CMOS/SOS and the generalized insulator in CMOS/SOI.

There are too many types of DI processes to give a complete summary. The more or less standard "DI tub" process used for hardened bipolar MSI circuits is the most mature. However, the process is difficult to extend to LSI and probably impossible to extend to VLSI. The so-called SOI processes, with the exception of SOS are immature, but appear to be adaptable to VLSI. The SOS process is used for MOS ICs but not for bipolar integrated circuits because bipolar devices require very high-quality silicon material. (See reference [6] for an excellent review of the rationale and techniques for making SOI structures.)

All of the total dielectric isolation processes appear to be effective for eliminating latch-up and reducing photocurrents by more than a factor of ten. The process may, however, introduce new problems such as γ-induced back-channel leakage in CMOS/SOS [3].

B Techniques Other Than Dielectric Isolation

Techniques other than DI for hardening against latch-up in CMOS include killing the lifetime of the parasitic bipolar by gold doping or neutron irradiation so that the parasitic bipolar gain is too small and using n on n^+ epi starting material. The lifetime killing technique is not applicable for LSI/VLSI; however, the epi material techniques appear promising.

Channel stops (heavy p^+ doping at the Si–SiO$_2$ interface) are used to prevent surface, side wall, and back-channel inversion. This technique does not appear to prevent the parasitic MOS from turning on if $\gamma > 10^5$ rads for relatively thick unhardened oxides. Component MOS hardening techniques can also be applied to the parasitic MOS. Effective component MOS hardening techniques include thin oxides and low temperature processing. Since component MOS devices must maintain prescribed electrical characteristics, techniques for hardening against parasitic MOS devices such as high p^+ interface doping (which raises the threshold voltage) are not effective.

It is clear from Fig. 1 that effective hardening techniques to displacement damage includes high majority carrier doping and low minority carrier lifetime. These techniques are consistent with γ and $\dot{\gamma}$ hardening techniques.

V SINGLE EVENT UPSET

A soft error or single event upset (SEU) is caused by ionization along the path of a single energetic particle passing through an integrated circuit. If the energetic particle can generate the critical charge within the critical volume of a digital device, then logic upset occurs. By definition, the critical charge is the minimum amount of charge necessary to change the state of a logic or memory cell. The critical charge can be deposited by direct ionization from cosmic rays and alpha particles and by secondary particles from nuclear reactions.

Soft errors caused by alpha particles emitted from IC packaging materials are of concern to commercial IC manufacturers [6]. Traces of thorium or uranium contained in the packaging material emit alpha particles of approximately 4 MeV that have a track length in silicon of about 20 μm. If all the energy of a single alpha particle is deposited in the critical volume, then $4 \times 10^6/3.6 = 1.1 \times 10^6$ electron–hole pairs would be generated (since 3.6 eV are required to generate one electron-hole pair) or $1.6 \times 10^{-19} \times 1.1 \times 10^6$ coulomb = 0.18 pC.

The critical charge [7] of typical LSI technologies is shown in Table IV.

TABLE IV

Critical Charge for Single Event Upset

Technology	Device	Feature size (μm)	Critical charge (pC)
CMOS	HM6508	4	0.3
CMOS	HS6508RH	4	0.8
CMOS	HS6508RH$^+$	4	20
CMOS	CD4061	15	3.5
CMOS	TC244	21	9
CMOS/SOS	CDP1821	5	1.1
CMOS/SOS	TC5416	4	0.2–0.6
CMOS/SOS	NBRC4042	2.5	12
NMOS	—	6	0.6
NMOS	—	2	0.08
NMOS	—	1	0.02
I^{2}L	SBP9900	4.5	0.4
I^{2}L	SBP9989	4.5	0.06

With the exception of the hardened CMOS device, the critical charge is roughly proportional to the square of the feature size. At a given feature size, however, it appears that the bipolar technology is more sensitive than the MOS technology. Even though CMOS/SOS in Table IV has a similar feature size and critical charge to other MOS technologies, the cosmic-ray induced error rate is about two orders of magnitude less, because the sapphire greatly reduces the effective critical volume for charge collection.

Applying constant field scaling, theoretical worst-case cosmic-ray induced error rates are shown in Fig. 7. It was assumed that the critical charge was proportional to the cube of the feature size. A square-law dependence of critical charge on feature size, which is now known to be realistic, would give similar curves but with much less dependence on memory size. The fact that the error rate does not always increase (even though the critical charge always becomes smaller with memory size) would be even more evident with a square-law dependence.

Methods of hardening against SEU include dielectric isolation and other techniques for hardening against $\dot{\gamma}$-induced photocurrents and latch-up. A very effective circuit hardening technique for CMOS involves using decoupling resistors in the cross-coupling lines of the inverter pair of a memory cell. This technique was used in the HS65008RH$^+$ device in Table IV, resulting in the 20-pC critical charge, which makes the device immune to SEU because the maximum charge deposition of any energetic particle is less than 10 pC if the feature size is less than 10 μm [8]. Hardening against alpha particles from IC packages is an active reliability

32. Radiation Hardening of VLSI Circuits

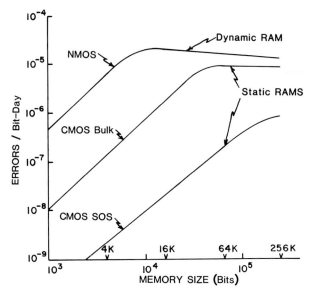

Fig. 7. Worst-case cosmic-ray induced upset as a function of memory size. (From reference [8].)

area. Parity checking is a very effective system software solution to most of the SEU problems.

RADIATION HARDNESS OF GaAs ICs   VI

The GaAs technology is often referred to as being inherently "radiation hard." The following discussion of the radiation hardness of GaAs devices relative to Si devices for each previously defined hardness constituent will put this in perspective and explain why.

Total Ionizing Dose   A

GaAs devices are inherently one to two orders of magnitude more resistant to γ than even the most γ-resistant Si devices. There are two reasons for this: (1) the GaAs component devices under consideration do not have gate insulators to trap charges; and (2) radiation-induced "turn on" of parasitic metal oxide semiconductor devices does not occur in GaAs ICs with or without passivating field insulators. The reason for this is that "surface inversion" is very difficult, if not impossible, in GaAs. This is the same reason why GaAs MOS devices are so difficult to make.

B Ionizing Dose Rate

GaAs is inherently more resistant to $\dot{\gamma}$ because of shorter minority carrier lifetime, larger activation energy for creating electron–hole pairs, and the fact that GaAs devices (JFET and MESFET) are built on semi-insulating substrates. The actual $\dot{\gamma}$ hardness of either a GaAs or a Si IC depends strongly on the circuit configuration. The complementary JFET GaAs RAMs have demonstrated a $\dot{\gamma}$ radiation hardness of 10^{11} rad/s. This is comparable to hardened Si complementary MOS fabricated on sapphire, which have shown the greatest upset hardness for a Si technology.

C Permanent ϕ Displacement Effects

The neutron-induced displacement damage constant relating to minority carrier lifetime and carrier removal rate are about the same for Si and GaAs. Because of higher doping levels and shorter minority carrier lifetime, GaAs appears to be hard to ϕ-induced displacement effects for MESFET, JFET, and bipolar devices. However, because of the low doping and initial high mobility, HEMT or MODFET GaAs devices may be very susceptible to displacement effects.

D SEU Effects

The first SEU test on GaAs RAMs gave disappointing results [9]. However, the more recent tests using SRAM design indicate a resistance to SEU equal to the standard Si CMOS RAM cell. Hardening techniques such as cross-coupled resistors can also be applied to the GaAs RAMs. The SEU effects do not seem to depend on whether the material is Si or GaAs, but rather depend on cell design.

VII HARDNESS TRENDS OF SILICON DEVICES

For a number of years, based on empirical data, it appeared that commercial NMOS ICs would become more susceptible to γ and $\dot{\gamma}$ radiation induced effects as the level of integration increased. This probably occurred because the component devices were not scaled in the vertical dimension and the threshold voltage margins were less. Recently this trend has turned around, and the NMOS γ hardness level has gone up from $\gamma \simeq 10^3$ rads to $\gamma \simeq 10^5$ rads for static HMOS III. The recent trend for advanced LSI bipolar, however, shows about two orders of magnitude decrease in γ hardness, caused by parasitic MOS devices introduced by side wall oxide isolation.

If constant field scaling is applied to the IC component devices, they theoretically increase in hardness as their size is scaled down. If MOS devices are scaled down by a factor k so that the feature size is scaled by k^{-1}, then it can be shown that the $\dot{\gamma}$ hardness increases by a factor k and that the γ hardness increases by a factor k^2. The ϕ hardness of bipolar transistors would probably also increase because of thinner basewidths resulting from scaling. New parasitic devices may be introduced during the scaling process that may dominate the overall hardness. It is not clear how the hardness of parasitic MOS devices already present would scale; however, parasitic bipolar devices already present would become much more of a problem relative to latch-up. The bottom line is that if the parasitic device effects can be suppressed, hardened VLSI appears feasible.

REFERENCES

1. J. R. Srour, "Basic Mechanisms of Radiation Effects on Electronic Materials, Devices, and Integrated Circuits," DNA-TR-82-20. 1982.
2. J. M. McGarity, *IEEE Trans. Nucl. Sci.* **NS-27,** 1739 (1980).
3. B. L. Buchanan, D. A. Neamen, and W. M. Shedd, *IEEE Trans. Electron Devices* **ED-25,** 959–970 (1978).
4. R. L. Pease, R. M. Turfler, D. Platteter, D. Emily, and R. Blice, *IEEE Trans. Nucl. Sci.* **NS-30,** 4216–4223 (1983).
5. J. P. Raymond, "Analysis and Testing of Radiation-Induced Transient Effects In Complex Microcircuits," DNA 6189T. 1982.
6. H. W. Lam, A. F. Tasch, Jr., and R. F. Pinizzotto, *in* "VLSI Electronics: Microstructure Science," vol. 4 (N. G. Einspruch, ed.), pp. 1–54. Academic Press, New York, 1982.
7. E. L. Peterson, P. Shapiro, J. H. Adams, Jr., and E. A. Burke, *IEEE Trans. Nucl. Sci.* **NS-29,** 2059 (1982).
8. E. A. Burke, *Technical Memorandum RADC-TM-81-ES-03* (1981).
9. S. A. Roosild, *Proc. 1984 ICCD* (1984).

Chapter 33
VLSI Imagers

RUDOLPH H. DYCK
Fairchild Camera and Instrument Corporation
Palo Alto, California

I. Introduction	581
II. Chip Architectures	582
A. General Remarks	582
B. Line-Scan Imagers	583
C. Television Imagers	585
D. Basic CCD Area Imager	588
III. Fabrication Technologies	589
A. Buried *n*-Channel CCD Register Structures and Processes	589
B. NMOS	592
C. Photoconductive Overlayer Technology	592
D. Color Technology	593
IV. Performance Variables versus Design Variables	593
A. Spectral Responsivity and Crosstalk	593
B. Nyquist Resolution and MTF versus Layout Variables	596
C. Dark Current and Dark Signal	598
D. Preamplifiers	600
Bibliography	601
References	601

INTRODUCTION   I

The material in this chapter provides the designers of imaging devices and of associated electronic cameras with some of the commonly needed information on various types of imagers, their designs and architectures, their processing, and their performance. The scope of the chapter is limited to silicon-based technology. It deals almost exclusively with the silicon chip itself, leaving out such important packaging topics as precision optical device packages, internal thermoelectric cooling, and engineering of precision hybrid focal planes. Also, there is a strong emphasis on issues related to large imagers, that is, those with the most photoelements (pixels), whether the array is of the line-scan type or the area type. Finally, some specialized topics, including preamplifier designs and blemish minimization, will receive only minimal attention.

II CHIP ARCHITECTURES

A General Remarks

An imaging device is made up of photosensing elements (pixels), scanning structures, and one or more charge-sensing preamplifiers. In some devices a charge coupled device (CCD) register provides both the sensing and the scanning functions. In MOS imagers the preamplification is usually off-chip.

VLSI imager technology is essentially all of the NMOS type; the various common subtypes are MOS, CCD, and CID (charge injection device), and some less common subtypes such as bucket brigade and resistive gate. The CCDs are (in most cases) of the buried channel type. Some interesting structures are combinations of these.

1 Photoelements

The photoelements themselves can be

(a) simple n^+p photodiodes,
(b) basic CCD type (gate-controlled and depleted np structure),
(c) virtual phase CCD type (p^+np structure with a depleted n region),
(d) Schottky barrier type (for infrared sensing), or
(e) upper-layer types in which a layer over the scanning structure can be a photoconductor (e.g., PbS) or a photodiode (e.g., an n^+i α-Si:H double-layer structure [1], or a ZnSe-Zn_xCd_{1-x}Te heterojunction structure [2]).

2 Charge Coupled Device

In a totally charge coupled device each scan removes, in principle, all free electrons from the n layer of the entire device, dumping them in one or more sink (or drain) diffusions. Then, after an integration period, the photoelectrons and any dark-current electrons are scanned out and sensed by a floating gate preamplifier. This totally CCD device can provide noise equivalent signal (NES) levels that are among the lowest, because the signal charge is converted to a voltage signal only once, at the minimum geometry input node of the preamplifier.

The following sequence of architectures is intended to be tutorial. Therefore, as the architectures proceed from simple to more complex, some of the comments that could be repeated will not be.

Line-Scan Imagers B

Although the most common line-scan imager architecture has a single row of pixels and two CCD registers, it is instructive to consider several types which are shown in Fig. 1.

Type 1: Photosensitive CCD Register 1

In this type, N-stages of CCD register provide N distinct photoelements (pixels). It produces good images with shuttered lens or with strobe-illuminated scene but has rarely been used since the earliest days of CCD technology. In operation, the locations of the pixels in one line of imagery can be shifted relative to other lines of imagery by selection of clock voltages and/or waveforms to provide staggered-pixel imagery, as shown.

Type 2: CCD Photoelements with One CCD Register 2

This type performs well with the scene imaged continuously on the device, provided that the CCD register is well masked. With one CCD register, variations are possible including

(a) use of photodiodes in place of CCD photoelements, providing higher quantum efficiencies but resulting in low-signal-level lag and sub-threshold leakage current effects, and

(b) use of two pixels for each stage of register, which must then be read out in interlaced fashion, that is, first the odd numbered pixels and then the even numbered.

Type 3: CCD Photoelements with Two CCD Registers 3

This is the most widely used architecture. A common variation has the registers extending on the output end and coming together into a single preamplifier. The variations listed for Type 2 are applicable.

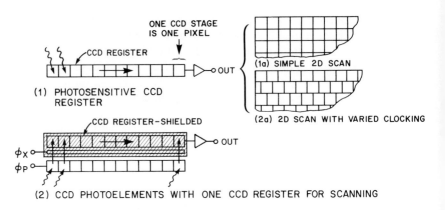
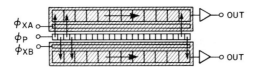
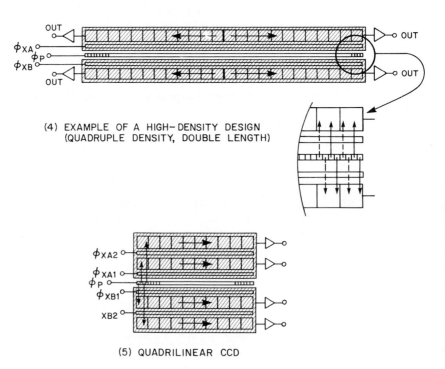

Fig. 1. Architectures of line-scan imagers (types 1–5).

Type 4: High-Density Design 4

As noted for a variation of Type 2 architecture, this example is shown with the density doubled as the result of incorporating interlace into the design.

Type 5: Quadrilinear CCD 5

This type requires careful design to avoid charge trapping due to narrow-channel effects during lateral charge transfer.

Television Imagers C

A television (TV) imager can operate in the staring mode, that is, it does not require a shutter. It should have at least 200 pixels in each direction, that is, at least 40,000 pixels total. Some TV imagers can be scanned a line at a time and thus do not require any signal storage sites, but the two most common CCD types, the inter-line transfer (ILT) type and the frame transfer (FT) type, both require signal storage sites for 50% of the pixels, that is, enough to store one entire TV field of signal. The common TV imager architectures are shown in Fig. 2.

Type 1: Frame Transfer CCD 1

This type can be made with a very simple CCD process. It requires high-powered clock drivers to transfer each entire field of the signal rapidly from the sensing array to the storage array. It has the problem of large chip size for standard TV sensing array sizes such as 11 mm diagonal and 16 mm diagonal (2/3-in. and 1-in. vidicon formats, respectively). The image smears somewhat during vertical transfer (typically 3%).

Type 2: Interline Transfer CCD 2

The typical design requires a very complex process that includes three poly-Si layers and three critical ion implantations. Exhibits horizontal aliasing. A variation uses a thin-film photoconductive array or photodiode array on top, which increases the active pixel area and decreases horizontal aliasing.

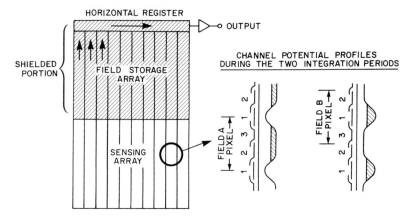

(1) FRAME TRANSFER CCD

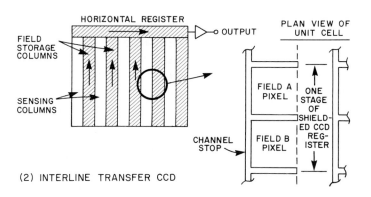

(2) INTERLINE TRANSFER CCD

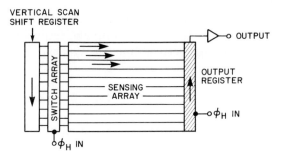

(3) LINE-BY-LINE TRANSFER CCD

Fig. 2. Architectures of TV imagers (types 1–6).

33. VLSI Imagers

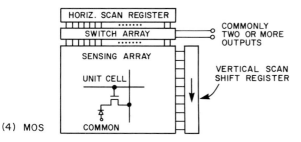

(4) MOS

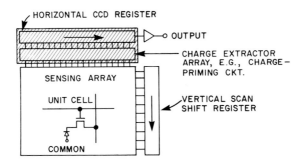

(5) CHARGE INJECTION DEVICE (CID)

(6) MOS ARRAY WITH CCD READOUT

Fig. 2. (*Continued*)

Type 3: Line-by-Line Transfer CCD

This type has only been demonstrated for quite small arrays. Because of the varying delay from top to bottom due to the CCD register at the right side of the device, as well as other problems, this architecture is not yet feasible as a TV imager. One interesting possible design uses metal gate virtual phase and backside illumination.

4 Type 4: MOS

This type can be made with a relatively simple NMOS process as used on some dynamic memories (one poly-Si layer, one metal layer). A variation, as in Type 2, uses a photosensitive layer on top.

5 Type 5: Charge Injection Device (CID)

This type is unique in that it has a nondestructive pixel-read capability.

6 Type 6: MOS Array with CCD Readout

This type eliminates the limitations of the horizontal MOS readout process. An example is the charge priming device (CPD).

D Basic CCD Area Imager

Figure 3 shows the simplest type of CCD area imager architecture. Its uses include snap-shot imagers, which require a shutter to give the best image, and time-delay-and-integration (TDI) imagers, which scan imagery in the fashion of a panoramic film camera. This is the most compact, most

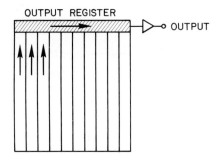

Fig. 3. Basic CCD area imager. Uses include snap-shot imaging and time-delay-and-integration (TDI) imaging.

efficient area imager, in that all of the unit cell is photosensitive and most of the area is available for CCD register potential wells.

FABRICATION TECHNOLOGIES III

Buried n-Channel CCD Register Structures and Processes A

Two-Phase, Implanted Barrier Types 1

One widely used high performance type of CCD is made with two layers of poly-Si and with a barrier implant of boron applied, without masking, after the first poly-Si layer is patterned. Figure 4 shows one stage of the 2-phase register in cross section together with a potential profile along the channel and with four clocking options.

High density CCD memories and TDI imagers have been made using this same phase-structure in an 8-phase ripple structure in which seven charge packets are controlled under eight phases of gate structure. This type of structure can be extended to a number of different variations using a third poly-Si layer. One such variation, providing a significantly smaller phase-length, is shown in Fig. 5. Compared with the 3-phase CCD, the big advantage of the 2-phase CCD is the built-in barrier feature that prevents the backspilling of charge. There is no problem with fast clock transitions as there is with the 3-phase CCD.

The Virtual Phase Type 2

The virtual phase type of CCD register [3] can be thought of as a modification of the 2-phase implanted barrier type, but a major modification: The two poly-Si gates of one phase of the structure are replaced by a single p^+ diffusion that is internally grounded; it must, of course, be clocked in the single-phase mode. The structure is shown in Fig. 6. In this version, five separate implants are required to form the pn structure, compared with two in the 2-phase design, but several advantages make it attractive:

(1) suppression of a major source of dark current (fast surface states),
(2) increased quantum efficiency, and
(3) flexibility in gate layout

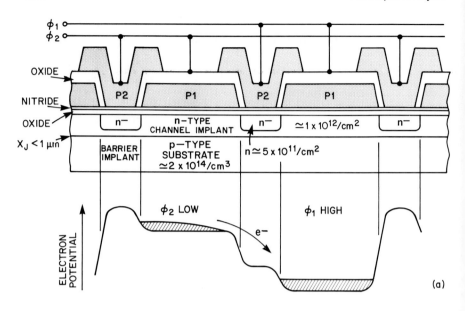

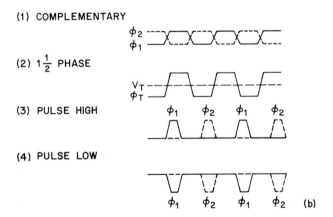

Fig. 4. Two-phase implanted-barrier CCD structure and a representative potential profile and (b) the clocking options. Typical concentrations and implant doses are shown.

The second advantage results when a sheet-type gate of some material is omitted above the p^+ diffusions. An example of the layout advantage is that in a very large array the many extremely narrow gate stripes can be replaced by a single large sheet, with none of the otherwise serious photolithographic problems.

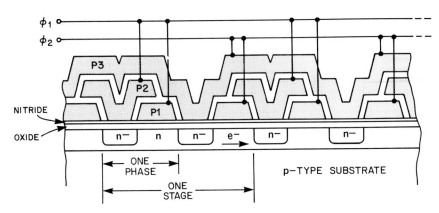

Fig. 5. A high-density, 2-phase implanted-barrier CCD register structure using three poly-Si levels. The poly-3 level can be patterned.

Multiphase Types

Three- and 4-phase CCD structures are also advantageous in some design situations. These do not require any implantations other than the one n implant that forms the buried channel region.

A *3-phase gate structure* made of three layers of poly-Si is particularly simple; each layer forms one phase. The simplicity is evident not only in the register region but also in the contact region, because in this structure it is not necessary to have contact openings to the metal buses opposite every stage of the register. Disadvantages of the 3-phase type are

(1) the sensitivity of the charge-transfer efficiency to the clock waveshape at high operating frequencies, and

(2) the sensitivity of the saturation charge level to the photolithography of both the first and second poly-Si levels, that is, this type is sensitive both to alignment and to amount of exposure and etch.

The *4-phase CCD gate structure* can be designed in either two-layer poly-Si or three-layer poly-Si, and the gates will appear just as in Figs. 4

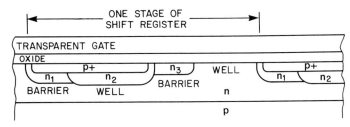

Fig. 6. The virtual-phase CCD structure.

and 5, respectively. The advantage of 4-phase over 2-phase, 3-phase, and virtual phase is high charge storage density per unit area of the CCD stage. This results for three reasons:

(1) Charge can be stored along at least 50% of the length of the stage at all times, that is, the CCD can be clocked to store a charge packet under either two or three adjacent gates at all times.

(2) This 50%-minimum fraction is independent of the critical dimensions of the gates.

(3) The height of the potential barriers is dependent only on clock amplitudes and can, in practice, be much greater than those of implanted barriers.

When high densities of charge are handled in multiphase CCDs, charge packets can include a significant amount of stored surface charge. Fortunately, this causes only a minor deleterious effect on charge transfer efficiency; for example, at frequencies near 20 KHz, efficiencies of more than 0.9999 are relatively easy to achieve while this effect is known to be occurring.

B NMOS

NMOS imaging arrays [4] are currently made by processes very similar to those used for some VLSI memories. A typical unit cell (see Fig. 2) has an area of 300 μm^2, 5–10 V operation is the norm, and a very important design objective is to minimize the vertical sense line capacitance because it has a major effect on the total random noise.

The process is considerably more complex for general purpose TV imagers, because these should have some antiblooming feature. A buried junction is needed so that when a photodiode becomes forward biased due to high light levels, it will inject into this buried junction. Also, an extra patterned implant is desirable to block injection to the drain (sense line) diffusion of the unit cell.

C Photoconductive Overlayer Technology

Perhaps the highest performance TV imagers are interline transfer CCD arrays with photoconductive or photodiode thin film photoelements formed on top. These require several additional layers: a lower electrode, the sensitive material (usually in more than one layer), and a transparent upper sheet electrode [1,2].

Color Technology D

A relatively simple technology for separation of red, green, and blue in image sensors makes use of two organic dyes in an overlapping structure to form three colors; photoelements not covered at all become a fourth type of pixel. The preferred dyes are yellow (high absorption in blue, high transmittance in green and red) and cyan (high absorption in red, high transmittance in blue and green); overlap of these produces a green filter.

Another possible color technology for imagers is the interference filter type. To make a satisfactory interference filter *on* an integrated circuit, it is necessary to form a highly planarized base layer. Planarizing can be done to a degree with a thick polyimide layer that forms an initial planarized surface and then by uniform etch-back with the right type of dry-etch conditions, forming a planarized surface in an underlying layer. This technology could in the future lead to higher quality color separation and greater stability in high stress environments such as space applications.

PERFORMANCE VARIABLES VERSUS DESIGN VARIABLES IV

Spectral Responsivity and Crosstalk A

Silicon as a sensor material for imagers is limited in the visible spectrum primarily by a high refractive index at the shorter wavelengths and in the infrared primarily by weak absorption (Fig. 7). The high refractive index and the absence of good antireflective coatings on most devices are largely responsible for the reduced quantum efficiencies in the 400–500 nm range (Figs. 8 and 9, upper spectra). The weak absorption in the infrared results in poor control of responsivity (Fig. 10) and poor resolution due to diffusion of deep carriers (Fig. 11). Of course, if infrared response is not of interest, these problems are avoidable by use of an infrared blocking filter. These subjects are treated in more detail in reference [5].*

---

* The author has taken a major portion of his "typical" performance information from Fairchild CCD Imaging product marketing literature. Reasons for making this heavy use of Fairchild device and technology data include (1) the large amount of information that is available about basically a single technology, and (2) the success of this technology as a manufacturing process for a comparatively long period of time. No value judgment is intended relative to some of the other technologies that have been shown to be quite successful in various respects.

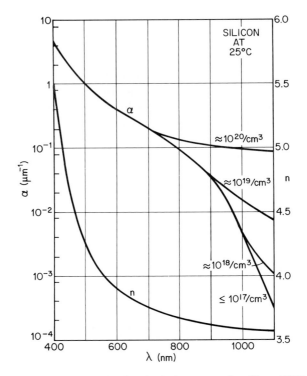

Fig. 7. Absorption and refractive index spectra for silicon (25°C).

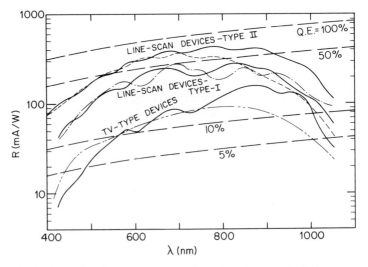

Fig. 8. Examples of response spectra from three types of VLSI imagers.

33. VLSI Imagers

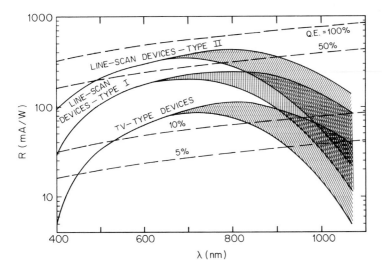

Fig. 9. Smoothed response spectra for the three types of VLSI imagers in Fig. 8.

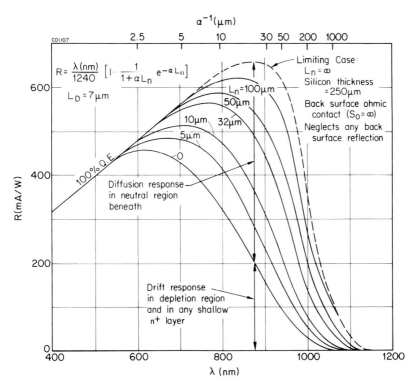

Fig. 10. Theoretical internal spectral response for one depletion depth and for several electron diffusion lengths (25°C).

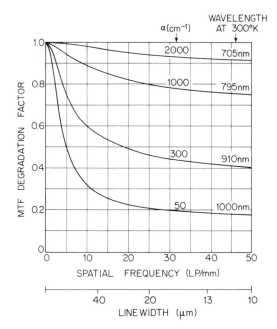

Fig. 11. Theoretical degradation in MTF due to diffusion of electrons below the depletion layer. Conditions = $L_D = 10$ μm, $L_n = 100$ μm.

B Nyquist Resolution and MTF versus Layout Variables

At least two different "ideal" image reproduction standards exist for solid state imagers: one simply requires the maximum possible information about the radiation intensity incident on each pixel, and the other requires that there should be no perceptible Moiré effect in the reproduced image due to the spatial frequencies of the array beating with those in the incident image. The latter is a goal for television devices. Array sizes that would meet the television standards of vertical line-count and also meet the second standard above, with equal horizontal and vertical resolutions, are 486 × 650 = 315,900 pixels, for NTSC, and 575 × 770 = 442,750 pixels, for PAL/SECAM.

Until such large arrays become practical, acceptable designs for smaller arrays are of considerable interest. Some acceptable designs simply have fewer columns. One convenient number of columns in NTSC television is 380, using a sampling clock operating at twice the chroma subcarrier frequency (455 cycles per line period and 16% blanking). Use of 380 columns in an interline transfer CCD gives the theoretical MTF behavior shown in Fig. 12b; the in-phase MTF is enhanced by the 50% opaque area (see Fig. 12a for reference). In an MOS array, it is practical to have a

33. VLSI Imagers

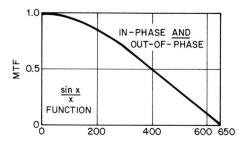

(a) "IDEAL" ARRAY WITH SQUARE PIXELS SAMPLED IN PAIRS

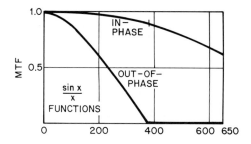

(b) BASIC 380-COLUMN INTERLINE TRANSFER CCD (50% OPAQUE REGION)

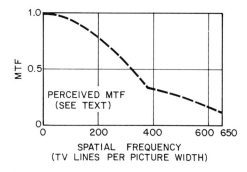

(c) MOS WITH 380 PIXELS PER ROW AND WITH STAGGERED PIXELS

Fig. 12. Theoretical horizontal MTF for some high resolution NTSC TV imager designs.

staggered-pixel design. With 380 pixels per row in a staggered-pixel MOS design, the theoretical perceived MTF behavior is shown in Fig. 12c. There is no distinction between in-phase and out-of-phase MTF because the viewer of the display tends to average what is seen in two adjacent lines; the curve in Fig. 12c is based on simple linear averaging and is intended to indicate approximately what is perceived. This theoretical curve neglects the effect of the finite width of the vertical metal sense lines in the MOS device.

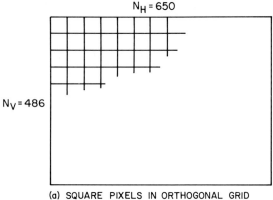

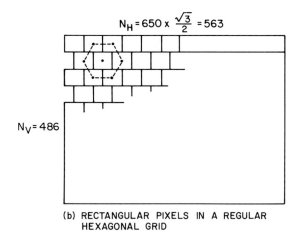

Fig. 13. Array sizes for "ideal" NTSC TV image sensors.

Perhaps the design in which the pixels are layed out in an hexagonal array as shown in Fig. 13b is optimal. Although it may be conceptually helpful to think of the staggered 380-pixel design of Fig. 12c as approximately hexagonal, it actually has 32% fewer pixels than the true hexagonal design.

C Dark Current and Dark Signal

Average dark current densities in VLSI imagers can vary widely depending on design and quality of processing. In a typical CCD imager, practically all the silicon surface is depleted, so the surface states make the dominant contribution to total dark current density. On the other hand, in a virtual-phase CCD imager, half of the active area never has a surface state component, and the other half needs a surface state component only during a short period needed for charge transfers. With current processing and gettering technology, a fairly typical CCD dark current density at room temperature is 3 nA/cm^2; but with virtual-phase technology, densities as low as 0.4 nA/cm^2 have been reported. Figure 14 shows the basic material parameters that determine dark current, along with their typical values and the typical temperature dependence.

Often, however, device performance is not limited by the average dark current but by shading nonuniformities such as a swirl pattern. Again, with current processing and gettering technology, these effects can be as low as 5 to 10% of the average dark current and possibly even less than that.

In MOS imagers, the reverse bias levels are lower and the depleted surface areas are less, resulting in room temperature dark current densities as low as approximately 0.2 nA/cm^2. Also, remarkably low nonuniformity levels have been observed.

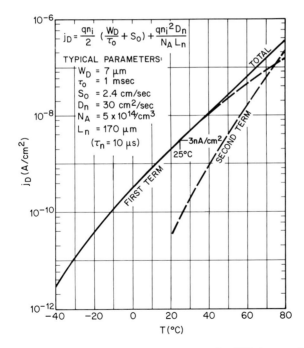

Fig. 14. Typical dark current versus temperature for CCD imagers in 1984.

TABLE I

Representative CCD Preamplifier Performance Characteristics

Type	N_{sat} (electrons/pixel)	NES at 5 MHz and 25°C (electrons/pixel, rms)	NES at 100 kHz, correlated double sampling, cooled (electrons/pixel, rms)
Resettable diffusion	1×10^6	400	—
	1×10^5	—	10 (6 best[a])
Floating gate	3×10^5	40 (30 best)	15
Resettable floating gate	3×10^5	60	—

[a] A large array made by GEC and used at very low charge levels has been shown to give this very low noise level [7].

D Preamplifiers

In CCDs, the preamplifier is a charge-to-voltage convertor. To minimize the noise equivalent input signal (NES), it is important to make the input capacitance as small as possible consistent with requirements for high-signal-level capability and with the noise properties of the small-geometry FETs that can be made with the given process. Typically, $1/f$-noise is a dominant noise in CCD preamplifiers; therefore, line-clamping or pixel-clamping (also called correlated double sampling) are highly desirable signal processing operations that can be done effectively off the CCD sensor chip. A good general discussion of the various types of CCD preamplifers is given by Kim [6]. These preamplifier types are

(a) resettable floating diffusion,
(b) resettable floating gate,
(c) true floating gate, and
(d) distributed floating gate

The latter two are apparently no longer used.

CCD preamplifers typically provide a saturation output voltage of 1 to 2 V; therefore, for a device that is required to have only a very small saturation charge (e.g., $\leq 1.5 \times 10^5$ electrons/pixel), not only can the input capacitance be made small but also a strong design effort should be made to achieve high gains. Table I shows the characteristics of some representative preamplifier designs.

BIBLIOGRAPHY

This bibliography is a listing of major publications that will provide the reader with general treatments of some of the more specific topics in the field of VLSI imagers, and which have not been referenced above.

33. VLSI Imagers

Barbe, D. F. (ed.), "Charge-Coupled Devices." Springer-Verlag, Berlin and New York, 1980.
Beynon, J. D. E., and D. R. Lamb (eds.), "Charge-Coupled Devices and Their Applications." McGraw-Hill, London and New York, 1980.
Chamberlain, S. G., and Kuhn, M. (eds.), "Joint Special Issue on Optoelectronic Devices and Circuits," *IEEE Trans. Electron Devices* **ED-25** (Feb. 1978).
Jespers, P. G., van de Wiele, F. and White, M. H. (eds), "Solid State Imaging." Noordhoff, Leyden, 1976.
McLean, T. P., and Schagen, P. (eds.), "Electronic Imaging." Academic Press, New York, 1979.
Morgan, B. L., and McMullan, D. (eds.), "Photoelectronic Image Devices." Academic Press, New York, 1979.
Sequin, C. J., and Tompsett, M. F., "Charge Transfer Devices." Academic Press, New York, 1975.
Weimer, P. K., and Cope, A. D., Image sensors for television and related applications, *in* "Advances in Image Pickup and Display," vol. 6 (B. Kazan, ed.). Academic Press, New York, 1983.

REFERENCES

1. O. Yoshida, Harada, N., Yoshino, T., and Ide, K., A CCD image sensor using a glow discharge amorphous Si photoconductor layer, *Eighth Symposium on Photoelectronic Image Devices, Imperial College, London, 1983 (Proceedings),* p. 51 (1983).
2. Y. Terui *et al.*, A solid-state color image sensor using a $ZnSe-Zn_xCd_{1-x}Te$ heterojunction thin film photoconductor, *Tech. Digest ISSCC* 34–35 (1980).
3. J. Hynecek, Virtual phase technology: a new approach to fabrication of large area CCDs, *IEEE Trans. Electron Devices* **ED-28,** 483 (1982). U.S. Patent No. 4,229,752 (J. Hynecek, Texas Instruments, Inc.).
4. M. Aoki *et al.*, $\frac{2}{3}''$ format MOS single-chip color imager, *IEEE Trans. Electron Devices* **ED-29,** 745. (1982).
5. R. H. Dyck, Design, fabrication, and performance of CCD imagers, *in* "VLSI Electronics: Microstructure Science," vol. 3 (N. Einspruch, ed.). Academic Press, New York, 1982.
6. C. K. Kim, The physics of CCDs, *in* "Charge-Coupled Devices and Systems (M. J. Howes and D. V. Morgan, eds.). Wiley, New York, 1979.
7. B. Thomsen and E. Sondergaard, Evaluation of the GEC 385 × 576 CCD image sensor for astronomical use, *Eighth Symposium on Photoelectronic Image Devices, Imperial College, London, 1983 (Proceedings),* p. 33 (1983).

Chapter 34
Noise in VLSI

A. VAN DER ZIEL

University of Minnesota
Minneapolis

I. Introduction 603
   A. Thermal Noise 604
   B. Shot Noise 604
   C. Burst Noise in Bipolar Transistors 605
   D. Flicker Noise or $1/f$ Noise 605
II. Various VLSI Circuits 606
   A. The CMOS Circuit 606
   B. The NMOS Circuit with Depletion Mode Load 607
   C. The NMOS Circuit with n-Channel Enhancement Mode Load 608
   D. CTL Bipolar Circuit 608
   E. The I^2L Bipolar Circuit 609
   F. ECL Circuit 610
III. Threshold Voltages 611
IV. Crosstalk 612
V. Alpha-Particle- and Cosmic-Ray-Induced Soft Errors in VLSI Circuits 612
   References 613

INTRODUCTION   I

The sources of spontaneous noise in VLSI circuits are thermal noise in MOSFETs, shot noise and burst noise in bipolar transistors, and flicker noise (or $1/f$ noise) in MOSFETs and bipolar transistors. Threshold voltages and circuit dimensions also affect the noise performance. Cross talk and the effects of α-particles and cosmic ray particles on the performance of the circuits are not spontaneous noise in the true sense of the word, but can be treated as such.

A Thermal Noise

The thermal noise of a conductance g kept at temperature T can, for a small frequency interval df, be represented by a current generator $(\overline{i^2})^{1/2}$ in parallel to g, where

$$\overline{i^2} = 4kTg\,df \tag{1}$$

The thermal noise of a MOSFET can, for a small frequency interval df, be described by an output current generator $(\overline{i_d^2})^{1/2}$, where

$$\overline{i_d^2} = \gamma 4kTg_{d0}\,\Delta f, \tag{2}$$

where $g_{d0} = (\mu w C_{ox}/L)(V_g - V_T)$. Here g_{d0} is the drain conductance for zero drain bias V_d, μ is the carrier mobility, w the channel width, L the channel length, C_{ox} the oxide capacitance per unit area, V_g the gate voltage, and V_T the turn-on voltage of the channel; the device is active for $V_g > V_T$. The parameter γ is a dimensionless parameter that depends on bias. Here $\gamma = 1$ for small drain bias V_d (linear mode) and $\gamma = \frac{2}{3}$ for long channels at saturation; γ can be somewhat larger than unity for very short channels at saturation (hot-electron effects).

B Shot Noise

The shot noise in a thermionic saturated diode that is carrying a current I can, for a small frequency interval df, be represented by a current generator $(\overline{i^2})^{1/2}$ in parallel with the device, where

$$\overline{i^2} = 2qI\,df, \tag{3}$$

and where q is the electron charge. It holds for any current I constituted by carriers that cross a barrier independently and at random; hence it also holds for currents in bipolar transistors.

If I_B is the base current and I_C the collector current, then

$$\overline{i_b^2} = 2qI_B\,df, \qquad \overline{i_c^2} = 2qI_C\,df. \tag{4}$$

Since the base conductance g_b and the transconductance g_m of the device are given by

$$g_b = qI_B kT, \qquad g_m = qI_C/kT, \tag{4a}$$

these expressions can also be written

$$\overline{i_b^2} = 2kTg_b\,df, \qquad \overline{i_c^2} = 2kTg_m\,df, \tag{4b}$$

corresponding to half thermal noise of g_b and g_m.

34. Noise in VLSI

If both the emitter and the collector are forward biased and if I_f is the forward current due to electron injection from the emitter and I_r is the reverse current due to electron injection from the collector, then $I_C = I_f - I_r$ and

$$\overline{i_c^2} = 2q(I_f + I_r)\,df. \tag{4c}$$

In the particular case of zero collector current,

$$\overline{i_c^2} = 4kTg_c\,df, \tag{4d}$$

where $g_c = qI_r/kT$ is the collector conductance.

C Burst Noise in Bipolar Transistors

Burst noise is caused by centers in the emitter-base space charge region that can absorb or emit an electron and thereby change the base and emitter current by an amount I, which is of the order of 10^{-8} A. The noise is, therefore, *pulse noise* with a duration of about 10^{-3}s. If the pulse signal is larger than the threshold voltage of the switching element, spontaneous switching can occur.

For current devices, it is unlikely that spontaneous switching can happen; but for very small devices, such an event could occur. Fortunately, burst noise can be prevented by proper emitter design. The burst noise is the result of heavy emitter doping, which produces dislocations in the emitter-base space charge region; these dislocations are caused by lattice distortion. By preventing lattice distortion, that is, by lower emitter doping (doping densities $<4 \times 10^{20}/\text{cm}^3$), one can eliminate the effect. For that reason the effect is not discussed further.

D Flicker Noise or 1/f Noise

Flicker noise in MOSFETs has a noise spectrum of the form $1/f^\alpha$ with α close to unity. It is caused by electron interactions with traps in the surface oxide, which causes fluctuations in the number of electrons; this, in turn, causes noise. At low drain bias, the noise can be represented by a current generator $(\overline{i_d^2})^{1/2}$ in parallel with the output, where $\overline{i_d^2}$ varies as V_d^2; at high drain bias, the noise saturates but is much larger than at low V_d.

In good bipolar transistors the noise comes from mobility fluctuations and is quite small; so its effect is negligible in comparison with the effect of shot noise.

To illustrate the effect of $1/f$ noise in circuits, we represent the white noise by $\overline{i^2} = A\,df$ and the $1/f$ noise by $\overline{i^2} = A(f_1/f)\,df$. Then the noise spectra are equal at the frequency $f = f_1$; for that reason f_1 is called the *corner frequency* of the $1/f$ noise. If we now assume that the computer circuits have a flat response for $10^{-3} < f < 10^7$ Hz and zero response outside, then the effects of the white noise and the flicker noise are equal if

$$\int_{10^{-3}}^{10^7} A \, df = \int_{10^{-3}}^{10^7} A(f_1/f) \, df. \tag{5}$$

Solving this equation, we find $f_1 \simeq 400$ kHz. For $f_1 \ll 400$ kHz, the effect of flicker noise is insignificant; for $f_1 \gg 400$ kHz, the flicker noise can be important.

For transistors, $f_1 < 1$ kHz; hence flicker noise is insignificant. For MOSFET circuits operating at zero drain current, $f_1 = 0$. But for MOSFET circuits operating at saturation, f_1 can be larger than 400 kHz; so the $1/f$ noise can give a significant contribution to the device noise.

II VARIOUS VLSI CIRCUITS

We now discuss the noise in various VLSI circuits. We shall prove the following theorem: Let the output of a particular circuit element feed into a capacitance C_0, then the mean square value of the noise voltage $\overline{V^2}$ developed across C is of the order of $\overline{V^2} = kT/C_0$ (see reference [1] for details).

A The CMOS Circuit

Figure 1 shows the CMOS circuit, where Q_p is a *p*-channel device, Q_n an *n*-channel device, and C_o the output load capacitance due to the input of subsequent stages. If the output feeds into m such stages and each state has an input capacitance C_i, then $C_o = mC_i$.

If the device is *on*, its input capacitance $C_i = C_i'$ is

$$C_i = C_i' = \varepsilon \varepsilon_0 w L/d. \tag{6}$$

Here ε is the relative dielectric constant, w the device width, L the device length, and d the oxide thickness. When the device is *off*, the oxide capacitance C_i' and the space charge capacitance $C_{ss} = \varepsilon \varepsilon_0 w L/d_s$ are in series; here d_s is the height of the space charge region. Hence

$$C_i = C_i'' = C_i' C_{ss}/C_i' + C_{ss}). \tag{7}$$

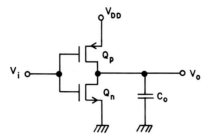

Fig. 1. CMOS circuit.

It is now easily seen that, when the output of a CMOS circuit feeds into m circuits in parallel, then

$$C_o = m(C_i' + C_i''), \tag{8}$$

since always one of the circuits is *off*. It is also easily shown that $\overline{V_o^2} = kT/C_o$ in each case. For example, if the p-type device Q_p is *on* but operating at zero drain bias, then

$$\overline{i_o^2} = 4kTg_{dpo}\,df, \quad \overline{V_o^2} = \int_0^\infty \frac{\overline{i_o^2}}{(g_{dpo}^2 + \omega^2 C_o^2)} = \frac{kT}{C_o}, \tag{9}$$

where $g_{dpo} = (\mu w C_{ox}/L)(V_g - V_{Tp})$ is the drain conductance at zero drain bias. The same holds when the n-type device Q_n is *on*.

As an example, take $w = 2\ \mu m$, $L = 0.5\ \mu m$, $d = 400$ Å, and $d_s = 800$ Å. Then $C_i' = 0.266 \times 10^{-14}$ F, $C_{ss} = 0.133 \times 10^{-14}$ F, and $C_o = 0.40 \times 10^{-14}$ F. If $m = 3$, then $C_o = 1.20 \times 10^{-14}$ F and $(\overline{V_o^2})^{1/2} = 0.62$ mV.

The NMOS Circuit with Depletion Mode Load   B

Figure 2 shows the NMOS circuit. For a *one* output, Q_1 is *off* and $V_o \simeq V_{DD}$; so Q_2 has zero drain bias ($V_{DD} - V_o \simeq 0$). Hence

$$\overline{i_{d1}^2} = 0, \quad \overline{i_{d2}^2} = 4kTg_{d2}\,df, \quad \overline{V_o^2} = kT/C_o, \tag{9}$$

where $C_o = mC_i'$ and C_i' is the input capacitance of Q_1 when Q_1 is *on*. For a *zero* output, Q_1 is operating at low drain bias ($V_o \simeq 0$) and Q_2 is saturated; thus,

$$\overline{V_o^2} = \int_0^\infty \frac{\overline{i_o^2}}{g_{d1}^2 + \omega^2 C_o^2} = \frac{kT}{C_o}\left(1 + \frac{\gamma g_{d2}}{g_{d1}}\right) \simeq \frac{kT}{C_o}. \tag{11}$$

Since the devices are usually designed so that $g_{d2} \ll g_{d1}$ and $C_o = mC_i''$, the device Q_1 looks into a circuit that is *off*.

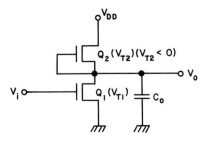

Fig. 2. NMOS circuit with *n*-channel depletion mode load.

As an example, let $w = 2\mu m$, $L = 0.5\ \mu m$, $d = 400$ Å, $d_s = 800$ Å, $C_i' = 0.266 \times 10^{-14}$ F, and $C_i'' = 0.089 \times 10^{-14}$ F, so that for $m = 3$, $C_o = 3C_i'$ when Q_1 is *off* and $C_o = 3C_i''$ when the device is *on*. Hence $(\overline{V^2})^{1/2} = 0.72$ mV and $(\overline{V^2})^{1/2} = 1.25$ mV, respectively.

C The NMOS Circuit with *n*-Channel Enhancement Mode Load

Figure 3 shows the NMOS circuit. For a *one* output, Q_1 is *off* and $V_o = V_{DD} - V_{T1}$; for a *zero* output, Q is operating at low drain bias $V_o \simeq 0$ and Q_2 is saturated.

This is approximately equal to the previous case; hence the values of $(\overline{V_o^2})^{1/2}$ are comparable [1].

D CTL Bipolar Circuit

Figure 4 shows the circuit that is the bipolar equivalent of the CMOS circuit. The input capacitance C_i of the device is $C_i = \varepsilon \varepsilon_o wL/d$, where w and L are the width and the length of the emitter contacts, respectively, and d is the width of the emitter space charge region. If the emitter donor

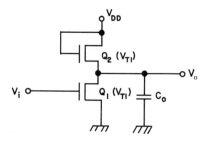

Fig. 3. NMOS circuit with *n*-channel enhancement mode load.

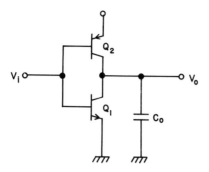

Fig. 4. CTL bipolar transistor circuit.

concentration N_d is large in comparison with the base acceptor concentration N_a, and V_{dif} is the diffusion potential of the emitter junction, then

$$d = \left[\frac{2\varepsilon\varepsilon_o(V_{\text{dif}} - V_{EB})}{qN_a}\right]^{1/2}, \quad (12)$$

where $V_{EB} = 0.70$ V when the device is *on* ($C_i = C_i'$) and $V_{EB} \simeq 0.20$ V [equal to the collector saturation voltage $V_{CE\text{sat}}$ ($C_i = C_i''$)] when the device is *off*. We then find

$$C_o = m(C_i' + C_i''), \quad \overline{V_o^2} = kT/C_o. \quad (13)$$

As an example let $V_{\text{dif}} = 0.90$ V, w = 2 µm, and $L = 1$ µm; then $d = 0.960 \times 10^{-1}$ µm when the device is *off*, and $d = 0.515 \times 10^{-1}$ µm when the device is *on*. Hence $C_i' = 0.41 \times 10^{-14}$ F and $C_i'' = 0.21 \times 10^{-4}$ F, so that for $m = 3$, $C_o = 1.89 \times 10^{-14}$ F and $(\overline{V^2})^{1/2} = 0.47$ mV.

The I²L Bipolar Circuit E

Figure 5 shows the circuit. Q_1 and Q_3 serve as current generators and Q_2 and Q_4 as high-speed transistors. The latter are usually provided with 4 or 5 separate collectors feeding into subsequent stages; we have drawn the circuit for only one of them. It goes without saying that the device has to be somewhat larger than in the CTL circuit to accommodate all the collectors; that is, the product wL must be larger. There are now two cases to be considered: *in* is a zero and *in* is a one.

(a) When *in* is a zero, Q_2 is off, Q_4 is on, and the base current I_B of Q_4 is equal to the current I_1 of Q_3. Hence,

$$\overline{i_o^2} = 4qI_1 \, df = 4kTg_1 \, df, \quad g_1 = qI_1/kT. \quad (14)$$

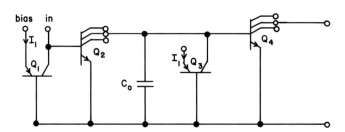

Fig. 5. I²L bipolar transistor circuit; Q_1 and Q_3 are low-frequency transistors (current generators) and Q_2 and Q_4 are high-frequency transistors with three collectors.

It is then easily seen that

$$\overline{V_o^2} = kT/C_i'. \tag{14a}$$

(b) When *in* is a *one*, the current I_1 of Q_1 flows into the base of Q_2; hence Q_2 is *on*. Since the forward current $I_f = \beta I_1$ of Q_2 is larger than I_1 if $\beta > 1$, a reverse current $I_r = \beta I_1$ must flow to the collector to make the net collector current zero. The device Q_3 contributes I to I_r. As a consequence $V_{C1} = V_{CEsat}$ and Q_4 is *off*.

We must now take into account that both the base and the collector current have noise. Since I_1 and I_B fluctuate independently, $\overline{i_1^2} = 4kTg_1\,df$, where $g_1 = qI_1/kT$. Since the current I of Q_3 flows into the collector, and this collector is saturated ($I_r = I_f = \beta I_1$), the output noise of Q_2 and Q_3 is

$$\overline{V_o^2} = \frac{kT}{C_i''}\left[1 + \frac{1}{2\beta} + \frac{C_i''/C_i'}{1 + C_i''/(C_i'\beta)}\right] \simeq kT\frac{(C_i' + C_i'')}{C_i'C_i''}. \tag{15a}$$

As an example, we take $wL = 10\ \mu m^2$. $V_{dif} = 0.90$ V and $V_{EB} = 0.70$ V when the device is *on*, and $V_{EB} = 0.20$ V $= V_{CEsat}$ when the device is *off*. Then $C_i' = 2.05 \times 10^{-14}$ F and $C_i'' = 1.10 \times 10^{-14}$ F. In the first case $(\overline{V_o^2})^{1/2} = 0.45$ mV, and in the second case $(\overline{V_o^2})^{1/2} = 0.77$ mV if $\beta = 6$.

F   ECL Circuit

Figure 6 shows the ECL circuit. It is easily seen that when Q_3 is *off*, then $\overline{i_1^2} = 4kTdf/R_c$; so $\overline{V_o^2} = kT/C_i$, where C_i is the input capacitance into which V_1 looks. When Q_3 is *on*, $\overline{i_1^2} \simeq 4kTdg(1/R_c + \tfrac{1}{2}g_m)$, where $g_m = qI_c/kT$ since I_c has full shot noise; so $\overline{V_o^2} = (kT/C_i)(1 + \tfrac{1}{2}g_mR_c)$, which is somewhat larger. It should be borne in mind that the C_i of the two cases may be different.

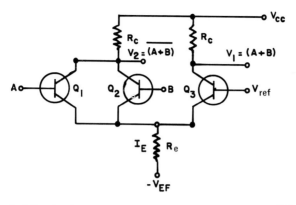

Fig. 6. ECL circuit showing the logic operation $(A + B)$ and $(\overline{A + B})$.

THRESHOLD VOLTAGES III

Since the noise presents a truly random signal, the instantaneous noise voltage V should have a normal distribution, or

$$P(V) = (2\pi \overline{V^2})^{-1/2} \exp[-\tfrac{1}{2}(V^2/\overline{V^2})]. \tag{16}$$

The probability p that the voltage V is positive and has a value $> V_1 \gg (\overline{V^2})^{1/2}$ is obtained by integrating Eq. (16). If $t = V_1/(\overline{V^2})^{1/2} > 3$, we have in good approximation

$$p = (2\pi)^{-1/2} \int_t^\infty \exp(-\tfrac{1}{2}u^2)\, dv \simeq \frac{\exp(-t^2/2)}{(2\pi)^{1/2} t}. \tag{16a}$$

We must now require that no circuit in the computer will switch during 1000 h. If there are N circuits in the computer and the computer operates at a pulse frequency f_p (e.g., 20 MHz), then the computer gives one error per 1000 h if

$$N = \frac{1}{p f_p \times 3.6 \times 10^6}. \tag{17}$$

A threshold $V_1 = 10(\overline{V^2})^{1/2}$ seems sufficient at present. This means, since $(\overline{V^2})^{1/2} \simeq 1$ mV, that the threshold voltage of the circuit must be at least 10 mV.

Since $\overline{V^2}$ is inversely proportional to the capacitance C_o, and this capacitance decreases when the device dimensions are scaled down, the minimum threshold for noise requirements goes up. But the actual threshold realized can also depend on the scaling; either it can be independent of the scaling down or it can decrease with scaling down. In either case the noise requirements and the actual thresholds achieved in the scaling are on a collision course. The limit has not been achieved as yet but may be reached by scaling down much further.

IV CROSSTALK

Crosstalk is defined as the undesirable coupling of energy between signal paths. In VLSI circuits, cross talk appears to come from two main sources:

(a) Capacitive or inductive coupling between lines and elements
(b) Common impedance coupling in the ground and battery lines

In these circuits the interconnection paths can be extremely dense, and cross talk can be a significant problem. (See reference [2] for details.)

V ALPHA-PARTICLE- AND COSMIC-RAY-INDUCED SOFT ERRORS IN VLSI CIRCUITS[†]

Soft errors induced by incident alpha particles are one of the most serious problems in VLSI development. Alpha-particle-induced soft errors have been observed in dynamic MOS RAMs and CCDs, in static MOS RAMs, and in bipolar RAMs. The failure mechanism is due to the collection of charge carriers produced in Si by low levels of alpha-particle radiation emitted from device packaging materials. The carriers are collected by nearby depletion regions in 25–50 ns and can cause a change of state of a memory cell. The effect is large enough to limit the mean time of failures to a few days or weeks for a 64K memory unless a design is chosen that minimizes the effects. Cosmic-ray bombardment has similar results, particularly at airplane altitudes or in space.

The resulting errors are called *soft* because there is no physical damage to the cell; only its logical contents have been changed. Soft errors are defined as random, nonrecurring single-bit failures in memory devices. The errors are not permanent; that is, no physical defects are associated with the failed bit. In fact, a bit showing a soft error is completely recovered by the following cycle with no greater change of showing an error than any other bit in the device.

† See reference [1] for a list of references on this topic.

A convenient way of expressing the magnitude of soft errors is the number of excess carriers involved. For alpha particles and cosmic rays they are typically 10^4–10^6 excess carriers, whereas the noise sources of Sections II and III are at the lower end of this range. For that reason, alpha radiation and cosmic rays are the main source of soft errors on present-day VLSI circuits.

REFERENCES

1. A van der Ziel and K. Amberiadis, Noise in VLSI, *in* "VLSI Circuits," vol. 7 (N. G. Einspruch, ed.), pp. 261–302. Academic Press, New York, 1984.
2. J. T. Wallmark, *IEEE Trans. Electron Devices* **ED-29,** 451 (1982).

Chapter 35

Limits to Performance of VLSI Circuits

R. T. BATE

Texas Instruments Incorporated
Dallas, Texas

I. Introduction	615
II. Fundamental Limits	616
III. Materials Limits	617
A. Maximum Achievable Velocity	617
B. Breakdown and Tunneling	617
C. Limiting Transit Time	617
IV. Device Limits	618
A. Barrier Lowering and Punchthrough	618
B. Tunneling through the Barrier	618
C. Power-Delay Product: The Minimum Energy Required per Logic Operation	620
V. Circuit and System Limits	622
A. Interconnections	622
B. Functional Throughput Rate	623
C. Error Correction	623
D. Fundamental System Limits	624
References	627

INTRODUCTION   I

The density and performance of integrated circuits have been increasing exponentially for many years. The primary reason for this is the exponential scaling of minimum geometries. In this chapter, we shall classify and discuss the types of limitations that can be identified. We consider only some of the limits that are thought to be fundamental, ignoring specific problems in the scaling of device technology such as contact techniques and interconnection materials. The rationale for this is simple: The nature of fundamental limits should make them immune to obsolescence, extending the useful life of this chapter. For the most part, these limits are general. However, the discussion of materials limits will be limited to semiconductor materials.

Because the emphasis is on fundamental constraints, the density and performance limits arrived at will be far beyond the present state of the art, and even far beyond the presently perceived practical limits of VLSI. They are also beyond the reach of transistor-based conventional architectures, and approaching them would require the invention of revolutionary architectures and devices.

Where approximations are made in the computation of limits, an attempt is made to err in the direction of higher performance–density to assure that the calculated limits cannot be exceeded.

A considerable body of literature exists on this subject, and a selected set of references can be found at the end of this chapter [1–6].

Various workers in this field have defined a hierarchy of limits [7,8], and we adopt this approach here, choosing the following simple scheme:

(1) fundamental limits,
(2) material limits,
(3) device limits, and
(4) Circuit–system limits.

II FUNDAMENTAL LIMITS

In the most general case, these are limits to performance, or to any other system parameter, that are expected to hold regardless of the kind of technology employed, be it VLSI, Josephson junctions, integrated optics, molecular electronics, or the human brain, and also regardless of the type of architecture. There are thought to be fundamental limits applying at all levels of the hierarchy, (i.e., materials, devices, and circuits/systems), but the very existence of such limits on computing is, at this writing, a subject of debate in the literature [9–13]. Some believe they can conceive of computers, very different from those that now exist, that could perform useful computations with essentially zero power dissipation. Since this issue is at present unresolved, we shall exclude this class of system of adopting the following definition.

The integrated systems under discussion have the following properties:

(a) They compute digitally by means of circuits composed of elementary logic gates and memory cells interconnected in architectures that may be serial, parallel, or combinations thereof.

(b) They are synchronous and are controlled by a single master clock of frequency f_c, and they are confined to a single chip.

(c) They do not employ resonant drive circuitry, and hence, all stored energy is dissipated during each clock cycle.

(d) The logic gates are not information-conserving (i.e., these gates retain no memory of their past inputs, outputs, or states, for more than one clock cycle.)

(e) The chip incorporates detection and correction circuitry that corrects single- and multiple-bit errors.

III MATERIALS LIMITS

In this case, we immediately specialize the discussion to semiconductor materials and assume that electron–hole conduction is the basis for device function [8].

A Maximum Achievable Velocity

The maximum current densities and switching speeds attainable in a semiconductor depend on the highest velocity to which a carrier can be accelerated. The only real limit is the velocity of light; but if we restrict ourselves to voltages insufficient for breakdown or interband tunneling, the maximum velocity for short times is determined by the maximum slope of the energy bands in wave-vector space. This velocity is about 1×10^8 cm/s for common semiconductor materials. For longer distances and times the velocity is limited by optical phonon emission to a saturation velocity v_s of about $1-3 \times 10^7$ cm/s, depending on the material.

B Breakdown and Tunneling

At electric fields above a critical field E_c of about 1 MV/cm, interband transitions that create electron–hole pairs begin to occur. At large distances, the mechanism for this is impact ionization, whereas for shorter distances, tunneling is responsible. If this occurs in a transistor that is turned off, the resulting leakage currents can adversely affect circuit performance.

C Limiting Transit Time

The existence of the above limits permits estimation of a minimum transit time τ of an electron between any two points differing in voltage by ΔV in terms of material parameters [8]. If d is the separation, then

$$\Delta V/d < E_c, \qquad (1)$$

or

$$d > \Delta V/E_c. \qquad (2)$$

But

$$\tau > d/v_s; \qquad (3)$$

so

$$\tau > \Delta V/(v_s E_c). \qquad (4)$$

For very short distances, less than 50–100 Å, v_s would be replaced by $v_l \sim 1 \times 10^8$ cm/s. The relationship discussed in this section is plotted in Fig. 1 as the line labeled *Intraband conduction*.

IV DEVICE LIMITS

A Barrier Lowering and Punchthrough

Nearly all known semiconductor switching devices employ a potential barrier between two electrodes at different potentials to inhibit current flow between those electrodes. This barrier is then lowered to switch on the device. In a 2-terminal device, the voltage on one terminal is increased sufficiently to lower the barrier, whereas a 3-terminal device uses a voltage on an intervening electrode to lower the barrier.

When the 2-terminal type of barrier lowering occurs in a 3-terminal device, we have *punchthrough*. The tendency for this to occur increases as the electrodes are moved closer together, and this is the primary limitation on scaling of source–drain separation and gate length in MOS/VLSI.

Various workers have studied this limitation and have predicted lower limits on gate length in the range 0.1–0.3 μm [7,14–19]. The existence of this limit implies a lower limit on transit time, because the carrier velocity cannot exceed the saturation velocity [8]. Using a gate length of 0.2 μm and a saturation velocity of 1×10^7 cm/s, we obtain a minimum transit time of 2 ps. This value is indicated as the transistor punchthrough limit in Fig. 1.

B Tunneling through the Barrier

Although punchthrough appears to be a fundamental limit on scaling of transistors, it is interesting to consider what the fundamental limit on the

35. Limits to Performance of VLSI Circuits

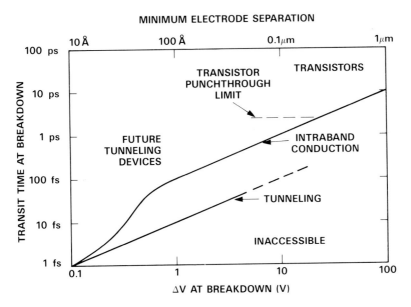

Fig. 1. Summary of materials limits for silicon. The minimum possible transit time of an electron between two electrodes differing in potential by ΔV is shown. The horizontal axis at the top of the figure indicates the minimum electrode separation to avoid breakdown or interband tunneling at this voltage. The curve labeled *intraband conduction* shows the limitation due to velocity saturation or, at small separations, the band structure limited velocity. The transistor punchthrough limit is indicated at 0.2-μm gate length, giving a limiting transit time of 2 ps. The curve labeled *tunneling* indicates the minimum transit times for interband tunneling, which could form the basis for future ultrahigh speed semiconductor devices. Minimum transit times set a lower limit on gate delay and an upper limit on system clock frequency.

ability of a potential barrier to inhibit current flow as it is made thinner would be if punchthrough could be eliminated. The stimulus for this inquiry is the advent of heterojunction technologies which permit the fabrication of very thin semiconductor barriers that cannot be lowered by an applied voltage [20].

The obvious limitation is intraband tunneling current, and 2-terminal semiconductor devices employing heterojunction barriers to hold off current should function with electrode separations of less than 0.01 μm. Calculations show that the effective tunneling carrier velocity is essentially the band structure limited velocity, $\sim 10^8$ cm/s. By the same argument used in Section III.C, this implies a minimum transit time that depends on material parameters and is essentially the same as the band structure-limited transit time. This relationship is plotted in Fig. 1 as the line labeled *Tunneling*. As one would expect, the minimum transit times are similar to those observed in superconducting Josephson junctions, because the current transport mechanisms are the same.

C Power–Delay Product: The Minimum Energy Required per Logic Operation

If a logic gate consumes an average power P_g, then the product of P_g and the duration t of a clock cycle is the energy E consumed by the gate in performing a single logic operation. (We define P_g and E to include all of the power, dissipated anywhere in the circuit, that is required to operate the gate.) The question as to whether or not a minimum amount of energy is fundamentally required to be dissipated in the performance of a logic or switching operation is a subject of continuing debate [9–13]. However, if we adopt the constraints of Section II, then we can assert that thermodynamics and quantum mechanics both place fundamental lower limits, which depend on the permissible error rate P_e on E [9,12,21–25]. The

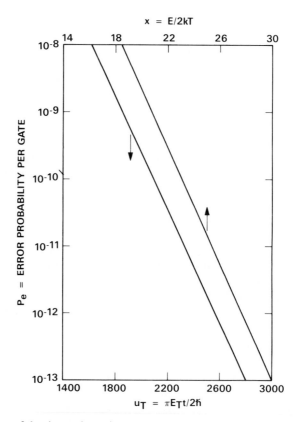

Fig. 2. Plots of the thermodynamic and quantum mechanical error probability functions defined in Table I for the range of error probabilities of concern.

35. Limits to Performance of VLSI Circuits

TABLE I

Summary of Limits

Limit	Value
Materials limits (Silicon, 300 K)	
Maximum achievable velocity	
Saturation velocity ($d > 100$ Å)	$v_s = 10^7$ cm/s
Band-structure limited velocity ($d < 100$ Å)	$v_l = 10^8$ cm/s
Breakdown–tunneling field	$E_c = 10^6$ V/cm
Limiting transit time	$\tau = \Delta V/(v_s E_c)$
Device limits	
Transistor punchthrough limit	Minimum transit time
	= minimum gate length/v_s
	= 0.2 μm/(10^7 cm/s) = 2 ps
Thermodynamic error probability	$P_e = \dfrac{1}{1 + \exp(E/2kT)}$
	$E = P_g t$ = gate power–delay product
	$kT(300$ K$) = 4.14 \times 10^{-6}$ fJ
Quantum error probability	$P_e = C/[u^{1/2} \exp(u/B)]$
	$u = (\pi^2/h)Et$
	$C = 1.735, \quad B = 105$
	$h/\pi^2 = 6.712 \times 10^{-20}$ fJ·s
System limits	
Chip power efficiency (CPE)	CPE = FTR/P
	FTR = $N_g f_c$ = (number of gates)
	(clock frequency)
	P = total chip power
Thermodynamic limit	FTR/$P < \dfrac{1}{2kT \ln[(\text{FTR}/R_e) - 1]}$
Synchronicity limit	(FTR/P) $\ln[(\text{FTR}/R_e) - 1] < 1.21 \times 10^{20}$
	gate·Hz/W (300 K)
	$A < (c/f_c)^2 = 9 \times 10^{20}/f_c^2$ cm^2
	A = maximum chip or subsystem area

minimal thermodynamic error probability is a universal function of E/kT, and the quantum error probability is a function of Et/h. Equations from reference [15] are displayed in Table I, and Fig. 2 shows these functions plotted over the relevant range of parameters.

We can use these plots to construct conventional power–delay plots, which show how absolute minimum power trades off with absolute minimum delay at constant error probability. This is done in Fig. 3. These limits should be independent of device technology within the constraints of Section II.

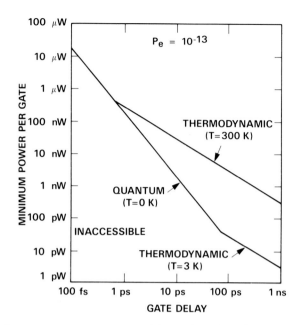

Fig. 3. Minimum gate power versus gate delay as determined from the thermodynamic and quantum limits for an error probability per gate of 10^{-13}. At room temperature, the quantum limits, which are essentially independent of temperature, become important for gate delays $< \sim 1$ ps, corresponding to a maximum clock frequency of 1 THz.

V CIRCUIT AND SYSTEM LIMITS

A Interconnections

Although the resistance and capacitance of contacts and interconnecting lines are currently a practical limitation on scaling of minimum geometries, we shall assume that these problems are soluble. The fundamental interconnection limitation, which would be encountered, for example, if optical data links were used on a chip, is the finite velocity of light. This places a fundamental upper limit on the length of links that can be used without encountering synchronization errors. Since the master clock must communicate with gates all over the chip, this implies an upper limit on the linear dimensions of a chip.

Quantitatively, the limitation can be stated as

$$A < (c/f_c)^2, \tag{5}$$

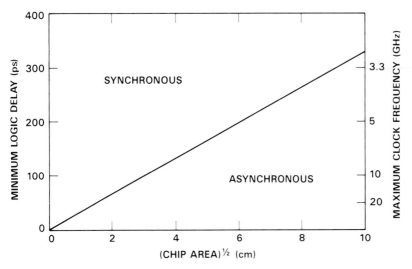

Fig. 4. The speed-of-light limitation on clock frequency to avoid timing errors in a synchronous system for a given chip area (synchronicity limitation). For self-timing asynchronous systems-on-a-chip composed of subsystems with local clocks, the area would be the maximum allowable for a subsystem.

where A is the area of the chip, f_c the clock frequency, and c the velocity of light. A plot of this relationship is shown in Fig. 4. Only asynchronous or self-timed [24] chips can have areas or clock frequencies greater than this limit. In the case of a self-timed chip, composed of subsystems controlled by local clocks, A can be interpreted as the area occupied by one of those subsystems.

B Functional Throughput Rate

A convenient measure of the processing capacity of a chip is the *functional throughput rate* (FTR), defined as $N_g \times f_c$, where N_g is the total number of gates. We can also define a specific FTR as FTR/A. The FTR, in units of gate·Hz, is essentially a measure of the maximum number of logic operations per second of which the chip is capable, while FTR/A, in gate·Hz/cm², normalizes for the chip area.

C Error Correction

As we have seen in Section IV.C, the fundamental limits on device performance involve a tradeoff of device power with error rate. Thus, it is highly desirable to take advantage of error correction techniques to minimize system power. The estimation of ultimate lower limits on system power must take this possibility into account.

As we reduce the power per gate, the error probability per logic operation will increase, and multiple-bit errors will become more probable. Although it is theoretically possible to detect and correct multiple-bit errors of any order by including a sufficient number of check bits in a word and adding the necessary correction circuitry, the overhead in chip area and power and the delay encountered in the correction circuitry must eventually degrade performance. Thus, there must be some error rate at which it becomes disadvantageous to reduce device power further and still maintain the frequency of occurrence of uncorrectable errors below the desired limit. For purposes of this chapter, we shall arbitrarily assume that this rate is given by $R_e = f_c$. Under this condition, a random single-bit error will occur, on the average, during every clock cycle. The probability of multiple-bit errors depends on the architecture, but errors in all of the bits in a word could occur under these conditions over the periods of time during which reliable system operation is required.

D Fundamental System Limits

We can use the fundamental device limits discussed in Section IV.C to derive limits on system performance. To begin, we note from Fig. 3 that for gate delays greater than 10 ps, we can neglect the quantum limits at room temperature. From Table I, the thermodynamic error probability is

$$P_e > 1/[1+ \exp(E/2kT)] \approx \exp(-E/2kT) \qquad P_e \ll 1. \tag{6}$$

Therefore,

$$E > 2kT \ln(1/P_e). \tag{7}$$

The total chip power is given by

$$P = N_g f_c E = \text{FTR} \times E, \tag{8}$$

and the chip error rate is

$$R_e = N_g f_c P_e = \text{FTR} \times P_e. \tag{9}$$

Therefore, we have

$$\text{FTR}/P < 1/[2kT \ln(\text{FTR}/R_e)]. \tag{10}$$

Thus, the maximum possible FTR is proportional to P/T times a slowly varying logarithmic function of FTR/(the permissible error rate).

35. Limits to Performance of VLSI Circuits

If we require, as in Section V.C, that $R_e < f_c$, then we have

$$\text{FTR}/P < 1/[2kT \ln(N_g)]. \tag{11}$$

To make this an absolute limit, we must place an upper limit on the total number of gates. We do this by considering the synchronicity condition and the ability to remove heat [26,27]. In general, the total heat that can be removed per unit chip area will be set by the method of cooling. If we denote this heat flux by F, then

$$F = P/A > 2kTf_c/AN_g \ln(N_g). \tag{12}$$

But the requirement for synchronicity is

$$A < (c/f_c^2). \tag{5}$$

These two inequalities can be satisfied only if

$$F = P/A > (2kT/c^2) \times f_c^3 \times N_g \ln(N_g). \tag{13}$$

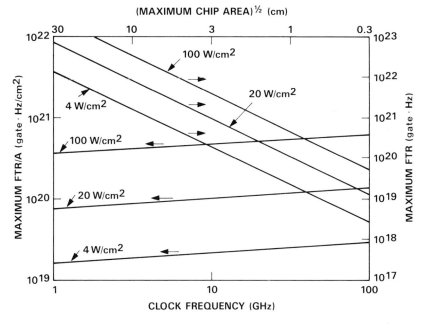

Fig. 5. The thermodynamically limited functional throughput rate as a function of clock frequency with heat dissipation per unit chip area as a parameter as determined from Eq. (17). Values of heat flux per unit area >4 W/cm² could not be achieved with forced air cooling. The maximum chip area shown on the top horizontal axis is determined from the synchronicity limitation, and it is assumed that the system error rate does not exceed the clock frequency.

The inversion of this inequality for $N_g \gg 1$ is

$$N_g < \frac{Y}{\ln[Y/\ln(Y)]}, \tag{14}$$

where

$$Y = c^2 F/(2kTf_c^3) = 1.1 \times 10^{41} \times F/f_c^3 \quad \text{at 300 K.} \tag{15}$$

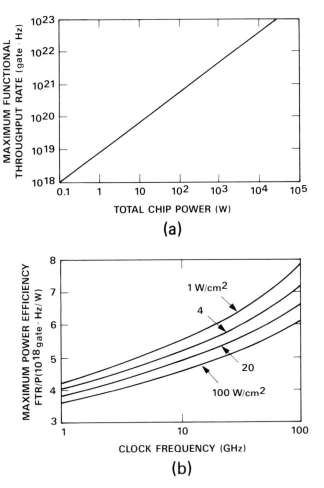

Fig. 6. (a) The thermodynamically limited chip functional throughput rate as a function of total chip power as determined from Eq. (16). This curve has a finite width which cannot be distinguished on this plot because of a weak dependence on heat flux per unit area and clock frequency [see part b]. (b) The limiting power efficiency, defined as the thermodynamically limited functional throughput rate divided by the total chip power, showing the weak dependence on clock frequency and heat flux per unit area. The chip error rate is assumed not to exceed the clock frequency.

Substituting this limit on N_g into Eq. (11), we find that

$$\text{FTR}/P < \frac{1}{2kT \ln[Y/\ln(Y/\ln Y)]}, \qquad (16)$$

with Y as defined in Eq. (15). Also,

$$\text{FTR} < \frac{1.1 \times 10^{41} \times F}{[f_c^2 \times \ln(Y/\ln Y)]}. \qquad (17)$$

Figure 5 shows plots of FTR and FTR/A versus f_c computed from Eq. (17) with F as a parameter, and Fig. 6 shows a plot of Eq. (16).

The maximum FTR for the total chip is a decreasing function of clock frequency, because the maximum allowable chip area to avoid synchronization errors decreases as the inverse square of the clock frequency. The FTR per unit chip area increases only logarithmically with clock frequency if the heat flux F per unit area is kept constant. The maximum achievable FTR and FTR/A increase approximately linearly with F.

The ratio FTR/P, which is equal to (FTR/A)/F, is a measure of the limiting power efficiency of the chip and also of the maximum performance per unit area obtainable at a given heat flux per unit area. It will increase logarithmically with clock frequency until the quantum limits are encountered at about 1 THz at room temperature, when it will be

$$[\text{FTR}/P]_{\text{max}} \approx 10^{19} \quad \text{gate} \cdot \text{Hz/W}. \qquad (18)$$

At higher frequencies, the limiting power efficiency will decrease as $1/f_c$. At lower temperatures, this maximum will occur at lower clock frequencies; but if we include the minimum power requirement of a Carnot refrigerator necessary to maintain the chip at low temperature [21,24], the value in Eq. (18) is unchanged. Since it will increase above room temperature, we must conclude that the quantum/thermodynamically limited power efficiency attainable by a system conforming to the constraints of Section II is about 1×10^{19} gate · Hz/W.

REFERENCES

1. Bledsoe, W. W. (1961). *IRE Trans. on Computers* **EC-10,** 530.
2. Chiba, T. (1978). *IEEE Trans. on Computers* C-27, 319–325.
3. Freiser, M. J. and Marcus, P. M. (1969). *IEEE Trans. on Magnetics* **MAG-5,** 82–90.
4. Keyes, R. W. (1977). *Science* **195,** 1230–1235.
5. Keyes, R. W. (1977). *IEEE Trans. on Computers* **C-26,** 1017–1025.
6. Keyes, R. W. (1981). *In* "VLSI Electronics: Microstructure Science" (N. Einspruch, ed.) Volume 1, pp. 185–230.
7. Chatterjee, P. B. et al. (1983). *Proc. IEE 130,* 105.
8. Meindl, J. M. (1983). *Tech. Digest, IEEE IEDM,* pp. 8–13.
9. Bekenstein, J. D. (1981). *Phys. Rev. Letters 46,* 623–626.
10. Deutsch, D. (1982). *Phys. Rev. Letters 48,* 286–288.

11. Many Authors (1982). "Proceedings of the Conference on the Physics of Computation" *Int. J. Theor. Phys. 21,* Nos. 3/4, 6/7, 12.
12. Porod, W. et al. (1984). *Phys. Rev. Letters, 52,* 232–235.
13. Robinson, A. L. (1984). *Science 223,* 1164–1166.
14. Bakoglu, H. B. and Meindl, J. D. (1984). *IEEE ISSCC Digest,* pp. 164–165.
15. Cooper, J. A. (1981). *Proc. IEEE 69,* 226–231.
16. Hoeneisen, B. and Mead, C. A. (1972). *Solid State Electronics,* 15, 819–829.
17. Lewyn, L. L. and Meindl, J. M. (1984). *IEEE ISSCC Digest,* pp. 160–161.
18. Pfiester, J. R. et al. (1984). *IEEE ISSCC Digest,* pp. 158–159.
19. Rideout, V. L. (1984). *In* "VLSI Electronics: Microstructure Science" (N. Einspruch, ed.) Volume 7, pp. 197–260.
20. Sollner, T. C. L. G. et al. (1983). *Appl. Phys. Lett. 43* 588–590.
21. Bate, R. T. (1982). *In* "VLSI Electronics: Microstructure Science" (N. Einspruch, ed.) Volume 5, pp. 359–386.
22. Keyes, R. W., and Landauer, R. (1970). *IBM J. Res. Develop. 14,* 152–157.
23. Landauer, R. (1961). *IBM J. Res. Develop. 5,* 183–191.
24. Mead, C. A. and Conway, L. A. (1980). "Introduction to VLSI Systems" Addison-Wesley, Reading, Mass.
25. Stein, K. U. (1977). *IEEE J. Solid State Circuits SC-12,* 527–530.
26. Grondin, R. O. et al. (1984). *IEEE J. Sol. State Circuits* (To be pub.).
27. Keyes, R. W. (1975). *Proc. IEEE 63,* 740–767.

Chapter **36**

Superconducting Integrated Circuits

THOMAS Y. HSIANG
University of Rochester
Rochester, New York

I. Model of a Josephson Junction	629
A. Constituents	629
B. Josephson Equations	630
C. Damped Pendulum Analogy	630
D. Control of Josephson Junctions	630
E. Threshold Characteristics	632
II. Digital Devices	632
A. Memory Cells and Junction-Controlled Gates	632
B. Loop-Controlled Logic Gates	635
C. Direct-Coupled Logic (DCL) Gate	636
D. Quiteron	637
III. Switching Speed of Josephson Junctions	638
A. The Pair-Breaking Rate	638
B. Scattering Rate of Electrons and Phonons	639
C. Other Limitations on Switching Speed	639
IV. Other Considerations	640
A. Wiring and Inductance	640
B. Noise in Josephson Junction Devices	642
References	642

MODEL OF A JOSEPHSON JUNCTION   I

Constituents   A

The basic unit of most superconducting electronic components is a Josephson junction, composed of two superconductors separated by an insulating barrier. The physical quantity that governs the electrical parameters can be described by the use of a phase parameter θ, which is the difference in the phases of the paired-electron wave functions describing the superconductors on the two sides of the barrier.

B Josephson Equations

The current due to the paired electrons and the voltage across the junction are related to θ by the pair of *Josephson equations* [1]:

$$J = J_0 \sin \theta, \tag{1}$$

and

$$(\partial/\partial t)\theta = (2e/\hbar)V, \tag{2}$$

where J_0, e, and $\hbar$ are the maximum dc supercurrent density, magnitude of electron charge, and Planck's constant, respectively.

C Damped Pendulum Analogy

In addition, single electron tunneling would contribute a conductance term G to the junction, and the displacement current can be expressed with a capacitive term C. The resulting differential equation of the junction is

$$I = I_0 \sin \theta + GV + C(\partial/\partial t)V. \tag{3}$$

Equation (3) can be put entirely in terms of θ by use of Eq. (2):

$$I = \frac{\hbar C}{2e}\frac{\partial^2 \theta}{\partial t^2} + \frac{\hbar G}{2e}\frac{\partial \theta}{\partial t} + I_0 \sin \theta, \tag{4}$$

which is the same as the equation that governs a damped pendulum with θ representing the angular displacement of the pendulum. Using this analogy, we find the natural frequency and critical damping coefficients to be

$$\omega_J = \frac{2e}{\hbar}\frac{I_0}{G}, \tag{5}$$

and

$$\beta_J = \omega_J C/G. \tag{6}$$

The equivalent circuit is shown in Fig. 1, and the current–voltage characteristic of an underdamped junction (a typical tunnel junction made with oxide barrier) is shown in Fig. 2.

D Control of Josephson Junctions

Josephson junctions are useful because the phase can be modulated by an external magnetic field, by an external inductively coupled control

36. Superconducting Integrated Circuits

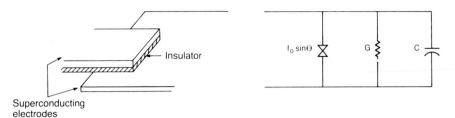

Fig. 1. Configuration of a Josephson tunnel junction and its equivalent circuit.

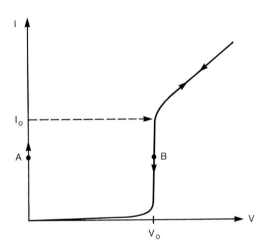

Fig. 2. Current–voltage characteristic of a hysteretic Josephson junction. Points A and B represent the switching points when the junction is used as a logic gate.

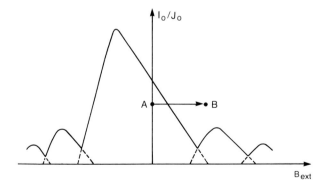

Fig. 3. Threshold characteristic of a Josephson junction backed by a ground plane. The switching of A → B represents a transition from zero voltage to finite voltage induced by external magnetic field.

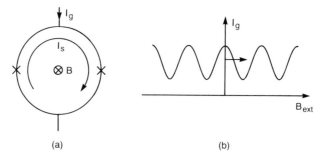

Fig. 4. Double junction configuration and threshold characteristic.

current, or by a directly injected current. The control equation is expressed as

$$\mathbf{J} = \rho(e/m)(\hbar\nabla\theta - 2e\mathbf{A}), \tag{7}$$

where $\mathbf{J}$, ρ, and $\mathbf{A}$ are the paired electron current density, paired electron density, and the vector potential, respectively.

E Threshold Characteristics

Equations (4) and (7), combined, generate a family of *control threshold characteristics*, such as those shown in Figs. 3 and 4 for single junctions and double junctions. These threshold characteristics separate the *superconducting* (zero voltage) and *normal* (finite voltage) states. A super-to-normal transition for fixed bias current (*gate current*) is indicated by an arrow in Figs. 2, 3, and 4. The asymmetry of the threshold characteristic in Fig. 3 is the result of the use of a ground plane that underlies the junction. Ground planes will be discussed in Section IV.A.

II DIGITAL DEVICES

A Memory Cells and Junction-Controlled Logic Gates

The single-junction switching property shown in Fig. 3 makes it useful as a logic gate, from which memory cells can be built. An example is a large-scale cache memory array made with double junctions [2].

Superconducting Quantum Interference Device

Two-junction switching also forms the basis for a nondestructive read–write memory cell [3], as shown in Fig. 5. The control current I_c flows through an overlay of one of the junctions that interrupts the superconducting loop, causing the reduction of I_0 in that junction, according to Fig. 3, and thus forcing the gate current I_g to flow preferentially in the other junction. This arrangement is known as a superconducting quantum interference device (SQUID). When two control lines and a third readout junction are incorporated into the SQUID, a memory cell is realized, as shown in Fig. 6, in which the simultaneous application of *write* current with either 1, 0, or *read* current accomplishes the *write* 1, *write* 0, or *read* function. The digital state is stored in the SQUID by the nondissipative nature of super-current, as shown on the right of Fig. 6.

Another feature of the double junction SQUID arrangement is that in either the 0 or the 1 state, the circulating current is such that the total magnetic flux is restricted to an integral multiple of a *flux quantum* ϕ_0, defined by fundamental constants:

$$\phi_0 = h/2e = 2.07 \times 10^{-15} \quad \text{Wb}. \tag{8}$$

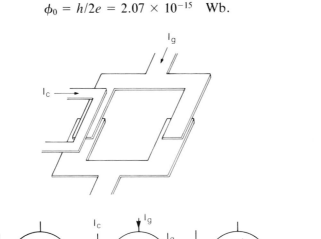

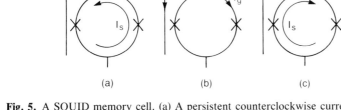

Fig. 5. A SQUID memory cell. (a) A persistent counterclockwise current I_s initially is stored in the SQUID. (b) Simultaneous application of the control current I_c and gate current I_g forces the left junction to become resistive by reducing its value of I_o. Current I_g is then directed through the left junction, reversing the magnetic flux inside the SQUID. (c) A clockwise current is now stored in the SQUID.

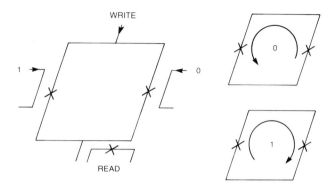

Fig. 6. Configuration of a NDRO memory cell made with a SQUID. Simultaneous application of *write* current with either *0*, *1*, or *read* accomplishes the *write 0*, *write 1*, or *read* function.

Invariably, the lowest state (flux = ϕ_0) is used, thus ensuring the uniformity of the quiescent currents of all cells in an array.

2 AND–OR Functions

The operational principle of the SQUID can be used to perform logical AND and OR functions with two or multiple inputs [4]. Shown in Fig. 7, the asymmetrical switching threshold is used for these logical functions for two control lines. The junction is initially biased below the zero-control critical current (point A). In the AND operation, the control current is so directed that a single control pulse is not enough to switch the junction into the finite voltage state (A → B). Only when both control lines are *on* does the junction switch out of the superconducting range (A → C), sending a voltage signal to the load. In the OR operation, the

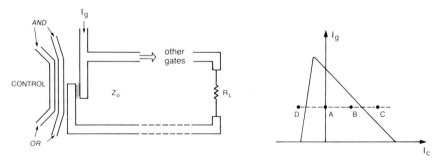

Fig. 7. Josephson junction as a logic gate for the logical AND and OR functions. For the explanation of the threshold characteristic, see text.

control current is in the reverse direction, allowing a single control to switch the junction (A → D).

This type of digital device is restricted in its usefulness by the low inductance of the junction which requires a large control current. They are used in memory devices, in wide-line circuits, and in drive gates.

B. Loop-Controlled Logic Gate

1. Three Junction Interferometer

By increasing the overlap between the control line and the SQUID loop and by controlling the SQUID with fields inside the loop rather than inside the junction, use of lower control current is possible. One such example is the three-junction interferometer [5] shown in Fig. 8. The three junctions have an intrinsic critical current ratio of 1:2:1, and the load resistance is chosen to maximize the output current ($R_L < V_g/4I_0$, where V_g is the order of millivolts). Extra resistors R_d are placed in the loop to damp out intrinsic oscillations.

2. Current Injection Logic Devices

SQUID gates can also be controlled via direct current injection, generating the family of current injection logic (CIL) devices. An example [6] is shown in Fig. 9. This device has two control currents I_A and I_B, in which I_A is asymmetrically fed to maximize the gain of the circuit.

Typically, the parameters of this circuit are chosen such that

$$L_1 I_{01} = L_2 I_{02} \quad \text{and} \quad (L_1 + L_2)I_{02} = \phi_0. \tag{9}$$

Since the entire loop (except for the junctions) is made with superconductor, I_B divides into the junctions according to the ratio of the inductances.

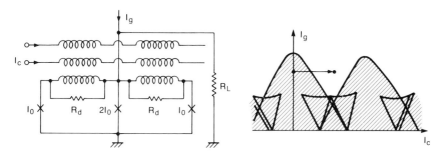

Fig. 8. Configuration and threshold characteristic of a three-junction gate.

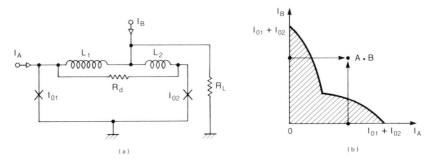

Fig. 9. Configuration and threshold characteristic of a current injection logic (CIL) gate.

Therefore, acting alone, I_B must reach $(I_{01} + I_{02})$ to drive the circuit into the state of finite voltage. The second condition in Eq. (9), restricting the circulating current in the loop, ensures that I_A alone is subject to the same constraint. However, when I_A and I_B are simultaneously supplied, a reduced total current is required to switch the gate, as shown in Fig. 9b. A minimization of the control current is achieved by making $I_{01}/I_{02} = 3$. This gate thus can perform the logical A · B function for the logical states represented by I_A and I_B. Other logical functions are also possible with variations on this circuit idea.

This logic family is simple to design at the cost of the poor isolation between input and output.

C Direct-Coupled Logic (DCL) Gate

A variation on the CIL logic family uses the concept of a resistive SQUID that consists of the multijunction SQUID interrupted by resistive components. Such gates, sometimes called JAWS (Josephson Atto–Weber switch) [7], operate on the principle that when feed current exceeds the critical current of a junction, a small bypassing resistor can preferentially draw the injected current.

An example of such a DCL gate [8] is shown in Fig. 10. Initially, the gate current I_g divides into the junctions according to the ratio of R_1 and R_2. The values of I_0 and R are chosen so that both junctions are superconducting. When the control current is injected, it acts antiparallel to the current in J1 while adding to the current in J2, driving J2 into the finite voltage state. The gate voltage is thus sent out to R_L as a logical signal.

The advantage of DCL gates lies in the fact that no inductive component is required for the operation of the gate, making even greater miniaturization possible.

36. Superconducting Integrated Circuits

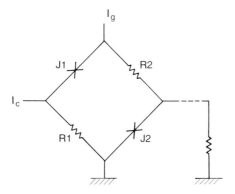

Fig. 10. Logic gate made with resistive SQUID: the direct coupled logic (DCL) gate.

D. Quiteron

A new superconducting, three-terminal switch has recently been proposed. It is named Quiteron (quasiparticle injection tunneling effect device) [9] by its inventors. It consists of three superconducting electrodes (S1, S2, and S3) separated by two tunnel barriers (J1 and J2). Junction J1 acts as an injector, and J2 acts as an acceptor, as shown in Fig. 11.

This device operates on a different principle from the Josephson device. It uses the fact that single electrons interfere with the formation of

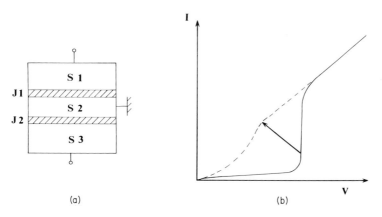

Fig. 11. (a) Configuration of a Quiteron. (b) Current–voltage characteristic of junction J2, at zero J1 injection (solid line) and maximum J1 injection (dashed line). The arrow indicates a logical switching between the two states. The actual locations of the switch points depend on the bias and on the load line (optimally, that of a nonlinear load for digital applications).

electron pairing in superconductors. Thus, if excess electrons are injected into one side of a tunnel junction, shown in Fig. 2, the values of I_0 and V_0 are both suppressed.

Junction J1 in Fig. 11 is biased at the resistive branch of its $I-V$ characteristic, causing extra electrons to be injected into S2, thus modulating the $I-V$ characteristic of junction J2. Power gain is achieved by making S2 thin to maximize the effect of electron injection and by making J1 a low current injector and J2 a high current acceptor. For digital applications, inversion power gain is further enhanced by using a nonlinear load (another tunnel junction).

Practical circuit ideas that make use of the Quiteron have not yet appeared in the literature. However, the versatility of Quiterons, especially for digital applications, will certainly generate may practical ideas.

III SWITCHING SPEED OF JOSEPHSON JUNCTIONS

A primary advantage of superconducting technology over semiconductor technologies lies in its superior speed. Only the 0.8 μm-line, hetero III–V compound devices that operate at liquid nitrogen temperature come close to a 2.5-μm-line superconducting technology in terms of speed [10].

The physical mechanism underlying the limitation on the speed of superconducting technology is similar to the limitation for semiconductor devices, which is the relaxation of carriers. This poses two intrinsic limitations: the pair-breaking rate and the inelastic scattering rate between electrons and phonons.

A The Pair-Breaking Rate

The pair-breaking effect occurs at a frequency above which the transport properties of a superconductor are completely dominated by normal, single electrons. Put differently, above this frequency the photon energy is capable of breaking electron pairs into their constituents, single electrons. The typical magnitude of this frequency is, therefore, the electron pair condensation energy divided by Planck's constant. This poses a switching constraint of the order of 0.1 ps and is usually a minor restriction on the switching time.

Scattering Rate of Electrons and Phonons   B

When a junction is switched into the finite voltage state, single electrons are injected into the electrodes. Analogous to the case for transistors, a normal electron charge imbalance is created inside the electrode [11]. The relaxation process is via the electron–phonon interaction, similar to that in III–V semiconductor devices. For the typical electrode material of Nb alloys or Pb alloys, this switching time is about 1 ps and represents the ultimate limit of superconducting technology (whether Josephson or Quiteron).

Other Limitations on Switching Speed   C

The actual device configuration and operating conditions pose other limitations on the speed [12]. The Josephson oscillation frequency, shown in Eq. (5), restricts the rate by which the phase θ can evolve. Since a phase angle of π is the minimum requirement for a finite voltage to set in Eq. (2), Josephson junctions can be turned on no faster than the delay time τ_D given by

$$\tau_D = (\pi C \phi_0 / 2 I_0)^{1/2}. \tag{10}$$

In addition, the gate current usually cannot be biased so close to I_0 that noise and fabrication uncertainties would spontaneously trigger a gate; thus an additional rise-time delay is needed to overcome the inertia [Eq. (4)]. This time is of the order of $CV_0/(I_g + I_c)$.

The turn-on delay and the rise time can both be reduced by optimization of the device fabrication, that is, by making the linewidth small and the barrier thin.

Table I summarizes the above discussion.

TABLE I

Limitation of Switching Time of Josephson Junctions

Mechanism	Determining factor	Typical magnitude (ps)
Pair breaking	$\hbar/4kT_c$	0.1–0.3
Inelastic scattering	Material dependent	1–10[a]
Josephson oscillation	$(\pi C\phi_0/2I_0)^{1/2}$	>5[b]
Turn-on delay	$CV_0/(I_g + I_c)$	>5[b]

[a] Values are for soft and transition metal superconductors.
[b] Lower limit of value quoted for 2.5-μm technology.

IV OTHER CONSIDERATIONS

A Wiring and Inductance

A superconductor exhibits perfect diamagnetism. A magnetic field **B** applied to a superconductor is screened off from the interior of the superconductor. This property, similar to the screening of a static electric field by normal metal, allows **B** to be expressed as the gradient of a scalar potential. As a result, electric current near a superconductor creates an image current inside the conductor, similar to the image charge effect near a conductor surface.

One consequence of the image effect is the distortion of current distribution in a superconducting circuit backed by a superconducting ground plane. The ground plane is usually made with a niobium film on the circuit substrate. Asymmetrical threshold characteristics such as shown in Fig. 3 are created in this way.

Furthermore, since the image current inside the ground plane is in an opposite direction to the external current, the self-inductance of external wiring is significantly reduced. The calculation of the reduction can be done with London's theory of superconductivity [13]. The result is given below.

First, it is noted that the screening of **B** (hence current) at a superconductor has a characteristic length, the *penetration depth* λ, given by

$$\lambda = \lambda_L(\xi_0/l)^{1/2} \quad \text{(impure superconductor)},$$

or

$$\lambda = 0.65\lambda_L(\xi_0/l)^{1/3} \quad \text{(pure superconductor)}, \tag{11}$$

where λ_L, ξ_0, and l are the London penetration depth, coherence length, and the electronic mean free path, respectively. The intrinsic properties of the superconductor determine λ_L and ξ_0. For typical materials used in superconducting electronics, $\lambda_L = 39$ nm (Nb), 37 nm (Pb), and $\xi_0 = 38$ nm (Nb) and 83 nm (Pb).

For a superconducting line of cross-sectional width w and thickness b, separated by a distance d from the superconducting ground plane, as shown in Fig. 12a, the self-inductance per unit length is given by (assuming $b_1 \gg \lambda$ and $w \gg d$)

$$L = \mu_0 D/w \quad \text{(superconducting)}, \tag{12}$$

36. Superconducting Integrated Circuits

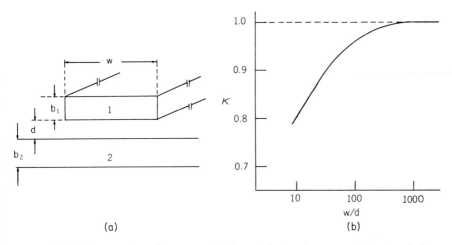

Fig. 12. (a) Configuration of a superconducting line backed by a ground plane. (b) Reduction factor of the line inductance due to the finite width to height (w/d) ratio.

where D, the effective thickness of the separation, is given by $D = d + \lambda_1 + \lambda_2$. This is to be compared with normal metal,

$$L = \mu_0 \quad \text{(normal)}. \tag{13}$$

The reduction ratio is then given by D/w.

For a superconducting loop of average radius r and wire width w, the inductance is given by

$$L = 2\pi\mu_0 r D/w \quad \text{(superconducting)}. \tag{14}$$

Compared with that for a normal ring,

$$L = \mu_0 r[\ln(16\, r/w) - 2], \quad \text{(normal)}, \tag{15}$$

once again there is a significant reduction.

For a finite ratio of w/d, Eqs. (12) and (14) must be modified by an extra factor κ, shown in Fig. 12b:

$$L = \mu_0 \kappa D/w \tag{12'}$$

and

$$L = 2\pi\mu_0 \kappa r D/w. \tag{14'}$$

The change in inductance due to the image current effect reduces significantly the cross talk in a circuit with high wiring density and also modifies the field propagation of the superconducting microstriplines used for signal transmission. These are the additional favorable factors for using superconducting electronics in high-speed digital circuits.

B Noise in Josephson Junction Devices

The sensitivity of Josephson junction devices is determined by the noise in the circuit. A simple shot-noise analysis has been developed [14] to calculate the noise property of the junction.

At frequencies much lower than the quantum mechanical limit (as usually in practice), the current noise density can be expressed as (in units of A^2/bandwidth)

$$S_I = 2eI_n \coth(eV/2kT) + 4eI_p \coth(2\ eV/2kT). \tag{16}$$

The terms I_n and I_p refer to the normal and pair electron currents. This expression reduces to the more familiar shot noise ($eV \gg kT$) or Johnson noise ($eV \ll kT$) in the respective limits:

$$S_I = 2eI_n + 4eI_p, \qquad eV \gg kT, \tag{17}$$

$$S_I = 4kT(I/V), \qquad eV \ll kT, \tag{18}$$

where $I = I_n + I_p$.

For digital devices, the tunneling shunt resistance G in Eq. (3) is small; hence the voltage limit, Eq. (18), is unimportant. In the finite-voltage state, the full expression, Eq. (16), has to be used. In the zero-voltage state, only shot noise due to the pair electrons needs to be considered, that is, $I_n = 0$ in Eq. (17):

$$S_I = 4eI_p, \qquad V = 0. \tag{19}$$

In a multicomponent Josephson circuit, the current noise not only limits the sensitivity (threshold control of the logic level) of the circuit but also restricts the biasing of gate current to be significantly less than I_0, thus limiting the speed of the device.

REFERENCES

1. B. D. Josephson, *Phys. Lett.* **1,** 251 (1962).
2. S. M. Faris, W. H. Henkels, E. A. Valsamakis, and H. H. Zappe, *IBM J. Res. Dev.* **24** (2), 143 (1980).
3. W. Anacker, *IEEE Trans. Magn.* **MAG-5,** 968 (1969).
4. D. J. Herrel, *IEEE J. Solid-State Circuits* **SC-9,** 277 (1974).
5. H. H. Zappe, *Appl. Phys. Lett.* **25,** 424 (1974).
6. T. R. Gheewala, *IEEE J. Solid-State Circuits* **SC-14,** 787 (1979).
7. T. A. Fulton, S. S. Pei, and L. N. Dunkelberger, *Appl. Phys. Lett.* **34,** 709 (1979).
8. T. Gheewala and A. Mukherjee, *IEDM Tech. Dig.,* 482 (1979).

9. S. M. Faris, S. I. Raider, W. J. Gallagher, and R. E. Drake, *IEEE Trans. Magn.* **MAG-19,** 1293 (1983).
10. H. Morkoc and P. M. Solomon, *IEEE Spectrum* **21** (2), 28 (1984).
11. K. E. Gray (ed.), "Non-Equilibrium Superconductivity, Phonons, and Kapitza Boundaries." Plenum, New York, 1981.
12. D. G. McDonald, R. L. Peterson, C. A. Hamilton, R. E. Harris, and R. L. Kautz, *IEEE Trans. Electron Devices* **ED-27,** 1945 (1980).
13. T. Van Duzer and C. W. Turner, "Principles of Superconductive Devices and Circuits," chapter 3. Elsevier, Amsterdam and New York, 1981.
14. A van der Ziel and E. R. Chenette, *Adv. Electron. Electron Phys.* **46,** 313 (1978).

Chapter 37

GaAs Digital Integrated Circuit Technology

T. GHEEWALA

GigaBit Logic, Inc.
Newbury Park, California

I.	Properties of GaAs	645
II.	GaAs MESFET Devices	647
III.	Planar GaAs Fabrication Process	649
IV.	GaAs MESFET Circuits	651
	A. Buffered FET Logic and Schottky Diode FET Logic Circuits	652
	B. Direct-Coupled Logic Circuit	653
	C. Static Random Access Memory	654
V.	Performance and Applications	654
VI.	Advanced GaAs Technologies	655
	A. High Electron Mobility Transistors	655
	B. Heterojunction Bipolar Transistors	656
	List of Symbols	656
	References	656

PROPERTIES OF GaAs I

The material properties of GaAs at 300 K are listed in Table I. The properties that lead to the significant interest in GaAs are

(a) the electron mobility μ_n, which is roughly seven times higher than that of silicon,

(b) the saturated drift velocity v_{sat}, which is about twice as high as in silicon, and

(c) the very high intrinsic bulk resistivity of the GaAs substrate, which minimizes parasitic capacitances and permits easy electrical isolation of multiple devices on a GaAs integrated circuit chip.

TABLE I

Properties of GaAs at 300 K[a]

Property	Value
Intrinsic mobility (cm/V · s)	
electrons	8500
holes	400
Saturation velocity of electrons (cm/s)	1.1×10^7
Dielectric constant	13.1
Energy gap (eV)	1.4
Barrier height (eV)	0.9
Effective mass of electrons, m^*/m_0	0.067
Minority carrier lifetime (s)	10^{-8}
Breakdown field (V/cm)	4×10^5
Intrinsic resistivity ($\Omega \cdot$ cm)	10^8
Thermal conductivity (W/cm · °C)	0.46
Linear coefficient of expansion (cm/cm · °C)	6.5×10^{-6}
Density (g/cm³)	5.32

From reference [1].

The mobility decreases as GaAs is doped with impurity atoms. For typical MESFET n-channel dopings ($N_D \simeq 10^{17}$ cm^{-3}) electron mobilities in the range of $\mu_n = 4000-5000$ cm/V · s are obtained.

At lower temperatures the mobilities in GaAs increase substantially because of reduced phonon scattering. Intrinsic electron mobilities of

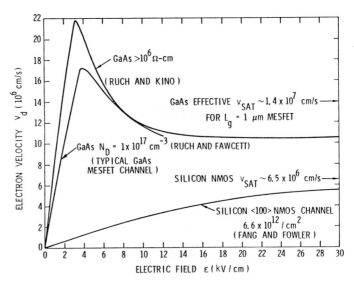

Fig. 1. Electron velocity as a function of applied electric field for silicon and GaAs.

$\mu_n \simeq 50{,}000$ cm/V · s at 77 K have been reported for the high electron mobility transistors (HEMT) which we shall discuss briefly in Section VI.

The drift velocity at which the electrons travel in the semiconductor determines the switching speed of the FET devices and is given by $v_s = \mu_n \times$ electric field E. In high electric fields, the electron mobility decreases because of increased scattering, and the drift velocity saturates as shown in Fig. 1. The electron velocity versus electric field dependence for GaAs is nonmonotonic with a peak velocity $v_s = 2.2 \times 10^7$ cm/s at around 3 kV/cm. Because of this velocity overshoot, GaAs devices switch faster than they would if the velocity rose monotonically to $v_{sat} = 1.1 \times 10^7$ cm/s [2]. This is especially true for devices with very small gate lengths ($L_g \leq 0.5$ μm).

II GaAs MESFET DEVICES

The predominant device used in GaAs integrated circuits is the metal semiconductor field effect transistor (MESFET) [1]. Some of the other promising device structures are discussed in Section VI. The GaAs MESFET *structure, energy band diagram,* and an *equivalent circuit model* are shown in Fig. 2. *Device parameters* for a typical GaAs MESFET with gate length $L_g = 1$ μm, gate width $W = 10$ μm, and pinch-off voltage $V_p = -1.0$ V are listed in Table II, and the dc I–V curve for the same device is shown in Fig. 3. The *saturation drain current* I_{DS} and the *transconductance* g_m of a MESFET for $V_{gs} > V_{ds} - V_p$ are given by Eqs. (1)–(3) below:

$$I_{DS} = K(V_{gs} - V_p)^2, \qquad (1)$$

$$g_m = 2K(V_{gs} - V_p), \qquad (2)$$

where

$$K = \varepsilon \mu_n W / 2aL_g. \qquad (3)$$

The *depletion layer thickness a* is indicated by the dotted region under the gate electrode in Fig. 2a. For large electric fields, I_{DS} and g_m are governed by saturation velocity as given by

$$I_{DS} = (\varepsilon W V_{sat}/a)(V_{gs} - V_p), \qquad (4)$$

$$g_m = \varepsilon W V_{sat}/a. \qquad (5)$$

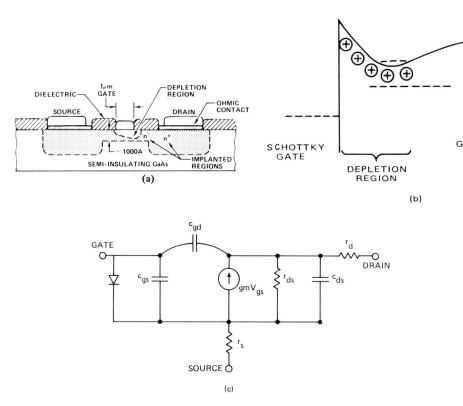

Fig. 2. (a) GaAs metal semiconductor field effect transistor (MESFET) device structure. (b) Energy band diagram. (c) Equivalent circuit model.

TABLE II

Typical Device Parameters for GaAs MESFETs[a]
1 μm and gate width W = 10 μm

Parameter		Value
Transconductance (saturated region)	g_m	1.2 mS
Gate-to-source capacitance	C_{gs}	12 fF
Gate-to-drain capacitance	C_{gd}	4 fF
Drain-to-source capacitance	C_{ds}	2 fF
Drain-to-source impedance (saturated region)	r_{ds}	10 kΩ
Drain and source series resistance	r_d, r_s	100 Ω
Schottky diode saturated current	I_s	1.3×10^{-15} A/μm²
Diode ideality factor	n	1.25

[a] Gate length $L_g = 1$ μm; gate width $W = 10$ μm.

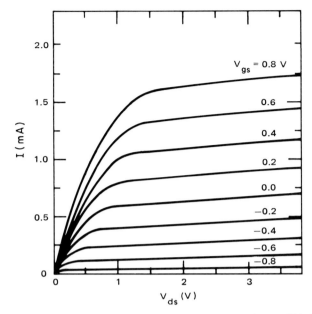

Fig. 3. GaAs MESFET I–V curve for gate length $L_g = 1$ μm and gate width $W = 10$ μm.

A figure of merit for the intrinsic speed of FETs is the *current gain-bandwidth product* f_τ given by

$$f_\tau = g_m/2\pi C_{gs}. \tag{6}$$

From Eq. (6) and the device parameters listed in Table II, we obtain $f_\tau = 18.6$ GHz for a GaAs MESFET with 1 μm gate length, which is about four times higher than that for a 1 μm gate length silicon MOSFET transistor.

The pinch-off voltage V_p is the value of V_{gs} at which the depletion layer thickness equals the thickness of the channel region a and is governed by the depletion layer thickness a and the doping level N_D. A MESFET is said to be a *depletion mode* (or normally-*on*) *device* if V_p is negative and an *enhancement mode* (or normally-*off*) *device* if V_p is positive.

III PLANAR GaAs FABRICATION PROCESS

The state-of-art GaAs integrated circuits are usually fabricated on 3-in. diameter liquid encapsulated Czochralski (LEC) wafers [2,3]. The active device regions are formed by ion implantation of dopant impurities into the semi-insulating substrates. Isolating regions are sometimes also formed by ion implantation of protons, or boron or oxygen ions, which create high-resistivity damaged regions. Typically, two implantation steps are used, a high-resistivity *n*-channel layer ($N_D \simeq 10^{17}$ cm^{-3}) and a low-resistivity n^+ layer ($N_D \simeq 10^{18}$ cm^{-3}) to form source and drain contacts. Silicon, selenium, or sulphur can be used as the donor impurity atoms. The Schottky gate electrode is formed by the deposition of titanium–platinum–gold alloys. The ohmic contact between the doped regions and the metal interconnects is achieved by the deposition of a thin layer of gold–germanium–nickel alloy. Plasma-deposited silicon nitride and silicon dioxide layers are used as insulators. Circuit elements are defined using high-resolution photolithography techniques followed by dry etching and enhanced lift-off.

The *cross-sectional view* of a GaAs MESFET integrated circuit is shown in Fig. 4, and the *sequence of processing steps* is shown in Fig. 5. It is a relatively simple process, requiring only six to eight mask levels.

Numerous enhancements to the basic GaAs IC process are being investigated to fabricate *self-aligned enhancement-mode ICs*. One generally accepted process enhancement is based on the use of a *refractory gate electrode* that can withstand the post-implant annealing temperature (850°C) [4]. In this case, as illustrated in Fig. 6, the source and drain n^+ regions are implanted and annealed for activation after the gate electrode is deposited, with the gate electrode acting as a mask to prevent n^+ implantation in the channel region beneath the gate electrode. Titanium,

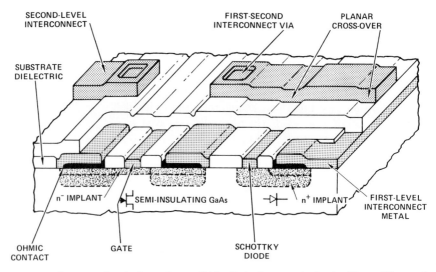

Fig. 4. Cross-sectional view of monolithic GaAs integrated circuit. (From Eden and Welch [2].)

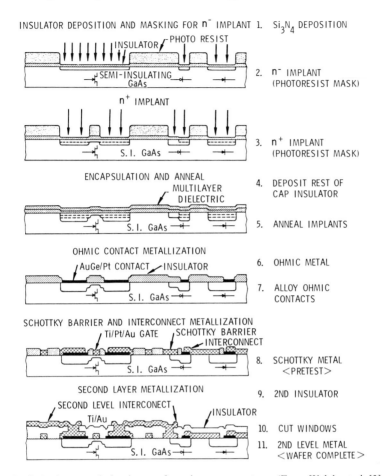

Fig. 5. GaAs integrated circuit manufacturing process steps. (From Welch et al. [3].)

tungsten, and molybdenum alloys as well as silicides have been used as the refractory gate materials.

IV GaAs MESFET CIRCUITS

The GaAs MESFET circuits are similar to their silicon MOSFET counterparts with the following main differences arising from the unique characteristics of GaAs MESFET devices:

(i) The maximum gate to source input voltage V_{gs} is limited to ≤ 0.8 V because of the presence of a forward biased Schottky diode between the gate and source electrodes.

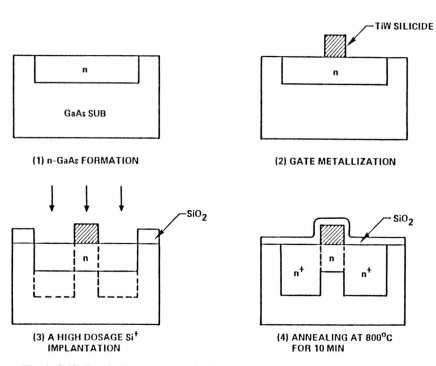

Fig. 6. Self-aligned enhancement-mode GaAs MESFET process steps. (From Yokoyama et al. [4].)

(ii) Complementary circuits are seldom used in GaAs technology because of the large difference between the electron and hole mobilities.

(iii) Additional power supply voltages and level shifting of signal voltages are necessary in GaAs circuits based on depletion-mode (normally-*on*) devices only.

A Buffered FET Logic and Schottky Diode FET Logic Circuits

Some examples of GaAs logic circuits are illustrated in Fig. 7. The circuits in Figs. 7a,b are known as buffered FET logic (BFL) and Schottky diode FET logic (SDFL), respectively, and they are characterized by the use of level shifting diodes to make the signal voltages compatible with the negative pinch-off voltages of normally-*on* devices ($V_p = -1.0$ V). The SDFL circuits also take advantage of the availability of ultrahigh-speed Schottky diodes on GaAs IC chips to perform the logic OR function. The BFL circuits use a depletion load device and a series and parallel combination of MESFETs to obtain the necessary logic function. In comparison with SDFL, the BFL circuits are characterized by higher

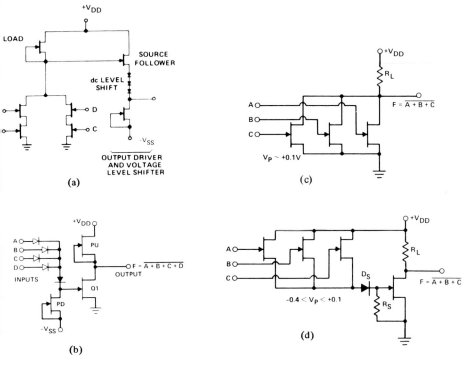

Fig. 7. Popular GaAs MESFET logic circuits.[5] (a) Buffered FET logic (BFL). (b) Schottky diode FET logic (SDFL). (c) Direct-coupled logic (DCL). (d) Quasi-normally-*off* logic. (From Eden [5].)

speeds, larger area and power, and larger fan-out driving capabilities. The SDFL circuits consume less area and power and, therefore, are more suitable for high density applications. The typical voltage swing in depletion-mode logic circuits is about 2 V, and a typical ratio of the pull-up to pull-down device widths is 0.6.

Direct-Coupled Logic Circuit B

The direct-coupled logic (DCL) circuit shown in Fig. 7c requires enhancement-mode devices with positive pinch-off voltages ($V_p = 0.1$); as a result they do not require level shifting diodes. A depletion-mode load device can be used in place of the load resistor R_L shown in Fig. 7c. The DCL circuits are characterized by the need for fewer power supplies, small area, small voltage swings ($\Delta V \simeq 0.8$ V), and low power. However, because of the smaller voltage swings, the DCL circuits demand tighter control of the pinch-off voltages than do BFL and SDFL circuits. For example, in LSI applications the acceptable standard deviation of pinch-off voltage V_p is about 25 mV for DCL circuits as opposed to about 50 mV for BFL and SDFL circuits. The quasi-normally-*off* gate shown in Fig. 7d is an in-between approach in which the pinch-off voltage can fall in either the enhancement- or the depletion-mode ranges (-0.4 V $< V_p < 0.1$ V), and it can tolerate larger variation in pinch-off voltages than can DCL.

C Static Random Access Memory

The GaAs IC technology is also extremely well-suited for ultrahigh-speed static random access memory (SRAM). The memory cell used in these applications is commonly a six transistor cell, similar to the ones used in silicon MOS RAMs.

V PERFORMANCE AND APPLICATIONS

The typical performance of some of the experimental circuits fabricated with 1-μm GaAs MESFET technology are highlighted below.

Ring oscillators	~50 ps delay at 1 mW/gate
Frequency dividers	~4 GHz
3-input NOR gate	~150 ps/gate
8 × 8 bit parallel multiplier	~6 ns multiply time (0.3 W power)
16 × 16 bit parallel multiplier	~10.5 ns multiply time (1 W power)
4K-bit SRAM	~3 ns access time (0.7 W power)

Applications of GaAs ICs are foreseen in the areas of communications, instrumentation, computers, and military. In communications the GaAs ICs are required for fiberoptic and satellite communications to provide low-noise signal detection and amplification as well as for gigabit rate data serializer or deserializers and processors. The instrumentation applications would include gigabit rate analog-to-digital and digital-to-analog converters, up to 5-GHz pulse and pseudo-random-bit pattern generators, as well as ultrahigh-speed shift registers, multiplexers, and demultiplexers. High-speed gate arrays and SRAMs will find applications in future very high speed computers. In addition to high speed and low power, the GaAs ICs offer very high radiation hardness which makes them attractive for some military applications.

ADVANCED GaAs TECHNOLOGIES VI

High Electron Mobility Transistors A

The high electron mobility transistor (HEMT) derives its performance from the extremely high electron mobility in undoped GaAs, especially at low temperatures. At liquid nitrogen temperatures (77 K) electron mobilities as high as 50,000 cm/V · s have been measured in undoped layers of GaAs. The salient feature of the HEMT structure (Fig. 8a) is the very thin undoped GaAs layer topped by a wide energy gap (E_g = 1.7 eV), n-AlGaAs donor level. Both these layers are grown by MBE techniques. The rest of the HEMT structure is similar to the MESFET. Isolation can be provided by either proton or boron implantation between adjacent devices or by mesa etching techniques. The active channel between the source and the drain is formed by a very thin layer of electrons in the undoped GaAs layer, contributed there by the n-AlGaAs layer. Ring oscillator delays of about 10 ps/gate and frequency divider operation at 5.6 GHz has been obtained at 77 K in HEMT circuits at 1-μm gate lengths. Frequency divider speed has been also improved to 8.9 GHz at 77 K by reducing the gate length to 0.5 μm.

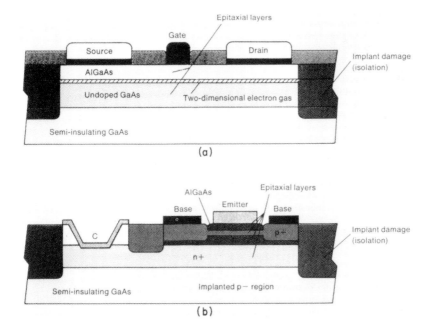

Fig. 8. (a) High electron mobility transistor. (b) Heterojunction bipolar transistor.

B Heterojunction Bipolar Transistors

The heterojunction bipolar transistor (HBT) is a very high-gain bipolar device structure in GaAs [5]. Its key feature (Fig. 8b) is the use of higher energy gap n-AlGaAs emitter ($E_g = 1.7$ eV) and GaAs base ($E_g = 1.4$ eV) materials. The difference in the energy gaps reduces the minority carrier current flow in the emitter to negligible levels resulting in very high current gain ($\beta = 1000$). The large energy gap of the emitter also allows higher base doping without adverse impact on the current gain, resulting in low base resistance and hence higher speeds. For an emitter size of 1 μm, $f_\tau \simeq 100$ GHz has been estimated for the HBTs. The HBTs are fabricated using MBE techniques and use ion implantation to achieve base contacts and isolation.

LIST OF SYMBOLS

a	Depletion layer thickness (cm)
f_τ	Current gain-bandwidth product (GHz)
g_m	Transconductance (S)
L_g	Gate length (cm)
N	Donor impurity density (cm)
v_{sat}	Saturation drift velocity (cm/s)
W	gate width (cm)
e	Elementary charge (C)
ε	GaAs permittivity (F/cm)
ϕ_B	Barrier height (eV)
μ_n	electron mobility (cm/V·s)

REFERENCES

1. S. M. Sze, "Physics of Semiconductor Devices," 2nd ed. Wiley, New York, 1981.
2. R. C. Eden and B. M. Welch, Ultra-high-speed GaAs VLSI: Approaches, potential and progress, *in* "VLSI Electronics: Microstructure Science," vol. 3 (N. Einspruch, ed.), chapter 4. Academic Press, New York, 1982.
3. B. M. Welch, Y. D. Shen, R. Zucca, R. C. Eden, and S. I. Long, LSI processing technology for planar GaAs integrated circuits, *IEEE Trans. Electron Devices* ED-27 (6) (1980).
4. N. Yokoyama, T. Ohnishi, K. Odani, H. Onodera, and M. Abe, TiW silicide gate technology for self-aligned GaAs MESFET VLSIs, *IEEE Int. Electron Devices Meeting Tech. Dig.*, pp. 80–83 (Dec. 1981).
5. R. C. Eden (ed.), Very fast solid-state technology, Proc. IEEE **70** (1), (1982). (This issue has various papers of interest on the GaAs technology.)

Chapter 38
VLSI in Personal Computers

DAVID J. BRADLEY

IBM
Boca Raton, Florida

I. Introduction	657
II. Microprocessor Evolution	658
A. First-Generation Microprocessors	658
B. Second-Generation Microprocessors	659
C. Third-Generation Microprocessors	661
D. The Next Generation of Microprocessors	662
III. The Evolution of a Personal Computer	662
A. The IBM 5100 Processor	663
B. The IBM Personal Computer	663
C. The IBM PCjr	664
IV. The Future of Personal Computers	665
A. The Two Main Challenges for Large-Scale Integration	665
B. Other Challenges	665
Bibliography	666

INTRODUCTION I

Large-scale integration of semiconductor technology created personal computers. Before powerful, single-chip computers were a reality, the personal computer was a toy devised by hobbyists and tinkerers. After the microprocessor was developed, however, the personal computer became an appliance that was within the reach of everyone. In particular, home computers owe their origin to large scale integration.

We shall look at the impact of VLSI on personal computers by looking at the history of single-chip microprocessors and the corresponding history of personal computers. In particular, we shall review the evolution of the Intel family of microprocessors and look at the progression of the IBM personal computer product line.

II MICROPROCESSOR EVOLUTION

In the late 1960s, the semiconductor industry was faced with the problem of designing chips to do specific tasks. Various techniques were devised to solve this problem, but the one of interest to use is the development of the microprocessor. The microprocessor allowed the semiconductor manufacturer to create programmable chips using large-scale integration; then system developers could take the chips and program them for specific applications. This relieved the logic designer from having to redesign a large-scale chip for each application. It also drove down the cost of the circuits, because a large number of systems could use identical parts. The microprocessor was at first intended to be a logic replacement, but it quickly became apparent that it was also a computer.

The original impetus for development of the microprocessor came from the calculator makers. More sophisticated calculators required more complicated logic circuits to implement the calculator algorithms. The Intel-inspired solution was to provide a programmable processor, with a simple instruction set. The sophisticated algorithms could then be done by programming the processor. The result was a processor that could handle all sorts of jobs.

A First-Generation Microprocessors

1 Intel 4004

The first programmable processor on a chip was the Intel 4004. This chip contained a CPU with a 4-bit data path, sixteen 4-bit registers, and a push-down stack. This chip made its first public appearance in November 1971. The 4004 went on to achieve success in a previously unknown area. However, the 4004 was just the first step toward a personal computer.

2 Intel 8008

Shortly after work began on the 4004, Intel responded to another customer request by designing an 8-bit processor. This microprocessor, the 8008, handled 8-bit data within its CPU. It had 45 instructions and could address 16K bytes of memory. With an 8-bit data path and an instruction set optimized for character string handling, the 8008 was a processor that could be used for general-purpose systems involving text manipulation as well as simple calculator functions.

Intel marketed the 4004 and 8008 as logic replacement devices rather than as computer systems. Because of this, and the fact that no one yet envisioned the wide range of personal computer applications, personal computer development was left to hobbyists. In 1974 a number of personal computers were designed and built by hobbyists (primarily radio amateurs, or *hams*). Some were even offered for sale. These kits, as the machines were offered, contained an 8008 and the supporting circuitry to make a complete system. Programming the system was accomplished by means of a set of front-panel switches and lights. The personal computer was in its infancy.

Other First-Generation Microprocessors 3

Intel was not the only company making microprocessors, although it had the market to itself for a year. Rockwell, Fairchild, National, Signetics, Toshiba, and AMI all introduced some type of microprocessor during the early 1970s. Texas Instruments became the largest-volume producer in the 4-bit market by announcing the *TMS-1000 processor;* this family of processors has provided the chips most often used in games, toys, and other low-end controller applications.

Second-Generation Microprocessors B

Intel 8080 1

The second generation of microprocessors was ushered in with the announcement of the Intel 8080. This 8-bit processor was an upgrade of the 8008. Introduced in April 1974, it improved on the 8008 in nearly every area. The 8080 offered a tenfold increase in throughput over the 8008; it could address four times the amount of main storage and ten times the number of I/O locations. It also required only six peripheral chips to complete an entire system, as compared with the 20 required for the 8008. The 8080 was a significant step forward in microprocessor design.

This step forward in processor design was quickly matched by a corresponding leap in personal computers. On the cover of the January 1975 issue of *Popular Electronics* was the *MITS Altair,* a personal-computer kit powered by an Intel 8080. The kit originally sold for $400, offering the person with limited finances the chance to have a real computer system. Demand for the system was overwhelming: The original forecast for sales was met in a single day following the publication of the story. That is usually recorded as the birth of the personal computer industry.

2   Motorola 6800

Although the 8080 started the personal computer industry, two other microprocessors that participated in the personal computer business were developed shortly thereafter. One was the Motorola 6800, noteworthy because it was the first microprocessor to run from a single 5-V power supply. A number of personal computers were designed using that chip.

3   Zilog Z80

The other microprocessor is the Zilog Z80. This processor was designed by Masatoshi Shima (who also directed the design of the 8080) and was an architectural improvement over the 8080. It could execute 8080 programs because it was object-code compatible, but it extended the 8080 instruction set in several ways. The Z80 was a more powerful processor than the 8080 and was easier to use in designing a system.

The Z80 became the processor of choice for personal computer systems. Its 8080 compatibility gave it a wide range of software to execute, and its system characteristics allowed it to implement the S100 bus, which had become a de facto hardware standard for the personal computer industry. The Z80 could also run the CP/M operating system, developed by Digital Research, which gave it access to the standard operating system for microprocessors. The capability of the Z80 can be measured by the fact that in 1983, seven years after its introduction, machines that used the Z80 as the central processor were still being introduced.

4   MOS Technology 6502

Another microprocessor that participated in the personal computer industry, and still does, is the 6502 developed by MOS technology. It was designed into the Apple system by Steve Jobs and Steve Wozniak. Similar to a 6800, this processor allowed the Apple to be produced at a low price, and the resulting mass distribution of the machine turned the personal computer into something other than a hobbyist's toy. Serious software, notably VisiCalc (TM), became available for the machine, which drove it into the offices of work areas of the country. Similarly, the 6502 became the standard for all home computer games with its use in the Atari Video Computer System (VCS).

C Third-Generation Microprocessors

1 Intel 8086 and Motorola 68000

The next generation of personal computers sprang from the next technological step of microprocessors. Spearheaded by the Intel 8086 and Motorola 68000, these processors managed 16-bit arithmetic and could address significantly more than the 64K bytes allowed by the earlier processors. The increased data flow resulted in faster program execution, and the larger address space allowed programmers to write more complicated and involved routines. Now, as before, the technology for personal computers came from large-scale integration of semiconductor technology.

2 Intel 8088 and the IBM Personal Computer

Perhaps the most notable of this generation of machines is the IBM Personal Computer. Driven by an Intel 8088, an 8-bit data bus version of the 8086, this machine has gained wide acceptance in the personal computer market. The IBM PC, along with its family members, the IBM Personal Computer XT and the IBM PCjr, has become a widely accepted architecture. Many new machines advertise IBM compatibility, which they achieve by using the same set of components that are used in the IBM machine.

3 Today's Personal Computers

Lest we overlook it, personal computer development has not been driven *solely* by processor development. The microprocessor is the most crucial element of the system and the most visible. However, the personal computer derives its desirability from its low cost and high power. Large-scale integration participated in all facets of the system design, not just that of the central processor.

Today's personal computers offer much more power at lower prices than personal computers of five, or even two, years ago. Part of this is because of the development of support chips for the microprocessors. Although the 8080, Z80, and 8088 provide the processing power, a number of other circuits are necessary in the system. The IBM Personal Computer uses the 8088 as the central processor, but it also uses several other LSI components to build the entire system. The system contains a direct memory access controller, an interrupt controller, a timer/counter, a floppy disk controller, an asynchronous communications element, and a CRT controller. Each of these is a single large-scale integrated chip that is devoted to the control of some particular element of the system. The combination of these components gives the system its tremendous power at minimal cost.

D The Next Generation of Microprocessors

The coming generation of personal computers will be controlled by the newer members of the microprocessor family. For Intel, these are the 80186 and 80286 processors.

1 Intel 80186

The 80186 is an advanced integration of the 8086. It combines the 8086 processor with a direct memory access controller, a timer/counter, and an interrupt controller, all in a single chip. This level of integration promises to lower the price of a system substantially; and, of course, the new processor offers greater performance than the 8086, as a result of improved chip architecture and faster circuitry.

2 Intel 80286

The Intel 80286 advances microprocessor design in a different area. Instead of combining functions, as in the 80186, the 80286 offers new functions. The 80286 has a protected mode of operation, in which each memory access is restricted to the memory region allowed for that task. The 80286 architecture also provides for virtual memory. The segment mapping of the 80286 has a *not-present* indicator. Combined with restartable instructions, this will allow automatic paging of programs. Thus, the 80286 has many abilities that were available only in a mainframe computer several years ago.

III THE EVOLUTION OF A PERSONAL COMPUTER

The effects of technology on personal computers can be illustrated by two machines: the IBM 5100, a personal computer introduced by IBM in 1976 using the best available technology at the time, and the IBM Personal Computer, announced and shipped in 1981. Between these dates

The IBM 5100 Processor   A

The IBM 5100 was the first personal computer offered by IBM. At the time of its introduction it was a state-of-the-art machine for a desktop environment.

The central processor for the 5100 was known as the *Program all logic in memory* (PALM). This name describes the situation that existed at the time the 5100 was designed. The processor and all of the logic in the system was implemented in bipolar technology. The processor consists of approximately 12 modules of 100 circuits per module master-slice technology. These circuits were expensive compared with the cost of FET memory then available. So the design goal of the 5100 was to put as much function as possible into read-only storage. The PALM processor is an 8-bit ALU design with the ability to address 64K bytes of memory. This relatively simple processor was backed by a large amount of memory to implement the system.

The 5100 used two different forms of ROM technology. A complete system required 64K bytes of high speed ROM. This ROM was available in chips with 36K bits per chip. The 5100 also had approximately 150K bytes of slower speed ROM (a cycle time of approximately 2.2 μs). This ROM was packaged with 48K bits per chip. Thus the system design traded processor circuits for a large number of ROM storage locations.

The 5100 also provided a *display adapter* that was composed almost entirely of standard TTL components. This display adapter gave a display of 16 rows of 64 characters each. The adapter required about 40 in.^{2} of board space to implement. An attachment to the keyboard and tape cartridge required a comparable space.

Finally, the 5100 used static memory technology. The RAM chips were 2K by 1 bit in size. They were combined in modules of four chips each, giving an effective package of 8K by 1 bit. This limited the 5100 memory to a maximum of 64K bytes.

The IBM Personal Computer   B

The IBM Personal Computer was introduced in August 1981. This system used the *Intel 8088* as the central processor. The 8088 is nearly identical to the Intel 8086, providing 16-bit arithmetic and a 1-MByte address space. Though it has only an 8-bit data bus and is restricted to transferring a single byte of information to and from memory each cycle, the 8088 is an order of magnitude more powerful than the PALM processor of the 5100 and has a much larger address space, more powerful instruction set (such as multiply and divide), and a faster clock rate.

The IBM Personal Computer uses large-scale integration technology to provide additional system functions. Instead of the bipolar logic of the 5100 there is a MOSFET processor. In addition, there are eight levels of prioritized interrupt. There are four channels of direct memory access and three 16-bit timer/counter channels. Each of these functions is available on the system board, along with up to 256K bytes of dynamic memory. The IBM Personal Computer uses 64K by 1 bit dynamic memory modules and 64K bit ROM modules. Even more illuminating is the fact that the Personal Computer requires only 40K bytes of ROM on the system board to make a fully functional system.

The adapter cards also have their share of large-scale integration. There are two different display adapters for the IBM Personal Computer. The one most similar to the adapter in the 5100 uses a single chip to generate all the horizontal and vertical timing information, as well as the display memory addressing. This adapter has a display capability of 25 rows of 80 characters each as well as several all-points addressable graphics modes. Besides the display function, the adapter has 16K bytes of memory to hold the display buffer. All of this is done in a board area of about 50 in.² This integration of function allows the IBM Personal Computer to offer much more display capability (such as color and graphics) than was offered on the IBM 5100.

C The IBM PCjr

The IBM PCjr is the lower cost version of the IBM Personal Computer and is intended for home and education environments. To appeal in those areas, it has to offer significant function at a moderate price. For the IBM PCjr, this was accomplished by building a machine that has function comparable to the IBM Personal Computer. The lower price was designed in by the use of additional large-scale integration.

To be sure, the PCjr does not have all the functions available in the IBM Personal Computer. Most noticeable is the absence of the direct memory access controller. However, considerable cost savings came about through the use of the video gate array. This large-scale integration CMOS chip contains most of the functions that are provided on the color/graphics adapter of the IBM Personal Computer. This gate array lets the PCjr display function fit on the system board without requiring another display adapter. Combining this display controller with the system memory made possible a considerable cost savings. The result is a system that is affordable and highly functional.

THE FUTURE OF PERSONAL COMPUTERS IV

Personal computers are just now beginning to enter the mainstream of life. Whereas five years ago it was unusual for a family or small business to own a personal computer, it is now common. However, personal computers are far from being pervasive. The personal computer will not be a part of every home and every business until it becomes simple and natural to use.

The Two Main Challenges for Large-Scale Integration A

Of course, most of the pressure for "user friendly" machines falls on the programmer. However, as the programming community strives to build systems more suitable for the unsophisticated user, two things will surely be needed.

(a) One is more memory. Each advance in system usability requires more system storage to hold the programs and data to drive it.

(b) The other requirement is for faster processors. As the programming system tries to do more for the user, it must execute more instructions; to be a natural system, it must execute them rapidly.

Other Challenges B

Advances in the functionality and performance of the *support circuits* are needed to build the future personal computers. Another area, just beginning to emerge from the laboratory and find its place in personal computing, is *speech synthesis*. Speech generation requires large amounts of storage and a high performance processor. Even more demanding will be the requirements for *speech recognition*. This and other forms of artificial intelligence will demand better performance from the system. These advances will be necessary for personal computer systems that can interact with people in a natural fashion.

The future looks just as bright for personal computers as the history has been. Continuing advances in large-scale integration will continue to drive the cost of personal computing down, while advancing the available functions.

BIBLIOGRAPHY

S. P. Morse, B. W. Ravenel, S. Mazor, and W. B. Pohlman, Intel microprocessors—8008 to 8086, *Computer* **13** (10), 42–60 (1980).

R. N. Noyce and M. E. Hoff, Jr., A history of microprocessor development at Intel, *IEEE Micro.*, 8–21 (1981).

D. A. Roberson, A microprocessor-based portable computer: The IBM 5100, *Proc. IEEE* **64** (6) (1976).

Chapter 39
VLSI in the Design of Large Computers

DONALD P. TATE
DENNIS G. GRINA
Control Data Corporation
St. Paul, Minnesota

I.	Introduction	667
II.	Performance	668
	A. Effect of Interconnection Delay	668
	B. Logic Gates	670
	C. Parallelism and Multiprocessor Systems	670
III.	VLSI Design Considerations	671
	A. Handling Complexity	671
	B. Minimizing Design Errors	671
	C. Locating and Correcting Errors	673
	D. Fault Tolerance and Array Yield	674
IV.	Design Tools and Staff	674
	A. Front-End Tools	675
	B. Back-End Tools	677
	C. Implementation of Tools	677
	D. Staff Productivity	678
V.	Summary	678
	References	678

INTRODUCTION I

The solution to a number of problems in the scientific and industrial fields constantly demand larger and faster computers. For more than 20 years supercomputer designers have been pressing state-of-the-art solid state technology and using the most advanced computer architectures to meet this challenge. The systems that have been designed during this period contain millions of transistors and took several years to develop. Because of the long development times and the need to compete in the marketplace, designers have had to choose a technology that they believe will still be viable when their product is finally introduced.

Today, designers of large-computer and supercomputer systems rely heavily on VLSI circuits to help meet the challenge in the marketplace. At any time, however, there still are choices to be made: MOS or bipolar, feature size, number of gates per array, gate array or custom design, and materials, to name a few. Those decisions will have to be made early in any development project, based on the designer's best judgment of current and future technology and on the market the product is designed to serve.

This chapter will not dictate choices but will look at system requirements and concentrate on the techniques and tools that will be generally applicable to any VLSI technology option.

II PERFORMANCE

A Effect of Interconnection Delay

High performance is almost always a major design goal for large state-of-the-art computer systems. The speed of early discrete-component computers was limited largely by the speed of the electronic components used to construct these systems. As the speed of electronic components increased, logic interconnection delay became an increasing barrier to higher machine performance. This trend is shown in Fig. 1.

The following table tabulates the effect on system delay of making these interconnections with coaxial wire, foil on a printed circuit board, and metal on a silicon wafer.

Logic element interconnection medium	Typical delay $s^{-12}/\mu m$	Typical length (μm)	Typical time (s^{-9})
On silicon chip	0.024	1,250	0.030
Foil on PC board	0.012	150,000	1.8
Coaxial wire	0.006	900,000	3.6
Light (reference)	0.0033		

Even though the delay per unit length is several times greater for silicon than for printed circuit boards or coaxial wire, the distance typically associated with making connections on silicon is such that the delay related to on-chip silicon connections is several orders of magnitude less than that associated with the other forms of connections.

39. VLSI in the Design of Large Computers

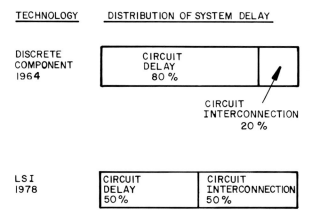

Fig. 1. Components of system delay versus technology.

As designers begin to use VLSI arrays in systems, the amount of logic that can be placed on a single chip increases, and two factors that can dramatically increase system performance begin to surface. First, a greater percentage of system interconnections occur on silicon (Fig. 2); hence interconnection delay decreases. Second, the system as a whole gets smaller: The connections between arrays, while still much longer

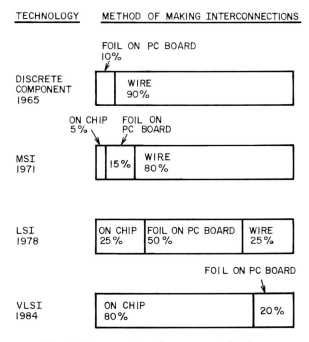

Fig. 2. Interconnect medium versus technology.

than those on silicon, become shorter than they would be in a non-VLSI system. Again, interconnection delay decreases, and system performance increases. VLSI systems are physically smaller, and, in general, smaller systems are faster.

B Logic gates

Other factors work together subtly to increase system performance in designs that use significant amounts of VLSI logic. The cost of a logic gate on a VLSI array is considerably less than it would be if other forms of logic were employed as shown in the following table.

Technology	Cost of gate (dollars)
Discrete component	
1965	1.50
LSI	
1978	0.15
VLSI	
1984	0.015

Furthermore, most VLSI arrays tend to be pin- rather than gate-limited (which means there are unused gates on most arrays). These two characteristics, combined with the fact that VLSI logic arrays contain so much logic that it is difficult to use an array type more than a few times in a system, allow the VLSI logic designer to use logic gates on the array rather freely. Since the designer does not need to be as concerned about part proliferation or the number of gates used, the design can be customized to perform a specific function at performance levels that are closer to the theoretical maximum.

C Parallelism and Multiprocessor Systems

As VLSI makes processor elements (networks and functional units) cheaper and physically smaller, parallelism (the use of multiple elements to perform different parts of a task simultaneously) becomes a much more viable design technique. Parallelism within an individual processor allows its performance to increase faster than the increase in raw gate speed. Since processors themselves are physically smaller and cheaper, multiprocessor systems can be fabricated practically without unreasonable cost or system interconnect delay penalties. These systems can share the tasks associated with a problem's solution and achieve system throughput that can be several times that attainable with a single processor system.

III VLSI DESIGN CONSIDERATIONS

Although VLSI technology has made possible some significant advances in supercomputer performance and some innovations in architecture, it has also posed some serious problems for the designer that lead to a longer, more complex design process.

A Handling Complexity

Since each array tends to be unique, there are many array types to be designed. Designing these individual VLSI arrays is a challenging undertaking. Several thousand gates must be managed. Design rules are complex. For example, if the designer is using MOS logic, circuit delay is a complex function of loading and interconnect distance. Even with emitter coupled logic (ECL), the designer must contend with the fact that transmission on the silicon wafer is *lossy*. As the VLSI array becomes larger, the resistive drop in transmission becomes significant. The loading and delay rules associated with these lossy interconnects become extremely complex.

VLSI design is simply too complex to be done by hand. A state-of-the-art electrical computer-aided design system (ECAD) is mandatory. The requirements for this system are discussed in Section IV.

B Minimizing Design Errors

Another challenge faces the designer using VLSI logic in large systems. It is critical that the VLSI design be virtually error free before it is committed to hardware. Errors that are not detected until prototype checkout can be disastrous.

With small-scale integration (SSI) technology, errors that escaped detection during design could usually be corrected by moving interconnect wires or by temporarily modifying a printed circuit (PC) board by cutting some foil and adding wire. Checkout could then continue. This is not true when VLSI is used. Since most logic interconnects are on the array, most errors can be corrected only by re-layout of one or more arrays. The re-layout and re-manufacturing cycle usually takes from six to twelve weeks.

The situation is made even worse by a phenomenon often referred to as the *layering effect,* which is caused by the fact that errors can mask each other. It is often necessary to correct one error to find others. Even though the CDC CYBER 205 design was extensively simulated, a typical functional unit (which would now fit on a VLSI array containing 20,000 gates) contained about three dozen logic errors that escaped detection until prototype checkout. Had this machine been designed with VLSI, it would probably have taken at least six array redesign cycles to eliminate all errors from a typical array.

One obvious way of dealing with this problem is to reduce the array redesign time. Unfortunately, this is very hard to do. The process of redesigning and fabricating a VLSI array is inherently time-consuming. The designer must seek out ways to minimize the number and impact of design errors if VLSI is to be used successfully.

1 Separate Control Logic From Data Logic

Most logic design errors occur in control logic rather than in data logic. Whenever possible, control logic should be implemented in a medium that can be changed without redesigning the array (PROMs, EPROMs, RAMs, and so on). Unfortunately, these devices are often not dense enough or fast enough to implement control in a high performance system. They should, however, be used wherever possible.

2 Minimize Array Types

Partitioning the system with a bit or byte slice design will usually help to minimize the number of array types in a design and, therefore, to minimize the number of arrays that must be redesigned because of an error. If bit or byte slice partitioning is not feasible, the designer should at least attempt to localize control to as few array types as possible. This will again tend to minimize the number of arrays that need to be redesigned because of a control error.

3 Make Array Interfaces Easy to Understand

Even if timing or other design considerations make it impossible to separate control from data logic, the designer employing VLSI should attempt to partition logic in such a way that the interfaces between arrays occur on natural logic boundaries that are easily comprehended by another designer. Most projects are organized so that one person or team is responsible for an array. Array boundaries, therefore, also represent interfaces between design teams. Errors caused by breakdown in human communications will be minimized if these array interfaces are easily understood and formally documented.

Use Top-Down Design 4

A good top-down design approach will give special attention to simplifying and documenting these interfaces early in the design process. The designer must, however, be careful that the low-level blocks postulated early in the design process are implementable within performance and cost goals.

Locating and Correcting Errors C

Even if the most careful design methodology is followed and extensive state-of-the-art simulation techniques are employed, some errors will escape detection during design. Provisions must be designed into the system to facilitate locating and correcting these problems quickly in prototype checkout. Some technique must be developed to minimize the impact of the layering effect.

Scan-In–Scan-Out 1

One of the most effective techniques for dealing with errors detected during prototype checkout is to implement a scan-in/scan-out approach, whereby data can be shifted serially into and out of the array. The IBM Level Sensitive Scan Design (LSSD) is an example of this approach [1,2]. It is well-known that the scan-in/scan-out feature allows the checkout technician to observe flip-flops that are not directly connected to array output pins and, therefore, makes it easier to isolate a design error to the failing element quickly. What is not as well-known is that scan-in/scan-out can be a powerful tool in overcoming the error layering effect. Scan-in/scan-out allows the checkout engineer to bypass failing logic or networks and test logic that is fed by or dependent on the failing logic. Correct data (obtained from simulation) can simply be forced into flip-flops that are fed by the failing logic. This technique can greatly increase the number of errors that can be detected in an array design before the array needs to be redesigned.

Fabrication Error Detection 2

Fabrication errors must also be addressed during the design phase. A method of detecting fabrication errors that is usable and implementable on the fabricator's test equipment must be designed into the array. Failure to provide this feature will result in VLSI arrays that cannot be tested and, therefore, cannot be manufactured. An on-chip maintenance system [3] that allows the array to generate its own test operands rapidly, combined with scan-in/scan-out capability, may prove to be the most effective way to provide this needed testing capability. This scheme has the advantage that it does not require high-speed testers, and it does not require test equipment to interface with a large number of array pins. Test operands can be generated using simple, well-known techniques that are appropriate for testing combinational logic.

D Fault Tolerance and Array Yield

Array yield per wafer during manufacturing is another factor that needs to be addressed during the logic design process. As the VLSI array die becomes larger, it may not be possible to obtain a satisfactory yield of 100% good parts from the array manufacturing process. State-of-the-art VLSI fabrication techniques yield between two and ten fabrication flaws per square centimeter of die area [4]. For an array that is 5 cm on each side, one can statistically expect between 50 and 250 defects on an average array. It is clear that this level of defects will not allow a sufficient yield of useable arrays, unless the designer takes steps to provide tolerance for some of these defects. It may be necessary to build some redundancy into the design to obtain sufficient yield from the manufacturing process. Techniques to produce this fault tolerance are currently being proposed by some manufacturers of large computer systems [5,6].

IV DESIGN TOOLS AND STAFF

Sophisticated ECAD tools are an absolute necessity in the VLSI design process, because the rules are so complex and the designs are so large. VLSI design that is done without state-of-the-art ECAD tools will not be completed as quickly or as cost-effectively as it could be. The time for getting the system operational will certainly be greater.

VLSI ECAD tools can be divided into two categories:

(1) Front-end tools are those routines that the designer uses to produce, document, and verify the design.

(2) Back-end tools are the programs that are used to translate a logical design to a physical design.

The two categories of tools have different criteria for judging their value.

39. VLSI in the Design of Large Computers

Front-end tools should be fast. They must provide quick feedback to the designer regarding the validity of his design. Feedback time should be measured in minutes rather than hours or days. The tools need to model accurately the physical design that will ultimately be produced. There should be a low probability that a back-end tool will discover an error in a design that successfully completed front-end tool verification.

The criteria for back-end tools are different. Since there is a very low probability that a design that appeared correct when analyzed with front-end tools will fail when exposed to the back-end tools, the back-end ECAD program will probably be run only once per array design. Speed and fast feedback are considerably less important than they are for front-end tools. Accuracy and thoroughness become key evaluation criteria. It is imperative that all documentation and test data needed to manufacture and test the array and array boards be automatically generated in a format that is compatible with manufacturing equipment and manufacturing procedures.

Front-End Tools A

Timing Verifier 1

One of the key front-end tools is an operand-independent *timing verifier*. This program should locate all potential long and short paths in a proposed design without the need for subjecting the network to test data. Timing performance of the physical circuits should be modeled as accurately as possible. Since the program need only be run once on any iteration of the design, it need not be particularly fast. Typically, when these programs are executed on a 20–25 MIP computer, they should be able to analyze a 200,000- to 300,000-gate design in half an hour of central processor time [7]. Once all the long and short paths detected by the timing analyzer program have been fixed, there should be no more timing problems in the proposed design.

Placers 2

To have accurate timing information on a VLSI array, it is necessary to have a reasonably accurate placement of gates or macros on the array. This requires a second kind of front-end tool, the *interactive* or *automatic placer*. For a relatively large VLSI array, this is one of the hardest tools to develop. The tool must be fast to give the designer timely feedback, and yet it must produce a design that can be implemented by back-end tools. The criteria for good placement are often ambiguous. (Should the average length of interconnect be minimized or should only certain critical nets be optimized? How important is minimization of resistive drop in lossy transmission paths?) It is critical that the placer be fast, use the right evaluation criteria, and produce a design that is routable with interconnect delays that are compatible with those assumed by the timing analysis program.

The designer also needs system-level timing information. This need dictates that a tool exist to assist in placing arrays on the PC board and placing PC boards in the system. The requirements for this placement tool are basically the same as those already cited for the tool that places gates or macros on the array.

3 Logic Simulator

Another critical front-end tool is a *logic simulator*. Since the timing verification program detects timing problems in the design, the simulation system should not model timing. It should be designed to help the designer determine if the design is logically correct.

This is a very challenging task. Most large systems have a large instruction set, which leads to a very large number of possible instruction sequences. Furthermore, to optimize performance of large machines, instruction execution is usually overlapped. This means that even though instructions A, B, and C may work individually, the sequence B, C, A may not work under certain conflict conditions.

To verify the logical validity of the design, it is theoretically necessary to simulate all possible instruction sequences. Almost all the problems found in prototype checkout of the CDC CYBER 205 involved logic problems in instruction sequences that had not been simulated. In a VLSI machine, each of these errors would probably have precipitated an array rework cycle, which would have delayed system checkout and shipment. Though for most large machines it is physically impossible to simulate every possible sequence, if unacceptable schedule delays are to be eliminated, it is critical that as many sequences as possible be simulated.

If the designer is to simulate a large number of sequences in a limited period of time, simulation programs must be capable of simulating a design quickly, even if it is large. One way to improve simulation speed is to use a hierarchical simulation system. Such a system has the ability to simulate a design at more than one level of complexity. It can simulate an area of primary concern in detail while simulating the rest of the system at a functional or block level. This allows the entire system to be simulated with minimum computer resources in a minimum amount of time. It also allows a detailed design of one functional area to be evaluated in a system environment before the detail designs of other areas are completed.

Other Front-End Tools 4

The VLSI designer needs a few more front-end tools to create and document his design. An *interactive delay calculator* that can accurately model the delay through a specific string of gates or macros and supply feedback to the designer in one or two seconds is imperative. A *schematic entry system* that allows the designer to produce a design interactively and automatically produces the master data base, which describes that design and provides input data for all other front-end and back-end tools, is mandatory.

Documentation 5

Front-end tools should produce all the documentation the designer needs to manage the design and successfully integrate it with other elements of the system. All design rules should be checked. No surprises should occur when the back-end programs are executed.

Back-End Tools B

The back-end ECAD tools needed for VLSI design do not differ greatly in principle from those needed for less dense logic, but they are more complex to develop and take longer to execute. Arrays and array boards must be routed. All arrays and documentation needed to manufacture and test the system must be generated. A system must exist to allow engineering changes to be documented, verified, and implemented quickly and effectively.

Implementation of Tools C

All front-end and back-end tools should share a common data base and should probably exist on a distributed processing system [8]. Most front-end tools should probably be implemented on an engineering workstation that has the ability to submit time-consuming, compute-bound tasks to a large traditional computing system. Back-end tasks should probably be implemented on the large computer system. Test equipment should have interactive access to the master data base.

D Staff Productivity

Though ECAD systems are critical to successful design with VLSI and should dramatically improve the productivity of the individual logic designer, there is a significant learning cycle associated with their use. The first VLSI design project attempted by a design team can take significantly longer to complete than would a comparable design project done with SSI or medium-scale integration (MSI). Project schedules must reflect this temporary decrease in productivity.

Because of the changes in design procedures, the successful VLSI logic designer must be an individual who can integrate traditional circuit design and logic design skills. Production of the floor plan for an array is not unlike the traditional logic designer tasks of placing integrated circuits on a PC board or placing PC boards in a system. Compensating for resistive drop and capacity loading effects are tasks that are routine to the circuit designer. ECAD tools have made this transition easier, but the VLSI designer must still master the basic principles of circuit and logic design.

V SUMMARY

VLSI has given the large-computer designer the speed and the packaging density needed for further advances in computer performance. It is the current chapter in the continuing development of more powerful and more sophisticated computing systems.

The large-computer and supercomputer systems of tomorrow will be fabricated largely with VLSI arrays. To work successfully with VLSI, logic designers must employ increasingly sophisticated techniques, processes, and tools to aid them in successfully completing their designs. The designer of tomorrow's supercomputer systems must have ready access to today's supercomputers coupled to a network of designer workstations and backed by a staff of competent programmers who can supply a steady stream of design tools to keep pace with the rapidly advancing VLSI technology of today—and tomorrow.

REFERENCES

1. E. H. Frank and R. F. Sproull, Testing and debugging custom integrated circuits, *Comput. Surv.*, December, **13**, 425–451 (1981).
2. T. W. Williams and K. P. Parker, Design for testability—A survey, *IEEE Trans. Comput.* January, **C-31**, 2–15 (1982).

3. D. R. Resnick, Testability and maintainability with a new 6K gate array, *VLSI Des.*, March, **4,** 34–38 (1983).
4. H. K. Dicken, Calculating the manufacturing cost of gate arrays, *VLSI Des.*, December, **4,** 51–55 (1983).
5. D. L. Peltzer, Wafer-scale integration: The limits of VLSI?, *VLSI Des.*, September, **4,** 43–47 (1983).
6. J. Schefter, Giant microcircuits for super-fast computers, **224,** 66–67, 155, *Pop. Sci.*, January (1984).
7. L. C. Bening, T. A. Lane, C. R. Alexander, and J. E. Smith, Developments in logic network path delay analysis, *Proc. 19th Design Automation Conf.*, pp. 605–615 (1982).
8. C. Barbour, System integration series, *Electron. Des.*, Sept. 1, **31,** 187–194 (1983).

Chapter 40

Electronic Warfare Applications of VLSI

CHARLES T. BRODNAX

E-Systems, Garland Division
Dallas, Texas

I. Introduction	681
A. Basic Definitions	681
B. Functional Divisions of Electronic Warfare	682
II. Electronic Warfare Support Measures	682
A. ESM Receiver Characteristics	683
B. Pulse Sorting Technology	683
C. Pulse Testing	685
D. VLSI Memory and Logic in Pulse Testing	686
E. Emitter Fingerprinting	687
F. Low Probability of Intercept Emitters	687
G. Design Considerations	687
H. Emitter Location	689
III. Electronic Countermeasures	690
A. VLSI for Deception	690
B. Simulated Signals	691
IV. Electronic Counter-Countermeasures (ECCM)	691
A. Training of Personnel	691
B. Equipment Characteristics	692
C. Filter Design with VLSI	692
List of Symbols	693
References	694

INTRODUCTION   I

Basic Definitions   A

Electronic warfare (EW) has been defined as "actions involving the use of electromagnetic spectrum energy to determine, exploit, reduce, or prevent hostile use of the electromagnetic spectrum and action which retains friendly use of the electromagnetic spectrum.''[1] EW can be used for both offense and defense. As an *offensive weapon,* the control information transmitted to an energy weapon can be altered to reduce the effectiveness of that weapon. When the purpose is to screen detection of friendly forces, EW is thought of as a *defensive weapon.*

Use of the electronic spectrum is crucial in modern warfare. We gather information about the enemy with electronic listening devices and radar. This information is sent from the collection platform to a station where it is processed, sorted, and sent to a central location. At the central location, the information is merged with other information from other sensors and other collection platforms. Based on the situation as perceived from the input information, responses are planned. The battle commander issues orders to his subordinates so that the planned response is implemented, assesses the effectiveness of the response, and plans the next action. Use of the electromagnetic spectrum is essential in this type of military operation. Not only is it important that we be able to communicate with friendly forces, it is equally important to deny the hostile force access to the information being transmitted.

B Functional Divisions of Electronic Warfare

We shall show by selected examples how application of VLSI can significantly impact EW capabilities in three functional areas:

(a) electronic warfare support measures (ESM);
(b) electronic countermeasures (ECM); and
(c) electronic counter-countermeasures (ECCM).

II ELECTRONIC WARFARE SUPPORT MEASURES

Electronic warfare support measures (ESM) is the division of electronic warfare involving actions taken to search for, intercept, identify, and/or locate sources of radiated electromagnetic energy for the purposes of immediate threat recognition. Before going into detail on how these functions can be performed and enhanced with VLSI, let's look at the environment in which ESM receivers must operate.

ESM Receiver Characteristics   A

The power density at the ESM antenna is significantly higher than that at the threat radar antenna. Power density at the receiving antenna of the radar is

$$P_R = \frac{P_t G_t \sigma}{(4\pi)^2 R_1^4} \qquad (1)$$

where P_t is the peak power transmitted by the radar, G_t the gain of the transmitting antenna, σ the radar cross section of the target, and R_1 the range from the radar to the target.

The power density at the antenna of the ESM system is

$$P_E = \frac{P_t G_t}{4\pi R_2^2} \qquad (2)$$

if R_2 is the range from the radar transmitter to the ESM antenna. If the ESM receiver is at the same range as the target, the power ratio will be

$$\frac{P_E}{P_R} = \frac{(4\pi)^2 R_1^4}{4\pi R_2^2 \sigma}. \qquad (3)$$

If $R_1 = R_2 = R$, then

$$\frac{P_E}{P_R} = \frac{4\pi R^2}{\sigma}. \qquad (4)$$

In Fig. 1 the ratio of P_E/P_R is shown. For a 100-m² target ($\sigma = 100$ m²) the relative power is 51 dB at 1 km and 91 dB at 100 km. This indicates that the ESM receiver is normally operating in a high signal-to-noise environment. Spread spectrum techniques that attempt to reduce this advantage are presented later in this chapter.

Pulse Sorting Technology   B

The ESM receiver must search, intercept, sort, and measure the parameters of all signals within the frequency range of the receiver and determine the direction of each emitter relative to the receiver. Due to the heavy use of signals in the electromagnetic spectrum, there are usually multiple transmitters on the frequency of interest. After the frequency selection, filtering, and pulse detection functions have been performed, the pulses must be sorted into pulse trains of constant pulse repetition

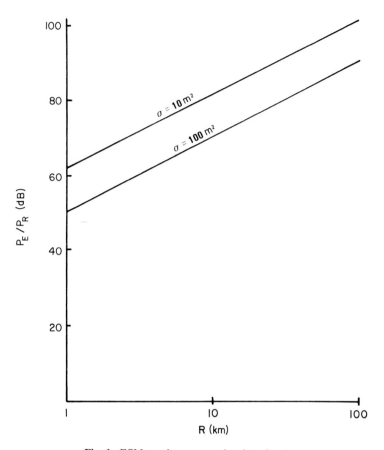

Fig. 1. ESM receiver power density advantage.

frequency (PRF). This sorting task is a natural one for VLSI implementation. Assume that the first pulse in the sequence to be sorted is received at time t_1 and that the leading edge of succeeding pulses are received at times $t_2, t_3,\ldots,t_i,\ldots,t_j,\ldots$, as illustrated in Fig. 2a. Let

$$\Delta t_{ij} = t_j - t_i \tag{5}$$

be the time interval between the leading edges of pulses at time t_i and time t_j. If

$$\Delta t_{ij} = \Delta t_{jk}, \tag{6}$$

we say that pulses at times t_i, t_j, and t_k are from the same transmitter, and we establish a track on this pulse train. When the third pulse has been

40. Electronic Warfare Applications of VLSI

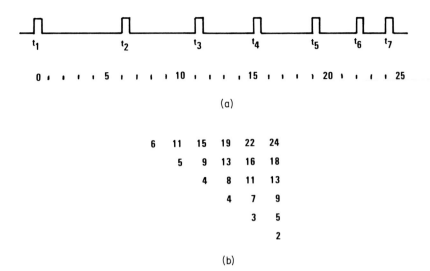

Fig. 2. Pulse sorting. (a) Pulse sequence. (b) Array of interpulse times.

received we can compute

$$\Delta t_{12}, \quad \Delta t_{13}, \quad \text{and} \quad \Delta t_{23}. \tag{7}$$

We compare Δt_{23} to Δt_{12} to see if these three pulses are from the same emitter. If $\Delta t_{12} < \Delta t_{23}$, we can say that pulses 1 and 2 are from different emitters, and we no longer need to test Δt_{12}. When $\Delta t_{12} = \Delta t_{23} = \tau_1$, we establish a track at this interpulse spacing and look for the next pulse in this string at time $t_3 + \Delta t_{23}$. As long as we continue to receive pulses at this rate, we say that that transmitter is still active. When we stop receiving pulses at that rate, we assume that the transmitter has turned off. When $\Delta t_{12} = \Delta t_{23} = \tau_1$, store τ_i and t_3 in the track file. Delete t_1 and t_2 from the transient file. If $\Delta t_{12} > \Delta t_{23}$, we say that at least one of pulses 1, 2, and 3 is not from a common pulse train and a common transmitter.

Pulse Testing C

As each new pulse is detected, it is tested to see if it is in a pulse train that is being tracked. This test consists of comparing the time from the last pulse in each train being tracked to the τ_i for that train. If it is one of these trains, the time of the last pulse is updated, and we wait for the next pulse. If it is not in one of these trains, it goes into the unsorted file to see if we can find a new train.

With $n - 1$ unsorted pulses, we have the array of all Δt_{ij}.

$$
\begin{array}{cccccccc}
\Delta t_{12} & \Delta t_{13} & \Delta t_{14} & \Delta t_{15} & \Delta t_{16} & \Delta t_{17} & \cdots & \Delta t_{1,n-1} \\
 & \Delta t_{23} & \Delta t_{24} & \Delta t_{25} & \Delta t_{26} & \Delta t_{27} & \cdots & \Delta t_{2,n-1} \\
 & & \Delta t_{34} & \Delta t_{35} & \Delta t_{36} & \Delta t_{37} & \cdots & \Delta t_{3,n-1} \\
 & & & \Delta t_{45} & \Delta t_{46} & \Delta t_{47} & \cdots & \Delta t_{4,n-1} \\
 & & & & \Delta t_{56} & \Delta t_{57} & \cdots & \Delta t_{5,n-1} \\
 & & & & & \Delta t_{67} & \cdots & \Delta t_{6,n-1} \\
 & & & & & & \vdots & \\
 & & & & & & & \Delta t_{n-2,n-1}
\end{array}
$$

When the nth unsorted pulse is received, we compute Δt_{jn} for each t_j in the unsorted file. We compare Δt_{ij} and Δt_{jn} for each ij pair. If a match is found, then $\tau_i = \Delta t_{ij} = \Delta t_{jn}$ is established as the interpulse spacing of a new pulse train, and t_n is stored as the time of the last pulse in this train. All rows and columns with a subscript i, j, or n are deleted from the array of spacings of unsorted pulses.

In Fig. 2b, $\Delta_{13} = 11$ and $\Delta_{36} = 11$; so we assume a train with PRI = 11 and a pulse at t_6. There is another apparent train with PRI = 4 indicated by Δt_{34} and Δt_{45}. Since another pulse with PRI = 4 has not been found at $t = 23$, $(t_5 + 4)$, we say that this emitter turned off at t_5.

The size of the array of all Δt_{ij} is $n(n - 1)/2$, where n is the maximum number of unsorted pulses expected in the time window of interest. The window must be at least twice the length of the longest interpulse spacing of the pulse trains of interest. The row of any pulse outside this window is deleted.

D VLSI Memory and Logic in Pulse Testing

Conventional computers will be very slow at this task because of the many comparison operations that have to be performed serially. Special purpose MSI/LSI based hardware can be built to perform this content addressable memory (CAM) function in parallel. Storing and comparing these data in a conventional machine would require extensive circuitry. With VLSI the memory and logic can be included on the chip with sufficient data paths to allow efficient implementation of this scheme. The result is expensive and bulky for memories of more than a few pulses. A custom VLSI chip design allows the straightforward integration of the required logic functions into the CAM array structure. The reduced size and power consumption that result are ideal for aircraft with complex EW threat environments.

Emitter Fingerprinting E

Characteristic Emitter Spectra 1

Emitters have characteristics that cause the pulses from one radar to differ from those of all other radars. The spectral content of a pulse from a given radar is unique. Radar identification is possible by analyzing the modulation on the pulse. A radar with a PRF of 1–10 kHz takes several hundred microseconds for the processing of each pulse.

VLSI Design Considerations 2

Assume a pulse train with a 10-kHz PRF is to be analyzed for purposes of fingerprinting the emitter. Furthermore, assume that one wants to form 32 frequency estimates on each pulse. Then 100 μs are available to form a 32-point FFT. A 32-point FFT requires 80 complex multiplications, or 320 real multiplications. With 2-μm design rules it is possible to design NMOS devices capable of performing 20 million real multiply-accumulates per second. One of these chips could perform the needed computation for a 128-point spectral analysis of pulses from the 10-kHz train. Similar techniques can be used to construct VLSI correlators capable of matching emitter fingerprints at these rates.

Low Probability of Intercept Emitters F

Radar and communications systems designers are attempting to spread energy in the frequency and time domains to defeat the advantage of the ESM receiver on a power density basis. The desire is to encode the signal so that it appears to be noise to anyone who does not have the code. By correlation techniques the received signal is integrated to appear to be a high power signal. Linear FM has been a popular code for spreading the energy in the CHIRP radars. Today Barker codes are gaining use. Other coding schemes are increasingly being employed.

Design Considerations G

Since the designer of the ESM receiver cannot assume that the codes will be known, the received signals must be processed in a way to detect the signals without knowing how the signal was encoded. One way to do this processing is to correlate over time. The output of the analog receiver is sampled and sorted for a period of time and then processed in a coherent manner. To achieve the dynamic range necessary to receive the low level signal, the bandwidth of the analog ESM receiver is limited. Channelized receivers are used where a number of narrowband high dynamic range receivers are arranged with contiguous frequency response. In this way the entire frequency band of interest is covered without giving up dynamic range.

A problem arises when the energy of the emitters is spread over the frequency range of several of these receivers. Correlating in time over the output of any of these receivers will not give the desired processing gain. Recalling that the correlation function is the Fourier transform of the power density spectrum, we can estimate the desired correlation function from the combined power density output of the receivers (Fig. 3). The FET is formed on the sampled output of each receiver. This FFT is then multiplied by its complex conjugate to form the power density estimate for the output of that bandpass receiver. What we have is a number of

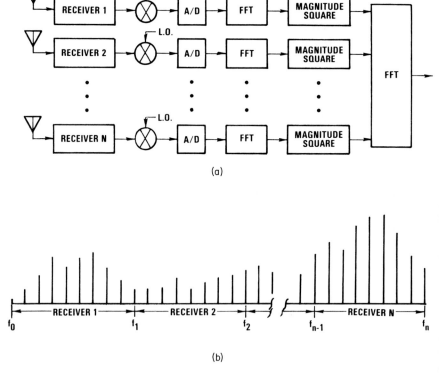

Fig. 3. Wideband correlation. (a) Block diagram. (b) Frequency estimates.

frequency estimates evenly spaced across the frequency band of each receiver as in Fig. 3b. These can be lined up to form the estimated spectral density for the entire bandwidth (or any subband of interest) of the receiver. Forming an FFT on this set of spectral samples gives an estimate of the desired correlation function. In this way, energy spread across several bandpass receivers can be coherently processed.

H Emitter Location

1 Estimates Needed

Location of an emitting system requires a significant amount of processing. The angle of arrival of the signal can be estimated, and from this an estimate of the direction of the emitter can be derived. If the direction of the emitter is known from two points, whose locations are also known, an estimate of the emitter location can be derived. If the ESM system is fixed, two receivers at different location are required. If the ESM system is in a moving platform, such as an aircraft, a single receiver suffices, and measurements can be made at successive points on the flight path.

2 Measurement Technique

Figure 4 shows the geometry for an emitter at location E and an ESM aircraft flying the path indicated. Assume that the direction to the emitter (relative to north) is ϕ_1 when the aircraft is at P_1 and ϕ_2 when the aircraft is at P_2. If P_1 has coordinates (X_1, Y_1) and P_2 has coordinates (X_2, Y_2), the range from P_2 to the emitter is

$$R_2 = \sqrt{(X_2 - X_1)^2 + (Y_2 - Y_1)^2} \sin(\phi_2 - \phi_1)/\sin(\phi_1 - \beta) \qquad (8)$$

where

$$\beta = \tan^{-1}(X_2 - X_1)/(Y_2 - Y_1). \qquad (9)$$

The estimated location of E can be computed as

$$X_E = X_2 + R_2 \sin \phi_2, \qquad (10)$$

$$Y_E = Y_2 + R_2 \cos \phi_2. \qquad (11)$$

Measurements of the angle ϕ are subject to significant error and the accuracy of the estimated position of E is quite sensitive to errors in the measurement of ϕ. Wegner [2] shows that the circular error probability (CEP) of the emitter location can be reduced by making successive estimates of X_E and Y_E and averaging these estimates.

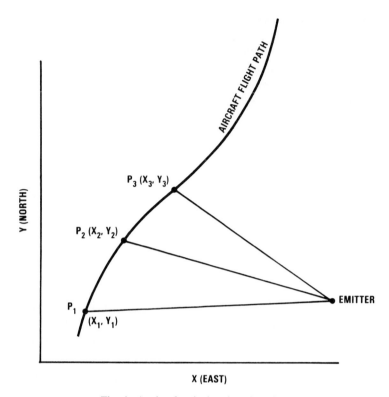

Fig. 4. Angle of arrival emitter location.

III ELECTRONIC COUNTERMEASURES

Actions taken to destroy or degrade the effectiveness of the enemies' electronic aids to warfare are classified as electronic countermeasures (ECM). These actions include jamming, deception, deployment of expendables, suppression, and destruction. Control functions associated with jamming, deployment of expendables, suppression, and destruction can benefit from application of VLSI. These control functions are system unique and are readily implemented with microprocessors. Guidance of munitions used in destruction of enemy emitters is a high-payoff application of VLSI but is not generally classified as an EW function.

A VLSI for Deception

The biggest opportunity for exploitation of VLSI in the ECM function is deception. For deception to be effective it must be transparent to the

40. Electronic Warfare Applications of VLSI 691

enemy so that he is unaware that any alteration of his transmission is taking place. Deception of radar usually requires that the radar pulse be received by the deception jamming system. The signal is then amplified, modulated, delayed or augmented, and retransmitted back to the radar signal. Intelligence in the jamming system is necessary so that alteration of the signal takes advantage of the characteristics of the radar. If done properly, the results can be the detection of false targets, range gate pull-off, or azimuth and elevation displacement. The deception jamming system depends on the ESM system for identification of the radar to be deceived. VLSI allows memories of sufficient capacity to store the characteristics and response for each radar encountered. Time is of the essence in deception jamming because the altered signal must appear at the radar receiver in time and of sufficient magnitude to appear to be the primary signal.

Simulated Signals   B

Another type of deception of increasing interest, now that the semiconductor technology can support the required storage and processing, is the injection of simulated signals into enemies' control systems that command actions desired by the jamming force. As an example, if the voice signals of a ground controlled intercept controller are simulated, the pilot can be commanded to take actions that cause him to cease to be a threat. He must not suspect that these are false signals. Only through the use of VLSI can this type of system be built, because the processing required to "learn" the sound of the controller's voice and then to simulate this sound would otherwise be prohibitive. However, through the use of the technology that gives us speaking toys and talking computers such deception capabilities are becoming possible.

ELECTRONIC COUNTER-COUNTERMEASURES (ECCM)   IV

Training of Personnel   A

ECCM has as an objective the retention of use of electromagnetic equipment despite the ECM employed by the opposing force. Equipments have been designed to operate in an ECCM environment. Of equal importance to the design features of the equipment is the training of personnel. The application of VLSI for ECCM is in implementation of equipment and in building ECM simulators and trainers.

B Equipment Characteristics

Traditional equipment characteristics for ECCM have included frequency agility, pulse compression, staggered PRF, side-lobe blanking, pulse-width discrimination, and a host of similar techniques intended to defeat the ECM system. All of these techniques increase the processing load. Many of the techniques mentioned above have been implemented in the analog portion of the system, but the trend today is to apply VLSI to a digital version of the technique of interest. The use of VLSI appears to be limited primarily by the speed of the A/D converters.

C Filter Design with VLSI

As systems become more complex, means of adapting to the electromagnetic environment will become increasingly important. A filter must be applied to the incoming time series that cancels the interfering signal and passes the desired signal. Since the spectrum of the interfering signal changes with time and location, the cancellation filter must be adaptive. One of the most powerful filter design techniques employed today is the maximum entropy method (MEM) [3] of determining filter coefficients.

If the sampled time series (sampled at constant rate Δt) of the environment, without the signal of interest, is

$$X_1, X_2, X_3, \ldots, X_{N-1}, X_N,$$

the autocorrelation function can be estimated by

$$\phi(\tau) = \frac{1}{N} \sum_{i=1}^{n-\tau} X_i X_{i+\tau}. \tag{12}$$

The maximum entropy spectrum is

$$P(f) = \frac{P_{N+1}/W}{\left|1 + \sum_{n=1}^{N} \Gamma_{n+1} e^{-i2\pi f n \, \Delta t}\right|^2}, \tag{13}$$

where W is the foldover frequency defined as $1/(2\,\Delta t)$, and P_{N+1} and Γ_{n+1} are obtained from the matrix equation

$$\begin{bmatrix} \phi(0) & \phi(1) & \cdots & \phi(N) \\ \phi(1) & \phi(0) & \cdots & \phi(N-1) \\ \vdots & & & \\ \phi(N) & \phi(N-1) & \cdots & \phi(0) \end{bmatrix} \begin{Bmatrix} 1 \\ \Gamma_2 \\ \vdots \\ \Gamma_{N+1} \end{Bmatrix} = \begin{Bmatrix} P_{N+1} \\ 0 \\ \vdots \\ 0 \end{Bmatrix}. \tag{14}$$

40. Electronic Warfare Applications of VLSI

Solving this matrix equation, we find that

$$\begin{Bmatrix} \Gamma_2 \\ \Gamma_3 \\ \vdots \\ \Gamma_{N+1} \end{Bmatrix} = \begin{bmatrix} \phi(0) & \phi(1) & \cdots & \phi(N-1) \\ \phi(1) & \phi(0) & \cdots & \phi(N-2) \\ \vdots & \vdots & & \vdots \\ \phi(N-1) & \phi(N-2) & \cdots & \phi(0) \end{bmatrix}^{-1} \begin{Bmatrix} \phi(1) \\ \phi(2) \\ \vdots \\ \phi(N) \end{Bmatrix}, \quad (15)$$

and

$$P_{N+1} = \phi(0) + \sum_{n=1}^{N} \phi(n)\Gamma_{n+1}. \quad (16)$$

The filter weights can be estimated either by inverting the matrix

$$\begin{bmatrix} \phi(0) & \phi(1) & \cdots & \phi(N-1) \\ \phi(1) & \phi(0) & \cdots & \phi(N-2) \\ \vdots & \vdots & & \vdots \\ \phi(N-1) & \phi(N-2) & \cdots & \phi(0) \end{bmatrix} \quad (17)$$

and solving for $\Gamma_2 \ldots \Gamma_{N+1}$ and P_{N+1} or by recursive techniques. Burg [4] presents one such technique.

$P(f)$ is the spectral estimate of the electromagnetic environment. Effective operation in this environment requires that interference be cancelled or at least reduced. From the estimated $P(f)$ a set of filter weights, which prewhiten the interference, is calculated and applied to the incoming time series for signal detection. This prewhitening of the environment is in addition to any matched filtering for the waveform of interest. Before the advent of VLSI, the processing required to design a filter based on maximum entropy techniques was prohibitive on all but large ground computers. Today, systems are being considered that readapt periodically in the field, particularly ground radar systems. With VLSI processing for this type of adaptation filter design can be performed on board an aircraft.

LIST OF SYMBOLS

σ	Standard deviation
π	Constant 3.14159...
Δt_{ij}	Interpulse spacing
τ_i	Interpulse spacing for pulse train from emitter i
ϕ_i	Azimuth to emitter from point i relative to north
$\phi(\tau)$	Autocorrelation function
β	Angle between two successive points on flight path, relative to north
$P(f)$	Power spectrum estimate

P_{N+1} Power output of $N + 1$ point prediction filter
Γ_j Prediction error filter coefficients
W Filter bandwidth

REFERENCES

1. AFP51-3. Department of the Air Force, 1 September 1978.
2. L. H. Wegner, "On the Accuracy Analysis of Airborne Techniques for Passively Locating Electromagnetic Emitters." U. S. Air Force Project RAND Report, June 1971.
3. J. P. Burg, Maximum entropy spectral analysis, *in* "Modern Spectral Analysis" (D. G. Childers, ed.), pp. 34–41. IEEE Press, New York, 1978.
4. J. P. Burg, A new analysis technique for time series data, *in* "Modern Spectral Analysis" (D. G. Childers, ed.), pp. 42–48. IEEE Press, New York, 1978.

Chapter 41
VLSI in Encryption Applications

CHARLES R. ABBRUSCATO
Racal-Milgo, Inc.
Miami, Florida

I. Introduction	695
II. Cryptography Overview	696
III. Data Encryption Standard	696
A. DES Basic Operation	696
B. Block Method	697
C. Stream Method	699
IV. Public Keys	702
V. Commercial Device Implementations	703
VI. VLSI Impact on Future	704
References	704

I. INTRODUCTION

The application of digital data communications is so widespread that it is difficult to imagine the world without it. As the flow of data increases, so does the vulnerability of that data to unauthorized disclosure and tampering. The ease and convenience of electronic funds transfer is a boon to the banking industry, but only if the message content remains secure. The ability to link data networks of geographically separated corporate facilities is of great value to the corporation, but possibly to an eavesdropper as well. Though the ability to make electronic financial transactions and to transport company secrets has created a new risk, the problems created by high technology can also be resolved by high technology. The use of VLSI in encryption devices has made the implementation of secure cryptographic algorithms not only possible but also practical.

II CRYPTOGRAPHY OVERVIEW

Cryptography is the art of encrypting or altering information into code or cipher, so that it is unintelligible to all but the intended recipient. The desire to encrypt information and the attendant desire to break the code have been with us since antiquity. What separates today's methods from methods such as Caesar's Cipher (a letter substitution cipher used by Julius Caesar), the ADFGVX Cipher (a simple product cipher used during World War I), and the Hagelin Machine (a mechanical rotor device used during World War II) is the ability to design and implement a cipher that is sufficiently complex to make brute force decipherment infeasible and cryptanalysis (i.e., shortcuts to decoding) virtually impossible. The ability to design such sophisticated devices is a result of the knowledge accumulated from previous methods and the capacity to use these techniques economically in high level integrated circuits.

One of the important characteristics of modern cryptography is the use of published algorithms. These algorithms encrypt data by means of user-generated digital words called *keys*. The best known and most widely accepted of these algorithms is the data encryption standard (DES) [1,2,3], which is approved by both the National Bureau of Standards (NBS) and the American National Standards Institute (ANSI). Over a half dozen manufacturers implement the DES on ICs. Two of the more popular versions are available from Advanced Micro Devices (Model No. 9518/Z8068) and Western Digital (Model No. 2001/2002). These are mature devices and represent the best of the first generation of DES ICs.

Maintaining the secrecy of the keys during key generation, distribution, and installation is an important task with ongoing expenses. Systems that can effectively transfer keys electronically, rather than physically, have an advantage. The strategies referred to as *public key methods* [1,2] have this ability. The best known and most widely scrutinized public key technique is the RSA algorithm, named after its co-inventors, Rivest, Shamir, and Aldeman. VLSI implementations of public key techniques have lagged behind the development of DES versions, primarily because of the relatively late start of public key, the lack of a standard algorithm, and the difficulty in attaining satisfactory data-throughput speeds.

III DATA ENCRYPTION STANDARD

A DES Basic Operation

The DES is essentially a complex series of permutations and substitutions to a block of 64 bits of data under control of a 64-bit key variable.

41. VLSI in Encryption Applications

The DES algorithm is structured so that each bit of the 64-bit output is dependent on each of the 64 input bits and each of the 56 information bits in the key (eight bits of the 64-bit key are used as parity bits). In addition to the obvious requirement that it be cryptographically secure, the DES was also designed with hardware implementation in mind. Encryption and decryption use the same key and the same algorithm, with only a minor variation on how the keys are manipulated internally.

The DES algorithm performs its encryption and decryption operations with a complete block of 64 bits at one time. However, the cryptographic system, the DES can be used with more flexibility. [1,3,4] The two basic categories of system operation for the DES are the *block method* and the *stream method*. In each of these two methods, there are some important variations.

B. Block Method

1. Electronic Code Book (ECB)

The basic block cryptographic method is called the electronic code book (ECB) mode of operation. In this mode, the plain text block of data to be encrypted is loaded into the DES input register, and the DES register yields the encrypted cipher text (Fig. 1). The same plain text always produces the same cipher text for a given key. Because of the complex

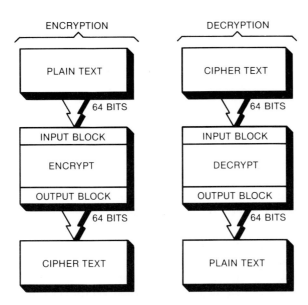

Fig. 1. Electronic codebook (ECB) mode.

function of the DES, a single-bit error in the cipher text block or in the nonparity key bits will propagate throughout the entire 64-bit block. However, that error will not affect any other blocks. These characteristics make ECB unacceptable for data communications with rigid protocols and format messages. On the other hand, ECB is a desirable method of providing secure system management messages (e.g., encoding keys for downline loading). In any case, it is imperative that the block boundaries be established and maintained between the encrypting and decrypting equipment.

All DES ICs support the ECB mode of operation. With the addition of buffer registers, this basic operation can be enhanced to create different modes of operation with critical performance characteristics.

2 Cipher Block Chaining (CBC)

To eliminate the shortcoming of block independence in the ECB mode, a feedback arrangement is added to create an operating mode variation called cipher block chaining (CBC). In this mode, the plain text block to be encrypted is exclusive-ORed with the cipher text from the previous block; the sum is then put into the DES for processing (Fig. 2). The AMD 9518 and the Z8068 (a higher speed but otherwise identical version) do this internally, eliminating the need for external registers. Note in Fig. 2 that an error during transmission will result in two receive blocks of data (128 bits) being in error. For error-prone transmission lines, this could produce an unacceptably high net error rate.

An important characteristic of CBC is that each block in the encryption process is dependent on every previous block, with the final block being dependent on every other block. This characteristic can be employed as a sophisticated checksum, providing a powerful method for authenticating or verifying that a message has not been altered., Using CBC, the transmitting end of the system routes the message to be authenticated through the encryption process. When the last block is processed, the most significant bits become the message authentication code (MAC). In the case of financial transactions, the MAC is 32 bits long. When the message is transmitted, the MAC is added to the end of the message. The receiving equipment similarly calculates a MAC from the received message and compares it with the MAC from the transmitting end. If they match, the received message is exactly the message sent, with no errors or alterations.

In addition to the requirement that both ends of the transmission system have the same key and be block synchronized, they must also have, for the first stage of encryption and decryption, the same initial block of bits. This block, known as the initialization vector (IV), must be distributed along with the cryptographic keys under a high level of security.

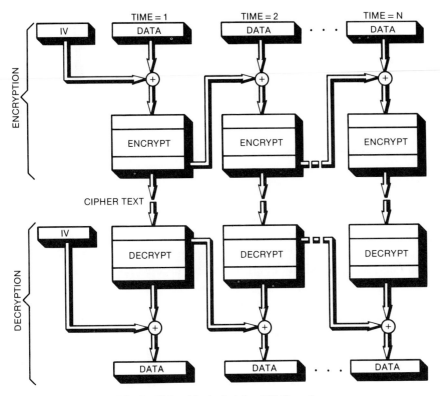

Fig. 2. Cipher block chaining (CBC) mode.

Stream Method C

In the stream cipher method, the DES algorithm (see Fig. 3) is used as part of a bit stream generator whose output is combined with the plain text (for encryption) or cipher text (for decryption).

Cipher Feedback (CFB) 1

An important type of stream cipher is the cipher feedback (CFB) mode of operation. Figure 4 shows a K-bit CFB arrangement in which the cryptographic bit stream uses the K most significant bits of the DES operation (where K can be any integer from 1 to 64). The K most significant bits of the DES output block register are exclusive-ORed with K plain text bits to produce K cipher text bits. The K cipher text bits are loaded back into the K least significant bits of the DES input block register, after the input register is shifted by K bits. The data throughput in this

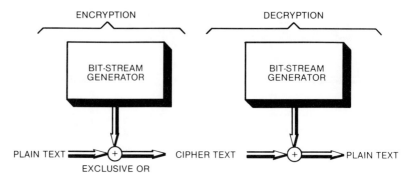

Fig. 3. Stream cipher.

mode is $64/K$ times slower than ECB or CBC because of the $64 - K$ unused bits. As with CBC, to start up error free, synchronization must be assured, and an IV must be used. However, unlike CBC, as long as the system is in synchronization, errors will propagate only until the error bits are shifted out of the input block register. Synchronizing the block bound-

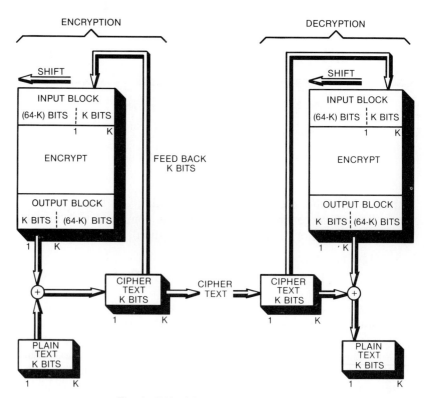

Fig. 4. K-bit cipher feedback (CFB) mode.

aries is just as critical with CFB as it is with ECB, but with CFB it is possible to use a one-bit wide feedback block. This single-bit CFB operation inherently requires no synchronization patterns or additional circuitry and is extremely valuable because it is protocol transparent. However, speed is sacrificed; even the 14-Mbit/s AMD Z8068 device can only process a data stream at just over 200 Kbit/s.

Output Feedback (OFB)   2

Another stream cipher similar to CFB is the output feedback (OFB) mode of operation (Fig. 5). OFB uses essentially the same logic elements external to the DES IC as CFB except that the feedback connection is slightly different. In this mode the feedback is taken directly from the DES output. Since the cipher text is not involved, errors are not extended. That is, one error in cipher text produces one error in decrypted plain text. Though this may be desirable for noisy transmission channels, it is vulnerable to undetected message modification. OFB is not self-

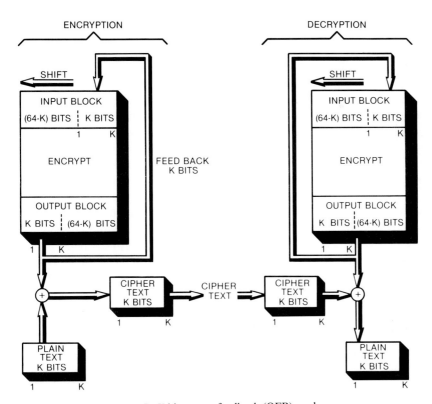

Fig. 5. K-bit output feedback (OFB) mode.

TABLE I

Typical Applications for DES Modes of Operation

Mode	Application
Electronic code book (ECB)	Secure service messages (e.g., keys, IV)
Cipher block chaining (CBC)	Authentication
	Data communication where error extension is desirable
Cipher feedback (CFB)	Data communications where protocol transparancy is required (single-bit CFB)
Output feedback (OFB)	Data communications over noisy channels

synchronizing; so if synchronization is lost, the system must be reinitialized with a new IV after synchronization of the block boundaries is attained. Table I shows typical applications for these modes of operation.

IV PUBLIC KEYS

As previously mentioned, the DES algorithm requires that both ends of the communications link have the same key. These keys must be distributed with a high level of security, in order not to undermine the security offered by the DES. This sometimes awkward, costly process can be avoided with a method called *public key cryptography*. With public key systems, two different but related keys are used: one for encrypting the data and one for decrypting the data. The encrypting key is made public, whereas the decrypting key is kept secret. Knowledge of the encrypting key alone is insufficient to determine the secret decrypting key except by brute force, which presents a prohibitive work factor. In Fig. 6, if A wishes to send a secret message to B, he obtains B's public encrypting key E_B from the public file; he then uses E_B to encrypt the message. This message can be decrypted only by the decryption key D_B, which is known only to B. Similarly, B uses A's public encrypting key E_A to send secure messages to A. Note that a public key file is not necessary in this simple case. The system can be designed so that each party merely asks the other party to send its public encrypting key over the communications link. In these arrangements, all communication is electronic, providing a convenient alternative to the DES requirement of manual key distribution.

In comparison with the DES, there are drawbacks to the public key at this time. There are no recognized standards, and commercially available LSI implementations have been slow in emerging. However, because

41. VLSI in Encryption Applications

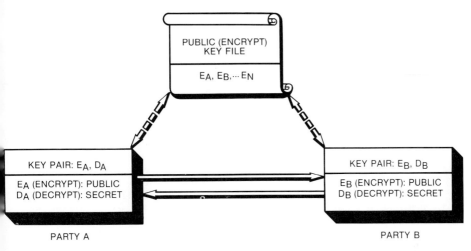

Fig. 6. Simplified public key system.

public key is computationally intensive, and data throughput is important, VLSI is the only hope to make higher speed public key systems commercially practical. At present, there are slower firmware implementations available, which are used in an efficient hybrid key management strategy. With these, the public key technique is used to transfer DES keys electronically, so that the DES can be used for the data communications.

V COMMERCIAL DEVICE IMPLEMENTATIONS

The DES algorithm is fairly new (1977), and the device implementations are varied. On one end of the spectrum are the low cost, low speed (3200 bits/s) firmware-driven devices such as Intel's 8294/8294A and TI's 99541. Other manufactures have taken the VLSI route to achieve higher speeds Motorola's model 6859, with its 400 Kbits/s capability, offers the lowest cost of the higher speed versions. Wester Digital's 2001/2002 higher speed (1.3 Mbits/s), higher cost devices represent a good value where high performance is required. Topping the field for high performance (14 Mbits/s) and cost are the AMD 9518 and Z8068. The AMD devices justify their high cost with extra internal registers for modes other than ECB; they can operate in the CBC mode an in an 8-bit CFB mode. Table II lists available DES encryption ICs in order of performance and price.

Higher speed operation can permit one device to perform both encryption and decryption, thus saving the cost of an extra device. In addition,

TABLE II

Available DES Encryption ICs[a]

Manufacturer	Model No.	Speed (ECB mode)	Comments
Intel	8294	640 b/s	Preprogrammed 8041A
	8294A	3.2 Kb/s	Preprogrammed 8042
Texas Instruments	99541	3.2 Kb/s	Preprogrammed TMS 7020
Motorola	6859	400 Kb/s	—
Western Digital	2001	1.3 Mb/s	
	2002		2002 has dual (8-bit) ports
Fairchild	9414	13.3 Mb/s	4-IC set
AMD	9518/Z8068	14 Mb/s	Operates in ECB, CBC, and CFB modes

[a] In order of increasing performance and price.

for K-bit CFB or OFB, the actual operating speed is 64/K times slower than the device rates stated above. Thus, single-bit CFB on the WD2001 yields a throughput of about 20 Kbits/s. If that device must also work in both directions of transmission, the throughput is reduced to about 10 Kbits/s.

Currently, no public key VLSI devices are commercially available, although several companies (such as RSA Security, Inc.) are working on such devices.

VI VLSI IMPACT ON FUTURE

VLSI has made the DES algorithm practical and available to those who need it. VLSI will also make public key methods practical for general use. Encryption and authentication will become more affordable by reducing the cost and size of the support circuitry necessary for a general purpose product with user-friendly features. Such features will include multimode DES capability, DES diagnostics, system diagnostics, alarms, clear mode, standby mode, interlock disable, displays, and so on. VLSI will make it reasonable to attain a high level of data security for almost any application from electronic funds transfer to personal computers.

REFERENCES

1. W. Diffie and M. E. Hellman, Privacy and authentication: An introduction to cryptography, *Proc. IEEE* **67**, 397–427 (1979).

2. C. H. Meyer and S. M. Matyas, "Cryptography: A New Dimension in Computer Data Security." Wiley, New York, 1982.
3. U. S. Department of Commerce, National Bureau of Standards, Federal Information Processing Standards (FIPS) Publication 46: "Data Encryption Standard." National Technical Information Service, Springfield, Virginia, 1977.
4. U. S. Department of Commerce, National Bureau of Standards, Federal Information Processing Standards (FIPS) Publication 81: "DES Modes of Operation." National Technical Information Service, Springfield, Virginia, 1980.

Chapter 42

Application of VLSI to Radar Systems

THOMAS K. LISLE
JOHN J. ZINGARO
Westinghouse Electric Corporation
Baltimore, Maryland

I. Overview	707
II. Functional Requirements and Radar Overview	708
A. System Requirements	708
B. Subsystems	708
III. Key VLSI Microelectronic Technologies	711
A. Digital VLSI	711
B. Analog VLSI	714
IV. Summary	719
References	720

I. OVERVIEW

Until the 1980s, microelectronic growth was principally limited to improving the density (gates per circuit) of standard LSI. However, with the advent of low-cost consumer items such as the single-chip calculator and the digital quartz watch, consumers tasted high technology and their appetite since has been insatiable.

With the resulting growth in the commercial industry, an acute awareness came to the defense community that the military could benefit from the new reaches of microelectronic growth, especially in high technology sensor systems such as radar. Historically, radar has been the military system that places the strictest demands on microelectronic technology. Wideband signals, high-speed data rates, extensive memory capacity, adaptive antenna control, and solid-state power generation are but a few of the radar requirements driving technology development under the general category of VLSI electronics. It is the purpose of this chapter to identify and illustrate briefly those key VLSI techniques that are having the greatest impact on radar systems.

II FUNCTIONAL REQUIREMENTS AND RADAR OVERVIEW

To clarify the relationship between technology and performance, an examination of the functional requirements of a future radar system is necessary. Radar is an electromagnetic system that is used for the detection and location of objects. It operates by transmitting a signal, usually a pulse-modulated sine wave, and then detecting and analyzing the echo signal. Radar does not have the resolution of the human eye but can "see" through conditions, such as darkness, haze, fog, rain, and snow, that are difficult or impossible for human vision. Radar also has the advantage of being able to measure the distance to the object that it sees.

A System Requirements

Radar systems have always measured parameters such as range to the target, target size, and the relative speed and direction of the target. Range, velocity, and azimuth require signal-processing techniques that use conventional LSI technology. However, as avionic systems require additional performance from radar, new radar modes emerge that accelerate the required technology (Table I). The relationship between requirements and technology indicate that wider signal bandwidths and hence high-speed signal processing requirements are emerging across all system specifications.

B Subsystems

The radar system comprises several subsections (Figs. 1 and 2). Each of these subsystems has unique technology requirements so that when the subsystems are connected by a digital data bus, the result is a system that performs the selected radar modes. In configuring a radar system, the commonality of VLSI technology across the boundaries of the radar subsystem becomes apparent.

In the simple radar system of Fig. 1, the transmitter serves as the source of microwave energy which is modulated to meet the radar mode requirement. This energy is coupled into free space and is focused and

42. Application of VLSI to Radar Systems

TABLE I

Radar Modes Drive Technology

Radar modes	Definition	Requirement that drives technology	Technology
Adaptive processing[a]	Responds to electromagnetic environment	Fast computation	VLSI, GaAs
Doppler processing	Extract doppler shift from radar return signal	Fast, complex arithmetic processing	VLSI
Target ID[a]	Extract target characteristics from radar signal return	Wide bandwidth, narrow pulse	VLSI, GaAs, fast ADC
Automatic terrain follow	Allows avionic system to follow and avoid terrain variations at rapid air speed	Measure scene variation and landmark changes	VLSI, memory, fast ADC
Automatic terrain avoidance			
Navigation	Accurate position of aircraft	Map matching to scene	Bulk memory, VLSI, high speed data rate
Monopulse MTI[a]	Tracking of moving targets	Multiple rf channels with accurate balance	VLSI, bulk memory, monolithic rf circuits
Track-while-scan	Ability to update position of targets while scanning radar beam for other modes	Rapid antenna beam agility, multiple data track files	VLSI, bulk memory, GaAs phase shifters
SAR map[a]	Synthetically generated large antenna aperture for high-resolution map functions	Wide bandwidth, narrow pulse	VLSI, CCD, high-speed ADC
Low probability of intercept[a]	Various techniques to deny interception of information by unwanted receivers	Wide bandwidth, coded waveform	SAW, CCD, monolithic, rf circuits

[a] Greatest impact drivers

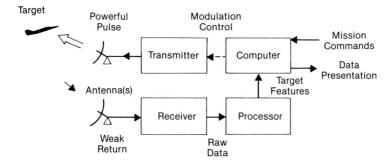

Fig. 1. How major radar subsystems relate.

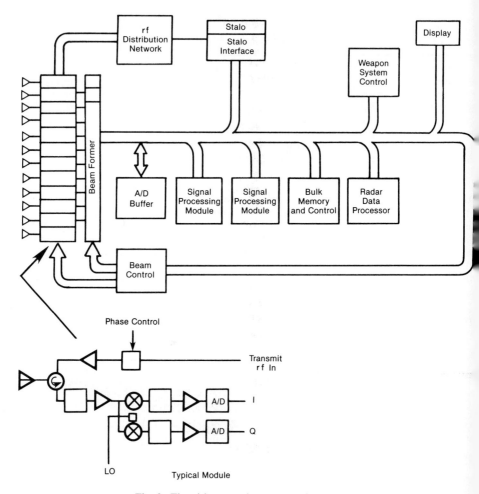

Fig. 2. The airborne radar system of the 1980s.

directed by an antenna. Targets (objects intercepting this energy) reflect a small part of this energy, which is coupled back through the antenna into a receiver that amplifies the weak return signal and removes the carrier frequency, leaving the modulated signal. This return signal is converted from analog to digital data and characterized for specific applications in a processor. Here the data are manipulated by mathematical algorithms to extract features that distinguish multiple targets and/or desired targets from unwanted targets (clutter). Target reports with their unique features such as direction, distance, radial velocity, and cross section (size) are fed into a computer that configures the processed data into a meaningful form.

Figure 2 is a generic block diagram of an advanced radar system that is

42. Application of VLSI to Radar Systems

expected to represent the system architecture of the late 1980s. The principal function blocks and their characteristics are as follows.

(a) A solid-state array antenna consisting of hundreds of rf transmit–receive modules and an electronic rf phase shifter used for beam forming and electronic scanning.

(b) An rf distribution network, usually a weighted network, required to distribute the rf energy with a minimum of amplitude and phase errors.

(c) Stalo (stable local oscillator), the source of the rf signal, requiring extreme precision in frequency and phase and amplitude stability.

(d) Stalo interface, a digital interface to the system control bus. Commands to the stalo are delivered along the bus and stalo modes, and bit status is returned.

(e) The A/D buffer, a large RAM that allows the digital outputs of the module to be stored, ordered, compiled, and reduced in speed.

(f) The signal processing module, a compact digital-processing architecture that uses common electronic functions to extract specific parameters of the radar signal. Consists of memory, multipliers, and adders which can be paralleled to form multiple signal-processing elements.

(g) Bulk memory and control, in which the program memory required for the control of the radar is stored. Mode changes can be instituted by external programming control.

(h) The radar data processor, which uses the data from the signal-processing module to execute algorithms that perform the required radar modes.

III KEY VLSI MICROELECTRONIC TECHNOLOGIES

Rapid growth in the microelectronics industry, as illustrated by memory growth trends in Fig. 3, has spurred radar developments. New radar modes that demand more signal bandwidth generate higher speeds and greater complexities in signal processing. Furthermore, the ability to perform the function in a smaller package has opened up a new spectrum of applications for radar systems. This section briefly explores those key VLSI microelectronic technologies that will most influence radar development.

A Digital VLSI

Digital VLSI can be broadly classified into three categories: *custom devices, memories,* and *gate arrays.* Radar signal processors combine

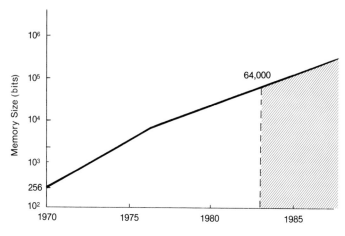

Fig. 3. Memory growth.

these VLSI functions with standard LSI and MSI functions. Figure 4 illustrates the effect VLSI will have on two typical radar signal processors as a function of time. However, the impact of VLSI will not be on size alone. Future systems will use dense functional blocks to increase the performance of the radar system and to incorporate new modes that were not previously feasible (Fig. 5). The constant-volume growth curve indicates the integration of new capabilities into a system. The constant-performance growth curve is indicative of new applications that are feasible because of reduced size and cost.

VLSI has created a new set of capabilities. New types of computer architectures such as those listed in Table II will be needed to address more complex scenarios. Many of the advances in digital VLSI will not result in strictly higher throughput. Some computing power will be used to ease the software designer's problems. Figure 6 shows the number of lines of software coding growing exponentially. While the cost per line has fallen, productivity improvements have not matched software growth. Higher-order languages will ease the software burden. In addition, some hardware will be dedicated to reliability, testability, and maintainability. A greater percentage of each integrated circuit will be dedicated to testability and fault tolerance (Fig. 7). Testability, combined with fault tolerance and programmability, will allow the signal processor to reconfigure under fault conditions and continue operation.

Though GaAs logic promises improvements over silicon logic (Table III), it remains an immature process and will require several years to enter the VLSI domain. A factor of 3 to 5 improvement in delay-power product is expected with equivalent lithography, and technologies such as the high electron mobility transistor (HEMT) could increase performance further.

42. Application of VLSI to Radar Systems

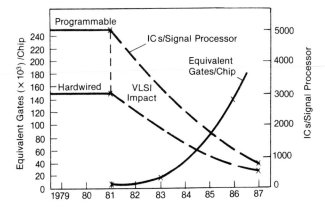

Fig. 4. VLSI shrinks system size.

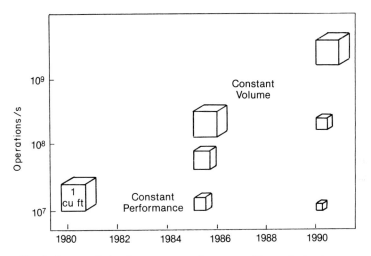

Fig. 5. Increased signal processor performance with constant volume.

TABLE II

Advanced Signal-Processing Architectures for Radar

Architecture	Functions
Programmable arrays	FFT, filters, convolution
Systolic arrays	Matrix operations, image processing
GP computers	Control, fixed algorithms
AI/expert systems	Addresses open ended or variable problems

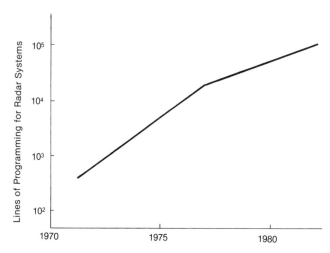

Fig. 6. Growth in software for radar systems.

LSI devices have been demonstrated in laboratories, and commercial SSI and MSI devices are expected in 1984.

B   Analog VLSI

Though the most visible impact of VLSI technology has been in digital circuitry, analog circuits have shown a vast improvement in complexity and accuracy. Differing demands for analog circuits, such as high dynamic range and low noise performance, prevent such circuits from ap-

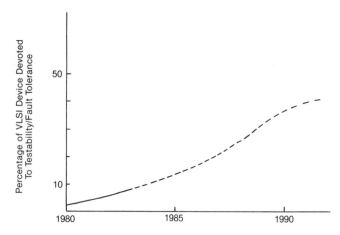

Fig. 7. Trend toward greater testability.

42. Application of VLSI to Radar Systems

TABLE III

Technology Comparison

Property	T²L	NMOS	Current technology CMOS Bulk	Current technology CMOS SOS	1985–1990 SOS	1985–1990 GaAs
Relative process maturity (1–10)	10	9	8	4	2	1
Process complexity (number of processing steps)	18–22	9–15	14–17	14–20	14–20	16
Logic complexity (number of components 2-input gate)	12	3	4	4	3–4	2
Packing density (gates/mm²)	10–20	100–200	40–90	100–200	200–500	300–1000
Propagation delay (ns) (typical value)	6–30 (10)	4–25 (15)	10–35 (20)	4–20 (10)	0.2–0.4 (0.3)	0.05–0.1 (0.07)
Speed–power product (pJ)	30–150	5–50	2–40	0.5–30	0.1–0.2	0.01–0.1
Typical supply voltage (V)	+5.0	+5.0	+10.0	+10.0	+20	+12
Signal swing (V)	0.2–3.4	0.2–3.4	0.0–10.0	0.0–10.0	0.0–2.0	0.0–0.8
Guaranteed noise margin (V)	0.3–0.4	0.5–2.0	3.5–4.5	3.5–4.5	0.2–0.8	0.2–0.3

proaching the density or complexity of microprocessor or memory circuits. But the complexity and the size of the circuits still classify them as VLSI circuits.

Radio Frequency Circuitry

Until recently reducing the physical size of microwave electronic assemblies was limited to hybrid stripline, using discrete devices. Now, monolithic circuits, such as the 3-W X-band power amplifier in Fig. 8, replace many discrete microwave elements. Radar applications using phased arrays and active apertures will incorporate monolithic GaAs and silicon blocks. Figure 9 illustrates a typical radar transmit–receive module. As this technology matures, multiple blocks in the T/R module will be integrated with control logic and interfaces to produce low-cost VLSI microwave circuits.

CCD Technology

Charge coupled device (CCD) technology blurs the distinction between the analog and digital domains. As shown in Fig. 10, a CCD acts as an

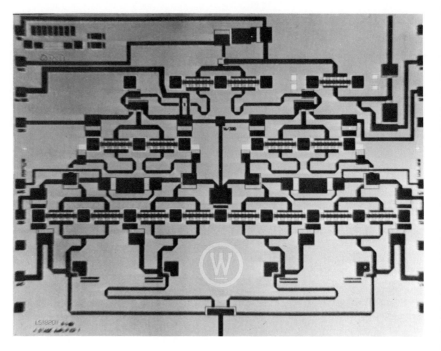

Fig. 8. 3-W, X-band amplifier.

analog sampling device and multiple delay lines. Table IV lists some applications for CCDs. While moderate-speed CCDs are mature, these wideband elements generally require operation at frequencies from hundreds of megahertz to more than a gigahertz. GaAs technology will be required to implement many of these functions.

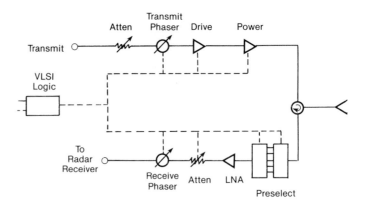

Fig. 9. Radar T/R module.

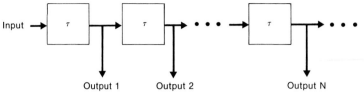

Fig. 10. CCD tapped-delay line structure.

TABLE IV

Radar Applications for CCDs

CCD	Application
Transversal filters	Preselection, ECCM
Adaptive filters	Postdetection integration
Time compression and expansion	Synthetic aperture radar and high resolution mapping
Correlation and convolution	LPI radar, ECCM

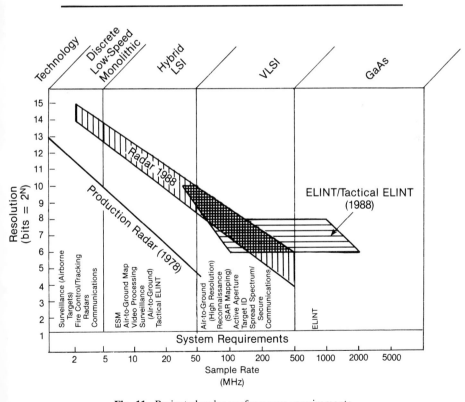

Fig. 11. Projected radar performance requirements.

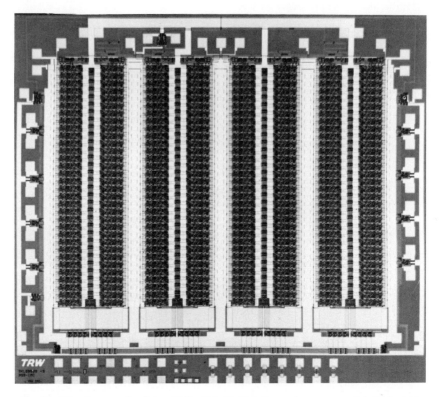

Fig. 12. Die photograph of the TDC 1025, TRW LSI products 1-μm, 60-MSPS, 8-bits flash A/D converter.

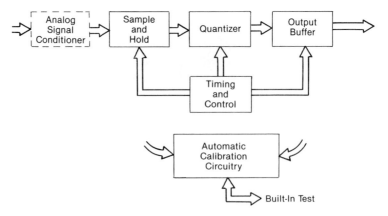

Fig. 13. Typical ADC subsystem block diagram.

3 Acoustic Technology

Acoustic technology performs filtering and signal synthesizing functions at intermediate frequencies (IF) of radar systems. Surface acoustic wave (SAW) and bulk acoustic wave (BAW) are used in a wide variety of applications. Some radar applications for acoustic devices are

(a) nondispersive delay lines,
(b) bandpass transversal filters,
(c) matched filters,
(d) local oscillators,
(e) correlations,
(f) pulse dispersive filters, and
(g) compressive filters.

These devices have not been compatible with IC technology. A new technique that is compatible with both silicon and GaAs integration has been developed [1]. With this monolithic film resonator technology, narrowband filters and resonators can be integrated with active circuitry. Decreasing filter size by orders of magnitude permits full monolithic integration of receiver subsystems. Such devices could significantly reduce the size of radar receivers and synthesizers.

4 Analog to Digital Converter Technology

Analog to digital converters (ADC) have historically been a choke point in limiting radar system performance. Increasing system requirements have steadily pushed requirements for ADCs (Fig. 11). However, the development of VLSI circuits such as the TRW 8-bit 60-MHz quantizer (Fig. 12) and the use of microprocessors in autocalibration circuits (Fig. 13) allows ADCs to meet system requirements up to 50 MHz. Performance beyond this level will require further development of GaAs or other advanced technology.

IV SUMMARY

In the 1980s the impact of microelectronics on radar systems is going to be dramatic. Radar system requirements demand more of technology than ever before. VLSI will allow the reduction of signal processor volume by allowing a significant number of processing functions to be placed upon a single chip. These chips, when combined with the gate-array technology, will permit an affordable, programmable, signal-processing module that can be reconfigured for each radar mode.

The emergence of monolithic technology in the analog and microwave domains will allow the development of an all-solid-state active aperture radar with the growth to a full digital beam formed radar by the end of the decade. [2] [3] [4]

REFERENCES

1. G. R. Kline and K. M. Lakin, 1.0 Ghz Thin Film Bulk Acoustic Wave Resonators on GaAs, *Appl. Phys. Lett.*, **43**, (8) 750–751, October 15 (1983).
2. M. I. Skolnik, "Introduction to Radar Systems," 2nd ed. McGraw-Hill, New York, 1980.
3. M. Shohat, Radars for the eighties, *Mil. Electron./Countermeasures,* **5,** (11) 34–40, November (1979).
4. W. R. Harden and T. K. Lisle, Impact of microelectronics upon radar systems, *in* "VLSI Electronics: Microstructure Science," vol. 5 pp. 297–328. Academic Press, New York, 1982.

Chapter 43

Medical Applications of VLSI Circuits

JACOB KLINE

Department of Biomedical Engineering
University of Miami
Coral Gables, Florida

I. Introduction	722
II. Dual-Chamber Programmable Implantable Pacemaker	722
A. Capabilities	722
B. Architecture	722
C. The CPU	723
D. Programmable Pacing Modes	724
III. Digital Hearing Aid	724
A. Advantage over Analog Aids	724
B. The I/O Subsystems	725
C. Technologies: 1981 versus 1985	725
IV. Computerized Tomography	726
A. Types of Scanners	726
B. Architectures of Scanners	727
C. Use of VLSI	728
V. Ultrasound Imaging	728
A. Method of Operation	728
B. Commercially Available Systems	728
C. Architecture of Ultrasound Systems	729
D. Obtaining Permanent Records	730
VI. A Computerized Local Area Network for an Intensive Care Unit	730
A. Topology	730
B. Data Transmission	730
VII. Evoked Potentials	731
A. VLSI versus Hard-Wired Averagers	731
B. Analysis of Evoked Potentials	732
VIII. Neural Stimulators	733
A. Applications	733
B. Types of Implants	733
C. Use of VLSI	733
D. Closed-Loop Controllers	734
References	735

I INTRODUCTION

VLSI circuits have made possible the inclusion of specialized parameters and modalities in medical instrumentation while effecting a reduction in cost, size, weight, and power consumption. As a result, highly sophisticated, implantable controlling devices such as pacemakers and neurostimulators, computerized critical care, and tomographic instrumentation have emerged. Commercial VLSI microprocessors are available for much of the medical instrumentation that is to be designed and extended in capability. However, for some devices, such as the pacemaker, implantable stimulator, and digital hearing aid, custom VLSI chips are usually required. This chapter presents a cross section of the utilization of VLSI circuits that has made significant adances in medical instrumentation possible. These applications are sufficiently varied and heterogeneous so that one could extract sufficient information to apply many of the circuits to the development of other types of devices and systems.

II DUAL-CHAMBER PROGRAMMABLE IMPLANTABLE PACEMAKER

A Capabilities

A pacer that incorporates a complex VLSI-based computer can provide many input–output capabilities that were unobtainable before. It can sense and pace both chambers of the heart in any complex sequence. Permanent programs in ROM and physician-alterable programs in RAM allow the pacer to function in many different operating modes [1].

B Architecture

The architecture of the pacer, with a single VLSI microprocessor of less than 200 mils on a side, is fabricated from CMOS technology (Fig. 1). The chip incorporates a 32 × 4 RAM, programmable logic array (PLA), arithmetic logic unit (ALU), and branch logic. The characteristics of the various sections that are incorporated in the VLSI chip are as follows:

(a) The *sense amplifier* has a sensitivity programmable from 0.5–2.5 mV, a 15–20 ms recovery time from a 160-mV overload, an overall gain of

43. Medical Applications of VLSI Circuits

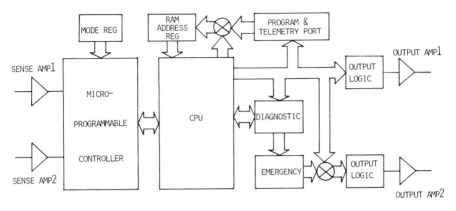

Fig. 1. Architecture of implantable programmable dual chamber pacemaker with CMOS VLSI chip that measures less than 200 mils on a side.

10,000, and a frequency response set by three external capacitors that ranges from 15 to 150 Hz.

(b) The *output amplifiers* are independently programmable to one of eight output amplitudes and pulse width combinations. Output current range is 1–12 mA. Output is capacity coupled to the stimulating electrode.

(c) The *CPU* shuts down when activity is zero. Average current drain is 6 μA.

(d) *Programming* of eight different programmable pacer modes is available. User programmable rates are 30–120 beats/minute; refractories and delays, 150–600 ms; and input sensitivities and atrial-ventricular delay times, 0 to 300 ms. Output currents are achieved with pulse-width modulated data telemetered under the supervision of a physician. The CPU echoes back the programmed data for verification. Illegal or unsafe programs are rejected.

The CPU C

The CPU provides a bidirectional function with the microprogrammable controller, which in turn offers multiple modes of response to the inputs of the sense amplifiers and to programmed modalities. The CPU provides timing, programming, telemetry, and diagnostic and output routines based on inputs from the controller. The indirect addressing of the RAM is done through a register connected to the telemetry port. The system clock, battery voltage, and fault conditions are checked and detected by the diagnostic circuitry [2].

D Programmable Pacing Modes

In addition to the above parameters and functions, a wide variety of single- and dual-chamber pacing modes are programmable, including algorithms that respond automatically to certain predefined cardiac arrhythmias. For example, a typical operating mode might require that the cardiac signal be monitored in both the atrium and ventricle simultaneously and that the pacer respond by either inhibiting an output or providing a pacing stimulus. The inhibited pacing responses may be independently determined for each chamber and the time delay between the responses may be a function of several other parameters. This type of operating mode is realized almost instantaneously and safely, using the computer capability of the single VLSI chip IC microprocessors.

III DIGITAL HEARING AID

A Advantages over Analog Aids

The block diagram of a unified digital hearing aid (DHA) with VLSI CMOS circuitry is shown in Fig. 2. The DHA offers the following advantages over an analog unit:

(a) lower signal-to-noise ratio,
(b) wider dynamic range,
(c) frequency selective gain,
(d) maximum power output over a range of 200 to 6000 Hz,
(e) real-time processing,
(f) reduced power consumption,
(g) small size,
(h) ease of fitting,
(i) rapid reprogramming to achieve arbitrary frequency responses, allowing a variety of patients to be serviced with a single design,
(j) less subject to degradation with age and mechanical shocks, and
(k) nonlinear amplification and control of temporary factors.

43. Medical Applications of VLSI Circuits

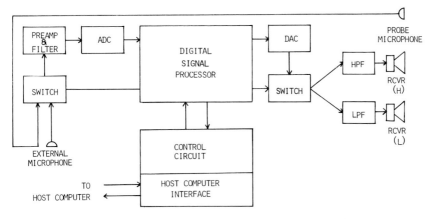

Fig. 2. Unified digital hearing aid with VLSI CMOS circuitry.

The concept of the VLSI DHA can be projected into other sound systems.

B The I/O Subsystems

The input subsystem consists of microphones that incorporate an antialiasing filter, preamplifier A/D converter, and a control channel that selects the microphone to be used. The audio output of the selected microphone is amplified and filtered by the preamp and antialiasing filter to emphasize signal frequencies between 200 and 6000 Hz and to attenuate signal frequencies above 6000 Hz. The digital signal processor with a single VLSI chip, TI TMS 320, provides independent gain and power limiting in 16 frequency bands, in real time at a sampling rate of 12,000 samples/s.

C Technologies: 1981 versus 1985

Chip size and power requirement for the necessary circuit complexity is available with 1981 VLSI technology using a speech synthesizer CMOS chip containing 60,000 transistors, operating at 1 Mhz, 5 V and consuming 30 mW. With 1985 VLSI technology, chips should be available that consume only 7 mW to perform the same function; with projected 1988 technology, power consumptions could be as low as 1.3 mW. Considering the

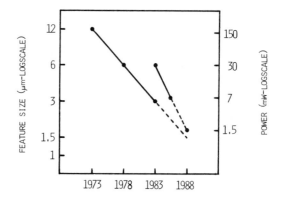

Fig. 3. Feature size of commercial ICs and estimated power consumption of 60,000-transistor speech-synthesis IC. (Courtesy of R. E. Morley, Washington University, St. Louis.)

1981 speech synthesizer device as a base and projecting along the feature size curve, Fig. 3, the 5 × 5 mm chip in 1981 is reduced to 3.3 × 3.3 mm in 1985 and to 2.5 × 2.5 mm in 1988 [3].

IV COMPUTERIZED TOMOGRAPHY

A Types of Scanners

The internal structures of the anatomy can be visualized and diseases diagnosed by applying imaging techniques such as

(1) the transmission and detection of X rays, computed axial tomography (CAT scanning);
(2) the detection of the decay products of an ingested radionuclide, positron emission tomography (PET scanning);
(3) the measurement of the relaxation times of protons subjected to a magnetic field, nuclear magnetic resonance (NMR).

Although the sources and characters of the radiation are different in each of the imaging techniques, the systems all reconstruct the final image from a sequence of measurements taken across the physiological sample volume or slice. Therefore, the storage, computing control, and display methods are fundamentally similar in each case [4,5].

43. Medical Applications of VLSI Circuits

Architectures of Scanners B

The block diagram in Fig. 4 shows a composite of the various scanner modalities.

(a) The *CPU* controls patient position, gantry motors, magnetic flux, and so on during the scan and supervises data flow during the reconstruction process.

(b) The *array processor,* in conjunction with the CPU, executes the reconstruction algorithm.

(c) The *disk memory* stores measurement and orientation data during the scan.

(d) The *ROM* contains the system firmware.

(e) The *RAM* stores intermediate results of computation and acts as disk buffer.

(f) The *keyboard* allows operator interaction with the instrument for selection of scan rate, section to be reviewed, display modes, and so on.

(g) The *display* is for viewing the reconstructed images.

(h) The *video cassette recorder* (VCR) can store the reconstructed images for record and subsequent reappraisal.

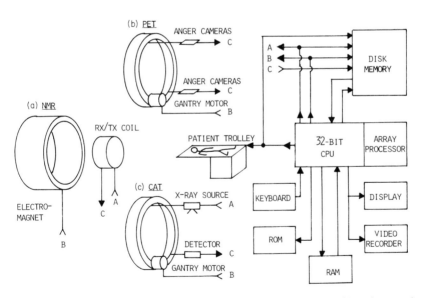

Fig. 4. Composite diagram of (a) nuclear magnetic resonance (NMR), (b) positron emission tomography (PET scanning), and (c) computed axial tomography (CAT scanning).

C Use of VLSI

Several VLSI, 32-bit microprocessors are commercially available for the CPU such as the Intel iAPX432, Motorola 69020, and National Semiconductor 32032. To maximize the benefits of the large direct memory addressing range of these units (16 Mbytes for the Intel), ROM chips such as the 256-Kbit Intel 27256 and devices such as the recently announced IBM 512-Kbit dynamic RAM will become increasingly important.

An architecture consisting of an array of hard-wired VLSI processing cells, employing parallelism and pipelining principles, can be used to replace the software implementation presently used for the image reconstruction. An example of the special-purpose processors that can be employed is the Texas Instruments TMS 320. The high-speed mass storage devices (hard disks) use VLSI controllers such as the Western Digital Corporation WD1010 Winchester disk controller. Finally, the high resolution monitor could be controlled by another VLSI chip, the NEC μD7220 graphics display controller.

V ULTRASOUND IMAGING

A Method of Operation

Tomographic images of internal anatomical details in any plane can also be produced by applying ultrasound pulse echo techniques. A transmitted beam is partially reflected wherever a change in the characteristic impedance of tissue occurs. Thus, a reflected image of the boundary between organs can be formed. Depth information is obtained from the time taken for the echo to return (speed of sound in soft tissue is approximately 1540m/s), and a measure of the impedance change is derived from the strength of the echo. Additional processing is required to compensate for the attenuation of the beam by body tissues by applying time gain control (TGC) [6].

B Commercially Available Systems

The Hewlett-Packard 77020 Ultrasound Imaging Systems are typical of the advanced state-of-the-art, real-time, phased-array scanners that use pulsed ultrasound techniques for viewing. A typical block diagram for

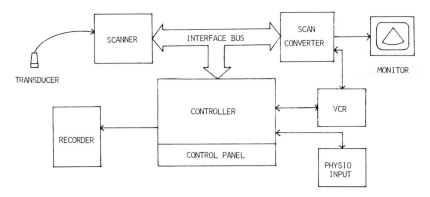

Fig. 5. Typical ultrasound diagnostic instrumentation system.

such a system is shown in Fig. 5. The transducer transmits and receives on 64 elements. For a 90° sector scan, the image consists of 121 acoustic scan lines. After analog to digital (A/D) conversion, each scan line consists of 396 data points. The intensity of each data point is represented by a 5-bit word that gives 30 shades of gray between peak black and peak white. The display has a high resolution of 480 lines with 640 pixels/line. (*Pixel* is a picture element.)

Architecture of Ultrasound Systems   C

The *controller* contains 2 VLSI 16-bit parallel SOS CMOS microprocessors with a 16-bit addressing range and transistor transistor logic (TTL) compatibility. One of the microprocessors regulates the operation of the whole instrument with a clock rate of 5 MHz, and the other is dedicated to controlling the interface bus. The controller contains 80 K of EPROM, DRAM and battery backed CMOS static RAM.

The *scanner* contains the same microprocessors as the controller board. The scanner incorporates the electronics to control beam steering, transmitting and receiving, pre-amplification, including time gain control (TGC), and the A/D conversion.

The *digital scan converter* (DSC) is controlled by the same pair of VLSI microprocessors as the controller board. The DSC also contains the display memory, output circuitry, VCR bit map, and the sector control and data functions. Information is shifted in the DSC on a 12.5-MHz synchronous data bus. The DSC receives data in polar form from the sector scan and converts it using a polar-to-cartesian algorithm. It then outputs high-speed noninterlaced video to the display and normal interlaced video to the VCR.

D Obtaining Permanent Records

The Physio Input unit of the HP 77020 can also display additional physiological data such as ECG, heart sounds, and pressure. These physiological parameters (e.g., R wave of ECG) can be used to trigger the scan after an operator-selected variable delay to freeze the image at a time of interest. Permanent records can be derived either by photographing the frozen display on an auxiliary TV monitor or writing time motion (TM) mode information to a high-speed chart recorder. Output is available for either of the common VCR formats [7].

VI A COMPUTERIZED LOCAL AREA NETWORK FOR AN INTENSIVE CARE UNIT

The system shown in Fig. 6 is a block diagram of a computerized local area network (LAN). It provides two-way data communication between patient monitoring instruments at a bedside in critical care areas (ICU and CCU) and remote points, such as a central nursing station and specialized computer facilities.

A Topology

The fundamental topology of the LAN system is that of a star. Each branch from the central controller is dedicated to one patient's bedside or to a central station or computer. A local BUS topology operates at each branch so that several instruments can be associated with each patient.

Each of the blocks contains an instrument–network interface (I/NI), at the heart of which is a custom serial distribution network (SDN) interface circuit chip (SIC). This VLSI chip, developed by Hewlett-Packard, is an NMOS, 5-μm geometry microprocessor containing over 10,000 transistors. The SIC has one-tenth the cost, one-fifteenth the power consumption and failure rate, and one-hundredth the area of the two-loaded, four-layer printed circuit boards it replaces.

B Data Transmission

The central controller acts as a rotary switch interrogating each branch of the star in sequence. This polling cycle has a fixed duration to allow real-time data transmission. At branch level, instruments can communi-

43. Medical Applications of VLSI Circuits

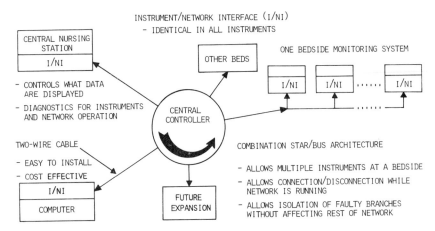

Fig. 6. Computerized local area network (LAH) for hospital critical care areas. (ICU, CCU, etc.)

cate with one another using a token passing protocol. The token signal is passed from instrument to instrument in a prescribed order. Possession of the token represents permission to transmit data. An instrument can claim the token as it comes by and then transmit its data while all other instruments listen. When transmission is complete, the token is passed on, allowing the next instrument to transmit. Since every instrument obeys the same protocol, only one type of interface is required to maintain compatibility with all instruments [8,9].

VII. EVOKED POTENTIALS

Evoked potentials (EP) refer to the recorded electrical activity of biological and physiological tissue, cells, and systems in response to a stimulus. The most common application in medicine of EPs is in brain activity, in which the response is usually obscured by EEG activity. Therefore, in this application, unique signal processing and response averaging techniques are employed to obtain EPs that are free from EEG interference and to improve the signal-to-noise ratio.

A. VLSI versus Hard-Wired Averagers

VLSI microprocessors made possible the development of a sophisticated, miniaturized, low cost EP system that obviated the problems inher-

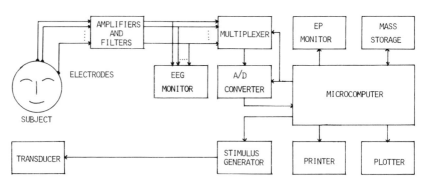

Fig. 7. Evoked potential instrumentation system.

ent in previous systems with bulky, expensive, inflexible, hard-wired logic averagers. Figure 7 shows a block diagram of such a system. The microcomputer contains all the functions of the controller, averager, and data processor. Data acquisition is achieved through the multiplexer and the A/D conversion modules. A graphic terminal functions as an EP monitor for display. Furthermore, microcomputer-based EP systems offer easy additions of mass storage devices and other peripherals such as digital plotters and printers.

B Analysis of Evoked Potentials

EP analysis requires extensive arithmetic calculations. Newly introduced VLSI devices for number crunching can be incorporated into the system for such calculations [10]. Coprocessors such as Intel 8087, AMD 9511, and AMI 2814 greatly reduce the time for calculations requiring multiplications and divisions. As a result, such VLSI devices will enhance the EP analysis by the implementation of fast-Fourier transforms and digital filter techniques. Electronics for providing the trigger pulse that coordinates the signals of the averager and stimulus generator can all be incorporated in one microprocessor. The analysis and display of multichannel EP information by means of topographic techniques are gaining importance in both theoretical and clinical applications [11]. Minicomputer and microcomputer systems are available that will rapidly generate color or gray-scale coded maps showing the distribution of signals derived from EPs. With more extensive use of VLSI devices, routine use of such topographic methods will be common in the coming years.

VIII. NEURAL STIMULATORS

A. Applications

Interactions between implantable electronic devices and the nervous system range from sensory stimulation to aid hearing or vision to peripheral nerve stimulation for movement disorders. In addition, stimulation of the central nervous system to control pain, suppress epileptic seizures, lower blood pressure by carotid stimulation, and to control respiration via stimulation of the phrenic nerve are also available. Stimulation of the surface of the cerebellum or deep brain structures can control psychiatric conditions. Also, electronically controlled neurosurgical implants that do not stimulate neural tissues include miniature pumps for the infusion of substances such as morphine for pain control and devices for the monitoring of intracranial pressure in hydrocephalics.

B. Types of Implants

Coupling between the stimulator and the target tissues ranges from body surface electrodes to electrodes implanted in the brain. Some of the devices, such as the transcutaneous nerve stimulators (TENS), are worn externally by the patient. These send relatively high-intensity currents to deep-lying tissues. Others include *passive* implants that receive their power and often even their control signals from an externally worn rf coupled transmitter. The highest levels of miniaturization and reliability coupled with very low current drain are required for the totally implanted devices, which carry their own primary batteries.

C. Use of VLSI

The application of VLSI technology to the design and construction of the various types of devices is most readily realized with externally worn devices. Commercially available CMOS microprocessors such as the 8-bit Motorola M146805, RCA 1802/1804, Rockwell 6502, and the 16-bit Harris 80C86 can be conveniently used in the systems. Development of flexible software that can be changed during the periods of clinical investigation has been facilitated by microprocessors. The devices are also needed to perform some self-diagnostic tests, such as verification of flow, reservoir level, and battery voltage. These require the addition of sensors that can be interfaced with the VLSI system. Some of these, such as battery voltage or accumulated discharge, can be included in the repertory of the functions of the VLSI.

1 Current Drain

Implantable stimulators pose a more severe problem of current drain because, though their supplies are smaller, they must function for years. All present implants use some lithium-type battery with 2.1–3.5 V open circuit voltage and 1–4 Ah capacities. These provide optimal space utilization and reliability when used singly; hence low-voltage VLSI circuits are desirable.

2 Custom VLSI Devices

Although noncustom chips have been used in cardiac pacemakers for several years, implantable neural stimulators with custom devices are still under development. The requirements for the implant include periodic two-way communication with an external device operated by the physician. Typically, the information is sent and received via binary coded rf bursts. Future requirements undoubtedly will include serial communication of several thousand bits at rates of about 500 bauds or above for each interaction. The security of the system against accidental reprogramming must be excellent. Also, sufficient static RAM and ROM will be resident for changing the function of the implant by noninvasive serial entry of new software and data. The device will have to keep track of certain events such as the dates of last activation and alteration of its operating parameters. It will be desirable to combine microprocessor function with memory and high current handling transmission gates (for powering the electrodes or a pumping mechanism) in a single VLSI custom processor–controller.

D Closed-Loop Controllers

At present the neurosurgical implants are open-loop controllers. The loop is periodically closed by the patient or his physician at intervals ranging from hours to months. Future developments can provide for a closed-loop mode of operation by incorporating a sensing feedback path, as is now virtually accomplished with all implantable pacemakers. Stimulators for the nervous system and pumps for drug delivery will also become more and more automatic. It is conceivable that monitoring the potentials at some points in the brain or on a single nerve bundle can provide the signals necessary for closing the loop.

Stimulation in response to the so-called aura that precedes epileptic seizures to prevent the seizures could become a possibility but would call for signal processing at a fairly fast rate, because the frequency range is up to 30 Hz; thus, more than simple filtering is necessary. Hence, the inclusion of amplifiers is desirable on a custom-made VLSI chip that "does everything." It may be necessary to have on-chip analog to digital converters as well as other special signal-processing functions. FFT is a common method of analyzing brain activity, but its use in implants seems to be remote. Finally, the infusion pumps will also be closed loop as sensors become available to measure physical parameters such as pressure or biochemical parameters such as pH or ion concentrations thus presenting further need for VLSI chips with multiple functions.

REFERENCES

1. N. V. Thakor and R. H. Kuhn, "Pacemaker Design with VLSI Technology," IEEE Frontiers of Computers in Medicine, pp. 78–83. IEEE, New York, 1982.
2. J. A. Berkman and J. W. L. Prak, Biomedical microprocessor with analog I/O, *IEEE Int. Solid State Circuits Conf.,* pp. 168, 169 (1981).
3. M. Engrebetson, R. E. Morley, and G. R. Popelka, An unified digital hearing-aid design and fitting procedure, paper presented at the *Association for Speech and Hearing Aids (ASHA) Conference, Cincinnati, Ohio* (1983).
4. *Proc. IEEE.* **71** (3) (March, 1983). Special issue on computerized tomography.
5. I. L. Pykett, NMR imaging in medicine, *Sci. Am.* 77–88 (May, 1982).
6. *IEEE Trans. Biomed. Eng.* **BME-30** (8) (August, 1983). Special issue on medical ultrasound.
7. Hewlett-Packard Company, Medical Products Group, "Ultrasound System Manual," October, 1983.
8. Hewlett-Packard Company, Waltham Division. "Design Criteria for a Local Area Network in an Intensive Care Unit." 1983.
9. M. D. Schwartz, "Applications of Computers in Medicine." IEEE Engineering in Medicine and Biology Society, New York, 1982.
10. B. A. Eisenstein and E. B. Morgan, "A Dual Processor System for the Analysis of Evoked Response," IEEE Frontiers of Computers in Medicine. IEEE, New York, 1982.
11. R. Coppola, M. S. Buchsbaum, and F. Rigal, Computer generation of surface distribution maps of measures of brain activity, *Comput. Biol. Med.* **12,** 191–199 (1982).

Chapter 44

Cardiac Pacer Systems

ROBERT D. GOLD
Cordis Corporation
Miami, Florida

I. Introduction	737
II. Cardiac Cycle and the Pacer Implant	738
A. Blood Circulation	738
B. The Natural Pacemaker	738
C. Applied Electrical Stimulation	739
D. The Electrocardiogram	740
E. Pacemaker Implantation	741
III. System Description	742
A. Method of Operation	742
B. Technology of Choice: CMOS	743
C. System Components	744
D. Other Support Instruments	746
IV. Pacemaker Design Considerations	747
A. Circuit Characteristics	747
B. Circuit Protective Measures	751
C. Battery Characteristics	752
D. Pacer Testing	754
V. Pacing Modalities	756
A. Pacer Mode Codes	756
B. Pacer Timing Cycles	759
VI. Future Trends in Cardiac Pacing	762
Bibliography	762
References	762

INTRODUCTION   I

The development of the totally implanted cardiac pacemaker is linked directly to the invention of the transistor. The first implant of a pacer in a human took place in 1958, just a few years after the junction transistor became commercially available. Since that time, the industry has grown to about $1 billion per year. Approximately 120,000 pacers are implanted per year in the United States, and worldwide the number is about 250,000 per year.

This rapid growth into a significant industry is the result of successful developments in three technologies: batteries, materials, and most importantly, microelectronics. The last includes semiconductor and hybrid integrated circuits, and microcomputers. Modern pacers and associated instruments use standard, semicustom, and full-custom integrated circuits. The successful development of these sophisticated medical devices requires close interaction between the biomedical and the microelectronics engineer. This chapter gives the microelectronics engineer the information necessary for effective communication with his biomedical counterpart in designing cardiac pacer ICs.

II CARDIAC CYCLE AND THE PACER IMPLANT

A Blood Circulation

Figure 1 shows a diagrammatic cross section of the heart. Oxygen-depleted blood flows through the venous system to the superior and inferior vena cava, which supplies the right atrium. Contraction of the atrium forces the blood into the right ventricle. After filling, the right ventricle contracts, forcing the blood through the pulmonary artery into the lungs. Here the carbon dioxide in the blood is exchanged for oxygen, and the enriched blood is returned to the left atrium, from which it then flows into the left ventricle. The ventricle contracts after being filled, pumping the oxygen-rich blood through the aorta into the arterial system. Backflow is prevented by valves located between each atrium and its corresponding ventricle, in the pulmonary artery, and in the aorta.

B The Natural Pacemaker

The heart is a special type of muscle. Muscle cells are electrically polarized. Each depolarization causes a contraction of the cell, followed by its relaxation and repolarization. The pumping action in the healthy heart is controlled by organized electrical activity that originates in a small group of muscle cells in the right atrium, the sinoatrial (SA) node, the heart's natural pacemaker. This activity spreads through the atria, causing them to contract in unison.

44. Cardiac Pacer Systems

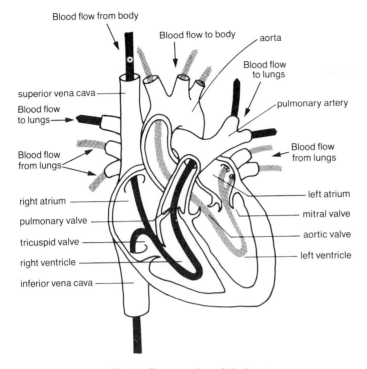

Fig. 1. Cross section of the heart.

At the base of the right atrium, just above the right ventricle, is another set of specialized cells, the atrioventricular (A-V) node. Electrical excitation is transmitted to the ventricles through the A-V node, which introduces a short delay (100 ms) to allow time for the atria to contract and pump their contents into the ventricles. The atrial and ventricular intracardiac signals differ in both amplitude and rate of change, or *slew rate*. Table I shows the normal range of values for these parameters [1].

Applied Electrical Stimulation C

Applied electrical stimulation can also cause the heart to contract. The excitation pulse amplitude that a pacer must deliver to stimulate the heart is an inverse function of its duration. This relationship, the *strength–duration curve*, is shown in Fig. 2 for constant-current pulses. Typical pacer output pulses are 5–10 mA in amplitude and 0.5–1.0 ms in duration. Voltage output pulses give a similarly shaped curve. Both curves are closely approximated by hyperbolic functions of the mean pulse ampli-

TABLE I
Comparison of Pacer Pulses and Natural Cardiac Signals

	Amplitude	Slew rate (V/s)	Pulse width (ms)
Intracardiac pacer pulse	1–8 V	≥40,000	0.1–2.0
Intracardiac atrial signal	0.5–10 mV (~4)	0.2–2	20–100
Intracardiac ventricular signal	0.5–40 mV (~15)	0.5–5	20–150
Surface electrode pacer pulse	0.5 mV–1 V	5–5000	0.1–2
Surface electrode atrial signal	0–0.5 mV	0.005–0.02	50–150
Surface electrode ventricular signal	0.5–4 mV	0.1–0.5	40–200

tude. The horizontal asymptote is called the *rheobase*, and the point at which the amplitude is twice the rheobase is called the *chronaxie*.

D The Electrocardiogram

The electrical activity that controls the cardiac cycle can be detected by appropriately placed chest electrodes. These result in the surface electrocardiogram (ECG). Figure 3 shows idealized and typical ECGs for a normal heart. The P wave originates from the depolarization of the atria,

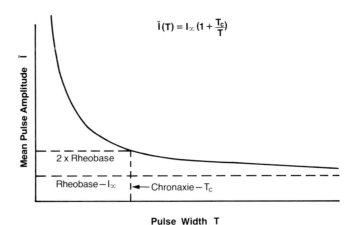

Fig. 2. Strength–duration curve relating mean current pulse amplitude required for stimulation as a function of pulse duration. A similar relationship exists for voltage pulses.

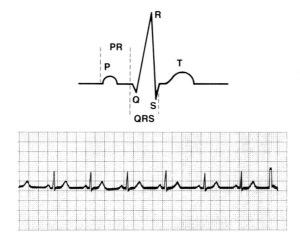

Fig. 3. Idealized and actual ECG for a typical normal heart.

which causes them to contract. The QRS complex corresponds to the depolarization and contraction of the ventricles. Finally, as the ventricles relax and return to their original electrical state (repolarization), the T wave is seen on the ECG. Any abnormality in the electrical conduction pathways, or any electrical activity originating spontaneously in other regions of the heart, will change the shape or timing of the ECG pattern. Modern pacers function by sensing the presence or absence of the P and R waves and by providing properly timed stimulation pulses to induce contraction when needed.

Pacemaker Implantation E

The pacemaker is usually implanted under the left or right shoulder, within a few centimeters of the skin (Fig. 4). A lead is inserted into a vein (usually the subclavian or the cephalic vein, but sometimes the jugular vein) and passed directly into the right atrium, and thence into the right ventricle, where it is lodged against the heart wall. Two leads are used for dual-chamber pacing of the right atrium and right ventricle. The entire implant procedure typically takes less than an hour and is done using local anesthesia. The pacer occasionally is implanted in the abdominal region with the lead electrode placed on the surface of the heart. This is a major surgical procedure, in which the chest is opened to expose the heart, and accounts for fewer than 5% of all implants.

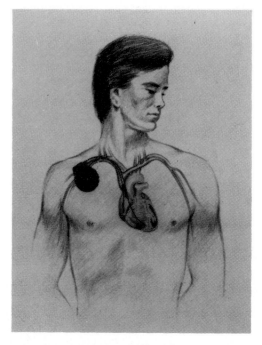

Fig. 4. Typical implant location and lead placement for a dual-chamber pacer.

III SYSTEM DESCRIPTION

The earliest pacers were simple pulse generators that delivered periodic electrical stimulating pulses to one chamber (usually the right ventricle) of the heart. Typically, the pulse rate was fixed at approximately 70 pulses per minute, although pacers were available at other rates.

A Method of Operation

Modern pacers are far more complex devices. Figure 5 represents a sophisticated two-chamber pacer based on hard logic timing and control circuits. These functions can also be software controlled in a microcomputer-based pacer design. The bandpass sense amplifiers detect normal or abnormal heart signals in two chambers (right atrium and right ventricle). The timing and output functions of the pacer are adjusted, on a beat-by-beat basis, depending on the presence or absence of these sensed heart signals. The output circuit regulates the pulses to the stimulating lead

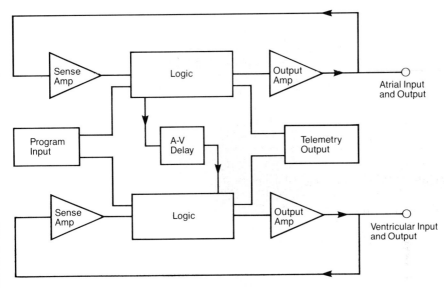

Fig. 5. Block diagram for a modern two-chamber pacer. The logic and delay functions can be provided by hard logic circuits or by a microcomputer. All functions shown, except for discrete passive devices, can be on a single chip.

electrodes. The logic controls all pacing functions, and a crystal oscillator (not shown) ensures accurate timing of these functions and parameters.

In addition, the modern pacer contains circuits that enable its parameters to be changed noninvasively by an external programmer and to telemeter information on pacer function, patient conditions, and overall system performance from the pacer to the programmer. A complete pacer and some of its components are shown in Fig. 6. Table II lists some typical pacer characteristics.

Technology of Choice: CMOS B

The principal microelectronics technologies used in implantable pacers are CMOS semiconductor ICs and thick film hybrid integrated circuits. The low current drain of CMOS circuits, as well as their low supply voltage operating characteristics (3 V or less) make these ideal choices for semiconductor ICs in pacers. Both metal-gate and silicon-gate devices are in use. Circuits have been based on semicustom and full-custom designs. They include commercial and custom microcomputers and hard-wired digital logic circuits. Earlier analog circuits (sense amplifiers) used a polysilicon bipolar technology to achieve the high impedance levels necessary for low current drain. Newer designs include CMOS analog circuits on the

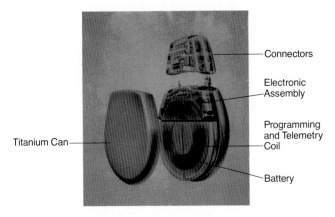

Fig. 6. Basic construction features of a modern pacer.

same chip as the custom microcomputer. Both hermetically sealed hybrid circuits and ceramic substrate circuit assemblies have been used to minimize the size of the pacer electronics.

C System Components

A complete pacing system consists of several components in addition to the implanted pacer. Those that directly influence the pacer design include the lead, which connects the pacer to the heart, a programming

TABLE II

Typical Pacer Characteristics

Output voltage amplitude	2–10	V
Output current amplitude	1–12	mA
Output pulse width	0.1–2.0	ms
Sense amplifier sensitivity	0.5–5	mV
Sense amplifier band pass	10–400	Hz
Sense amplifier frequency peak	50–90	Hz
Pacer rate, programmable range	30–150	bpm
Pacer rate, most commonly used	50–90	bpm
Pacer atrioventricular delay	75–250	ms
Pacer refractory period	200–500	ms
Current drain @ 70 bpm, 510 Ω load	10–30	μA
Power source voltage	2–7	V
Power source charge capacity	0.5–3.5	A·h
Pacer size (height × width)	47 × 48–65 × 56	mm
Pacer thickness	8–20	mm
Pacer weight	30–100	g

instrument for changing the pacer parameters noninvasively, and an instrument for interrogating and receiving telemetered information between the implanted pacer to the user (this can be a separate unit or an integral part of the programming instrument).

The Lead 1

The lead is made of a conductor, an insulating sheath, a connector pin which is inserted into the pacer, and an active electrode which is placed in the heart. The conductor consists of one or more helically coiled wires, usually either an alloy of cobalt, nickel, and other metals, or stainless steel. The wires are encased in an insulation sheath of silicone rubber or a special polyurethane. Bipolar leads consist of two separately insulated wires, within a common insulating sheath, and two electrodes. The electrode can be of the same material as the wire or can be titanium, platinum, or carbon. The electrode surface can be smooth or can be made porous to improve its electrical characteristics and its mechanical stability when placed against the heart wall. Immediately behind the electrode, fins or tines on the lead sheath are used to help anchor the electrode in the heart. Metal fingers, screws, or other devices also can be used to improve electrode stability.

Two electrical aspects of the lead have an important bearing on pacer circuit design.

First, in unipolar leads, the electrode tip can serve as one electrically active terminal (the cathode), with the pacer case serving as the other, indifferent electrode (the anode). Alternatively, in bipolar leads, a metal band located about 2–5 cm behind the tip can serve as its indifferent electrode (anode). The unipolar lead is usually better suited to sensing small heart signals; but it is also more susceptible to sensing electromagnetic interference and spurious muscle potentials.

Second, a polarization potential between the electrode and the heart tissue arises from the difference in charge carriers in the electrode (electrons) and the tissue (ions). This potential is analogous to that of a metal–semiconductor junction in that equilibrium requires the presence of a space charge layer at the electrode surface. Several equivalent circuits have been proposed to model the electrode–tissue interface impedance. The most common consists of a resistor in series with the parallel combination of the polarization capacitor and resistor. The space charge capacitance can vary from 0.01 to 25 μF, and its parallel resistor ranges from 1 to 50 kΩ depending on electrode material and size. The series resistor is on the order of a few hundred ohms and includes both tissue and lead wire resistance. The long relaxation time of the capacitor, on the order of tens to hundreds of milliseconds, effectively disables the sense amplifier immediately following a pacer output pulse. This time is called the *repolar-*

ization period. Modern pacers have additional circuits that accelerate the discharge of this capacitor.

2 Programming and Telemetry

The first noninvasive, electronically controlled programmable pacing system was introduced in 1972. A hand-held programmer transmitted a string of magnetic pulses to the pacer, which contained a miniature reed switch to sense the pulses. A simple pulse counter decoded the signal, providing 24 parameter combinations (6 rates and 4 output currents). Protection against spurious programming was provided by pulse width and pulse frequency discrimination and by an initial minimum pulse count. This basic system is still widely used. It can provide up to 128 parameter combinations in a single programming sequence. By programming eight parameters independently, where each parameter has four possible values, over 65,000 parameter combinations have been provided in a pacer.

However, subsequent improvements have been made in the signal transmission means, the modulation methods, and the coding schemes. Carriers include continuous and modulated magnetic pulses, and low frequency rf (10–200 kHz). Pulse width and pulse position modulation schemes are used, in addition to the simple pulsed *on–off* method. Digital codes are typically based on patterns ranging from 14 to 32 bits and use a variety of self-checks and identification codes. Both reed switches and pickup coils are used in the pacer to sense the programming signal. As a result, modern programming systems are almost completely impervious to spurious programming. Moreover, many millions of programmable parameter combinations are available, allowing far more sophisticated pacers to be designed.

Telemetry of information from the pacer to the programmer (or other external receiver) was introduced in 1978. The energy transmission, modulation, and encoding are similar to the programming methods. The telemetered information includes programming verification, measured parameter values, ECGs taken directly in the heart itself (intracardiac electrograms), essential patient information, and various additional checks to reduce further the possibility of misprogramming. A block diagram of a programmer with telemetry capability is shown in Fig. 7.

D Other Support Instruments

Two other frequently used instruments are the *transtelephonic transmitter* and the *pacer systems analyzer* (PSA).

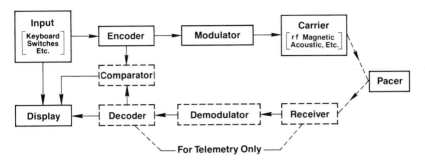

Fig. 7. Basic programmer elements.

The transmitter is used by the pacer patient at home to send a simple ECG-like record to the doctor's office or to a special monitoring center. The patient places an electrode on each wrist, or under each armpit, or on a finger of each hand. The transmitter, which resembles a small acoustic modem, transmits a signal artifact, in some cases, rather than the actual detected signal from the pacer.

The implanting physician uses the PSA to perform three functions.

(1) The measurement of the patient's electrical signals (i.e., the P-wave, which denotes atrial contraction and the R-wave, which denotes ventricular contraction). This information ensures that the pacer sensing characteristics are adequate for the patient.

(2) The measurement of the pacer itself, to verify that it is functioning normally.

(3) The PSA provides external simulation of the implantable pacer during lead placement and maintains pacing in a pacer-dependent patient during a replacement implant.

PACEMAKER DESIGN CONSIDERATIONS IV

Circuit Characteristics A

The principal factors in the design of an integrated circuit chip include:

(1) operating voltage range,
(2) operating currents, with and without output pulses,
(3) timing accuracy,
(4) stability with temperature and battery voltage,
(5) input and output impedance levels,

(6) sense amplifier for the detection and amplification of low level signals (less than 1 mV), and

(7) end of battery life detection.

The requirements for each factor are discussed below. Because of the diversity in designs, the values should be taken only as guidelines. Note that most modern pacers are designed with crystal controlled microcomputers.

1 Operating Voltage

The lower limit (2 V) is set by the decrease in battery voltage as it discharges, for the IC operating from a single cell power source. The upper limit (12 V) applies to circuits that have more than one cell and include a voltage multiplier.

2 Operating Currents

The normal circuit operating current, exclusive of the output pulse, should be as low as possible to maximize the battery life of the implanted pacer. Typical values for modern dual-chamber pacers are on the order of 5 to 15 μA. The output pulse amplitude and duration ranges are given in Table II. The effective additional average current is a function of pulse amplitude, pulse width, pacer rate (i.e., duty cycle), and pulse waveshape. For example, a dual-chamber pacer operating at 60 beats per minute with constant-current output pulses of 5 mA and 1 ms would add 10 μA to the circuit current.

3 Timing Accuracy

The resolution accuracy required for pulse rate, atrioventricular (A-V) delay, and refractory period is on the order of several milliseconds. For example, one popular dual-chamber pacer series (Gemini, manufactured by Cordis Corp.) has a microprocessor cycle time of 7.8125 ms. The most widely used crystal frequency of 32.768 kHz, corresponding to a 30.517-μs clock pulse, provides the timing resolution necessary for the narrow pulse widths and also for the programming and telemetry pulses.

4 Stability with Temperature and Battery Voltage

The pacer characteristics should remain essentially constant over the temperature range of 20 to 42°C. The pacer is routinely tested by the manufacturer at room temperature, and many physicians use a PSA to verify these values just prior to implant. The implanted pacer temperature can range from 34 to 42°C for a nominal body temperature of 37°C, depending on the environment and the patient's health. Similarly, the pacer characteristics should not change as the battery voltage decreases and its output resistance increases with time. Some pacers have been designed with a pulse width that increases as its amplitude decreases due to decreasing battery voltage. The purpose of this design is to provide a more nearly constant output energy as the battery is depleted. However, analysis of the strength–duration curve shows that this has relatively little effect in ensuring the continued stimulation of the heart [2].

5 Input and Output Impedance Levels

The sense amplifier *input impedance* should be greater than 20 kΩ to avoid attenuation of the cardiac signals. Since the cardiac signal source impedance can range from 300 to 3000 Ω, the signals will be attenuated by less than 15%; for the most typical source impedance level of 500 to 1000 Ω, the attenuation will be less than 5%.

The *output impedance* level depends on whether a constant-voltage or constant-current output circuit is used. Although these are the terms commonly used to describe their corresponding output pulses, more accurate terms would be capacitor-discharge or capacitor-charge circuits, as shown in Fig. 8. In Fig. 8a, the output capacitor is discharged through the heart when the switching transistor is closed, after charging between output pulses to the full battery voltage; the capacitor produces an output waveform into a resistive load, as shown in Fig. 8b. The circuit of Fig. 8c shows a constant current circuit, in which the output capacitor is charged through the heart when the transistor is turned on. In this circuit, the transistor is not driven into saturation but operates in the active region to maintain a constant current through a resistive load, as shown in Fig. 8d. Between pulses, the capacitor discharges through the heart and the large transistor load resistor (20–40 kΩ). For both circuits, the output capacitor is typically about 5–10 μF.

6 Sense Amplifier Requirements

As shown in Table I, the cardiac signals typically are a few tenths to several millivolts in amplitude. The amplifier bandpass characteristics are selected to optimize the sensitivity to P and R waves and to be relatively insensitive to the T wave and to spurious signals generated by various muscles (myopotentials). Although different manufacturers have optimized their amplifiers somewhat differently, most amplifiers peak be-

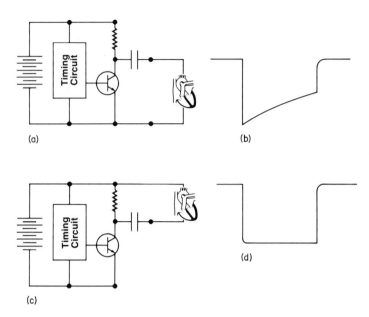

Fig. 8. Commonly used output circuits and their corresponding waveforms into a resistive load. (a) Capacitor discharge, or constant voltage circuit. (b) Output waveform for (a). (c) Capacitor charge, or constant current circuit. (d) Output waveform for (c).

tween 50 and 100 Hz and roll off at 6 to 12 dB/octave on both the low- and high-frequency sides (Fig. 9). For some pacers, the sensitivity differs for positive and negative signals.

Provision must be made to minimize the effect of the output pulse on the amplifier input because they share the same lead. One common approach is to include a transmission gate at the amplifier input that disables the input during an output pulse. Also, the sense amplifier is effectively disabled following an output pulse, until the lead–tissue polarization charge is dissipated. The sense amplifier can include a gate that provides a low impedance discharge path for the polarization capacitance, or it even can include an active charge pump or sink to accelerate the discharge further. These techniques have led to blanking times as low as 10–15 ms.

7 End of Battery Life Detection

It is essential that the pacer provide some end-of-life (EOL) indication as the battery becomes depleted. For lower cost and older pacers with *RC* oscillators, the effect of a change in battery voltage on oscillator frequency is usually used to alert the physician when battery EOL is near.

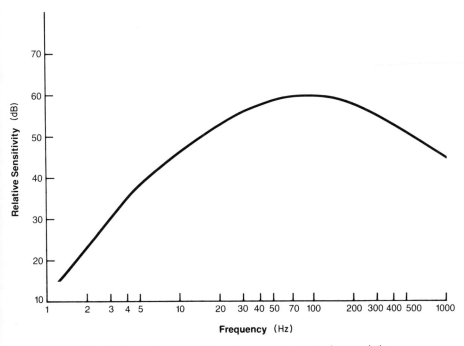

Fig. 9. Typical sense amplifier frequency response characteristic.

With crystal oscillators, the pacer must include a battery voltage level detector that produces a discrete change in a measurable pacer parameter (most frequently, pacer rate) when the battery nears its end of life.

Circuit Protective Measures B

The pacer includes circuits that are designed to allow continued operation of the pacer in certain external or internal conditions that otherwise might be harmful to the patient.

Defibrillation Protection 1

Pacer patients sometimes require defibrillation to break an otherwise fatal ventricular arrhythmia. The several thousand volts at up to 300 J that is applied across the chest to stop fibrillation is sufficient to damage or destroy the internal pacer circuit. To protect the circuit, zener diodes are placed across the output to limit the voltage at the pacer terminals.

2 Noise Window

External electromagnetic interference (EMI) signals can interfere with the function of the pacer. Depending on the type of pacer and the nature of the EMI, the pacer can be either inhibited or caused to pace at a dangerously high rate. To protect against these occurrences, a frequency counter is activated for a small portion of each pacer cycle. If an excessively fast signal is sensed, the pacer automatically reverts to a preset rate. Noise filters and other means can also be employed.

3 Rate Limit

The failure of some circuit components can cause the pacer to operate at dangerously high rates. Also, certain pacers are designed to stimulate the ventricles at a rate determined by their sensed atrial signals (atrial synchronized pacers). Patients with such pacers who experience high atrial rates because of some disease or defect in their atrium or in their SA node, or even in response to stress, can have their ventricles paced at excessive rates. Depending on the patient's condition, ventricular rates of 160 beats per minute (bpm), or even lower, can be very hazardous, and rates greater than 300 bpm can be fatal. Pacers, therefore, include circuits that limit the rate to a predetermined value. Both hardware circuits (e.g., simple one-shots) and software protection can be used. In addition, countdown circuits can be used to decrease the ventricular rate either gradually or abruptly in the event of a high atrial rate.

4 Backup Oscillator

As further protection against certain component failures, the pacer can include a backup *RC* oscillator in addition to its primary crystal oscillator. The backup oscillator is activated by any failure in the primary oscillator and, in some designs, by low battery voltage. The latter feature alerts the physician by pacing at a specified rate other than the programmed rate.

C Battery Characteristics

Table III lists the principal battery chemistries that have been used in pacers. Mercuric oxide–zinc (HgO/Zn) was used for the first 15 years of implantable pacers. However, because of its relatively low energy density, low output voltage, and internal self-discharge current (hence poor life), it has since been replaced by cells based on lithium compounds. The

TABLE III

Battery Characteristics by Cell Chemistry

Chemistry	Charge density (A · h/mL)	Energy density (wh/mL)	Open circuit voltage (V)
Li/CuS	0.40	0.82	2.1/1.7[a]
Li/I$_2$	0.32	0.80	2.8 → 2.2[b]
Li/Ag$_2$CrO$_4$	0.19	0.573	3.0
Li/MnO$_2$	0.26	0.77	3.0
Li/SOCl$_2$	0.31	1.1	3.6
Li/BrCl in SOCl$_2$	0.31	1.2	3.9/3.6[a]
Li/PbI$_2$	0.10	0.42	1.9
HgO/Zn	0.30	0.41	1.36

[a] (Initial voltage level)/(second voltage level). See text.
[b] The arrow implies continuous change in voltage throughout battery life.

two most widely used of these are lithium iodide (LiI) and lithium cupric sulfide (LiCuS). These have higher energy densities, higher voltages, and no self-discharge current. As a result, modern pacers have a typical life expectancy of five to ten years under normal conditions.

The most important characteristics of a pacer battery, in addition to those shown in Table III, are the changes in voltage and internal resistance as the battery is discharged. The LiI cell shows a monotonic decrease in voltage and an increase in resistance from about 1000 to 40,000 Ω as it discharges. Some pacer parameter values, therefore, degrade continuously with time (e.g., output amplitude and pacer sensitivity). The LiCuS cell, on the other hand, exhibits a constant output voltage of 2.1 V and resistance below 50 Ω for 85 to 90% of its total expected life. The voltage then drops very rapidly to a new plateau of 1.7 V, with an increase in resistance to about 1000 Ω, for the remaining 10–15% of its life. Regardless of battery type, an end-of-life indicator is essential.

Two other power sources that have been used are nickel–cadmium (NiCd) rechargeable cells, and nuclear batteries. The NiCd cells proved to be relatively unreliable, and for some patients the need to recharge the battery once or twice a week for an hour or more proved to be a significant psychological problem. Pacers with these batteries are no longer being implanted. The nuclear battery, using plutonium 238, emits alpha particles that produce heat as they bombard its container. A thermopile converts the heat to a voltage of about 0.2 to 0.7 V, and a dc-to-dc converter steps it up to 5 or 6 V. The battery life is estimated to be 25 to 30 years; nuclear-powered pacers, in fact, have had the best record of pacer longevity. Their higher cost, need for additional precautions, the additional federal controls imposed by the Nuclear Regulatory Agency, and the advent of the lithium cell prevented this battery from becoming widely used.

D Pacer Testing

The principal parameters that require testing, on both the integrated circuit pacer chip and the completed pacer, are input sensitivity, output amplitude, and various timing functions (rate, delay, refractory, etc.). Although a generalized standard method exists for these parameters (see e.g., *AAMI Pacemaker Standard,* listed in the bibliography) there is great latitude in test methods. Some of the variations in use by different manufacturers are described below.

The initial IC development tests include device characterization tests and qualification acceptance tests. The IC manufacturer performs on-line tests of logic functions and parametric functions. To optimize testing with respect to completeness and test time, it is essential that the IC be designed for testability as well as for basic functionality. In addition to the tests performed by the IC manufacturer, the pacer manufacturer carries out an extensive program of initial device performance characterization and on-line functional tests. Testing the analog functions usually consumes substantially more time than testing the digital functions, both in device characterization and on-line tests.

1 Initial Characterization and Qualification Tests

These tests include the characterization of the sense amplifier, output pulse, rate, A-V delay, refractory period, current drain, and all other functional parameters. The tests are performed at room temperature and nominal voltage, and also as functions of temperature and voltage. Logic pulse wave shapes and timing are evaluated, as are all logic paths. Where applicable, measurements are compared with computer simulation values. The IC qualification tests include burn-in for up to several thousand hours at 125°C. Numerous additional tests common to high reliability military specifications, including burn-in of the completed electronic assembly, also are performed. Acceptance criteria include minimal parametric shifts as well as a very low failure rate.

2 On-Line Tests and Design for Testability

A modern microcomputer-based pacer IC can contain several thousand logic gates, as well as analog functions, on a single chip. High-speed automatic IC test equipment (e.g., Fairchild Series 20 or Sentry test system) can exercise all logic gates and logic paths, typically in less than a second. The analog functions can also be tested on such equipment, but test time can be up to several minutes per chip. The amount of testing can be increased and the test time reduced if the IC is designed with testing in mind. For example, the address bus and data bus may not be brought out to pads in a custom design. Adding a dedicated test bus requires only a few pads and provides indirect access to the logic functions to be tested. The addition of test transmission gates allows the interconnect pattern to be reconfigured to simplify the test procedures and reduce test time for a more thorough chip test.

Input Sensitivity Test 3

A test signal is applied to the pacer, and its amplitude is adjusted to produce a change in output in accordance with the pacer design type. For example, a pacer may provide periodic output pulses, if no signal is sensed, but be inhibited (i.e., provide no output) when a signal is sensed. The sensitivity value is that which just inhibits the pacer output. The procedure is simple but usually time-consuming, because the test signal amplitude cannot be incremented more rapidly than the pulse rate.

A variety of test signal wave shapes are in use among pacer manufacturers. Figure 10 shows some of the more common wave shapes used. The rectangular wave is typically 5–10 ms, but it may be substantially longer to distinguish sensitivity to positive and negative signals. The haversine is 25 Hz (40 ms) for ventricular sense amplifiers and can vary from 25 to 100 Hz for atrial amplifiers. The half-sine wave and triangular wave can vary over a similar range.

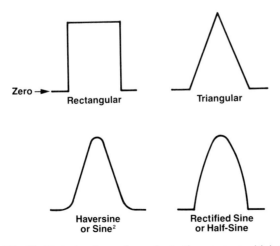

Fig. 10. Test signal waveforms for testing pacer sensitivity.

4 Output Amplitude Test

This measurement is standardized for a 510-Ω resistive load, although other values may be used. The measured value depends on the precise point on the output pulse at which the measurement is made. Common points include the absolute peak, a point 90 μs into the peak (measured from the start of the pulse), an average value over the first 90 μs, the mean value, and the value at the midpoint of the pulse. Some measurement circuits can produce grossly erroneous readings for output pulses with a relatively slow rise time.

V PACING MODALITIES

A wide variety of pacing modalities is possible, depending on which chambers are paced, which are sensed, and whether the response to a sensed signal is inhibition of the output in the same chamber or a triggered output in the same or other chamber. The advent of IC microcomputer chips, sophisticated programming, and telemetry schemes has made it possible for a single pacer to offer most or all of these modalities. The design of a pacer IC requires an understanding of these modalities and the terminology used to describe them.

A Pacer Mode Codes

The most widely used mode code, the three-position code proposed by the Intersociety Commission for Heart Disease (ICHD), is shown in Fig. 11.

(1) The first position indicates which chamber is paced: ventricle (V), atrium (A), or both chambers (D, for double).

(2) The second position indicates which chamber is being sensed; it uses the same three letters, or the letter (O) to indicate that the pacer does not include a sense amplifier.

(3) The third position indicates the response to a sensed signal: triggered (T), inhibited (I), atrial triggered and ventricular inhibited (D), or no response (O).

The three examples shown in Fig. 11 are commonly used pacing modes.

This code recently was extended to five positions to identify multiprogrammable pacers and pacers with certain functions intended to restore

44. Cardiac Pacer Systems

Position	I	II	III
Category	Chamber(s) paced	Chamber(s) sensed	Mode of response(s)
Letters	V- Ventricle	V- Ventricle	T- Triggered
	A- Atrium	A- Atrium	I- Inhibited
	D- Double	D- Double	D- Double*
		O- None	O- None

*Atrial triggered and ventricular inhibited.

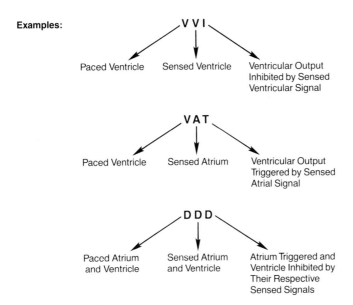

Fig. 11. Three-position ICHD pacemaker code. (From Gold [1]. © 1983 The Association for the Advancement of Medical Instrumentation, and PW Communication, Inc., reprinted by permission.)

very rapidly beating hearts to a more normal rhythm (the latter are called antitachyarrhythmia pacers).

(4) The fourth position indicates programmability as simple programmable (P, for one or two programmable parameters), multiprogrammable (M, for more than two programmable parameters), or multiprogrammable with telemetry features (C, for communication).

(5) The antitachyarrhythmia terms in position five do not apply to the types of pacers discussed in this chapter but are described fully elsewhere [3].

Another modification of the three-position code is shown in Fig. 12. The key to this code is the use of the basic three-position ICHD code successively to describe atrial activity, atrioventricular activity, and ventricular activity. It reduces the ambiguity of the three-position code in

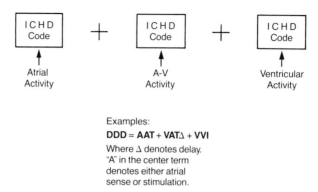

Fig. 12. Modified three-position pacemaker code. (From Gold [1]. © 1983 The Association for the Advancement of Medical Instrumentation, and PW Communication, Inc.; reprinted by permission.)

describing dual-chamber pacers and describes these pacers more completely, while retaining the basic structure of the simple code. If it becomes necessary to describe ventricular-to-atrial activity, a fourth three-letter code term is easily added.

An alternate to the ICHD code has been accepted by the North American Society for Pacing and Electrophysiology (NASPE). This code uses terms in the format of a numerator and denominator to describe atrial and ventricular activity, respectively, and also cross-chamber activity. The terms are defined in Fig. 13.

Figure 14 is a schematic representation of how the pacer output circuits and sense amplifiers are connected to the heart to produce various pacing

$$\text{Code} = \frac{\text{Terms Describing ATRIAL Activity}}{\text{Terms Describing VENTRICULAR Activity}}$$

Examples:

$$\text{Atrial Inhibited} = \text{AAI} = \frac{\text{PSI}_a}{\text{O}}$$

$$\text{Ventricular Inhibited} = \text{VVI} = \frac{\text{O}}{\text{PSI}_v}$$

Dual Chamber Pacing and Sensing, with Atrial Triggered and Ventricular Inhibited Response
= AAT + VATΔ + VVI

$$= \frac{\text{PS T}_a \text{ I}_v}{\text{PS I}_v \text{ T}_a}$$

Definition of Terms

O = No Activity in Chamber
P = Pacing
S = Sensing
I_a = Inhibited Response to Sensed Atrial Event
I_v = Inhibited Response to Sensed Ventricular Event
T_a = Triggered Response to Sensed Atrial Event
T_v = Triggered Response to Sensed Ventricular Event

Fig. 13. NASPE specific mode code. (From Gold [1]. © 1983 The Association for the Advancement of Medical Instrumentation, and PW Communication, Inc.; reprinted by permission.)

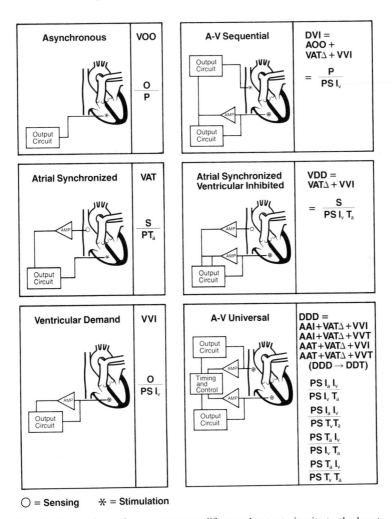

Fig. 14. Connections of pacer sense amplifiers and output circuits to the heart.

modes [1]. Relatively simple single- and dual-chamber pacing modes are shown in the three diagrams on the left. The more complex dual-chamber modes are shown in the three diagrams on the right. Note that the A-V universal mode (DDD) actually has four subsets, depending on which chambers inhibit and which are triggered in response to a sensed signal.

Pacer Timing Cycles   B

Pacer operation can be described in terms of the pacer timing cycle. The duration of one timing cycle, in seconds, is equal to 60 divided by the programmed rate in beats per minute (bpm). Pacer outputs occur at the end of the programmed minimum rate period (Tp) if

(1) cardiac signals are not sensed,
(2) electromagnetic interference or noise is detected,
(3) the pacer is programmed to asynchronous pacing, or
(4) a test magnet is applied.

The timing cycle diagram (Fig. 15) shows each channel divided into two parts: a *refractory period* of programmable duration (shaded area) and an *alert period* (unshaded). During the refractory period, the pacer does not respond to cardiac activity. During the alert period, the two channels respond to sensed cardiac activity according to the programmed mode.

In single-channel operation, the timing cycle resets either at the end of the pulse-to-pulse interval or when a sensed event occurs during the alert period. In two-channel operation, paced or sensed events in the atrial channel initiate an A-V delay. The timing cycles for both channels reset for either a paced or a sensed event in one of the channels, with the controlling channel depending on the pacer design.

To prevent the ventricular channel from sensing the output pulse in the atrial channel, a *blanking* period is included in the sensing function of the ventricular channel. Blanking is initiated by the atrial output pulse. During the blanking period, which may range from 10 to 100 ms, the ventricular sense amplifier is disabled.

Timing cycle diagrams, in conjunction with their corresponding ECGs, are often used to describe or analyze specific pacing modes. To illustrate, Figs. 16a–d show the operation of one type of DDD pacer (Gemini model, manufactured by Cordis Corp.) under four conditions.

(1) In Fig. 16a, no cardiac signals are sensed, and dual-chamber pacing occurs at the programmed rate with a programmed A-V delay between the atrial and ventricular stimulus pulses.

(2) If P waves are sensed during the alert period of the atrial channel (Fig. 16b), the atrial channel output is inhibited, and an A-V delay is

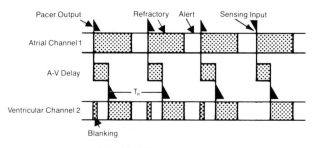

Fig. 15. Timing cycle diagram.

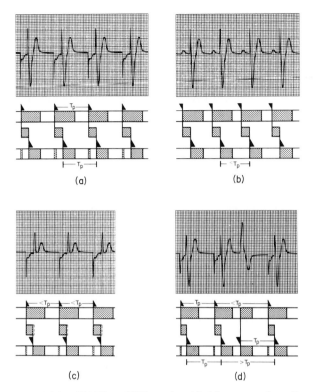

Fig. 16. DDD (AAI + VATΔ + VVI) mode with (a) no sensed cardiac signals; (b) P waves sensed during the alert period of the atrial channel; (c) R waves sensed during the A-V delay; (d) a PVC sensed during the ventricular alert period.

initiated. If an R wave does not occur, the ventricular channel produces an output pulse at the end of the A-V delay period, and timing cycles are reset for both channels.

(3) If no P waves are sensed, but R waves are sensed (Fig. 16c), the ventricular channel output is inhibited, and the timing cycle is reset. The atrial pulse-to-pulse interval is shortened by the difference between the programmed A-V delay and the patient's natural A-V conduction period.

(4) Fig. 16d shows the effect of a premature ventricular contraction (PVC) that occurs during the ventricular alert period. The timing is reset to the end of the A-V delay, causing a lengthened atrial and ventricular pulse-to-pulse interval.

Note the changes in both the ECGs and the timing diagrams. An understanding of the timing cycle diagrams for all pacing modes under all possible sensed conditions is essential to the proper design of a complex pacer IC.

VI FUTURE TRENDS IN CARDIAC PACING

Pacers and pacing systems are becoming increasingly complex. New implantable devices that terminate tachyarrhythmias and ventricular fibrillation are under clinical investigation. Pacers have been or are being developed that sense and transmit intracardiac electrograms (internal ECGs), the presence and frequency of occurrence of certain types of arrhythmias, and even patient information records. Sensors for other physiological functions will also be used to control pacer functions. The fundamental characteristics of the next generation of pacers will be determined by down-loading their controlling software from a master instrument. The accompanying instruments will be more complex and sophisticated than they are today. They will require more complex ICs, including greater RAM and ROM capacities in a single chip. Pacer technology is inextricably linked to VLSI.

BIBLIOGRAPHY

AAMI Pacemaker Standard, issued by the Association for the Advancement of Medical Instrumentation, Arlington, Virginia. Contains definitions for some pacing terms, and includes a section on function, test methods, and procedures.

S. S. Barold and J. Mugica, "The Third Decade of Cardiac Pacing. Futura Publishing Company, Mount Kisco, New York, 1982. Contains numerous chapters on advances in pacer technology and clinical applications of pacers.

R. D. Gold, Cardiac pacing—From then to now, *Med. Instrum.* **18**:15 (1984); describes the history of cardiac pacing, with emphasis on the technological basis for advances in pacing.

Pace, the official journal of the North American Society of Pacing and Electrophysiology and the Cardiac Pacing Society, published by Futura Publishing Company, Mount Kisco, New York. Contains papers of both clinical and engineering importance.

REFERENCES

1. R. D. Gold, The State of the Art in Pacer Technology, AAMI Technology Assessment Report, Cardiac Monitoring in a Complex Patient Care Environment, 1982. [Also published in revised and updated form in *Cardiology Product News* **3**, (6), 1; **3**, (7), 38, (1983).]
2. A. D. Bernstein and V. Parsonnet, Implications of constant-energy pacing, *Pace,* **6**:1229 (1983).
3. A. D. Bernstein, R. R. Brownlee, R. D. Gold, *et al.,* Report of the NASPE mode code committee, *Pace* **7**, 395, (1984).

Chapter 45

VLSI in a Complex Medical Instrument

JOHN FOSTER

Coulter Electronics
Hialeah, Florida

I. System Architecture	764
II. Sample Handling and Preparation	765
A. Example: Aspirate and Dispense an Aliquot	765
B. Motor Control	767
C. I/O Interface Device	769
D. Interval Timer	769
III. A/D Conversion and Data Accumulation	770
A. A/D Conversion	770
B. Data Accumulation	771
IV. Central Data Processor	772
A. Software	773
B. Hardware	773
C. Error Detection and Correction	773

Effective medical diagnosis and treatment for hospital patients can be achieved only if the laboratory can provide the physician with timely, accurate test results. In the areas of hematology and clinical chemistry alone, dozens of individual tests on the patient's blood, serum, and urine are routinely required. The smaller hospital performs, therefore, hundreds of tests daily and the larger hospital performs thousands. Automated instruments that minimize labor of the medical technician make it possible for the laboratory to achieve this work. Advances in electronics, especially in computer systems, microprocessors, RAM and ROM memory, and peripheral support chips, have made possible the performance and cost effectiveness of these instruments.

I SYSTEM ARCHITECTURE

The system diagram of a typical, automated instrument is given in Fig. 1. The instrument system can be divided into two parts. analyzer and cental data processor.

(a) An operator enters into the *central data processor* (CDP), through the keyboard and display screen or from a remote computer (by modem), a listing of the tests that are to be performed on each patient sample. The CDP holds in disc memory the detailed and extensive control information needed to define all the tests the analyzer can perform. After preprocessing all the test requests, the CDP outputs the required control data to a master microcomputer in the analyzer, which then directs the required processing activities.

(b) Patient samples (e.g., blood, serum, or urine) are put into a *sample handling and preparation* section where each sample is automatically

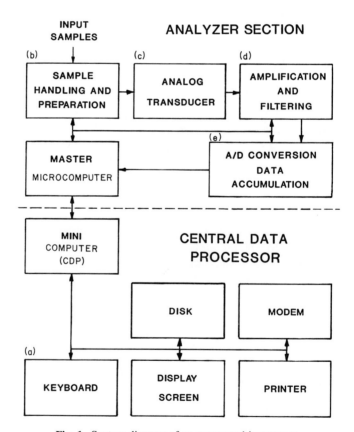

Fig. 1. System diagram of an automated instrument.

partitioned, diluted, treated with reagents, incubated, and so on, according to the requirement of the tests that are to be made.

(c) After sample handling and preparation, the test specimens move to an *analog transducer* section in which physical, chemical, optical, and other parameters representing the ultimate test results are translated into electronic signals.

(d) After amplification and filtering,

(e) These signals are passed to a *A/D conversion and data accumulation* section. The raw data, now in the form of digital bytes, are passed through the microcomputer (for editing and checking) on to the CDP.

Under joint control of the microcomputer and the CDP, a continuous process take place, producing several hundred test results per hour. As the data for each test are received at the CDP, it is processed according to stored mathematical algorithms to give the test result. As results for all tests requested on a given patient are achieved, they are output to the printer to form a hard copy report.

SAMPLE HANDLING AND PREPARATION II

The most difficult design problems of this section are mechanical and fluid dynamic. The actions of motors, valves, pumps, stirring devices, blowers, heaters, and cooling devices must be coordinated precisely and accurately. This is accomplished by one or more local microcomputers under the direction of the master microcomputer of Fig. 1. This distributed processing approach, made feasible by the low cost of microprocessor, ROM and RAM chips, simplifies the timing problems of these real-time systems, in which any delayed response from the controlling microcomputer might abort an entire group of tests. Distributed processing also partitions the software development, the cost of which can equal or exceed that of the hardware development.

Example: Aspirate and Dispense an Aliquot A

To illustrate how the microcomputer can aid a fluid mechanical problem, consider the requirement to aspirate a small volume (e.g., a few microliters) of fluid from one vial and dispense it into another vial with volumetric imprecision of less than 1%, the operation to be accomplished in a few seconds. A positive displacement mechanism, depicted in Fig. 2, can be used. A small probe is connected by a length of tubing to a glass syringe, which is connected to an elevator that is driven up and down by

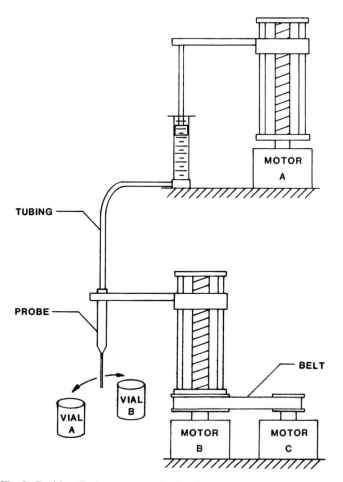

Fig. 2. Positive displacement mechanism for metering an aliquot of fluid.

the action of a lead screw turned by stepping motor A. A similar elevator and screw driven by stepping motor B can raise or lower the probe arm, and motor C can rotate the probe arm to position it over vial A or vial B. To aspirate fluid from vial A and dispense it into vial B, motor C positions the probe over vial A, motor B lowers the probe into vial A, motor A pulls the syringe piston upward a precise distance from its home position, aspirating fluid, motor B raises the probe above vial A, motor C rotates the probe over vial B, motor B lowers the probe into vial B, motor A returns the syringe piston to the home position, dispensing the fluid into vial B.

Since the mechanical movements must be performed accurately and precisely, the stepping motors must make a few thousand steps per second. For a volumetric resolution of 0.1% of the syringe volume, a full stroke requires 1000 steps of motor A. To aspirate (or deliver) full stroke volume in one second, the motor must make the 1000 steps in one second. Each motor, moreover, has four windings that must be properly sequenced for each step. The control of motors A, B, and C in real time can be done only by a microcomputer system.

Motor Control   B

Let us detail the control of motor A for a single, full stroke of the syringe piston from fully extended to closed position, which requires 1000 motor steps. To achieve rapid response, a 5-V motor is driven from 24-V supply (Fig. 3). This allows current (and thus torque) to rise very quickly

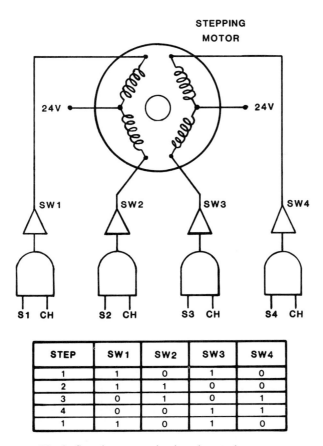

STEP	SW1	SW2	SW3	SW4
1	1	0	1	0
2	1	1	0	0
3	0	1	0	1
4	0	0	1	1
1	1	0	1	0

Fig. 3. Stepping motor circuit and control sequence.

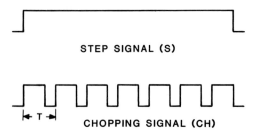

Fig. 4. Relationship of step signal and chopping signal.

in the windings as they are sequenced by step control signals (S1, S2, S3, S4) applied to one input of the AND gate that controls the drive to each power switch (SW). The second input to each AND gate allows a high-speed chopping signal (CH) to limit maximum winding current (Fig. 4). The S and CH signals are supplied by a microcomputer system (Fig. 5). The 8085 microprocessor uses a 27128 ROM, a 2016 RAM, an 8255 programmable peripheral interface device, and an 8253 programmable interval timer. The 8255 creates the step signals, and the 8253 controls the chopping signal.

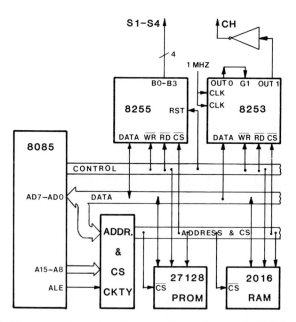

Fig. 5. Microcomputer system for generating motor signals S1–S4 and CH.

45. VLSI in a Complex Medical Instrument

I/O Interface Device C

The 8255 is a general-purpose programmable I/O device with 24 I/O pins. Its function is to interface peripheral devices to the microcomputer system. The 24 I/O lines are grouped into three 8-bit ports. In Fig. 5, lines B0–B3 of the B port provide the step signals S1–S4. This one chip could control up to six motors of four windings each. To create the required sequence for the S1–S4 signals, the following steps must be performed by the microprocessor.

(1) Selecting the 8255 chip, set address lines A0 and A1 to write a control byte (from the data bus), which programs port B as an output port.

(2) Set lines A0 and A1 to output data on bus through port B. Data lines D0–D3 contain the 1010 code for Step 1 (Fig. 3).

(3) After the proper time interval is allowed for the motor to achieve Step 1, the next data (1100) are output to cause Step 2.

This process is continued until all 1000 steps are achieved. The time interval for each step is chosen to allow the motor to accelerate and then decelerate back to rest.

Interval Timer D

The 8253 programmable interval timer has three independent 16-bit counters. Two, C0 and C1, are used here to create the desired CH signal. The 8085 is programmed to

(1) select the 8253 chip, set address lines A0 and A1 to write a control byte (from data bus) which programs counter C0 to operate in a square wave mode;

(2) in similar manner, program counter C1 to operate as a retriggerable one shot;

(3) load each counter in turn with a count value (one or two bytes) which will be down counted to zero.

The operation of the 8253 is then as follows (assume C0 and C1 have count values of 100 and 60, respectively).

(1) C0, counting the 1 MHz clock, has first a high output for 50 μs and then a low output for 50 μs (i.e., a square wave). This establishes the period T (Fig. 4) at 100 μs.

(2) C1, the gate of which is driven by the output of C0, will be low, going high when a count of 60 (or 60 μs) is reached. This low output

passes through the inverter, creating a CH signal that is on (high) 60 μs of each 100 μs.

As the step signals S1–S4 are shortened (lengthened) the duty cycle of CH is increased (decreased). If both the torque-speed curve of the motor and the load-position curve are known, each motor step can be given the required drive current. Motor performance can be optimized without overheating.

III A/D CONVERSION AND DATA ACCUMULATION

As shown in Fig. 1, samples, after suitable preparation, pass to an analog transducer function which converts certain chemical or physical parameters into electrical signals. After amplification and filtering, these signals pass to an A/D conversion–data accumulation function, both of which are techniques for producing digital data.

A A/D Conversion

A/D converter chips are essentially IC implementations of classical A/D concepts. An analog signal, usually ranging between a few millivolts and a few volts, is first sampled by a sample-and-hold circuit, the output of which is then converted to its digital value in binary or BCD. The two key specifications are *conversion rate* and *number of bits output*. Other specifications of interest (e.g., accuracy, linearity, and drift) usually relate to the value of the least significant bit (LSB). Though great advances have been made at the low end of the spectrum, A/D converters above 12 bits and 1-MHz conversion rates presently remain hybrid circuits. Inexpensive A/D chips in panel meters and digital multimeters have displaced analog meters. Even with all the advances in solid state and IC technology, however, single chip, high precision, high speed A/D converters have not yet been achieved. Most likely, higher conversion rates will be achieved more readily than higher precision: speed of response is easier to obtain than greater stability. The LSB of a 16-bit A/D converter, for example, is 153 μV when the full signal is 10 V. The TRW 1025, an 8-bit, 50-MHz chip sells for under $500 whereas the more recent Sony CX 20052, an 8-bit, 20-MHz chip, is reportedly in the $50 range. Achieving a low cost (i.e., $20) 16-bit chip at 1–5 MHz remains a challenge to the industry.

Data Accumulation B

Data accumulation, as defined here, involves the counting of electrical pulses having amplitudes between certain limits. Signals from Coulter® counters, for example, often fall into this category. The limits can be fixed, operator selectable, or adaptive by microcomputer control. VLSI technology has greatly simplified this counting task. Assume, for example, it is necessary to count the pulses occurring in each of four amplitude ranges during a time interval of one second. The circuit of Fig. 6 can be used. The AM9513 system timing controller chip is a flexible peripheral support chip. A complete description is obtainable from Advanced Micro Devices, Inc., P.O. Box 453, Sunnyvale, CA 94086. For our purpose, we note that the 9513 has five general-purpose 16-bit counters with a variety

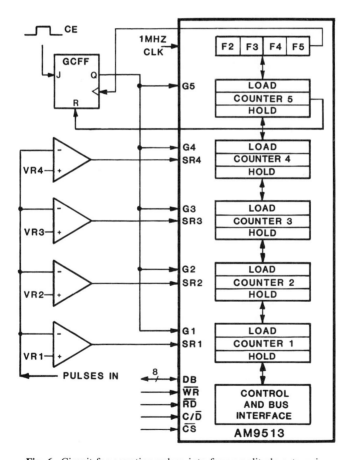

Fig. 6. Circuit for counting pulses into four amplitude categories.

of programmable operating modes. Associated with each counter are a *load register* and a *hold register*. These registers can be used to preload the counters to any count and to hold accumulated values for inspection by or transfer to the microcomputer without disturbing the counter's content. The chip has also a programmable frequency scaler circuit that divides by 10 or 16 at each of four stages, F2–F5.

In Fig. 6, a 1-MHz signal is provided to the X2 input, and the frequency scaler is programmed to divide by 10. The output of F5 is thus a 100-Hz clock signal (i.e., one million divided by ten thousand). Counter 5 is programmed to accept the output of F5 as its count source and to count down with an active high pulse at terminal count (TC). When preloaded with 100 BDC, therefore, this counter establishes the desired one-second interval for pulse counting. The output of F5 also clocks a gate control flip-flop (GCFF) which is discussed below.

The positive-going pulses to be counted are input to the negative inputs of four comparators, the positive inputs of which are referenced to four different reference voltages, VR1–VR4, as shown. As a pulse exceeds the reference voltage of a comparator, its output goes negative true. Counters 1–4 are programmed, therefore, to count up on falling source edges and are preloaded with zero. After programming and initialization of the 9513 by the microcomputer, the counters are armed and ready to count when their gates go high.

The counters' gates are held low by the gate control flip-flop until a *count enable pulse* (CE) and clock F5 turns on the GCFF. After one second, counter 5 outputs a positive pulse, resetting the GCFF and interrupting the microprocessor (connection not shown) which can then interrogate the count accumulated in each of the four counters, 1–4.

The above example was chosen to illustrate the concepts of data accumulation; it is easy to extrapolate this basic approach to a powerful, adaptive configuration. The reference voltages VR1–VR4 could be produced, for example, by the microcomputer outputting to D/A converters. The microcomputer could, moreover, monitor the accumulation of counts as they occur and take several alternative actions as appropriate, such as changing the thresholds or extending the count time.

IV CENTRAL DATA PROCESSOR

The requirements for the CDP are not notably different from other systems requiring a minicomputer or the latest generation of super microcomputers. The system designer can form the CDP system from individual components (e.g., the blocks shown in Fig. 1) or purchase the complete system from one supplier.

Software

The software capabilities of a fully featured instrument system require the use of multitasking operating system, a high level language, and often a hard disk. Compiled code of a high level language executes more slowly and requires more memory than machine or assembly code. In fact, had IC technology not lowered the cost of memory, the software (and hence the instrument system) would not be practical. It would not be acceptable either to write several hundred man years of software in assembly language to hold down memory cost or to write in a high level language with an extremely high memory cost. This trend, to accomplish all software (even motor control) with the highest possible language, will continue as VLSI reduces the cost and increases the speed of memory.

Hardware

The Texas Instruments 990/10A processor is representative of the current hardware available to the system designer. The 990/10A, a full function processor on a single printed circuit board, uses the 16-bit TMS 99000 microprocessor chip, contains up to 512 kilobytes of error-correcting memory (using a memory mapping chip to convert 15-bit logical addresses into 20-bit physical addresses), and has extensive I/O capability.

Error Detection and Correction

As memory size increases, it is worthwhile to consider the need for error correction; for as VLSI technology has packed more bits into each chip, memory requirements have increased. The key factor in evaluating the relative merits of error correction is the *mean failure rate* (MFR) of a single memory chip. Today, MFR figures vary from about 0.2 to 0.02% per 1000 h depending on process and process control, memory type (e.g., static or dynamic), whether the chip is new or has been in production several years, and the type of prestressing and testing done. One should thus expect a chip to fail in some manner, on average, between 5×10^5 and 5×10^6 h. To the present time, the MFR has been, moreover, largely independent of the number of bits on the chip.

Consider a 512-Kbyte memory having a 16-bit word. Error detection requires a 17th (parity) bit, and single-bit error correction and double-bit error detection require six extra bits per word. The increase in chip cost is 29.4%. Including the additional costs of error correction logic, mounting, power supply, testing, and so on, a cost penalty of 40% for error correction could be estimated. If we use 64K chips, our 512-Kbyte memory requires 68 chips for error detection and 88 chips for error correction.

Optimally organized, each bit of each word of an error-correcting memory is contained in a separate chip. If a single bit of one word fails, the probability that the next single-bit error will occur in that same word is negligibly small. Thus, though single-bit errors cause failure in a memory having only error detection, they can by and of themselves be ignored when error correction is used. The argument just made for single-bit errors, moreover, is only less valid for double-bit errors (same chip) by a factor of about 2, triple-bit errors by a factor of about 3, and so on. We may reasonably make the dichotomy that "small" errors involve up to 1% of the total bits on the chip and that the entire device has failed if more than 1% of the bits fail. We note that these "small" errors, occurring by themselves, have only a very small probability of causing memory failure. The error-correcting memory fails, then, by having a total device failure and, additionally, a "small" error or a second device failure. Total MFR can be broken into two parts: *mean entire device failure rate* (MEDFR) and *mean small error failure rate*. If MEDFR can be estimated, the value of error correction can be established. Consider two cases:

Case (1) MEDFR = MFR
Device failures only.
Error detection: system fails at MFR.
Error correction: system fails at $\frac{1}{2}$ MFR.
Enhancement factor = 2.
Net memory improvement: $(2 \times 68/88) = 1.55$ or 55%.

Case (2) MEDFR = 1/10 MFR
Failure of an entire device is infrequent; small errors dominate.
Error detection: system fails at MFR.
Error correction: system fails at 1/10 MFR.
Enhancement factor = 10.
Net memory improvement: $(10 \times 68/88) = 7.73$ or 673%.

Error correction with a 16-bit word thus always gives a mean reliability improvement of 55% and an improvement of several hundred percent when small errors are the dominant mode of chip failure. At a cost increment of 40%, error correction should be justifiable, unless the improvement in reliability is not required. Using a MFR of 0.2% per 1000 h, our 512-byte memory without error detection would fail (on average) every 7340 h or every 10 months at 24 h per day. Error correction would extend the failure period to 15.5 months minimum (Case 1), or to 77 months (Case 2). A final argument for error correction can be made if system up-time is very important. A warning (red light) can alert the operator or serviceman that the first error has occurred. Replacement or repair of the memory before the second error occurs is then probable.

Chapter 46

Impact of VLSI on Speech Processing

RICHARD V. COX

AT&T Bell Laboratories
Murray Hill, New Jersey

I. Introduction	775
II. The Nature of Speech and Speech Processing	777
A. Speech Production	777
B. Main Components of Speech Processing	778
C. Speech Analysis and Synthesis	779
III. Spectral Estimation: Algorithms and Hardware	780
A. Filter Bank Estimation and Linear Prediction Coefficient Analysis	780
B. VLSI Digital Signal Processor Chips	781
IV. Speech Recognition	781
V. Speech Coding and Speech Synthesis	782
A. Comparison of Coding and Synthesis	782
B. The Vocoder	783
C. Current Research	783
VI. The Future	783
References	784

INTRODUCTION   I

Our purpose in these few pages is to acquaint the reader with some of the fundamental problems of speech processing and how VLSI technology has made an impact on them. What we seek to show by our examples is that the increased computation speed and memory storage of VLSI devices are opening up this field. Previous to VLSI, the known algorithms that could be implemented in real time performed too poorly or were too expensive to build, the existing hardware was not fast enough to allow the better algorithms to be implemented in real time, even in the laboratory. VLSI devices are capable of 100 or more times the speed of the laboratory computers that were used to develop the algorithms.

With VLSI, many systems that rely heavily on speech processing such as Texas Instruments' *Speak and Spell,* automobiles that talk to the driver, and speech recognizers that can be added to personal computers have already reached the marketplace. Table I shows the impact of the VLSI breakthrough on three speech processing examples. The simple speech recognition task progressed from a laboratory computer with a dedicated array processor, to 5 boards of hardware, and soon can be expected to be available as a chip set. As a second example, the ADPCM (adaptive differential pulse code modulation) speech coder progressed from a single board to implementation on a multi-purpose chip in 1980, and is now available from NEC on a custom chip. Furthermore, the ADPCM algorithm has become more sophisticated. The third example in Table I is a low bit rate coder. In 1979 a real time implementation would have required a rack of equipment. By 1983 the same coder could be implemented in a chip set produced by Motorola [4].

A common feature of these examples is that they were previously technically feasible. The advent of VLSI made them economically feasible. The ADPCM example is particularly instructive because it illustrates how VLSI can reduce the cost of a product and simultaneously improve its quality and capability by providing increased computational complexity at low cost. When this chip is in widespread use, it will double the capacity of all digital telephone lines. These developments relied on a theory that unifies the areas in speech processing. In this chapter we shall outline this theory and describe some of the basic problems of speech processing and how this theory can be applied to solve them. Then we shall describe how VLSI has made existing solutions economically feasible and is making new solutions possible.

TABLE I

Real Time Hardware Implementations

Task	1/1/79	1/1/83	Near Future
Isolated word recognizer	Laboratory minicomputer + array processor	5 boards, special-purpose hardware	VLSI chip set
ADPCM	1 board	1 DSP chip	Custom chip
LPC-10 vocoder	Rack of hardware	VLSI chip set	Single chip

THE NATURE OF SPEECH AND SPEECH PROCESSING   II

Speech Production   A

To understand the nature of speech, one must begin with a discussion of how speech is produced. Figure 1 is a simplified mechanical model of the speech producing mechanisms of the vocal tract, which is terminated at one end by the vocal cords and at the other end by the mouth. The nasal tract is acoustically linked to the vocal tract when the velum is lowered for production of nasal sounds. All vocal sounds are generated by acoustical excitation of the vocal tract and can be classified into three categories: *fricative* sounds such as /sh/ in sugar; *plosive* sounds such as /b/ in both, and voiced sounds such as vowels. The period of a voiced sound is called the *pitch period* and its fundamental frequency is called the *pitch*. The most noticeable feature of most speakers is their pitch. If we say someone has a "low voice" we mean that they have a low pitch.

Frequency Response of the Vocal Tract   1

The frequency response of the vocal tract conveys the information in speech. Figure 2 shows the short-time spectra of three sounds. Included with each spectrum is a smoothed estimate of the short-time spectrum. This estimate can be considered to be an estimate of the vocal tract acoustic filter frequency response. Note that each estimate contains regions, called *formants*, with high concentrations of speech energy corresponding to the resonant frequencies of the vocal tract. The reason that we can distinguish between different sounds is the different placements of the formants of each sound.

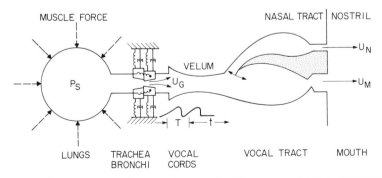

Fig. 1. Engineer's diagram of the vocal tract (after Flanagan *et al.* [6], © 1970 IEEE).

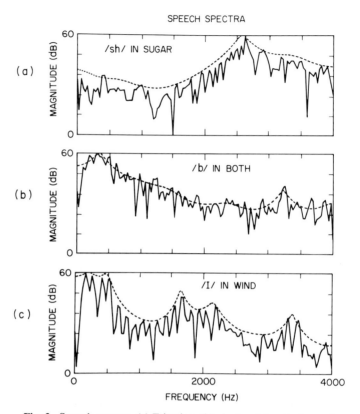

Fig. 2. Speech spectra. (a) Fricative, (b) plosive, (c) voiced sounds.

The Source–Filter Model of Speech Production

These properties of the vocal tract give rise to the source–filter model of speech production (Fig. 3). The excitation source for the vocal tract can be classed as *quasi-periodic pulses* for voiced sounds or *broad-spectrum white noise* for plosives and fricatives. The vocal tract acts as a linear filter on the excitation waveform to produce the sounds we recognize. This model provides a mathematical framework for speech processing. With this framework, a speech waveform can be separated into an *excitation signal* and a *time-varying linear filter* that models the vocal tract.

Main Components of Speech Processing

Speech processing can be divided into three main areas: speech synthesis, speech recognition, and speech coding.

46. The Impact of VLSI on Speech Processing

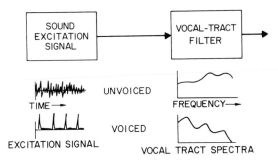

Fig. 3. Source-filter model of speech production.

(a) *Speech synthesis* is the representation of the linear filter by a few parameters and its combination with an artificial excitation signal to produce synthetic speech. The naturalness of the speech is determined by how well these two approximations match their natural counterparts.

(b) *Speech recognition* is based on the extraction of the linear filter parameters from a real speech waveform. The degree to which the extracted parameters match stored parameters of known speech sounds determines which sounds the computer can recognize.

(c) Speech coding is the representation of an original speech waveform with less information than is required for reproducing the exact waveform values. The object of speech coding is to do this without degrading the perceived quality of the reproduced waveform.

The basis for the solutions of the problems in each of these three areas is the source–filter model.†

Speech Analysis and Synthesis C

Figure 4 shows the block diagrams for speech analysis and synthesis subsystems. The important elements in these diagrams are the blocks labeled *Digital speech analysis* and *Digital speech synthesis*. These are the processes of estimating the source and filter parameters for the vocal tract (for the analysis subsystem) and synthesizing speech using these parameters (for the synthesis subsystem). By choosing to implement these processes digitally we require the analog signal to be digitized in the analysis subsystem and the resulting digital signal to be changed to analog form in

---

† For readers interested in the nature of speech and speech processing, reference [1] is a comprehensive treatment of this entire area, including all of the fundamental theory and important algorithms.

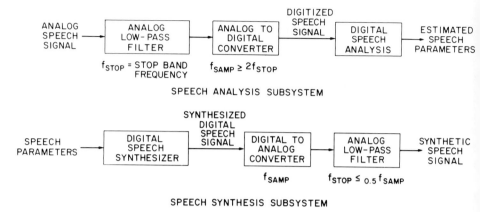

Fig. 4. (a) Speech analysis subsystem. (b) Speech synthesis subsystem.

the synthesis subsystem. The extra hardware needed for these tasks is more than offset by the great flexibility of digital processing. This is the area in which VLSI has had, and will continue to have, an impact on speech processing by providing high-speed, flexible, inexpensive digital processors. As an example of all three areas of speech processing, we shall consider spectral estimation.

SPECTRAL ESTIMATION: ALGORITHMS AND HARDWARE

Filter Bank Estimation and Linear Prediction Coefficient Analysis

One of the central problems in speech processing is the estimation of the vocal tract filter in the source–filter model. Estimates of the physical parameters of the vocal tract are too difficult; estimates of the frequency response must suffice. Two methods have received the most attention: filter bank estimation and linear prediction coefficient (LPC) analysis. In *filter bank estimation,* the frequency response is estimated as the output of a series of bandpass filters covering the range of the speech spectrum. The filter bank can be implemented by using digital filters on VLSI devices. *LPC analysis* is based on the assumption that the vocal tract filter can be modeled as an all-pole digital filter. As we shall see, LPC analysis and synthesis are used as a foundation for a number of VLSI speech-processing implementations.

One measure of the complexity of algorithms is the number of instructions per second that they require to be performed in real time. The resolution of the filter banks determines the complexity of the filter bank method. For speech recognition about 15 bands are needed. The rough number of multiplications and additions per second needed to implement such a filter bank is one million instructions per second (Mips). Similarly, the number of LPC coefficients computed and the frequency of their update determines the complexity of the LPC algorithm. For many applications, we need about 0.25 to 0.5 Mips to obtain LPC coefficients in real time.

VLSI Digital Signal Processor Chips B

The first VLSI digital signal processor chips, such as the Bell Labs' DSP [2], were designed to do digital filtering. They have been used for speech applications including digital filter bank and LPC analysis implementations. These chips typically contain a hardware multiplier and accumulator, a separate arithmetic processor for addressing, a small amount of on-chip RAM for data storage, and a larger amount of on-chip ROM for program instructions and coefficient storage. The performance measure for such chips is the maximum number of simultaneous multiplications and additions that can be performed per second. Most of the available DSP chips have speeds of from 1 to 5 Mips. However, this number represents 100% utilization of the arithmetic processor. Such rates cannot be achieved in practice because of the nature of the algorithms.

A recently designed Motorola LPC analyzer chip [3] performs 400,000 multiplications, 800,000 additions, and 80,000 divisions per second for implementation of real time LPC analysis. Divisions usually take more time to compute than multiplications. The advent of VLSI has made fast dividers possible. A number of the papers in references [3–5] are about VLSI chips or chip sets that can implement LPC analysis and/or synthesis.

SPEECH RECOGNITION IV

Figure 5 is a simplified diagram of a speech recognition system that is based on conventional pattern recognition. Speech is initially processed by a spectral analysis system, usually one of the methods mentioned above. The purpose of the analysis system is to extract feature data based on the spectral information of the speech. A set of *labeled reference patterns* or *templates* must be provided to the recognition system. The preparation of this set is referred to as *pattern training*. The templates are

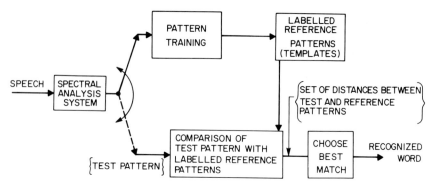

Fig. 5. Elements of a speech recognition system.

stored as a sequence of feature data. When an unknown word (*test pattern*) is presented to the system, it is also converted into a time series of feature data. Next, the test pattern for the unknown word is compared with the templates of all words in the vocabulary. To do this, a *spectral distance metric* is required to measure the similarity between the test word and each of the word templates. The vocabulary word whose template most closely matches the test pattern is chosen as the recognized word.

SPEECH CODING AND SPEECH SYNTHESIS

Comparison of Coding and Synthesis

Speech coding and speech synthesis are different, but many of the processing procedures are applicable to both. The input to a speech coder is speech, while the input to a synthesizer is text. At the transmitter, the coder analyzes the speech and forms a string of digital codewords to represent it. At the receiver, the codewords are transformed back to a speech waveform and played out. A synthesizer must translate text into a similar string of digital codewords that can be input to a receiver to be played out as speech. The synthesizer and coder can use the same hardware for the receiver/synthesizer. In addition, the off-line process of forming and storing a digital codeword vocabulary for the synthesizer can use the same hardware as the transmitter for the coder.

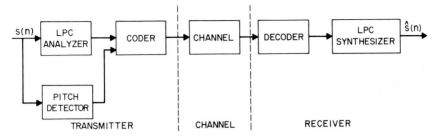

Fig. 6. LPC vocoder system.

The Vocoder   B

The source–filter model of speech production gives rise to a mathematical model that can be used for either coding or synthesis. The simplest example of using this model as a method of coding speech is the classical *vocoder* (from the contraction of the words *voice coder*). The most popular form of vocoder today is based on LPC spectral analysis (Fig. 6). The government standard vocoder is called LPC-10, because it uses a 10-pole filter to model the vocal tract. A number of special-purpose speech processing chips have been developed with VLSI technology to implement LPC analysis and synthesis for the government's LPC-10 vocoder.

Current Research   C

Current research on speech coding and synthesis has two goals: to improve the quality of synthetic speech without increasing the data rate and to reduce the data rate without degrading the quality. Algorithms that achieve either objective usually require more memory or higher computational speech or both. Without VLSI technology, such improvements would be difficult, if not impossible.

THE FUTURE   VI

The Texas Instruments LPC speech synthesizer chip and the programmable digital signal processor chips were the first devices to demonstrate the economic feasibility of speech processing. For any speech processing task, there is a tradeoff between quality and the computational complexity of the algorithm performing the task. In the past, the complexity of many algorithms made them economically unfeasible. With the increased digital signal processing capabilities of VLSI devices, these algorithms are now feasible. In addition, VLSI also makes even larger tasks possible. For the future, we can expect technical developments in both these directions.

REFERENCES

1. L. R. Rabiner and R. W. Schafer, "Digital Processing of Speech Signals." Prentice Hall, Englewood Cliffs, New Jersey, 1978.
2. *Bell System Tech. J.* **60** (7), Part 2 (September, 1981). Special issue on the digital signal processor.
3. *IEEE Trans. Acoust., Speech, Signal Process.* **31** (1), Part II, pp. 247–348 (February, 1983). Joint special issue on integrated circuits for speech.
4. *Proc. 1982 IEEE Int. Conf. Acoustics, Speech and Signal Processing,* pp. 510–528, 1049–1091.
5. *Proc. 1983 IEEE Int. Conf. Acoustics, Speech and Signal Processing,* pp. 471–522.
6. J. L. Flanagan, C. H. Coker, L. R. Rabiner, R. W. Schafer, and N. Umeda, Synthetic Voices for Computers, *IEEE Spectrum,* **7**, (10), 24, (October 1970).

Chapter 47

Application of VLSI to Pattern Recognition and Image Processing

TZAY Y. YOUNG
PHILIP S. LIU
HSI-HO LIU

Department of Electrical and Computer Engineering
University of Miami
Coral Gables, Florida

I. Introduction	785
A. Computer-Vision Levels	786
B. System Components	786
II. Image Processing and Pattern Recognition	787
A. Image Transforms and Enhancement	787
B. Image Segmentation and Description	788
C. Pattern Recognition	788
III. VLSI Architectures for Pattern Recognition	789
A. Arrays for Matrix Operations	789
B. Arrays for Statistical Pattern Recognition	791
C. Arrays for Syntactic Pattern Recognition	793
IV. LSI/VLSI Image Processing: Architectures and Systems	795
A. VLSI Arrays for Discrete Fourier Transform and Image Filtering	795
B. Arrays for Spatial Domain Operations	796
C. LSI and VLSI Image Processing Systems	797
References	799

INTRODUCTION I

Pattern recognition and image processing have applications in many areas, including speech recognition, character recognition, remote sensing, fingerprint analysis, biomedical signals and image processing, industrial automation, and robotics. For image processing and recognition, the primary motivations are to develop and construct computer-based vision systems. The goals and requirements of computer vision systems depend on specific applications.

A Computer-Vision Levels

Computer vision can be divided into low-level, intermediate-level, and high-level vision.

(a) Low-level vision techniques are primarily concerned with segmenting or partitioning images into line boundaries and regions.

(b) Intermediate-level techniques include analysis of two-dimensional shapes, texture, perspective, shadows and occlusions, and relationships between regions. In this way, regions obtained by segmentation are transformed into more descriptive forms.

(c) High-level vision techniques are concerned with the recognition of objects and the interpretation of relevant components of an image in the context of the goals and prior knowledge.

B System Components

A block diagram of a computer vision system is shown in Fig. 1. The system consists of four major components—*preprocessing, segmentation, description,* and *recognition*—and there may be feedback paths between the blocks. The last three blocks correspond roughly to the three levels of vision. The purpose of preprocessing is to enhance the image and restore any degradation caused by the sensor and the environment of the scene; it may or may not be necessary in a computer vision system.

A digital image consists of the gray levels of a large number of pixels (picture elements), for instance, $256 \times 256 = 65,536$ pixels. Many image processing operations are performed repeatedly over the pixels. Both preprocessing and segmentation are high-throughput operations; processing speed can be improved significantly by parallel processing and pipelin-

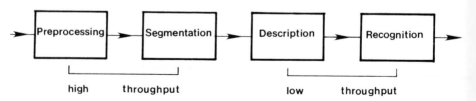

Fig. 1. Block diagram of a computer vision system.

ing, using special VLSI architecture [1]. Description and recognition require sophisticated techniques to perform feature extraction, region description, scene analysis, and object identification by statistical or syntactic methods. These tasks can be shared by a set of processors in a multiprocessor system to improve computation speed.

IMAGE PROCESSING AND PATTERN RECOGNITION II

Image Transforms and Enhancement A

Transforms 1

Let $\mathbf{Y} = [y_{ij}] = [y(i,j)]$ represent the gray levels of the pixels of an $n \times n$ image. A two-dimensional discrete Fourier transform (DFT) can be defined as

$$z_{k\ell} = \sum_{i=0}^{n-1} \sum_{j=0}^{n-1} y_{ij} \exp[-j2\pi(ki + jl)/n], \qquad (1)$$

where we have used j to represent both the unit imaginary number and an index. Clearly, we can define a matrix

$$\mathbf{A} = [a(k,i)] = [\exp(-j2\pi ki/n)] \qquad (2)$$

and can express Eq. (1) in matrix form,

$$\mathbf{Z} = \mathbf{AYA}. \qquad (3)$$

The inverse transform can be expressed in a similar manner.

Enhancement 2

The two-dimensional DFT is a frequency-domain representation of the image. Frequency-domain methods for image enhancement modify the DFT of an image by two-dimensional filtering followed by inverse DFT. For example, noise in an image usually appears in a salt-and-pepper fashion and contains significant high-frequency components. *Low-pass,* or *smoothing, filters* can be used to diminish spurious effect caused by the noise or sometimes by inappropriate sampling of the image. On the other hand, the edges of objects in an image can be highlighted and sharpened by high-frequency emphasis filters. These two-dimensional filters can be easily specified in frequency domain.

B Image Segmentation and Description

1 Segmentation

A popular approach to segmentation is the detection of the edges of an object by template matching. To match a 3×3 template or window, $\mathbf{W} = [w(k,l)]$, we need to compute

$$\hat{y}(i,j) = \sum_{k=-1}^{1} \sum_{l=-1}^{1} w(k,l)y(i+k, j+l), \tag{4}$$

and then compare the result with a predetermined decision threshold. By selecting the weighting coefficients $w(k,l)$ properly, template matching can be used for various spatial tasks including line and edge detection.

2 Description

Region description characterizes a region by its measured properties and its relationships to other regions. *Two-dimensional shape* is an important characteristic of a region, which can be described by Fourier descriptors. An image plane can be regarded as a complex plane with real and imaginary coordinates. A point in the image becomes a complex number in the complex plane. The edges of a region form a closed contour; thus, moving around the contour generates a complex-valued periodic function that can be expanded into a Fourier series. Fourier descriptors are defined in terms of the Fourier coefficients in such a way that they are invariant to translation, rotation, and the size of the region.

Another set of invariant shape descriptors are the *invariant moments*. The size and orientation of an object can be treated as additional properties. For texture analysis, a set of statistical texture measures has been developed. The descriptors and measures describe a region or an object with a set of numerical values that are suitable for machine recognition using well-developed statistical pattern recognition techniques.

C Pattern Recognition

In statistical pattern recognition, a set of n measurements of an object is regarded as an n-dimensional vector in a vector space. The vector space is divided into several decision regions based on statistical analysis or decision theory.

Let $\mathbf{x}$ be an n-dimensional column vector. If each class of the pattern vectors is Gaussian distributed, the discriminant function for class k may be expressed as

$$D_k(\mathbf{x}) = -2 \log p_k(\mathbf{x})$$
$$= (\mathbf{x} - \boldsymbol{\mu}_k)^T \mathbf{R}_k^{-1}(\mathbf{x} - \boldsymbol{\mu}_k) + \log(2\pi)^n |\mathbf{R}_k|, \qquad (5)$$

where $\mathbf{R}_k$ and $\boldsymbol{\mu}_k$ are covariance matrix and mean vector, respectively. This is a quadratic discriminant function, and the pattern vector $\mathbf{x}$ is assigned to class k if $D_k(\mathbf{x})$ is minimum among the values of discriminant functions of all classes. The computation of $D_k(\mathbf{x})$ requires that $\mathbf{R}_k$ and $\boldsymbol{\mu}_k$ be known. If they are unknown, $\mathbf{R}_k$ and $\boldsymbol{\mu}_k$ must be estimated from a set of pattern vectors that are known to belong to class k.

If the covariance matrices are identical, the pairwise discriminant function for class j and class k is linear,

$$D_{jk}(\mathbf{x}) = D_j(\mathbf{x}) - D_k(\mathbf{x})$$
$$= 2\mathbf{x}^T \mathbf{R}^{-1}(\boldsymbol{\mu}_k - \boldsymbol{\mu}_j) - \boldsymbol{\mu}_k^T \mathbf{R}^{-1} \boldsymbol{\mu}_k + \boldsymbol{\mu}_j^T \mathbf{R}^{-1} \boldsymbol{\mu}_j. \qquad (6)$$

Class j is preferred if $\mathbf{D}_{jk}(\mathbf{x}) < 0$.

The *syntactic pattern recognition approach* describes complex patterns by means of a set of *pattern primitives*, which could be line segments, regions, or components of objects, and *structure rules*, which define the relationships between the primitives. The syntactic approach is based on formal language theory. The primitives are regarded as alphabets; together with the structure rules they form the grammar of a language. String grammars, tree grammars, and graph grammars have been proposed for pattern recognition purposes. The recognition of an object is accomplished by parsing techniques that compare the relationships between the primitives with the structure rules of the grammar. It is noted that the complex patterns of the objects must have an underlying structure for the syntactic method to be successful.

VLSI ARCHITECTURES FOR PATTERN RECOGNITION   III

Arrays for Matrix Operations   A

It is clear from Section 2, that matrix multiplication and covariance matrix inversion are essential steps in statistical pattern recognition. VLSI systolic and wavefront arrays have been proposed that are particularly suited to matrix operations. These arrays have the desirable features of simple processing elements (PEs), simple communication and control structures, and regular data-flow patterns. One of the simplest array structures is the orthogonally connected square array [1].

1 Matrix Multiplication

Consider the multiplication of two $n \times n$ matrices,

$$\mathbf{Z} = \mathbf{AB}. \tag{7}$$

If we assume the matrix A is already inside a square array, as shown in Fig. 2, **B** can be piped in to interact with **A** to produce the multiplication result. During each computation cycle, all the cells in the array perform the same multiplication step: Each cell PE_{ij} takes the sum of partial products from its left neighbor, adds it to its partial product of $a \times b$, and then passes the sum to its right neighbor for the next multiplication step. The computation time required is $(4n - 2)$ units, including n units of matrix **A** loading time.

2 Matrix Inversion

Matrix inversion involves several computation steps that can be implemented on a reconfigurable array [2].

(a) The first step in the computation of the inverse of a symmetric nonsingular covariance matrix $\mathbf{R} = [r_{ij}]$ is to decompose it into an upper triangular matrix $\mathbf{U} = [u_{ij}]$ and a lower triangular matrix $\mathbf{L} = [l_{ij}]$. In Fig. 3a, the orthogonally connected square array is reconfigurated along the diagonal to become an upper triangular array for $\mathbf{L} - \mathbf{U}$ decomposition

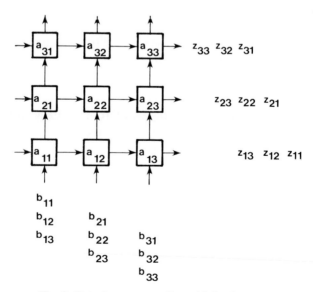

Fig. 2. Data-flow pattern of a multiplication array.

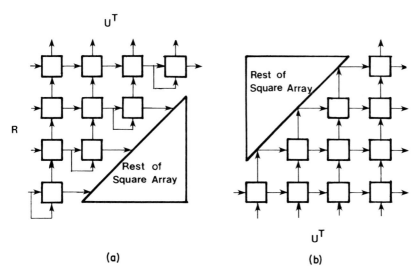

Fig. 3. (a) Reconfigured triangular array for $L - U$ decomposition. (b) Computation of $V = U^{-1}$ on a reconfigured lower triangular array.

[2]. The skewed rows of the matrix R are piped into the triangular array in parallel as inputs. Each PE in the array has two inputs and two outputs that are buffered or latched internally and an internal register to store l_{ij}. The PEs operate in two modes, the multiplication mode and the division mode; it is assumed that each operation takes one unit of time to complete.

(b) The next step for covariance matrix inversion is the computation of $V = U^{-1}$ using a reconfigurated lower triangular array, as shown in Fig. 3b. The matrix U is piped back without delay to the bottom of the array for the computation of V; the latter remains in the triangular array after the computation.

(c) Step 3 is the computation of the inverse of L, which is denoted by $M = [m_{ij}]$. Since R is symmetric, $m_{ji} = v_{ij} u_{jj}$, and M can be easily computed from V. The last step is the computation of $R^{-1} = U^{-1} L^{-1} = VM$, using the square array as a multiplication array. With the computation overlap of the steps accounted for, the total computation time of the four steps for covariance matrix inversion is $(13n - 4)$ units.

Arrays for Statistical Pattern Recognition B

A statistical pattern analysis array that consists of a square array and a linear array is shown in Fig. 4a. The proposed array processor [3,4] is

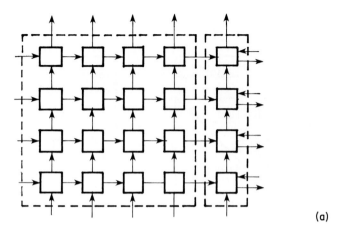

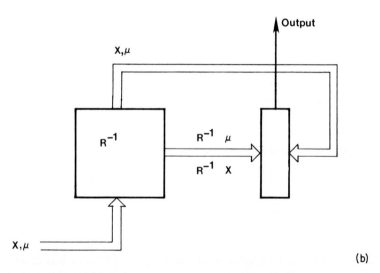

Fig. 4. (a) A reconfigurable pattern analysis array processor. (b) Computation of quadratic discriminant function.

reconfigurable in the sense that data paths can be changed by modifying the communication structure between the two arrays and/or the structures among the PEs.

1 The Square Array

Each PE of the square array has two inputs and two outputs, and a PE can have several internal storage registers. It can perform three types of operations:

47. Application of VLSI to Pattern Recognition

(a) data transfer and loading,
(b) multiplication and addition (or subtraction), and
(c) division.

The square array can be reconfigured into upper and lower triangular arrays as discussed in Section 3.A. Thus, $\mathbf{R}^{-1}$ can be computed for a given covariance matrix $\mathbf{R}$ and then piped back and stored in the array. The block diagram for computing the quadratic term of the quadratic discriminant function is shown in Fig. 4b, where $\mathbf{X} = [\mathbf{x}_1, \mathbf{x}_2, \ldots, \mathbf{x}_m]$ represents m sample vectors being classified. The quadratic term can be expressed as $(\mathbf{x} - \boldsymbol{\mu})^T(\mathbf{R}^{-1}\mathbf{x} - \mathbf{R}^{-1}\boldsymbol{\mu})$. With $\mathbf{R}^{-1}$ stored in the square array, the array essentially performs matrix multiplication.

The Linear Array 2

For the linear array, each PE has three inputs and two outputs, and its operations are very similar to those of the PEs of the square array. The skewed outputs of the square array, $\mathbf{R}^{-1}\boldsymbol{\mu}$ and $\mathbf{R}^{-1}\mathbf{x}$, are piped into the linear array; at the same time, the skewed data streams $\boldsymbol{\mu}$ and $\mathbf{X}$, having passed through the square array, enter the linear array from the other side. The vectors $\boldsymbol{\mu}$ and $\mathbf{R}^{-1}\boldsymbol{\mu}$ are stored in the internal registers of the linear array, before the array executes the appropriate multiplication and subtraction operations. The computation time is $(2n + m)$ units. It is noted that by changing data paths between the square array and the linear array, the array processor can also be used to compute the linear discriminant function and estimate $\boldsymbol{\mu}$ and $\mathbf{R}$ from a set of sample vectors belonging to the same class [3,4].

Arrays for Syntactic Pattern Recognition C

A formal string grammar is a four-tuple $G = (N, \Sigma, P, S)$, where N is a set of nonterminals, Σ is a set of terminals, P is a set of productions or rewriting rules, and S is the start symbol. A language $L(G)$ is a set of sentences or strings composed of terminals only; each string can be derived from S by proper applications of productions from the set P. Parsing or syntax analysis is the procedure for finding a derivation of a given string. It could be a top-down or a bottom-up procedure.

Triangular Array 1

The Cocke–Younger–Kasami (CYK) parsing algorithm can be implemented by an upper triangular array, as shown in Fig. 5 [3,4]. The input string of length n enters the array from the diagonal of the triangle. Each

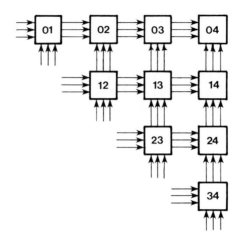

Fig. 5. An upper triangular array for syntactic pattern recognition. (From Y. P. Chiang and K. S. Fu, Proc. Workshop CAPAIDM, pp. 220–226, © 1983 IEEE.)

PE_{ij} computes and stores the parsing results of the substring consisting of the symbols i through $j - 1$ ($0 \leq i < j \leq n$). More specifically, PE_{ij} computes recurrently the set $\{A \mid \text{for some } k, i < k < j, A \rightarrow BC \text{ in } P, B \text{ in } PE_{ik}, C \text{ in } PE_{kj}\}$, where A, B, C, are nonterminals. The PEs with the same value of $(j - i)$ compute in parallel. Three channels—fast belt, slow belt, and control lines—are connected between adjacent PEs. Data in the fast belt move from one PE to another in one unit of time, whereas data in the slow belt take two time units. Data in the fast belt are copied to the slow belt at certain stages determined by the control signals; $2n$ time units are required before the final results are available at PE_{on}. Earley's parsing algorithm can be implemented by using a similar triangular array.

2 Calculation of String Distances

When pattern grammars are not available or when a complete description is not required, decision theoretical approaches based on string distances can be applied in a manner similar to statistical pattern recognition. With syntactic representation, the distance between two strings is defined as the smallest number of insertions, deletions, and substitutions of string symbols required to derive one string from the other. For two strings x_1, x_2, ..., x_n and y_1, y_2, ..., y_n, the partial distance at the (i,j)th step is $\delta(i,j) = \min\{\delta(i,j - 1) + I(y_j), \delta(i - 1, j - 1) + S(x_i, y_j), \delta(i - 1, j) + D(x_1)\}$, where I, S, and D stand for insertion, substitution, and deletion, respectively. An orthogonally connected, $m \times n$ systolic array can be used to compute string distances; data in $PE_{i-1,j-1}$ are passed to PE_{ij} via $PE_{i-1,j}$.

LSI/VLSI IMAGE PROCESSING: ARCHITECTURES AND SYSTEMS IV

VLSI Arrays for Discrete Fourier Transform and Image Filtering A

The choice of DFT or FFT (fast Fourier transform) for future VLSI image processing is a tradeoff between computation costs and communication and control costs.

Discrete Fourier Transform 1

The block diagram for the computation of the DFT, $\mathbf{Z} = \mathbf{AYA}$, is shown in Fig. 6. It is assumed that both $\mathbf{Y}$ (the image) and $\mathbf{A}$ (the Fourier matrix) have been loaded into the orthogonally connected square array. The matrix $\mathbf{A}$ is taken out from the top of the array and piped back from below, so that it interacts with $\mathbf{Y}$ to form a matrix product $\mathbf{YA}$. The next step is to feed $\mathbf{YA}$ back into the array to interact with the stored $\mathbf{A}$ for the computation of the two-dimensional DFT. The inverse DFT can be calculated in a similar manner. Thus, with a frequency-domain filtering function loaded into the array, two-dimensional image filtering can be implemented [1,3].

One-Dimensional Fast Fourier Transform 2

Fast Fourier transform (FFT) promises a significant reduction in the number of multiplications required for calculating DFT. LSI processors

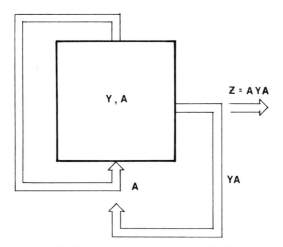

Fig. 6. Computation of DFT using a square matrix multiplication array.

for one-dimensional FFT are available commercially [1]. The heart of a FFT processor is a high-speed multiplier–accumulator (MAC) that performs the basic butterfly computations in FFT. A single-chip MAC has been developed by TRW, which is capable of performing a 16 × 16-bit multiplication in 100 ns. With a single MAC, the total computation of a one-dimensional 1024-point FFT takes less than 6 ms.

3 Two-Dimensional Fast Fourier Transform

A two-dimensional FFT can be computed by first calculating one-dimensional row FFTs and then column FFTs. If $n^2/2$ PEs are used, a two-dimensional FFT requires only $2 \log n$ units of time. The FFT algorithm sends the outputs of its PEs back to the inputs of the same set of PEs, and it requires rather complicated branching or switching circuits. The proposed DFT array is slower, but it has a regular data-flow pattern and simple communication and control structure.

B Arrays for Spatial Domain Operations

There are several possible approaches to VLSI implementation of template matching.

(a) The first approach [5] for a 3 × 3 template uses $(n - 2)$ kernel cells, each kernel cell consisting of nine basic PEs to form three linear arrays, plus a row interface PE to add the results emanated from the linear arrays. Three data streams enter a kernel cell, two of which exit from the cell after necessary computations and then enter the next kernel cell. If the template size is large, computation speed can be improved by using a single kernel cell. At Jet Propulsion Laboratory, a 35 × 35-element pipelined convolution kernel has been fabricated with VLSI chips, each of which contains five PEs [4]; 245 such chips are connected together to represent a 35 × 35 template.

(b) It is desirable to implement frequency-domain and spatial-domain operations on the same VLSI array. It turns out that template matching can be implemented with matrix multiplications, and hence the computations is similar to computing DFT. As many as four matrix multiplication operations are needed for a 3 × 3 template [1]. The values of the matrix elements must be precomputed from the given weighting coefficients, possibly at the host computer.

(c) In the third approach, the image is stored in an $n \times n$ orthogonally connected square array, and the array is conceived as overlapping templates [3,4]. With 3 × 3 templates, the (i,j)th template computation involves nine PEs, in the following order: $PE_{i-1,j-1}$, $PE_{i-1,j}$, $PE_{i-1,j+1}$, $PE_{i,j+1}$, $PE_{i+1,j+1}$, $PE_{i+1,j}$, $PE_{i+1,j-1}$, $PE_{i,j-1}$, $PE_{i,j}$. All n^2 PEs perform multiply and add operations at the same time, and it takes only 9 units of time to complete the template-matching computation of the image. Each coefficient is presented to all PEs simultaneously, and all PEs are required to be active from the beginning. This implies the broadcasting of weighting coefficients and control commands to the PEs. Thus, the method is based on a combination of systolic and cellular array concepts.

C. LSI and VLSI Image Processing Systems

1. Cellular-Array Processor

Most existing parallel image processing systems are cellular-array processors which are basically SIMD (single instruction, multiple datastream) machines. The array processor consists of a number of identical processing cells that may have some form of topological interconnections. Each cell has a PE and a local memory. The interconnection allows data to be passed from one PE to a number of other PEs. There is a single control unit that broadcasts instructions one at a time to all PEs. All active PEs must execute a broadcast instruction synchronously and operate on their own data stream respectively. Data or pixel values can enter the system either directly through a processing cell or through the control unit. By connecting a processing cell to its 4, 6, or 8 nearest neighbors, spatial-domain operations for image enhancement and segmentation can be executed in parallel for each pixel in the array processor.

2. Massively Parallel Processor

The massively parallel processor (MPP) of Goodyear Aerospace is developed to process large amounts of image data at high speed [4,6]. The array unit (ARU) provides the parallel image processing power and is under the control of microprograms that execute inside the array control unit (ACU). The structure of the ARU is shown in Fig. 7. The processor plane consists of 16,384 PEs arranged as a 128 × 128 square array. Each PE can transfer single-bit data or communicate under program control with one of its four nearest neighbors: up, down, right, and left. Eight PEs are fabricated into a VLSI CMOS chip. In each PE, there are six one-bit registers, a shift register with depth up to 30 bits which plays the same role as an accumulator in a conventional computer, an internal data bus, a one-bit full adder, and some combinatorial circuits. A PE can execute arithmetic operations and calculate any of the 16 Boolean functions of two

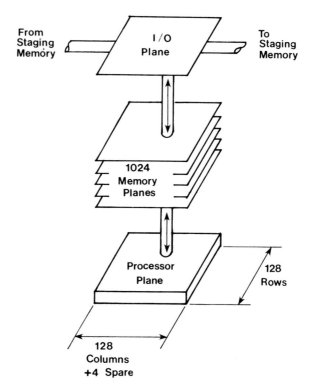

Fig. 7. Array unit of the MPP. (From K. E. Batcher, Proc. Workshop CAPAIDM, pp. 170–174, © 1983 IEEE.)

variables. All PEs execute in the SIMD mode, and an image is processed bit-plane by bit-plane. Each PE has 1024 bits of memory as provided by the 1024 memory planes. The I/O plane sends and receives 128×128 bits of image or subimage column by column to and from a staging memory.

The basic cycle time of the MPP is 100 ns. Special high-level language is available to program parallel processing operations of the ARU. Table I shows the estimated execution speed with respect to often-used image processing algorithms, assuming a frame size of $512 \times 512 \times 8$ bits.

3 Future Processors

In the future, hierarchical and reconfigurable general-purpose image processing and recognition systems will be implemented with VLSI components.

(a) The lowest level would be cellular arrays and/or systolic arrays to handle low-level parallel-image processing tasks.

TABLE I

The Estimates of Algorithms Executed on MPP[a]

Algorithm	Time (μs)
Median filter	282 per frame
Histogram	205,712 per frame
Threshold	37 per frame
Region growing	528 per frame
Coordinate search	230 per frame
Two dimensional cross correlation	128 per target

[a] From J. L. Potter, *Proc. Workshop Computer Architecture for Pattern Analysis and Image Database Management*, pp. 51–56 (1981). © 1981 IEEE.

(b) The second level could be a number of multiprocessors that handle description and recognition tasks and call upon the first-level processors to perform the necessary low-level operations.

(c) The top level could be a master processor that handles scheduling tasks to optimize system throughput and also I/O functions with the outside world.

REFERENCES

1. T. Y. Young and P. S. Liu, Impact of VLSI on pattern recognition and image processing, in "VLSI Electronics: Microstructure Science, vol. 4 (N. G. Einspurch, ed.), pp. 319–360. Academic, New York, 1982.
2. P. S. Liu and T. Y. Young, VLSI array design under constraint of limited I/O bandwidth, *IEEE Trans. Comput.* **C-32,** 1160–1170, (December 1983).
3. K. S. Fu (ed.), "VLSI in Pattern Recognition and Image Processing," Springer-Verlag, New York, 1984.
4. *Proceedings Workshop on Computer Architecture for Pattern Analysis and Image Database Management*, IEEE Computer Society Press, Silver Spring, Md, 1981 and 1983.
5. H. T. Kung, Special-purpose devices for signal and image processing: An opportunity in VLSI, *Tech. Rep. CS-80-132*. Dept. Computer Science, Carnegie-Mellon University, Pittsburgh, Pennsylvania, 1980.
6. *Computer Magazine* **16** (1), (January, 1983). Special issue on computer architectures for image processing.

Chapter 48
VLSI Approach to FM Detection

SHARBEL E. NOUJAIM
General Electric Company
Schenectady, New York

I. Introduction	801
II. Frequency Modulation	802
A. Fundamental Concepts	802
B. FM Detection: Classical Approach	802
III. VLSI Approach to Frequency Modulation Detection	803
A. Mean-Frequency Estimation	804
B. Digital Double-Correlation Mean-Frequency Detector (Estimator)	804
IV. System Implementation Example: An Implantable Pulsed Doppler Flowmeter	807
A. Doppler Ultrasound	807
B. System Description	809
C. Implementation of System Major Blocks	811
D. System Integration	815
E. In-vivo Data	819
V. Conclusion	820
References	820

INTRODUCTION   I

In the past ten years the emergence of high speed, high density digital electronics has had significant impact on systems previously implemented with analog signal processing architecture [1–5]. The critical enabling technologies to this transition have been the analog-to-digital converters and digital filtering techniques.

This chapter is an outline of the major steps in applying the VLSI technology to a system previously realizable only with discrete analog components. Included in the analysis is a discussion of a new architecture for a highly accurate mean-frequency detector that is suitable for monolithic VLSI implementation.

II FREQUENCY MODULATION

In contrast to linear modulation (AM), frequency modulation (FM) is a nonlinear process. The baseband signal (message waveform) varies the frequency of the carrier while the carrier amplitude remains constant [6,7].

A Fundamental Concepts

In FM modulation, the modulated waveform is expressed as

$$x_c(t) = A_c \cos[\omega_c t + 2\pi f_\Delta \int_{-\infty}^{t} x(\lambda) \, d\lambda], \tag{1}$$

where A_c, ω_c are the amplitude and frequency of the carrier, respectively, f_Δ is the frequency-deviation constant, $x(t)$ is the message waveform, and the instantaneous frequency deviation of the FM signal is defined as

$$f(t) = f_\Delta x(t). \tag{2}$$

It is assumed that the message has no dc component; that is, $\bar{x} = 0$. Otherwise, the integral in Eq. (1) would diverge as $t \to \infty$. Practically, any dc message component is usually blocked in the modulator circuits.

B FM Detection: Classical Approach

The receiving system of Fig. 1 is universally used to recover a frequency-modulated signal. The input signal is amplified in an *rf amplifier* and passed onto a *mixer* where the modulated rf carrier is mixed with a waveform generated by a local oscillator that operates at a frequency f_{osc}. The process of mixing generates sum and difference frequencies. The sum is rejected by a filter, and the difference, named the *intermediate frequency* (IF = 10.7 MHz), is passed through the IF filter to the *demodulator*, in which the baseband signal is recovered.

In the FM *discriminator* of the demodulator, (Fig. 1b) the transfer of the signal through a network that has a transfer-function magnitude $|H(f)|$ that varies linearly with frequency over some range, is essential to the

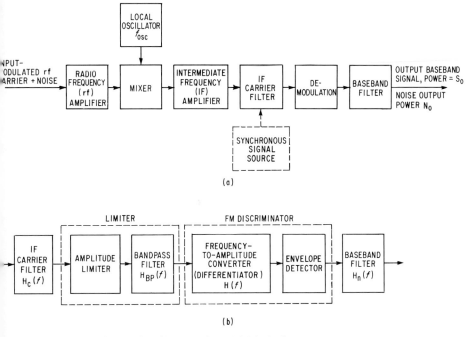

Fig. 1. Conventional FM receiver. (a) Block diagram. (b) Demodulator.

recovery of the baseband. There are many networks that can provide such linear variation [6,7]. One such network is a *differentiator*. In addition, for the discriminator to respond only to change in the instantaneous frequency of the input signal, the signal has to pass through an amplitude *limiter* to remove any amplitude variations.

Commonly, the IF filter and the limiter consist of a transistor-amplifier stage operating into a tuned resonant collector circuit. This tuned circuit requires coils and capacitors, which makes it unsuitable for VLSI implementation.

To design a highly integrable single-chip FM demodulator, a new approach based on the advantages offered by VLSI techniques had to be taken. This approach relies on the fact that the mean frequency of the FM signal at any instant is a function of the modulating baseband signal.

VLSI APPROACH TO FREQUENCY MODULATION DETECTION   III

The relationship between the mean-frequency estimation of a given spectrum and FM demodulation is given by a short explanation of the fundamental concept.

A Mean-Frequency Estimation

The mean frequency of a random or deterministic process is defined as the first moment of the power spectrum normalized by the zero moment [8,9] and is

$$\bar{f} = \int_0^\infty fS(f)\,df \bigg/ \int_0^\infty S(f)\,df, \quad (3)$$

where $S(f)$ is the real power-spectrum density. When the signal is complex, the frequency-domain limits include both the positive and negative frequencies, for which

$$\bar{f} = \int_{-\infty}^\infty f\tilde{S}(f)\,df \bigg/ \int_{-\infty}^\infty \tilde{S}(f)\,df, \quad (4)$$

where $\tilde{S}(f)$ is the power-spectrum of the complex envelop of the signal. In the time domain, mean frequency is defined [8,9] as

$$\bar{f} = -j\dot{\tilde{R}}(0)/2\pi\tilde{R}(0). \quad (5)$$

Here, $\tilde{R}(0)$ is the autocorrelation function of the complex envelope of the signal at $\tau = 0$, and $\dot{\tilde{R}}(0)$ is its derivative. Equations (4) or (5) can be approximated by a different number of frequency estimators [8–11], but in this analysis only the double-correlation mean frequency detector is discussed because of its superiority over others [4,10].

B Digital Double-Correlation Mean-Frequency Detector (Estimator)

Linear and normalized first-moment processors have stimulated great interest in the field of frequency estimation because of their theoretical accuracy for all input spectra [8–11]. The problems associated with their implementation, however, include the necessity for

(a) a high-quality linear phase differentiator,
(b) a high-speed linear multiplier, and
(c) a high-speed divider for normalization

These and others have complicated the design of accurate linear first-moment processors using analog signal processing techniques [4].

The block diagram of the digital double-correlation mean-frequency detector is shown in Fig. 2. The quadrature detector generates $x(t)$ and $y(t)$, which are the in-phase and quadrature components of $e(t)$, and the analog-to-digital converter converts them into a sequence of samples $x(n)$ and $y(n)$, respectively. The estimate of the mean frequency $\bar{f}$ is equal to [4,8]

48. VLSI Approach to Frequency Modulation Detection

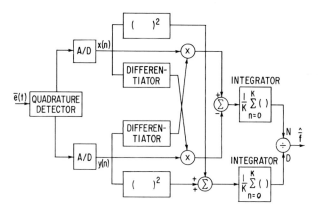

Fig. 2. Digital double-correlation mean-frequency detector.

$$\bar{f} = k \frac{E[x(n)\dot{y}(n) - \dot{x}(n)y(n)]}{E[x^2(n) + y^2(n)]}, \qquad (6)$$

where k is a constant, $E[.]$ is the expectation, $\dot{x}(n)$ and $\dot{y}(n)$ are the time derivatives of $x(n)$ and $y(n)$ respectively. For an ergodic signal, it is possible to write [12]

$$\bar{f} = k \sum_{n=0}^{K} [x(n)\dot{y}(n) - \dot{x}(n)y(n)] \Big/ \sum_{n=0}^{K} [x^2(n) + y^2(n)], \qquad (7)$$

where K is the number of samples for smoothing. The above mean-frequency estimator is especially well suited for signal recovery in a frequency modulated system, doppler ultrasound, frequency-shift keyed systems, and many others. The circuit required (Fig. 2) includes linear multipliers, adders, subtractors, and a divider, which makes it highly suitable for digital implementation and very difficult to design with analog components. A more detailed description of the different blocks is given in Section IV. The use of such a frequency estimator to demodulate FM signals is different from the classical approach [6,7], and some examples are given below.

Frequency Modulation (FM) Detection 1

An exact description of the FM spectra is difficult [6] except for certain simple modulating signals. Therefore, instead of attempting the analysis with arbitrary $s(t)$, a tone modulation example is highlighted.

With tone modulation, the instantaneous frequency of an FM signal varies in a sinusoidal fashion about the carrier frequency. Specifically [6], if $s(t) = A_m \cos \omega_m t$,

$$\tilde{e}(t) = A_c \cos\left(\omega_c t + \frac{2\pi f_\Delta A_m}{\omega_m} \sin \omega_m t\right), \tag{8}$$

where A_c, ω_c are the amplitude and frequency of the carrier, respectively. f_Δ is the frequency deviation and ω_m is the modulating frequency. To simplify the notation let

$$\beta = \frac{2\pi f_\Delta A_m}{\omega_m} = \frac{A_m f_\Delta}{f_m}, \tag{9}$$

so that

$$\tilde{e}(t) = A_c \cos(\omega_c t + \beta \sin \omega_m t). \tag{10}$$

The input signal $\tilde{e}(t)$ is multiplied in the quadrature detector (Fig. 2) by $\cos \omega_c t$ and $\sin \omega_c t$ and is then low-pass filtered to generate $x(t)$ and $y(t)$ [8], which can be expressed by

$$x(t) = A \cos(\beta \sin \omega_m t), \tag{11a}$$

$$y(t) = A \sin(\beta \sin \omega_m t). \tag{11b}$$

The derivatives of $x(t)$ and $y(t)$ are expressed as

$$\dot{x}(t) = -A\beta\omega_m \cos \omega_m t \sin(\beta \sin \omega_m t), \tag{12a}$$

$$\dot{y}(t) = A\beta\omega_m \cos \omega_m t \cos(\beta \sin \omega_m t). \tag{12b}$$

Substituting Eqs. (11) and (12) in Eq. (7) and evaluating the output instantaneously (i.e., $K = 0$), we obtain the mean frequency $\bar{f}$,

$$\bar{f} = k \frac{A^2\beta\omega_m \cos \omega_m t [\cos^2(\beta \sin \omega_m t) + \sin^2(\beta \sin \omega_m t)]}{A^2[\cos^2(\beta \sin \omega_m t) + \sin^2(\beta \sin \omega_m t)]}$$

$$= k\beta\omega_m \cos \omega_m t. \tag{13}$$

Substituting Eq. (9) in Eq. (13) results in

$$\bar{f} = k \frac{A_m f_\Delta}{f_m} 2\pi f_m \cos \omega_m t = A_m \cos \omega_m t, \tag{14}$$

with the proper selection of the constant k. Equation (14) shows that, indeed, the output of the double-correlation mean-frequency detector is the input modulating tone signal $s(t) = A_m \cos \omega_m t$. This approach is valid for any other input spectra.

2 Frequency-Shift Keying Detection

A frequency-shift keying signal consists of two frequencies ($\pm \Omega$) and a carrier (ω_c) and is represented by [7]

$$\tilde{e}(t) = A\cos(\omega_c \pm \Omega)t. \qquad (15)$$

Following the same steps of Section III.B.1. we can write

$$x(t) = A\cos(\pm\Omega t), \qquad (16a)$$
$$y(t) = A\sin(\pm\Omega t) \qquad (16b)$$

and

$$\dot{x}(t) = \mp\Omega A\sin(\pm\Omega t), \qquad (17a)$$
$$\dot{y}(t) = \mp\Omega A\cos(\pm\Omega t). \qquad (17b)$$

Substituting Eqs. (16) and (17) into Eq. (7) gives the mean frequency

$$\bar{f} = \pm k\Omega\, \frac{A^2[\cos^2(\pm\Omega t) + \sin^2(\pm\Omega t)]}{A^2[\cos^2(\pm\Omega t) + \sin^2(\pm\Omega t)]} = \pm k\Omega, \qquad (18)$$

which is a simple recovering scheme compared with the synchronous approach [7] in which a local carrier with frequencies ($\omega_c \pm \Omega$) has to be generated.

IV SYSTEM IMPLEMENTATION EXAMPLE: AN IMPLANTABLE PULSED DOPPLER FLOWMETER

The need for long-term chronic studies in cardiovascular research, without the danger of infection or loss of instrumentation caused by damaged percutaneous leads, led the way to totally implantable instrumentation. Such devices are also significant in fetal and neonatal, toxological, and hepatic research [13]. The principal parameters to optimize in the design of these devices are

(a) small size,
(b) low power consumption,
(c) reliability, and
(d) reduced bandwidth of the transmitted signal.

This example will emphasize the impact that VLSI-based architecture has on system design. The use of an alternative detection approach (double-correlation mean-frequency estimation) selected for its suitability to monolithic implementation in a high density digital technology will be described.

A Doppler Ultrasound

The doppler ultrasound system shown in Fig. 3 has a transmitted signal $\tilde{e}(t)$ in the form of [8–11]

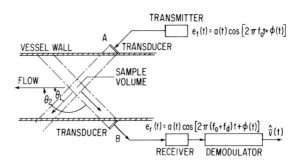

Fig. 3. Block diagram of the CW ultrasonic flowmeter.

$$\tilde{e}_t(t) = a(t) \cos[2\pi f_0 t + \Phi(t)], \quad (19)$$

where f_0 is the carrier frequency (on the order of 10 Hz), $a(t)$ is the amplitude function, and $\Phi(t)$ is the phase function. Usually $a(t)$ and $\Phi(t)$ are slowly varying functions relative to the carrier frequency; therefore, $\tilde{e}(t)$ can be considered to be a narrowband signal. In the *continuous-wave (cw) doppler ultrasound flowmeter*, $a(t) = E_0$ and $\Phi(t) = 0$; therefore,

$$\tilde{e}_t(t) = E_0 \cos 2\pi f_0 t. \quad (20)$$

The backscattered signal from the moving particles is a doppler-shifted version of $\tilde{e}_t(t)$ and is

$$\tilde{e}_r(t) = E_0 \cos 2\pi (f_0 + f_d)t, \quad (21)$$

where f_d is the doppler frequency. In the receiver, this signal can be synchronously distinguished from the excitation frequency by processing it through the quadrature detector and the mean-frequency estimator, as described above, which gives

$$x(t) = \frac{E_0^2}{2} \cos 2\pi f_d t, \quad (22a)$$

$$y(t) = \frac{E_0^2}{2} \sin 2\pi f_d t, \quad (22b)$$

and finally,

$$\bar{f} = k f_d. \quad (23)$$

By using the doppler equation,

$$f_d = 2 \frac{v}{c} f_0 \cos \theta, \quad (24)$$

where f_d is the doppler-frequency shift, v is the particle velocity, c is the sound velocity, f_0 is the transmitted carrier frequency, and θ is the angle between transducer beam and particle direction, we can calibrate the output of the mean-frequency estimator to measure the mean velocity of the particles inside the vessel, which is given, from Eqs. (23) and (24), by

$$\bar{v} \alpha \left(2 \frac{f_0}{c} \cos \theta \right)^{-1} \bar{f}, \qquad (25)$$

which is of great value in many applications, especially the medical ones. The above technique is not restricted to a CW system but can also be applied to a pulsed doppler system [4].

System Description   B

In a *pulsed-doppler flowmeter*, a short burst of sound is transmitted. A single transducer serves as both a transmitter and a receiver. By observing the doppler shift of the scattered return at a certain time after each transmitted ultrasonic burst, we can measure the velocity at a fixed range (Fig. 4). Demodulators 1–n have to be identical if the velocity profile ($\bar{v}_1$, $\bar{v}_2$, ..., $\bar{v}_n$) is to be reconstructed accurately. The design of identical demodulators using analog techniques is difficult because high-precision components, matched offset, and other features are required.

A simplified block diagram of an implantable pulsed-doppler flowmeter designed by Gill [14] is illustrated in Fig. 5. It can be seen that the signal processing and velocity profile are computed external to the body. For a 10-range velocity profile, the bandwidth of the transmitted signal is approximately 1 MHz [13,14], which makes it a wideband signal. On the other hand, the mean frequency $\bar{f}_i$ of the doppler spectrum at each range r_i has a bandwidth depending on the rate of the heart beat (5–20 Hz), compared with the doppler bandwidth of 20 KHz. Therefore, the transmitted bandwidth can be reduced by a factor of 1000 if monolithic digital mean-frequency estimators are used to compute the average velocity $\bar{v}_i$ at each

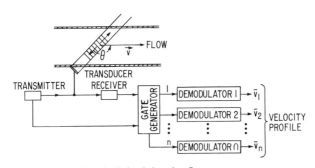

Fig. 4. Pulsed-doppler flowmeter.

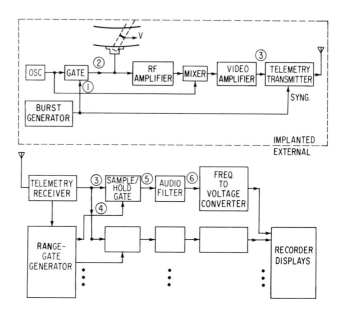

Fig. 5. Simplified block diagram of implantable pulsed-doppler flowmeter.

range r_i [4]. Such a narrowband system is illustrated in Fig. 6. The digital demodulator designed for range r_i (dashed area) can be easily duplicated for the other ranges, and therefore an accurate velocity profile can be established.

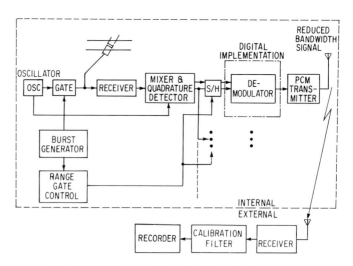

Fig. 6. Digital implantable pulsed-doppler flowmeter.

Implementation of System Major Blocks C

The monolithic implementation of the analog front end (oscillator, burst generator, range gate control, receiver, and mixers) is explained in details in the work done by Gill [14]. The highlight of this subsection is the digital implementation of the demodulator which allows on-board mean frequency estimation.

A detailed architecture of this demodulator is shown in Fig. 7; some of the blocks are explained next.

A Simple Analog-to-Digital Converter 1

It was pointed out earlier that linear mean-frequency estimators are more efficiently implemented in a digital form. Therefore, the problem is how to convert the analog signal into digital words with a minimum number of analog components.

Linear delta modulation (LDM) is a simple method of analog-to-binary conversion [15–17]. The LDM system consists of a two-level quantizer and a feedback path containing a single integrator (Fig. 8). It acts as a converter with an analog input $y(t)$ and a binary output $\{b_n\}$. The delta

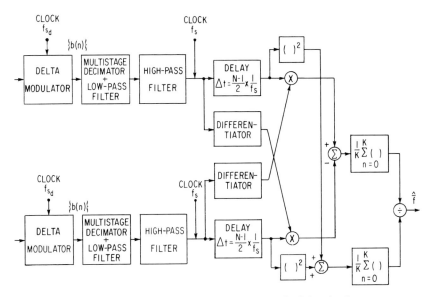

Fig. 7. Digital demodulation architecture for the pulsed-doppler flowmeter.

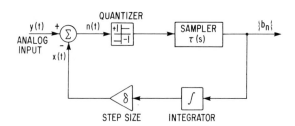

Fig. 8. Basic delta-modulator system.

modulator transforms the continuous signal into the binary sequence

$$(b_n) = \ldots b_{-1}, b_0, b_1, \ldots, \quad (26)$$

in which b_i can have the value of ± 1.

Figure 9 is a block diagram of the binary-to-digital converter operating on the binary sequence produced by the delta modulator [16]. This converter contains a digital replica of the delta-modulator feedback loop and a linear filter that is considered to have an ideal low-pass transfer function. This filter processes the integrated DM signal $x(n)$ and rejects the quantization noise occurring outside the band of the original analog signal.

With low-pass filtering, the equations governing the granular noise are [18–21]

$$N_G = \tfrac{1}{8} K_q \frac{d_1}{f_c^2} \times \frac{s^2}{F^3}, \quad (27)$$

where s is the slope-loading factor and is equal to

$$s = \delta f_{sd}/\sqrt{d_1}; \quad (28)$$

F is the bandwidth expansion factor;

$$F = f_{sd}/2f_c \quad (29)$$

and K_q is a constant that for many applications is $\tfrac{1}{3}$.

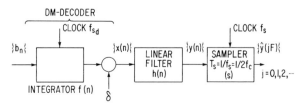

Fig. 9. Binary-to-digital converter.

48. VLSI Approach to Frequency Modulation Detection

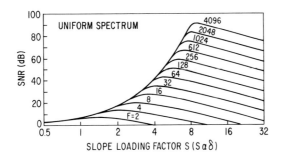

Fig. 10. Signal-to-noise ratio of the linear delta modulator with a uniform spectrum signal as input.

The signal-to-noise ratio at the output of the analog-to-digital converter is defined as the ratio of signal power S to total quantization noise N_Q, or

$$\text{SNR} = S/N_Q. \tag{30}$$

The SNR as a function of the slope-loading factor s is illustrated in Fig. 10 for a uniform-spectrum signal and at several values of F. It can be seen that, unfortunately, quantization noise in linear DM systems is sensitive to any variations in s (i.e., any change in step size δ or signal amplitude); as a result, the range of s over which the SNR is near maximum is limited.

When no filtering is required, the converter reduces to an *up–down counter*, and the encoder consists only of a DM and a counter (Fig. 11). This structure is simple to implement but results in aliasing of the high-frequency noise components of $x(n)$ into the signal bandwidth. Because of multiple aliasing, the total granular noise power N_G in the baseband $[0, f_c]$ becomes

$$N_G \approx \delta^2/3. \tag{31}$$

This is a severe limitation in the LDM system and would have a serious effect on the estimate of the mean frequency [4]. Noise reduction techniques [11], however, help resolve this problem and improve the dynamic range of the system, as explained later.

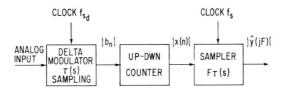

Fig. 11. Delta modulation to digital encoder with a single coefficient.

2 Anti-aliasing Filter

A note should be added on the design of the anti-aliasing filter, which is required before any sampling operation can be performed. The DM sampling frequency is usually much higher than the signal bandwidth ($F > 32$ for reasonable SNR), which simplifies the amplitude response specifications of this filter and can be implemented by just an *RC* circuit [4].

3 Linear Phase Differentiator

Equation (7) is based on the assumption that the differentiator used in the implementation of the double-correlation mean-frequency estimator described in Fig. 2 is ideal. This implies that the amplitude and phase response of such filters should be [22–24]

$$M(\omega) = \omega, \tag{32}$$

$$\phi(\omega) = \omega \, \Delta t, \tag{33}$$

where the linear phase response gives a constant group delay at all frequencies. The design of the above filter is achieved by using finite impulse–response technique (FIR), which is simple to implement using digital VLSI approach [25,26]. Figure 12 illustrates a mean-frequency

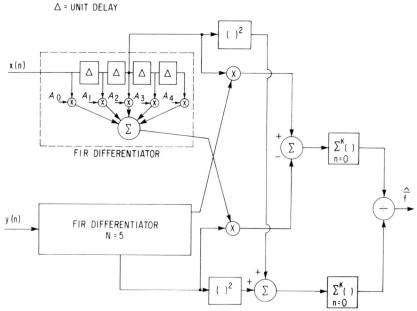

Fig. 12. Double-correlation mean-frequency estimator. FIR implementation of the differentiators.

estimator in which the linear phase can be compensated by tapping the data at the center of the delay line $[(N - 1)/2]$ used to build the FIR differentiator [4]. The response of the mean-frequency estimator to a monochromatic input signal with and without phase compensation is illustrated in Fig. 13.

The results show that compensation of the phase of the differentiator is important in reducing the error of $\bar{f}$.

D. System Integration

From a monolithic point of view it is advantageous to design the FIR differentiator with a minimum number of coefficients and to use the simple A/D converter described in Section IV.C.1. Techniques are available for the optimization of digital filters [27], but what can be done about the quantization noise introduced by the A/D?

The answer to this question is given by a short explanation of the effect of input noise on the estimation of mean frequency.

1. Effect of Input Noise

The expectation and variance of $\bar{f}$ are used to describe the relative error and the SNR in the estimation of the mean frequency as a function of the SNR at the input and are given by [11]

$$\varepsilon = \frac{E[\bar{f}] - \bar{f}_s}{\bar{f}_s} = \frac{(\bar{f}_n/f_s) - 1}{1 + \text{SNR}_{in}}, \tag{34}$$

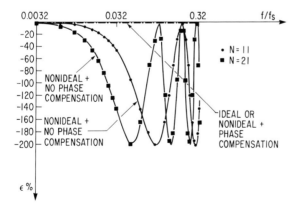

Fig. 13. Relative error as a function of input frequency for a monochromatic signal.

and

$$\frac{1}{\mathrm{SNR_{out}}} = \frac{\mathrm{Var}\,\bar{f}}{\bar{f}_s^2},\tag{35}$$

where $\bar{f}_s$ is the true mean-frequency of the signal.

For a wideband noise with mean frequency zero (i.e., $\bar{f}_n = 0$) Eqs. (34) and (35) become [11]

$$\varepsilon = \frac{-1}{1 + \mathrm{SNR_{in}}},\tag{36}$$

and

$$\frac{1}{\mathrm{SNR_{out}}} = \frac{1}{\bar{f}_s^2 T}\left[\frac{1}{12}\left(B_s + \frac{B_n}{\mathrm{SNR_{in}^2}}\right) + \frac{1 + TB_n}{B_n \mathrm{SNR_{in}^2}}(\bar{f}_s - \bar{f}_n)^2\right],\tag{37}$$

where B_n is the noise bandwidth, B_s is the signal bandwidth, and $\bar{f}_n$ is the mean frequency of noise. From the preceding equations one can see that, when $\mathrm{SNR_{in}} = \infty$, the relative error is equal to zero, and when $\mathrm{SNR_{in}} = 0$ (signal power $\ll$ noise power), it is as high as 100%.

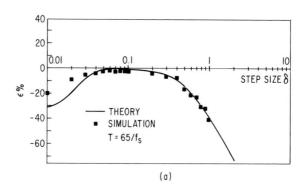

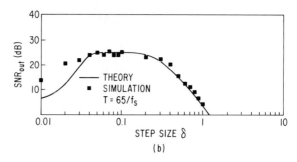

Fig. 14. Estimation of mean frequency as a function of step size for a band-limited gaussian signal. (a) Relative error in percent ($\varepsilon\%$). (b) $\mathrm{SNR_{out}}$ in decibels.

48. VLSI Approach to Frequency Modulation Detection

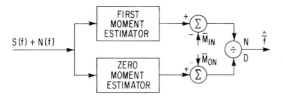

Fig. 15. Centroid detector with noise cancellation.

It is shown [4] that the quantization noise introduced by the DM in the A/D conversion can be approximated by a flat spectrum with $B_n = 2f_c$ and $\bar{f}_n = 0$, and the dynamic range can be expressed as a function of the DM step size. Figure 14 shows the relative error and the SNR in the estimation of $\bar{f}$ for a band-limited gaussian signal as a function of δ (dynamic range) when no sophisticated filtering is introduced and when a simple A/D, as described in Fig. 11, is used.

It can be seen that as δ increases or, in other words, as the SNR$_{in}$

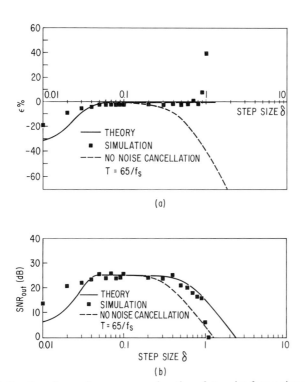

Fig. 16. Estimation of mean frequency as a function of step size for a noise-compensated centroid detector having a band-limited gaussian signal as input. (a) Relative error in percent ($\varepsilon\%$). (b) SNR$_{out}$ in decibels.

decreases, the error in the estimate increases, and the SNR$_{out}$ decreases. To increase the dynamic range and keep the A/D simple, some other tricks had to be used, as explained below.

2 Noise Cancellation

It is shown [4,11] that when the nature of the input noise can be determined, its zero- and first-moment estimates are subtracted from the zero- and first-moment of the normalized mean-frequency detector to improve the response. This is illustrated in Fig. 15.

For a DM noise (with no filtering), $\overline{M}_{1n} = 0$ and $\overline{M}_{0n} \approx \delta^2/3$ [4] which, when substituted back into the system described in Fig. 15, gives the results illustrated in Fig. 16. For the same band-limited gaussian signal (Section IV.D.1.), it can be seen that the range of step size δ for which $\varepsilon\%$ is below 10% has increased compared with Fig. 14a (dotted line). The same argument applies for the SNR in Figs. 14b and 16b.

It can be concluded that noise canceling has improved the dynamic range of the system without a noticeable increase in the hardware. If more improvement is required, filtering can be added (Fig. 9) at the expense of

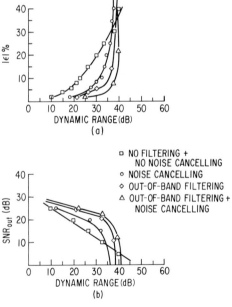

Fig. 17. Dynamic range of the centroid detector for different values of error in the estimate of mean frequency. (a) Relative error ($\varepsilon\%$) versus dynamic range. (b) SNR$_{out}$ in decibels versus dynamic range.

more complexity in the hardware. A comparison of the different techniques on the bases of relative error $\varepsilon\%$ and SNR_{out} versus input signal dynamic range is summarized in Fig. 17.

In-vivo Data E

In this experiment, the data from the hepatic artery of a dog were recorded via a cw doppler velocity meter. The received doppler signal was quadrature detected and low- and high-pass filtered before recording the in-phase and quadrature baseband signals on a two-channel audio recorder. These signals were then input into the digital double-correlation mean-frequency estimator, using the delta modulator for analog-to-digital conversion. The estimated mean frequency as a function of time is illustrated in Fig. 18, where the input signal is scaled so that the mean power is unity.

The results in Fig. 18a were obtained after digitizing the analog signal through a perfect A/D converter; this plot serves as reference, assuming that the A/D noise is negligible. In Fig. 18b, the signal was digitized by a delta modulator at a sampling frequency 32 times the Nyquist rate and with a step size of $\delta = 0.1$. The SNR_{in} under these conditions should be

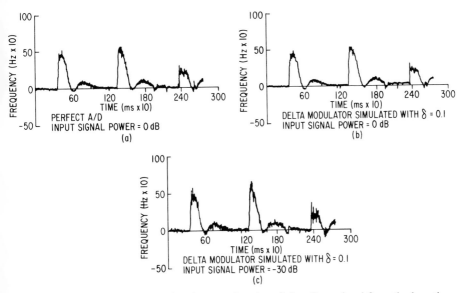

Fig. 18. Mean-frequency estimation as a function of time for a signal from the hepatic artery of a dog. (a) Doppler system using a perfect analog-to-digital concerter. (b) Doppler system using the delta modulator with an optimal step size. (c) Doppler system using the delta modulator with the same step size as in (b) but with input power reduced to -30 dB.

high (25 dB), and the effect of DM noise on the estimate should be minimum, as can be observed by comparing the qualitatively similar curves in Fig. 18a and b. If the input signal power drops by 30 dB, and the step size of the DM remains constant, the output SNR of the A/D will be approximately equal to -3 dB, which would result in a poor mean-frequency estimation. It was shown (Section IV.D.2.) that noise cancellation improves the system response under such conditions. Figure 18c illustrates the results obtained by this technique; this plot is very similar to the other two.

V CONCLUSION

The success of applying VLSI to system applications is in the determination of architectural approaches that are well matched to the technology. In the application discussed, doppler flowmeters, the mean-frequency estimator, which is the critical system requirement, was reexamined in light of a high performance, high density, digital technology and then designed within this context. Simulation results presented demonstrate the performance capability of the system. Monolithic design of a mean-frequency detector in an advanced CMOS technology is now underway.

REFERENCES

1. W. C. Knight, R. G. Pridham, and S. M. Kay, Digital signal processing for sonar, *Proc. IEEE* **69**(11), 1451–1506, November, (1981).
2. M. I. ElMasry, "Digital MOS Integrated Circuits." IEEE Press, New York, 1981.
3. R. J. Apfell, H. Ibrahim, and R. Ruebush, Signal-processing chips enrich telephone line-card architecture, *Electronics,* (1982).
4. S. E. Noujaim, "Digital Signal Processing for Doppler Systems." Ph.D. dissertation, Stanford University, December 1982.
5. R. Gregorian, Glenn Wegner, S. D. Flanagin, and J. G. Gord, A CMOS adaptive delta modulation CODEC chip for PABX applications, *ISSCC Digest of Technical Papers* (1983).
6. A. B. Carlson, "Communication Systems." McGraw-Hill, New York, 1975.
7. H. Taub and D. Schilling, "Principles of Communication Systems." McGraw-Hill, New York, 1971.
8. W. R. Brady, "Theoretical Analysis of the Ultrasonic Blood Flowmeter," Ph.D. dissertation, TR No. 4958-1, Stanford Electronics Laboratories, Stanford University, Stanford, California, 1971.
9. L. Gerzberg and J. D. Meindl, SSB mean doppler shift detector for simultaneous forward and reverse flow-velocity measurement, *Proc. 28th ACEMB, New Orleans, Louisiana, Sept. 1975,* p. 448 (1975).
10. H. V. Allen, "Totally Implantable Bidirectional Pulsed Doppler Blood Flow Telemetry: Integrated Timer–Exciter Circuitry and Doppler Frequency Estimation." Ph.D.

dissertation, TR No. 4958-4, Stanford Electronics Laboratories, Stanford University, Stanford, California, May 1977.
11. L. Gerzberg, "Monolithic Power-Spectrum Centroid Detector." Ph.D. dissertation, TR No. G557-2, Stanford Electronics Laboratories, Stanford University, Stanford, California, May 1979.
12. A. Papoulis, "Probability, Random Variables, and Stochastic Processes." McGraw-Hill, New York, 1965.
13. J. W. Knutti, "Totally Implantable Bidirectional Pulsed Doppler Blood Flow Telemetry: Integrated Ultrasonic Receiver, Diameter Detection, and Volume Flow Estimation." Ph.D. dissertation, TR No. 4958-3, Stanford Electronics Laboratories, Stanford University, Stanford, California, July 1977.
14. R. W. Gill, "Am Implantable Pulsed Doppler Ultrasonic Blood Flowmeter Using Custom Integrated Circuits," Ph.D. dissertation, TR No. 4958-3, Stanford Electronics Laboratories, Stanford University, Stanford, California, May 1975.
15. J. C. Balder and C. Kramer, Analog-to-digital conversion by means of delta modulation, *IEEE Trans. Space Electron. Telem.* pp. 87–90, (1964).
16. D. J. Goodman, The application of delta modulation to analog-to-PCM encoding, *Bell Sys. Tech. J.* **48**(2), pp. 321–343 (1969).
17. R. Steele, "Delta Modulation Systems." Wiley, New York, 1975.
18. R. Steele, SNR formula for linear delta modulation with bandlimited flat and RC-shaped Gaussian signals, *IEEE Trans. Communications* **COM-28** (1980).
19. J. E. Abate, Linear and adaptive delta modulation, *Proc. IEEE* **55**(3), 298–308 (1967).
20. L. J. Greenstein, Slope overload noise in linear delta modulators with Gaussian inputs," *Bell Sys. Tech. J.* **52,** 387–421 (1973).
21. D. J. Goodman, Delta modulation granular quantizing noise, *Bell Sys. Tech. J.* (1969).
22. A. Antoniou, "Digital Filter Analysis and Design." McGraw-Hill, New York, 1979.
23. A. V. Oppenheim and R. W. Schaffer, *"Digital Signal Processing."* Prentice-Hall, Englewood Cliffs, New Jersey, 1975.
24. R. W. Hamming, "Digital Filters." Prentice-Hall, Englewood Cliffs, New Jersey, 1977.
25. A. Peled and B. Liu, A new approach to the realization of non-recursive digital filters, *IEEE Trans. Audio Electroacoust.* **AU-21,** 477–484 (1973).
26. A. Peled and B. Liu, A new hardware realization of digital filters, *IEEE Trans. Acoust. Speech, Signal Process.* **ASSP-22,** 456–462 (1974).
27. J. H. McLennan, T. W. Parks, and L. R. Rabiner, A computer program for designing optimum FIR linear phase digital filters, *IEEE Trans. Audio Electroacoust.* **AU-21,** 506–526 (1973).

Chapter 49
VLSI Impact on Modem Design and Performance

LEROY D. YOUNG, JR.
EDWIN J. HILPERT

Racal–Milgo, Inc.
Miami, Florida

I. Introduction	824
II. Modem Overview	825
A. Low-Speed Modems	825
B. Medium-Speed Modems	825
C. High-Speed Modems	825
III. Incentives for VLSI Use in Modems	827
A. Pricing	827
B. Data Rates	828
C. Error Rate Performance	829
D. Throughput	829
E. Reliability	829
IV. Custom Implementations	830
A. Transmitter Implementation	830
B. Multiple-Chip Implementations	831
V. Commercial Devices	833
A. Bit-Slice Processors	833
B. FSK Modem Chips	833
C. DSP Chips	834
VI. VLSI and Value-Added Features	834
A. Nondata Features	834
B. Parametric Measurements	834
C. Diagnostics	834
VII. Future Trends	835
References	835

I INTRODUCTION

Although the theoretical basis for modem design has existed for several decades, only in recent years have the necessary IC building blocks for effective implementations become available. The dramatic changes taking place in VLSI circuits continue to accelerate modem development in every area, from low-speed modems on-a-chip to sophisticated high-speed modems with digital signal processing (DSP) integrated circuits. Voice-grade modems that use these chips are now cost effective, reliable, and offer high performance. Additionally, VLSI circuits incorporate many value-added features for diagnostics and system control into the modem. This paper describes some design features and available VLSI components that are useful for digitally based modems.

II MODEM OVERVIEW

A modem (acronym for *mo*dulator–*dem*odulator) is used to condition digital data for transmission and error-free recovery over a communication channel. For this discussion, consider that the communication channel is a telephone line (3002 voice-grade or equivalent) with a bandwidth of 300 to 3300 Hz (Fig. 1). The type of conditioning required for digital data is a function of the data speed, the quality of the telephone line, and the acceptable complexity of the modem itself. Data rates can be several times the bandwidth of the telephone line because of VLSI implementations.

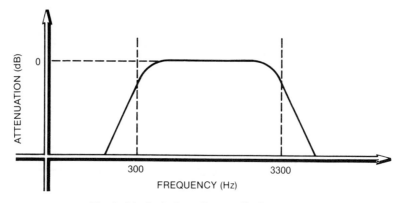

Fig. 1. Ideal telephone line amplitude response.

49. VLSI Impact on Modem Design and Performance

Voice-grade modems can be broadly classified as follows: low speed (up to 1200 bps), medium speed (up to 4800 bps), or high speed (greater than 4800 bps). Modems can be further classified by modulation type (frequency, phase, amplitude, or combinations), synchronization method (synchronous or asynchronous), or transmission method (simplex, half duplex, or full duplex). Classifications such as these will be defined further as we describe VLSI implementations of representative modems.

Low-Speed Modems A

Although the concepts of amplitude, frequency, and phase modulation are not unique to voice-grade modems, certain specific implementations are predominant in modems. For example, most low-speed modems use a form of discrete frequency modulation in which the transmitted signal is one of two discrete frequencies depending on the binary data input. This technique, frequency shift keying (FSK), is used in several commercially available LSP chips. Unlike higher-speed modems, FSK modems encode only one data bit per signaling element, or *baud*.

Medium-Speed Modems B

Medium-speed modems typically use a form of phase modulation, either phase shift keying (PSK) or differential phase shift keying (DPSK). The distinction between the two methods is that DPSK phase changes are made with reference to the phase of the preceding signaling element, as opposed to absolute encoding. The number of phases used is generally two or four, with multiple data bits encoded for each baud. These medium-speed modems can be implemented with commercial processors or custom devices.

High-Speed Modems C

Higher-speed modems use a combination of amplitude modulation (discrete levels) and DPSK (discrete phases), causing the carrier to have multiple states in each baud. The modulation method most commonly used for 9600-bps operation is the grouping of four data bits per baud to generate two amplitudes and eight phases. This technique is called quadrature amplitude modulation (QAM). Designs for such high-speed modems use integrated bit-slice processors or custom-designed VLSI chips.

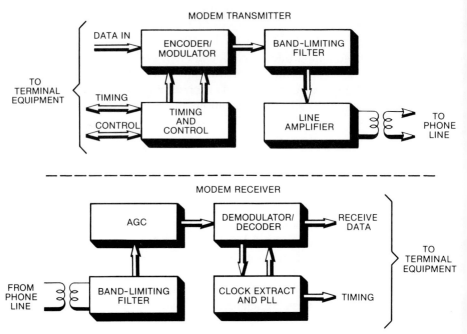

Fig. 2. Simplified block diagram of a voice-grade modem.

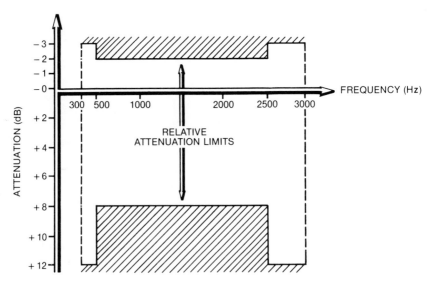

Fig. 3. 3002 unconditioned line attenuation characteristics referenced to 1004 Hz.

TABLE I

Selected Bell Standard Modems

Modem type	Speed (bits per second)	Transmission method	Modulation method
Bell 103	300	Full-duplex	FSK
Bell 201	1200	Half-duplex	PSK
Bell 202	1200	Half-duplex	FSK
Bell 212A	300	Full-duplex	FSK
Bell 212A	1200	Full-duplex	PSK

Figure 2 is a block diagram of a medium-to-high speed voice-grade modem. The major elements are the transmitter (modulator) and receiver (demodulator). Within these elements are shown the various DSP functions that may be required to perform the transmit and receive tasks. VLSI has enabled more and more of these functions to be performed digitally. The result is repeatable, highly accurate data processing within the modem. These high-speed modems operate over worst-case telephone lines, such as shown in Fig. 3. This diagram shows electrical characteristics for a 3002 telephone line without conditioning.

Table I shows a cross reference of several Bell Standard modems with some of the pertinent technical information such as speeds and modulation types. There are also international standards specified by the International Consultative Committee for Telegraphy and Telephone (CCITT) [1].

III INCENTIVES FOR VLSI USE IN MODEMS

Significant benefits in cost, speed, reliability, and performance are well-known incentives for generalized use of VLSI in modem product development.

A Pricing

Pricing incentives for VLSI circuits in modems are dictated to a large degree by data rate considerations. A pricing indicator widely used in the 1970s was "a dollar per bit" for certain higher-speed modems. That is, in past years a 9600-bps modem might have cost about $10,000. This pricing structure for all modems in all speed ranges has been steadily driven downward with the increased incorporation of LSI. Current pricing indicates that high-speed modems are now available for less than $0.40 per bit [2]. This downward trend in pricing tells only a portion of the story. Along with the lower pricing, modem manufacturers have been able to offer increased reliability, lower power dissipation, and better performance by using VLSI circuitry in major portions of the modem.

B Data Rates

Incentives for achieving higher data rates at acceptable levels of error performance are being realized by both custom and commercial DSP VLSI devices. The higher data rates are achieved only by using equalization methods to compensate for line impairments (such as amplitude and delay distortion) and by implementing more sophisticated modulation and demodulation techniques. These compensating and data conditioning tasks are computationally intensive, especially for high-speed modems, and hardware intensive for lower-speed modems. Both areas have been addressed successfully by VLSI, as described in the implementation discussions to follow.

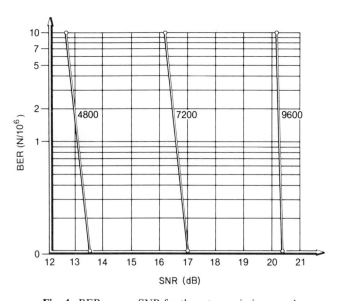

Fig. 4. BER versus SNR for three transmission speeds.

Error Rate Performance C

Performance incentives for the use of VLSI have focused primarily on error rates and throughput. The cost-per-bit analysis previously described is valid only if the transmitted bits are received error-free and if they are customer data bits, not modem overhead bits.

Bit error rate (BER) is a relative indicator of the quality of modem operation over a given line. This is measured as the number of error bits received per million bits transmitted. BER is usually plotted versus the SNR to show clearly the system noise characteristics required for various performance levels. Figure 4 shows BER versus SNR for a representative modem at three data rates. VLSI-based modems now exhibit acceptable error performance at high data rates (less than one error per million bits at 9600 bps) over unconditioned 3002 lines.

Throughput D

Throughput for modems is a function of the data rate less any required modem overhead transmissions. Most medium-to-high speed modems transmit a preamble prior to sending the user's data. This *training sequence* permits the receiver to adapt itself to existing line conditions. Shortening this training period, of course, allows throughput to increase proportionally. By incorporating automatic and adaptive equalizers into VLSI firmware-based machines, modem designers have shortened training sequences from several hundred milliseconds to less than 20 milliseconds in products that run at 9600 bps. In cases of repeated transmissions over the same line, the time for retraining can be made less than 10 milliseconds by having the receiver use certain parameters "learned" from the previous data transfer.

Reliability E

A low-profile but very important incentive for use of VLSI is the positive impact on the mean time between failures (MTBF). High levels of integration now produce MTBFs as high as 60,000 h. This is a dramatic increase in reliability, with improvements still continuing.

IV CUSTOM IMPLEMENTATIONS

A Transmitter Implementation

In the early 1970s volume production became available for custom-designed LSI chips. Several progressive companies embarked on designs that were innovative and cost competitive, although the up-front costs and time penalties were severe. One modem function that was successfully converted to LSI by Milgo, Inc. (now Racal–Milgo, Inc.) was a transmitter coder and filter circuit for use in low-to-medium speed modems (Fig. 5). The chip was used to effect the data conditioning necessary for control of intersymbol interference (ISI) [3]. The resulting technique proved effective and was used through several generations of Racal–Milgo's modem development.

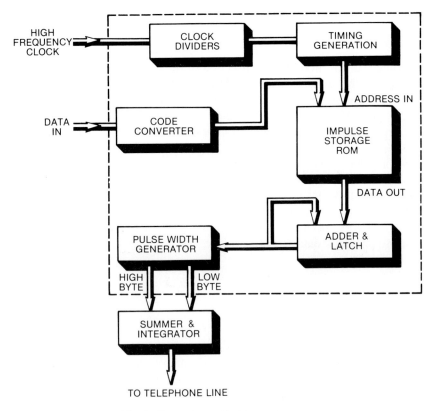

Fig. 5. Transmitter-chip block diagram.

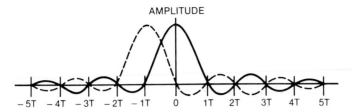

Fig. 6. Impulses with zero ISI.

Refer to Fig. 6 to appreciate the need to control ISI. Notice that when impulses are sent over the transmission media at a given baud rate, *smearing* occurs because of various line imperfections such as envelope delay distortion and frequency distortions. These imperfections cause the impulse to spread over several succeeding baud times and interfere with the receiver's ability to extract the data properly. Appropriately conditioning the impulses at the transmitter and/or receiver reduces the interference to very close to zero at all subsequent baud sample times. This effectively reduces the ISI.

Figure 6 also shows two impulses with properties that minimize that ISI. When the second impulse occurs, its amplitude will be detected without error by the receiver, if sampled as the preceding impulse passes through zero. To generate impulses with these ISI properties, the transmitter must produce accurate, repeatable impulses with predictable zero crossings. Racal–Milgo's custom LSI chip accomplished these goals using a fully integrated arithmetic look-up table approach. This table contains sampled versions of all the required transmitted impulses (all legal combinations of phase and amplitude) stored in on-chip ROM. A timing chain generates the necessary clocking information to step through selected ROM segments as dictated by phase and amplitude coding of the user's data.

An adder–accumulator on the chip generates a running summation of individual past and present samples and outputs a pulse-width modulated version of the sums to an external integrator for transmission to the phone lines. Thus, a look-up table approach allows LSI to perform precisely the filtering necessary for low ISI.

Multiple Chip Implementations   B

As design rules were relaxed and the chip geometries available for custom designs approached the low micrometer range, several modem manufacturers became more aggressive in their design efforts with multi-

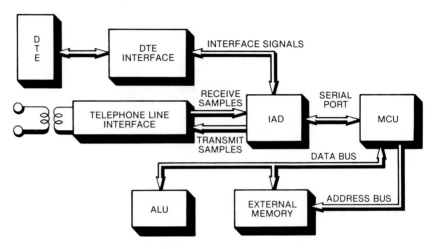

Fig. 7. A 4800-bps custom modem implementation.

chip solutions to data transmission requirements. Numerous partitioning approaches were used as designers sought the best tradeoffs for hardware and performance [4,5].

One multichip solution to this partitioning dilemma is demonstrated in Racal–Milgo's LSI-based Omnimode series of medium-to-high speed modems. The designs allow for 2400, 4800, 9600, and higher-speed operation in a variety of configurations. Figure 7 shows the block diagram for 4800-bps architecture. For speeds higher than 4800, the system is configurable for multiple use of the arithmetic unit and memory control chip.

In Figure 7, the ALU is a high-speed arithmetic unit (5-μm NMOS) that uses pipelining and multitask instructions to implement specialized modem-dependent DSP operations. It performs a 16 × 16 multiply, for example, in three clock cycles of 2.5 MHz and produces a full 32-bit output available in two 16-bit bytes.

The MCU is a memory control unit chip (5-μm NMOS) that performs the necessary queuing for the memory-fetch pipeline as well as controls loop counts, table pointing, and interrupt processing.

The IAD chip is the interface chip (5-μm NMOS) to the outside world, comprising an 11-bit A/D converter, an 11-bit D/A converter, plus all clock generation, computational processor communication, and phase correction for data clocks. By using the specialized instructions available in this custom-designed chip set, all modem tasks (except certain analog functions and line interfacing) can be handled under software control [6].

V COMMERCIAL DEVICES

A Bit-Slice Processors

In the mid-1970s, microprocessors with the necessary computational speeds and configuration capabilities for bit-slice processing became available. The processors were bit-slice building blocks and were incorporated into a limited number of products during this period. The AMD 2900 is one of the bit-slice family, with speeds of less than 100 ns for certain arithmetic operations. These designs were effective for speed considerations, even at the expense of high power dissipation and the large number of support chips required for full implementations.

B FSK Modem Chips

Commercial LSI manufacturers gave their earliest attention to low-speed FSK modems. Now with the explosive growth of personal computers and the anticipated need for the transfer of data over dial-up telephone lines, renewed efforts are being seen in the area of FSK chip sets and modems on a chip. Exar, for example, offers more than a half dozen chips to support a number of modem configurations. Other manufacturers also offer versions that allow flexible single-chip modem designs, as shown in Table II.

TABLE II

Commercial Modem Processors

Modem type	Available devices	Manufacturer
Bell 103	AMD7910	AMD
	MC6860	Motorola
	MC11412	Motorola
	TMS99532	TI
Bell 202	TCM3101	TI
Bell 203	TP3320	NSC
Bell 212A	CH1760	Cermetek

C DSP Chips

Noncustom high-speed modems have been implemented with discrete devices or bit-slice processors. Texas Instruments is changing this trend with a new low-power, high-speed DSP chip, the TMS 320. Designed with a modified Harvard architecture, the TMS pipelined processor can calculate the $a \times b + c = d$ type operation with sufficient speeds to allow a 4800-bps modem to be designed around a single chip and a 9600-bps unit to be designed around two of the chips. Although certain modem algorithms such as receive bit "slicing" do not fit the $a \times b + c = d$ format, the overall algorithm computation tends to be as efficient in terms of speed as that of some custom-designed chips.

VI VLSI AND VALUE-ADDED FEATURES

A Nondata Features

Value-added options are enhancements to modem performance that result from the interaction of several types of VLSI devices. Features such as auto dial, looping capabilities, secondary channels, and diagnostics allow users to configure modems to suit the needs of their systems.

B Parametric Measurements

The same VLSI-dependent processing power that produces high-performance data transmission and recovery can be used effectively to measure certain performance parameters such as signal quality, SNR, phase jitter, delay distortion, and amplitude distortion. Some of these parameters can be measured during actual data transmission (noninterruptive), yielding approximations of actual conditions.

C Diagnostics

Many high-speed modems offer enhancements for diagnostics and control by providing low-speed secondary or tertiary channel capabilities. These channels typically operate from 75 to 150 bps on a carrier near the edge of the main channel and are actually modems-within-a-modem.

Newly available switched capacitor implementations can provide for the notching of the main channel spectrum to allow the bandpassing and insertion of the secondary channel. Custom and commercial single-chip switched capacitor devices are used to provide all necessary secondary channel functions with minimal external support circuitry.

VII FUTURE TRENDS

As VLSI technology becomes faster and more efficient, modem designers are turning their attention toward the commercial aspects of some of the theoretically proven algorithms for speed, performance, and feature enhancements.

Available voice-grade modems that break the 9600-bps barrier are again near the $1 per bit price; however, the phone line savings due to increased throughput still make these units economically attractive. VLSI will cause the same downward pricing trend for these higher speed products as it has demonstrated during the past ten years.

Other VLSI circuits such as large RAMs, nonvolatile memories, enhanced DSP chips, and single-chip, high-speed modems will make specialized modems practical. Tailoring of such modems will be accomplished by down-line loading of computational programs and protocol-dependent options. Long-term performance data can even be stored for later retrieval and analysis.

REFERENCES

1. C. Barney, *Electronics* **56,** 95–97 (1983).
2. *Datapro Reports on Data Communications* **3,** C33-010-101–C33-010-124. Datapro Research Corporation, 1983 Delran, New Jersey.
3. A. P. Clark, "Advanced Data-Transmission Systems." Wiley, New York, 1977.
4. K. Murano, Y. Mochida, F. Amano, and T. Kinoshita, *ICC Conf. Rec.* **3,** 37.3.1–37.3.5 (1979).
5. H. Suzuki, S. Honda, H. Yahata, T. Nose, A. Nakano, S. Kawakami, and H. Amano, *ICC Conf. Rec.* **3,** 5F.1.1–5F.1.5 (1982).
6. Racal-Milgo staff, *Digital Design* **13,** 90–92 (1983).

Chapter 50

Impact of VLSI on Distributed Communications

DANIEL HAMPEL

RCA Government Communications Systems Division
Camden, New Jersey

I. Introduction	837
II. VLSI Technology and Trends	838
A. VLSI Performance	838
B. Types of VLSI	840
III. Distributed Communications	841
A. Local Area Networks	841
B. Signal Processing for Communications	847
IV. Future Trends	849
References	850

INTRODUCTION I

This chapter relates VLSI technology to distributed communications systems. The technology is described in terms of state-of-the-art and trends; comparisons of alternative transistor types are drawn; and important relationships regarding chip densities and storage capacities versus minimum feature definition and year are shown.

The meaning of distributed communications is given and focused on a present-day example of such systems, mainly local area networks (LANs). Architecture, transmission and interconnection alternatives, and protocols are discussed and examples of each given. Then VLSI chips for use in a specific LAN are listed.

No exposition of distributed communications or VLSI is complete without discussing signal processing. Functions and applications are given for voice, video, and graphics.

A discussion of future trends summarizes the underlying principles and highlights one of the latest uses of commercially distributed communication.

II VLSI TECHNOLOGY AND TRENDS

A VLSI Performance

Five basic VLSI technologies are now being used and are under further development. These are based on the following transistor types:

(a) Bipolar,
(b) NMOS (single-channel n-type),
(c) CMOS (complementary n- and p-types),
(d) CMOS–SOS (complementary, but fabricated on an insulating saphire substrate),
(e) GaAs.

Each of these technologies has advantages for particular applications. Their relative performance with typical calibration data are given in Table I. This data must be interpreted with respect to given geometries or minimum feature definitions.

VLSI (very large scale integration) pertains to chips containing more than the equivalent of 1000 gates. VHSIC (very high speed integrated circuits), as defined by the Department of Defense, are chips with 1.25-μm minimum feature-size, with a figure of merit of at least 5×10^{11} gate $\cdot$ Hz/cm^2. This is obtained by multiplying the total gates by the maximum clock rate and dividing by the chip area in square centimeters.

TABLE I

Comparisons of Basic VLSI Technologies

VLSI technology	Speed	Power dissipation	Packing density
Bipolar	High	Relatively high (400 μW/gate)	Moderate (40 gates/mm^2)
NMOS	Lowest	Moderate (150 μW/gate)	Highest (150 gates/mm^2)
CMOS	Moderate	Low	Moderate
CMOS/SOS	Highest for MOS (1–2 ns)	Very low (40 μW/gate)	High (50 gates/mm^2)
GaAs	Highest (sub-ns)	Relatively high (1 mW/gate)	Lowest

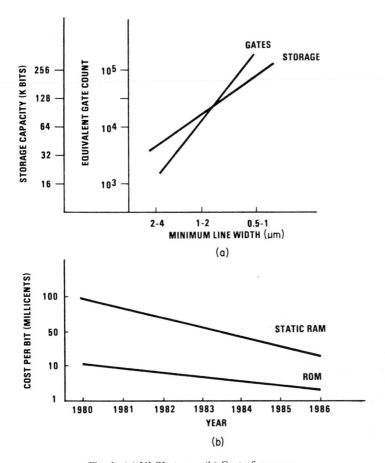

Fig. 1. (a) VLSI stages. (b) Cost of memory.

The trends [1] of gate count and storage capacity per chip as a function of minimum features sizes are shown in Fig. 1a. Some 1–2 μm devices are now in use, which implies the availability of chips with 10,000 gates (upward of 40,000 transistors). The availability of memory devices with feature sizes close to 1 μm enables the design of chips with 256 Kbits of RAM. The cost trend for memory chips, both RAM and ROM, based on these memory device advances is shown in Fig. 1b.

Examples of currently available VLSI chips are given in Table II for different transistor types and geometries. Note that the transistor count for a given size chip is a strong function of the regularity of the device patterns. Memory chips, with symmetric repeating patterns, have a low relative cost whereas random logic has a high relative cost.

TABLE II

Examples of State-of-the-Art VLSI Chips

Manufacturer and type		Technology and geometry	Chip size (mm²)	Number of transistors	Clock	On-chip memory
HP	HP 9000 CPU	NMOS, 2.5 μm	44	450,000	18 MHz	9K × 38 bit
TI	TMS 320 signal processor	NMOS, 2.7 μm	39	55,000	—	3K × 8 bit, (16 × 16 mult
Toshiba	T6386 microprocessor	CMOS, 2 μm	48	66,000	—	2K × 8 bit
RCA	Correlator	CMOS/SOS, 1.25 μm	75	72,000	—	—
RCA	TA 12905 microprocessor	CMOS/SOS, 3 μm	82	20,000	20 MHz	

B Types of VLSI

The VLSI chips that can be used in distributed communications systems, as well as other applications, fall into one of three categories.

1 Catalog or Off-the-Shelf Parts

Some of these chips are described in Table II. They will fulfill the microprocessor and memory functions of most systems. Other standard parts are made for high volume applications for commonly encountered functions in specific systems such as bus interface units.

2 Gate Arrays or Semicustom VLSI

Chips with up to 6000 uncommitted gates in a variety of bipolar and MOS technologies are available for use in special applications. By interconnecting these gates with CAD routing programs and having the stocked wafers metallized accordingly, one can rapidly and efficiently implement special random logic functions chips. These are sometimes referred to as "glue" chips in a system, because they interconnect with standard VLSI parts. Such parts do not have the packing density associated with fully customized designs; but for limited volume, production economics and turnaround are in their favor. They can be cost effective (including nonrecurring as well as recurring changes) in volumes as low as 50 parts when considering their IC component counterparts.

Fully Custom VLSI 3

Fully custom technology is used for the high volume production (tens of thousands or greater) to maximize packing density and minimize the per-gate cost. This is the technology that is used to create chips such as those in Table II. Here, the gates and other logic and storage cells required in a design are applied as needed and called up from a cell library with design automation techniques. The higher nonrecurring costs for design and verification is amortized by the resulting smaller chip sizes, yielding more chips per wafer and low chip cost. Also, fully customized designs provide chips with substantially more functionality, generally replacing several alternatively designed or available chips. On an end product, this results in lower cost, smaller size, and reduced power requirements.

DISTRIBUTED COMMUNICATIONS III

The best example of existing distributed communications systems is the local area network (LAN), which serves one or more communities of processors and terminals. These networks use packet switching techniques pioneered by the ARPANET computer communication network, which provide cost-effective data communications among information processing systems [2].

Early LANs were based on radio broadcast systems (ALOHA network, Hawaii). Subsequent systems have been based on coaxial cable transmission systems using baseband or broadband modulation schemes.

The technology of loop and ring structures has taken advantage of these access schemes and has led to LANs. Because LAN's now involve virtually all of the concepts of distributed communications and in the future will be meshed with wide-area networks, embracing similar principles, they will be described here.

Local Area Networks A

The LAN is archtypical of a distributed communications system. It embodies all the characteristics and features of distributed communications. The LAN represents an extension of data networks, making packet-switching widely available. It is a communications system that connects various computers to terminals, enabling each of them to share peripherals and data, that is, it permits several users to share resources.

LANs can provide data processing, word processing, voice communications, video and graphics service, and electronic mail service, largely as a result of VLSI technology and its applications to microcomputers, storage, and signal processing.

A LAN, a distributed communications system of its own and perhaps the technology with the greatest proliferation, represents one aspect of communications networks in general. LANs interconnect through switches and wide-area networks as shown in Fig. 2.

1 LAN Architectures

Three of the major architectures [3] for implementing LANs are shown in Fig. 3: the star, loop (ring), and bus structures. In these basic architectures there are choices of interconnection media, transmission or modulation type, and operational protocols. Figure 4 lists the predominant technological alternatives for LAN realizations. Technological choices are optimally paired with other alternatives, as will be described. The transmission alternatives relate to the modulation of the information, and the protocols involve the methods for resolving conflicts in a shared resource environment.

(a) The *star configuration* is essentially that of private branch exchange (PBX) systems for voice communications based on twisted pairs of telephone wire. These systems can be upgraded for data and have a concentrated centralized control; the operation of each station is independent. Compared with the other alternatives they require more wire or cable for interconnection and may be more vulnerable to failure because of over-concentration of the control resources. In more recent star architectures, the control is distributed and the central interconnect is passive.

(b) The *bus configuration* is becoming widely used, particularly for advanced office automation or local data processing networks. There is no unique control resource, and it represents a truly distributed approach to communications. It is quite compatible with wire or coaxial cable as well as with broadband or baseband techniques. Cabling is minimized, reliability is high, users can be readily added or removed, and VLSI has dramatically reduced costs.

(c) The *ring* system, in general a more structured busing approach, can take advantage of simple token protocols and has a predictable and stable performance even under heavy loading. On the other hand, average delays are long even under light load.

2 Transmission

Baseband operation is the simplest approach with the media transmitting bits generated at the various stations. Rates of 10 Mbits/s are com-

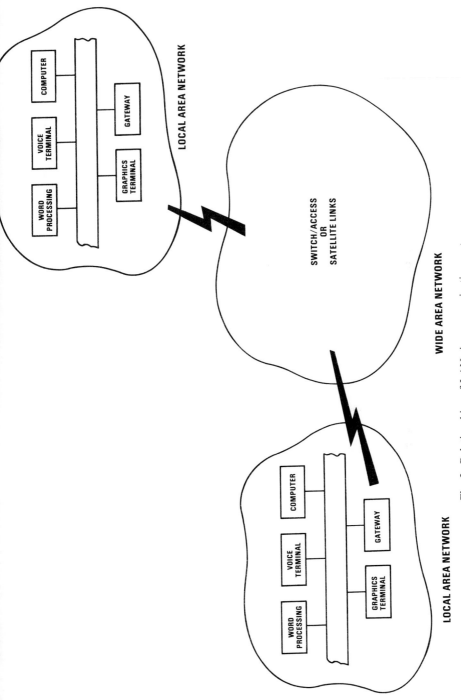

Fig. 2. Relationships of LANs in communication systems.

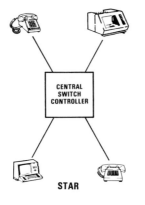
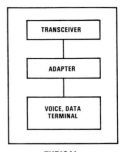
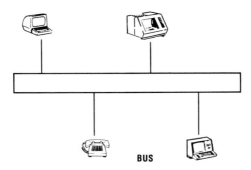
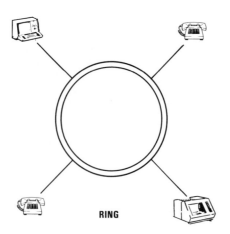

Fig. 3. LAN architectures.

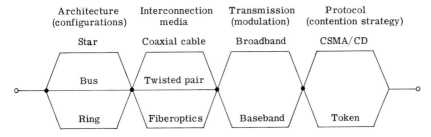

Fig. 4. Technological alternatives for LANs.

monly achieved with low cost coaxial cable, and fiber optics promise data rates of 500 Mbits/s or more. In broadband operation, multiple high-frequency carriers (frequency division multiplexed) are used and modulated by the voice, data, or video information. Higher bandwidths are possible, with more expensive cable, and broadband interface transmitter–receivers are more complex. Broadband systems can use the entire radio spectrum bandwidth available between roughly 10 and 500 MHz. Just like different channels in cable television, local subnetworks can operate in different frequency bands.

Most existing systems use twisted-pair wire or various capacity coaxial cable. Fiberoptic transmission is the choice of more recent designs because of inherent immunity to electromagnetic interference, absence of ground loop and RF leakage problems, and broad bandwidth compared with copper circuits. Presently, fiberoptics are more compatible with star configurations due to optical tapping and coupling considerations. To take maximum advantage of fiberoptic bandwidth capacity, the very high speed VLSI technologies will be necessary. Logic circuits capable of processing hundreds of megabits with all necessary functions of modulation, buffering, and coding will directly impact and accelerate the trend to fiberoptics.

Protocols 3

The best known random-access scheme for bus systems is carrier-sense multiple access with collision detection (CSMA/CD). Under a CSMA protocol, every station ready to send must listen, before transmitting an information frame, to detect if a transmission is already in progress. If another transmission is already in progress, the station (or terminal) defers sending. Due to propagation delays, collision cannot be completely avoided; however, if sensed, a transmission can be aborted and rescheduled at a later, random retransmission interval.

The prominent polled or controlled-access scheme (in a ring) uses a permission token. In a bus system the station receiving the token is free to send an information frame. At the end of the transmission, the station frees the token and passes an addressed token to the next station. In a ring architecture there is sequential ordering so that addressing is not required.

For data rates of 1 Mbit/s, token ring and CSMA/CD bus systems have the same delay-throughput characteristic, whereas for data rates of 10 Mbits/s, the token ring clearly shows better performance over a wide range of parameters. Also, the problems in ring systems are more amendable to digital solutions which benefit from VLSI.

4 Security and Protection

By its very nature, information in a distributed communications system is vulnerable to interception, insertion, and disruption. Wiretapping, spoofing, and failure of a component in a multiple access system must all be dealt with. Encryption is commonly used to prevent interception and insertion. Message data are multiplied by an encoding key; then, to decrypt the cipher text, the receiver must have the agreed-on key to divide the transmitted data.

A data encryption standard (DES) developed by the U.S. Bureau of Standards is commonly employed for link, end-to-end, and public key encryption. VLSI so used provides a high degree of economical protection for distributed communications networks.

Component and cable failure problems can be dealt with by redundancy, self-checking features, and automatic disconnect features.

5 LAN Examples

Ethernet of Xerox Corporation is a well-known example of a CSMA/CD bus system, and the Cambridge Ring has been implemented by Racal–Milgo in PLANET. Ethernet systems of coaxial cable of 1 km in length can connect up to 256 devices transmitting data at 3Mbits/s.

Four dual-chip sets have been, or are, under development for implementing an inexpensive link between work-station nodes and the transceivers that feed into Ethernet. Such VLSI will cut the cost of Ethernet connections that would otherwise require about 100 ICs on several printed circuit boards. The characteristics [4] of each chip set, composed of an encoder–decoder and a controller, are summarized in Table III. A new ring system, announced at Telecom '83 is the IBM Zurich Ring, which uses baseband transmission and token passing. When outfitted with fiberoptics, data rates of 30 Mbits/s will be possible at a distance of over 1 km.

A single NMOS VLSI chip, the Western Digital WD2840, has been

TABLE III

VLSI Chips for a LAN

Development team	Controller	Encoder–decoder[a]	VLSI technology for controller
AMD, DEC, MOSTEK	LANCE AM 7990 MK 68590	SIA AM 7991 MK 3F91	NMOS
INTEL	82586	ESI 82501	HMOS[b]
SEEQ Technology	EDLC 8001	—	NMOS
Ungerman–Bass, Fujitsu	DLC MB61301	MB502	CMOS

[a] All encoder–decoder chips are bipolar devices.
[b] High performance NMOS.

designed as a token access controller. This chip has three major elements: a serial communications subsystem, a two-channel DMA controller, and a microprocessor with internal ROM and RAM.

B Signal Processing for Communications

Nowhere is the impact of VLSI more apparent than in the signal processing used in distributed communications systems. The lower costs of implementation and the speeds at which signal-processing chips can operate permit more functions and flexibility than alternate realizations and make possible new uses of distributed communications.

The digital processing of communications signals is defined as the calculations performed on digitized samples of signals to enhance or transform their values for improved transmission or detectability. Important communications applications of signal processing include bandwidth reduction of voice and image data for improved transmission efficiency. This involves a tradeoff between transmission and terminal resources. With higher performance, low cost VLSI, terminals can reduce the bandwidths of audio and video signals by factors of ten or greater while still preserving intelligibility.

1 Voice

Specifically for voice, standard 8-bit samples at 8 KHz (64 Kbits/s PCM) can be reduced to 16 Kbits/s data by using subband coding techniques or to 2.4 Kbits/s by using linear predictive coding (LPC). A very small, flexible, high quality, full-duplex 2.4-Kbits/s LPC system has been implemented with commercially available VLSI circuits [5]. Using three Nippon Electric Company PD7720 single processing signal chip microcomputers and an Intel 8085-based microcomputer, this LPC system requires a total of 16 ICs, dissipates 5.5 W, and occupies 18 in.^{2} of circuit area.

2 Video and Graphics

Standard TV broadcast bandwidths dictate the use of 80 Mbits/s channel capacity, considering necessary sampling rates and digitization resolution. With digital signal processing techniques, digital video rates of 1.5 to 18 Mbp/s can be attained. These techniques reduce quality somewhat but maintain the primary image communications intelligibility, especially for graphics and slow-changing imagery. Systems employing as few as 56 Kbits/s are being developed, which degrade by losing resolution when overwhelmed by too much motion [6]. Slow scan systems do not use video compression but instead store all frame information in a buffer, transmit the bits to the receive buffer, and display a new frame every few seconds.

Real-time graphics are being applied to work stations in Ethernet communications networks. These workstations provide color display and manipulation of characters, two- or three-dimensional vectors, parametric curves and surfaces, and solids with shading and hidden surfaces removed [7]. Silicon Graphics has produced a VLSI chip with about 75,000 transistors to perform the basic operations of matrix transformation, geometric clipping, and mapping to output device coordinates.

Table IV lists many of the functions normally performed in digital signal-processing equipment. The VLSI chips that usually perform these functions are signal-processing architecture microcomputers, general-purpose microprocessors, multipliers, and memory. Besides digital signal processing, analog signal processing is now available and compatible with

TABLE IV

Functions Performed by Digital Signal Processing Chips

Application	Function
Digital filtering	Both finite impulse and infinite impulse response
Spectral analysis	Both discrete and fast Fourier transform
Modulator–demodulator (Modem)	Tone generation, detection, and synchronization
Adaptive equalization	Echo cancellation
Transmultiplexing	Frequency division to time division transformations

VLSI technology. Through the use of *switched-capacitor arrays,* many communications functions can be optimally implemented in the sampled analog data domain (rather than sampled digital data). For example, a variety of switched-capacitor filters can be integrated with varying numbers of poles. Also, digital-to-analog converters, integrators, and amplifiers can now be realized along with all-digital logic nets in both NMOS or CMOS technologies. The principle behind switched-capacitor arrays is that of using an MOS (single channel or complementary) operational amplifier, a feedback capacitor, and an input capacitor that is switched at a much higher clock rate than the signal bandwidth, which effectively makes it perform as a resistor equivalent.

FUTURE TRENDS IV

While the cost of communications is decreasing, the cost of computing, because of VLSI, is decreasing even more rapidly. As a result, to minimize communications costs it pays to use systems that tend to maximize digital processing. Distributed communications systems accomplish this tradeoff. Instead of communicating with strictly centralized resources, such as switches and computers (as when hardware was a relatively expensive commodity), it is now more economical to distribute the resources, placing more burden on user terminals where VLSI is most effective. This is not to say that all centralized systems will be replaced. Instead they will be augmented with distributed systems.

This trend to economical, high performance VLSI will lead to improvements in communications services as varied as those in the past have been for moving vehicles (via radio packet switching) local area networks. Terminals in offices, shops, homes, and factories will be used for voice, video, and so on. The television set will be part of a bidirectional communication channel that provides shopping, banking, entertainment, information, home management, security, utility telemetry, and distributed office service. ITT has recently announced BIGFON in West Germany. This *Broadband Integrated Glass Fiber Optic Network* incorporates all available video and audio services and ties in business facilities such as data communications, facsimile transmission telex, and teletext. The telephone of the future will be a home station operated with an infrared control unit, as easy to handle as a TV control but providing not only videophone but also a variety of business services. The VLSI is inserted into these products either from standard parts (particularly memory and processors) or from a form of custom integration such as those already developed for some types of LAN functions.

As the cost of VLSI products decrease, distributed communications will become more pervasive and will be part of a substantial percentage of all homes and offices by the end of this decade.

REFERENCES

1. D. J. McGreivy and K. A. Pickar, "VLSI Technologies Through The 80's and Beyond." IEEE Computer Society Press, Silver Springs, Maryland, 1982.
2. L. Kleinrock, On resource sharing in a distributed communication environment, *IEEE Commun. Mag.* **17** (1), 27–34 (1979).
3. R. D. Rosner, "Distributed Telecommunications Networks." Lifetime Learning Publications, Belmont, California, 1982.
4. H. J. Hindin, *Electronics,* Oct. 6, 89–103 (1982).
5. J. A. Feldman, E. M. Hofstetta, and M. L. Malpass, *IEEE J. Solid-State Circuits* **SC-18,** 4–9 (1983).
6. T. Ishiguro and K. Linuma, *IEEE Commun. Mag.* **20** (6), 24–30 (1982).
7. J. H. Clar and T. Davis, Work station unites real-time graphics with Unix, Ethernet, *Electronics* **56,** 113–120 (1983).

Chapter 51

Applications of VLSI to the Automobile

FRANK S. STEIN

Delco Electronics Division
General Motors Corporation
Kokomo, Indiana

I. Introduction	851
II. Driving Forces	852
A. Government Controls on Emissions and Fuel Economy	852
B. Benefits of Electronic Implementation	852
III. Automotive Electronics Market	853
IV. Reliability	854
V. Engine Controls	855
A. I/O Structure of Engine Control Module	856
B. Microprocessors in Automobiles	857
VI. Body Computers	859
VII. Entertainment Systems	861
VIII. Cellular Telephones	862
IX. Conclusion	863
References	864

INTRODUCTION   I

Electronics, in the form of vacuum tube radios, first appeared in the automobile around 1930. In the mid-1950s, just a few years after the transistor was invented, solid state electronics was introduced into the car in the form of germanium power and small-signal transistors for the radio.

The first control application appeared about 1960, as silicon rectifiers found their way into the alternator system; and at the end of the decade, the first automotive custom IC was designed and used in the passenger car voltage regulator, along with discrete output devices. High voltage silicon transistors and ICs were used in the high energy automotive ignition system a few years later. The microprocessor was invented and marketed in the early 1970s, and the first automotive microprocessor was introduced a few years thereafter.

This trend toward applying electronic innovations to the car within a short period after their development continues, even as the complexity of ICs increases. The extensive use of ICs, and of VLSI in particular, is having a large effect on the automobile. Not as apparent is the fact that the automobile industry, with its high production base and its goals of increased reliability, higher performance, and cost effectiveness, in turn is having a large effect on the VLSI industry.

Some recent applications and future trends of VLSI in the automobile will be discussed in this chapter. Only a few applications are addressed in detail. But, first, let us examine the impetus for VLSI application to the automobile.

II DRIVING FORCES

A Government Controls on Emissions and Fuel Economy

Federal emission controls imposed in 1968 and tightened in succeeding years have called for reduced emissions of hydrocarbons, carbon monoxide, and oxides of nitrogen in the automobile exhaust. Legislation passed after the oil embargo established fuel economy requirements that imposed further constraints on the automobile's performance. To meet these conflicting requirements of reduced emissions and increased fuel economy, several decisions were made: to develop a catalytic converter that would reduce gaseous emissions (and emissions of lead as well, because lead-free gas must be used to avoid poisoning the catalyst); to target the air/fuel ratio for stoichiometry ($A/F = 15$); and to utilize electronic engine control systems. The microprocessor had recently been developed, and electronic innovations had already been used extensively and successfully in the car. Thus, automotive electronic systems engineers were ready, willing, and able to develop the necessary systems; and they proceeded to do so.

B Benefits of Electronic Implementation

Indeed, the microprocessor offered many major benefits to the auto manufacturer besides facilitating compliance with federal mandates.

51. Applications of VLSI to the Automobile

These benefits include:

(a) higher system reliability that resulted from the reduced component count and fewer interconnections;
(b) the opportunity to offer a host of new features and higher performance in the areas of safety, entertainment, convenience, instrumentation;
(c) improved buildability, testability, and serviceability; and
(d) lower system cost.

In view of the federal mandates, on the one hand, and these new opportunities, on the other, the auto industry quickly latched onto the microprocessor. Today, with fuel economy and emission requirements achieved, microprocessor control systems are continually being refined and expanded to provide new levels of performance, economy, and customer satisfaction.

III AUTOMOTIVE ELECTRONICS MARKET

It is interesting to examine the extent to which electronics pervades the automobile. Excluding radios and similar entertainment products, the domestic automotive electronics market for 1984 is estimated at $2.3 billion. At 8 million cars per year, this averages about $300 per car. This market is expected to double in 1988 (Fig. 1) for a projected base of 9.5 million cars, corresponding to a 60% increase, or $480 per car [1]. Entertainment features may add perhaps another $200 per car, on the average. It is apparent

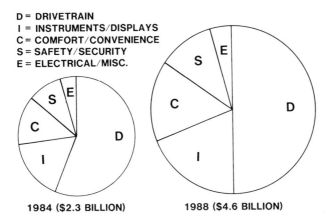

Fig. 1. Projected growth of the nonentertainment automotive electronics market (Frost & Sullivan Inc. [1]).

from these figures alone that the automotive market has a major impact on the semiconductor supplier. Perhaps the biggest impact is on quality requirements and reliability, as discussed in the following section.

IV RELIABILITY

Due to the very large volumes used, semiconductors must be of the highest quality to avoid an excessive number of failures. Elaborate testing programs have been developed to qualify the IC for the hostile environments seen in and by the automobile. For example, tri-temperature testing is a necessity: high temperature (65 to 150°C), room temperature, and low temperature (usually −40°C). It is apparent that in the field four-seasons testing will require long lead times.

TABLE I

Environmental and Other Tests for Automotive Devices[a]

Test	Passenger compartment requirements	Under-the-hood requirements
Temperature cycle (1 h/cycle, 1000 h)	−40−+125°C	−40−+150°C
Bias humidity (1000 h)	90–98% RH 30–65°C cycle	90–98% RH 30–65°C cycle
Power temperature cycle (1000 h)	−40−+85°C	−40−+125°C
Storage life (1000 h)	150°C	150°C
Autoclave (121°C, 14.5 psig)	48 h	48 h
Mechanical shock	500 g peak (1 ms) or 1500 g peak (0.5 ms); 5 drops	500 g peak (1 ms) or 1500 g peak (0.5 ms); 5 drops
Thermal shock (100 cycles, 5 min. soak)	−65−+100°C	−65−+150°C
Salt spray (5% solution)	35°C (48 h, ICs)	35°C (48 h, voltage regulator; 96 h, ignition)
Additional Tests		
	Voltage extremes	Lead pull
	Load dump	Thermal resistance
	Transient voltages	Operating life
	Input protection	Vibration
	Resistance to solvents	Solderability
	Lead bend	Solder temperature

[a] Stein, F. S. (1982). *International Electronics Packaging Society, Proceedings*, 400–03.

TABLE II

Transient Stress Testing to Simulate the Automotive Environment[a]

Transient added to B⁺ line (V)	Source impedance (Ω)	Test description
±32	60	10 pulses; 40-μs sine wave; 1% duty cycle
±250	225	10 pulses; 1-μs sine wave; 1% duty cycle
±450	200	5-MHz rate; pulse burst of 20 μs duration; 10 bursts
±125	10	Exponential pulses; 1-ms decay time constant; 10 pulses
−125	10	Exponential pulses; 20-μs decay time constant; 10 pulses
±60	1000	Exponential pulse; 1-ms decay time constant
+60 (load dump)	0.1	10 pulses; 5-ms rise time; 115-ms decay time constant; 1% duty cycle

[a] Motz, P. R., and Vincent, W. A. (1983). *IEEE Proceedings of the Custom Integrated Circuits Conference*, 392–98.

Table I is a list of environmental requirements that electronic systems must withstand in automobiles. Other parameters and considerations, any or all of which must be applied to ensure high reliability, are also listed. Transient stress testing (Table II) is another type of test to ensure survivability. Fault verification or test grading of new IC designs is also a must, even though it is very time-consuming and costly for VLSI.

The trillion miles that are driven annually by cars in the United States make it apparent that the automobile manufacturer must demand high quality and reliability. It is equally apparent that VLSI vendors must meet these high standards if they wish to participate in this major market.

V ENGINE CONTROLS

The engine control module (ECM) was the first major automotive application of the microcomputer, and the first ECM appeared in the 1977 Oldsmobile MISAR system. Its sole function was to control electronic spark timing (EST). Further functions were added to ECMs in succeeding years by various manufacturers, as shown in Table III.

Adaptive fuel control in later ECMs enables errors in the fuel flow system to be corrected continuously. Corrections can be made to compensate transient engine operating conditions. Even as functions are being added to the ECM, up-integration through the expanded use of VLSI is leading to smaller, more rugged systems with higher reliability and reduced cost.

TABLE III

VLSI in Automobile Engine Control Functions[a]

Year	Functions
1977	Control of electronic spark timing with engine control module (ECM), Oldsmobile MISAR system
1978	Linear exhaust gas recirculation (EGR)
1979	Closed loop carburetor control (O_2 content of exhaust controls A/F ratio)
	Control of idle speed
	Secondary air control
	Canister purge
1980s	Open- or closed-loop throttle-body (port fuel) injection
	Self-diagnostics
	Early fuel evaporation
	Modulated displacement
	Spark knock control
	A/F ratio control by mass airflow sensing
	On–off and proportional exhaust gas recirculation
	Transmission control
	Integrated cruise control
	Air management
	Fuel shut-off during deceleration

[a] From reference [2].

A   I/O Structure of Engine Control Module

An early ECM required three densely packed 5.5×8.0 in.^{2} circuit boards with perhaps a dozen LSI ICs and several dozen SSI ICs. Present systems provide three times as many functions but occupy only a single circuit board, thanks to the reduced IC count resulting from greater use of VLSI (Fig. 2).

Before describing the complexity of the microprocessors that drive these systems, it is instructive to look at the I/O structure of a typical ECM (Fig. 3). The functions of the outputs have been described. The inputs, which must be sensed prior to controlling the outputs, usually include most of the following [3]:

(a) battery voltage
(b) crankshaft angle position
(c) engine rpm
(d) manifold absolute pressure
(e) barometric pressure
(f) air flow
(g) fuel flow
(h) partial pressure of oxygen

(i) partial pressure of oxides of nitrogen
(j) engine knock
(k) coolant temperature
(l) air temperature
(m) throttle position
(n) exhaust gas flow

Figure 4 shows a block diagram of an engine control module.

Microprocessors in Automobiles   B

The microprocessors or microcomputers themselves vary widely. Most are NMOS but some are CMOS; most are customized but some are standard; they use 4-, 8-, 10-, 12- or 16-bit operation, depending on performance requirements. The memory capacity in ROM, calibration PROM, RAM, nonvolatile RAM, and EPROM or EEPROM varies according to need, with growth projected to over 250 Kbytes. Future systems will

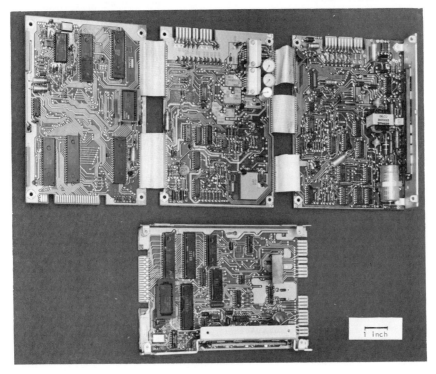

Fig. 2. Size reduction in engine control modules through application of VLSI: 3-board system, 1982; 1-board system, 1984 (GMC).

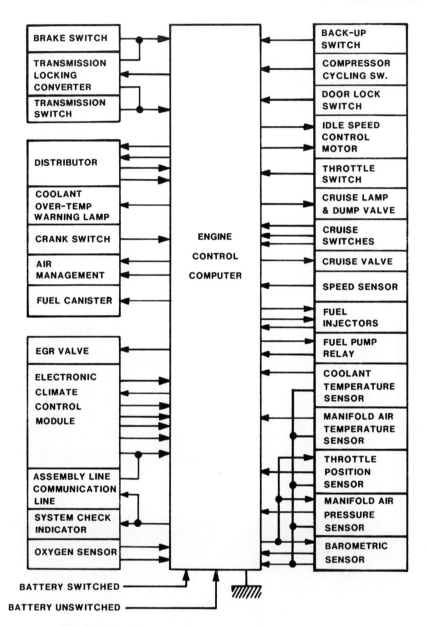

Fig. 3. Typical I/O of an engine control microcomputer (GMC).

51. Applications of VLSI to the Automobile

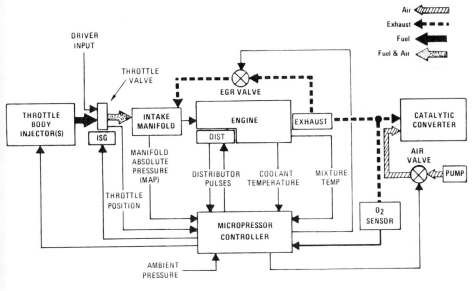

Fig. 4. Block diagram, throttle-body electronic engine control system (GMC).

require mass data storage, greater I/O, faster data transmission, and greater computer throughput capability. The impact on the automobile of tomorrow will be further improvements in both performance and reliability.

VI BODY COMPUTERS

Vehicle body computers (as contrasted with engine computers) have been developed to provide cost-effective instrumentation functions and features other than engine controls, while minimizing complexity, maximizing reliability, and improving serviceability. The first body computer combined a microcomputer-controlled automatic heating and ventilating control system, a fuel economy display, and a trip computer. It controlled four major areas: fuel display, climate control, communication with actuator controllers, and integration of other body interior functions; detailed functions are listed in Table IV. Figure 5 illustrates a version for a top-of-the-line car. The system used a 6801 microcomputer containing a 2-Kbyte ROM and a 128-byte RAM. External memory resided in two ICs: a custom IC combining a 4-Kbyte ROM, a 128-byte RAM, and an 8-bit discrete I/O port, plus a standard IC with additional program and calibration in a 4 Kbyte EPROM. A custom 8-bit analog-to-digital converter was used for

TABLE IV

Body Computer Functions

Fuel Display Panel
  Instantaneous fuel economy
  Average fuel economy
  Range
  Fuel consumed
  Fuel remaining
	Communication with Actuator Controllers
	  Ventilation mode actuators
	  Continuously variable ventilation fan
	  Continuously variable coolant fan speed with
	    diagnostic feedback

Climate Control Panel
  Ventilation mode control
  Fan speed control
  Temperature control
  Outside temperature display
  Diagnostic moding and display
	Integration of Other Body Interior Functions
	  Instrument panel dimming
	  Powertrain controller communications
	  Control of accessory power
	  Automatic twilight control of exterior lamps
	  English/metric moding of readouts
	  Enhanced ventilation system diagnostics

11 analog signals and 10 discrete I/O signals. The 6801 was used for serial data link ports, the pulse-width modulated signal for the ventilating fan, and other discrete outputs. A 16-V switching power supply was also included for the vacuum fluorescent displays in the radio, fuel data panel, and climate control panel; alternative types of displays (LCD, LED, CRT, EL, etc.) would, of course, require different power supplies.

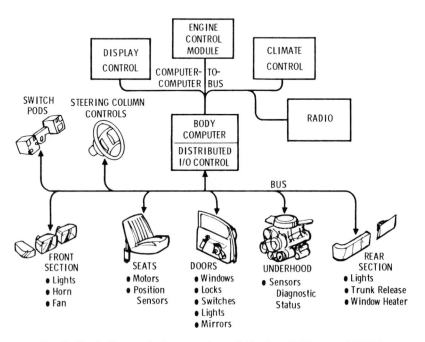

Fig. 5. Block diagram, body computer and distributed I/O control (GMC).

A recent body computer controls a CRT display system featuring a touch-sensitive screen and four display modes: radio controls, climate controls, trip monitor, and a summary of all three. It retains the previously mentioned features plus a time-of-day clock and introduces a broadcast bus using the built-in bidirectional serial data link (UART) in the 6801 type microcomputer. Additional features may be offered in CRT displays, such as navigation or route guidance, wherein maps are displayed, scale factor is selected, routes to desired destination are indicated, and a cursor tracks the car's position on the map.

Body computers of the future will be designed to provide the central computing resource for more fully integrated vehicle electrical systems, including more extensive system diagnostics. Some power control options may be included, but the trend is to keep the power control near the actual load to minimize the power distribution network. Another trend relates to broadcast buses: Each bus is placed in the vehicle to provide communications between systems with similar requirements. Vehicle data communications will play an increasingly important role in the design of vehicle electrical systems as the system requirements grow and the cost effectiveness of VLSI circuits continues to improve.

ENTERTAINMENT SYSTEMS   VII

The electronically tuned radio (ETR) constitutes a major product improvement which the advent of VLSI has made cost effective. Conventional mechanically tuned radios either rotate interleaved capacitor plates or shift the position of magnetic cores in inductance coils to achieve the desired channel frequency. Fine tuning is required, and drift may occur. With ETRs, on the other hand, all AM and FM stations can be tuned directly to their nominal frequency, thereby obviating the need for fine tuning; and no drift is observed. This is achieved through the technique of frequency synthesis, which is derived from a microprocessor-controlled digital phase-locked loop circuit. A quartz oscillator is used to provide a precise frequency with which the local frequency oscillator of the radio is compared. Tuning is effected through small pushbutton switches; no other mechanical movement occurs. The switch signals a microcomputer to provide a new frequency comparison ratio. The electronic circuitry then converts this into a digital readout of the channel frequency.

The microcomputer in question contains up to 1.5 Kbytes of ROM, a 56-byte RAM including nonvolatile RAM to store the stations selected, and time-of-day data that is software-controlled. A block diagram of a typical ETR is shown in Fig. 6. Entertainment systems of the future may

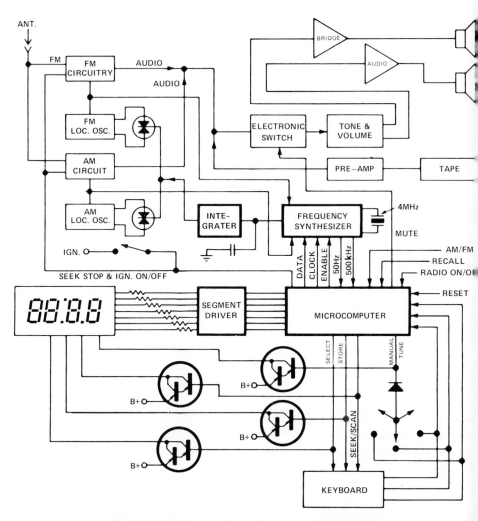

Fig. 6. Block diagram, electronically tuned radio (GMC).

use digital signal processing (DSP), which will make even more extensive use of VLSI.

VIII CELLULAR TELEPHONES

We shall close with a future application, one whose volume is projected to grow from only 40,000 in 1984 to 800,000 by 1988 [1]. This is the cellular mobile radio telephone (CMRT). Cellular phones make extensive use of digital logic to control a network of land-base and mobile transceivers. Cities are divided into small cells, to each of which is assigned a pair of frequencies, one to transmit and one to receive. Adjacent cells have different frequency pairs, and calls are automatically handed off by a land-base computer network as the car leaves one cell and enters another. This computer exercises control via a serial data stream integrated with the user audio. In addition, logic circuitry is required to interface the handset and control head with the transceiver unit, normally located in the trunk. The complex nature of the digital circuitry in cellular communication demands the use of microcomputers and other VLSI devices to meet the minimal weight and space allocations in today's car.

VLSI is expected to lead to the eventual integration of CMRT with heating and ventilating, entertainment, and driver information systems into a single display and control unit. Up-integration of this sort is a trend. Although some of today's luxury cars contain as many as a dozen microprocessors, more and more systems will become integrated in the automobile of the future in the interests of performance, reliability, and cost effectiveness.

CONCLUSION IX

A listing of further applications would include, but not be limited to, the following [4–6]:

(a) electronic climate control, which can control passenger compartment temperature to within 1°F;

(b) voice information systems, in which voice synthesis can be used to announce any number of abnormal circumstances, identify the problem, and evaluate its urgency, even suggesting actions to alleviate it;

(c) voice recognition, with the car responding to its "master's voice";

(d) diagnostic systems that can provide code numbers to identify over 100 possible malfunctions or performance deficiencies if they occur;

(e) electronic power steering that can be made speed sensitive;

(f) automatic electronic leveling or air-suspension system, especially useful in today's cars to provide constant headlamp aim and consistent handling characteristics irrespective of vehicle-loading conditions;

(g) proximity-warning and collision-avoidance systems, which can be made interactive;

(h) multiplexed wiring to eliminate bulky wiring harnesses;

(i) air cushion restraint systems;

(j) transmission control;

(k) antiskid or antilock braking;
(l) electronic seat-position control; and
(m) theft-deterrent and keyless-entry systems [4,5,6].

This list omits more applications than it includes. It is intended only to stimulate the imagination, which, in view of the increasing capabilities of VLSI, will prove to be the limiting factor for applications of VLSI in the automobile of the future.

REFERENCES

1. Frost & Sullivan, Inc., "The Automotive Electronics Market in the U.S." New York, 1983.
2. J. Bereisa, *IEEE Trans. Ind. Electron.* **30** (2), 87–96 (1983). (This entire issue is devoted to automotive applications of microprocessors.)
3. R. K. Jurgen, *IEEE Spectrum* **20** (10), 33–39 (1983).
4. *Int. Congress on Transportation Electronics, Proc.* (P-111). SAE, Warrendale, Pennsylvania (1982).
5. "An Update on Automotive Electronic Displays and Information Systems" (P-123). SAE, Warrendale, Pennsylvania, 1983.
6. "Electronic Engine/Drivetrain Controls" (SP-540). SAE, Warrendale, Pennsylvania, 1983.

Chapter 52

How to Protect VLSI Intellectual Property

HAROLD LEVINE
Sigalos & Levine
Dallas, Texas

I. Patent Protection	866
A. What Is a Patent?	866
B. Why Obtain Patents?	866
C. How Is a Patent Obtained?	867
D. How Much Does It Cost to Obtain a Patent?	869
E. To Litigate or Not to Litigate Patents?	870
F. Negotiated Settlements versus Litigation: Pros and Cons	871
G. Other Economic Realities of Patents: Patent Program Effectiveness Measurement	872
II. Copyright Protection	873
A. What Is a Copyright?	873
B. How Is Copyright Protection Obtained?	874
C. Scope of Copyright Protection and Commercial Practical Realities for VLSI	874
III. Trade Secret Protection	876
A. What Is a Trade Secret?	876
B. Scope of Trade Secret Protection	876
C. Advantages and Disadvantages of Trade Secret Protection	876
D. Practical Do's and Don'ts to Establish and Maintain Enforceable Trade Secret Protection	878
IV. Other Key Aspects in the Protection Program	880
A. Employee Relations Agreements: Exit Interviews	880
B. Visitor Agreements	880
C. Vendor Agreements	880
D. Use Log Books	881
V. Summary	881
References	882

I PATENT PROTECTION

A What Is a Patent?

Although approximately 60,000 patents are granted and issued by the United States Patent and Trademark Office each year in the United States, very little is actually known about them by the general public.

A patent, when granted and issued, is in the form of a booklet having the Patent Office seal on the front of it. But it is much more than a simple booklet or disclosure of an invention. It is a statutory grant from the United States Government to the patentee or patent owner, giving the right to exclude others throughout the United States from making, using, or selling the patented or protected development for a period of 17 years [1]. Thus, the patent is a legal statutory monopoly which allows the patentee to be the sole source of the patented product or service for a period of 17 years.

The patent is also a property right inasmuch as the patented invention must be new and unobvious to one skilled in the art and is, therefore, the exclusive property of the patent owner. As such, the patent owner has personal property rights [2] in the patent, as he does in any other type of personal property. Thus, it can be sold, hypothecated, licensed, pledged, assigned, or otherwise used.

Further, if the patent is an improvement patent, although the patent owner may have the right to exclude others from making, using, or selling his invention, the patentee himself may not be able to use it if such use, in fact, would infringe earlier patent(s), held by others, upon which it is an improvement.

B Why Obtain Patents?

Since patents have the attributes of property and provide a statutory monopoly, one can understand how patents can be valuable business assets and why inventors and invention owners should obtain patents.

If the invention is not an improvement patent over an existing device but is a basic innovation in the art (such as the transistor), and if that basic invention is or will be in demand by the public, there are many advantages to be obtained by having a patent.

(1) The inventor or his company, to whom the patent is assigned, or patent owner will be willing to invest the necessary time and capital in research and development to perfect the invention so that it can be used

52. How to Protect VLSI Intellectual Property

by the public. The patentee or his assignee will be able to have the opportunity to recoup their investment if the product or service is one that is or will be in demand by the public. Thus, the patent is an incentive for the inventor to disclose his invention to the public so that the public will benefit from the use of the product or service and so that others seeing it will be able to make improvements and thus improve the quality of life.

(2) The patents can be used in a defensive manner inasmuch as they can be used as negotiable assets in licensing arrangements with other companies. For instance if the inventor of one invention needs to use an invention of another, he may be able to enter into a license agreement with the other whereby he allows the other to use his patented invention in return for the granting to him of the right to use the other's patented invention. Thus, patents are property that can be sold, negotiated, or traded and thus can prove valuable in the conduct of business.

(3) Because patents are property and because a great deal of money and labor goes into the research and development of those inventions, a patent that is obtained has an asset value to the company. Thus, a patent portfolio may increase the value of the company by several millions of dollars, depending on the size and nature of the portfolio.

(4) Because patents have the attributes of property and because of laws that have been passed to encourage the development of inventions, they can be used to obtain income tax benefits such as capital gains treatment.

How Is a Patent Obtained? C

The patent laws of the U.S. and the various countries around the world all require certain formalities to be observed in the filing of a patent application [3]. The application for a patent consists of a specification that includes

(a) an abstract of the disclosure,
(b) cross references to related patent applications, if any,
(c) a brief summary of the invention,
(d) a brief description of the several views of the drawings,
(e) a detailed description of the drawings,
(f) claims that describe the invention for which protection is sought.

In addition, drawings must be submitted when necessary [4], and the specification must refer to the drawings by numerals and letters. The claims are of particular interest inasmuch as they are, in fact, the "metes and bounds description" of protection for the invention. Further, because the inventions usually involve some technical art or science such as engineering, physics, chemistry, and the like, they must be described by one who understands that particular art or science. Thus, it can be seen that patent applications are technical in nature and should be prepared with care by a professional who is knowledgeable not only in the technical nature of the invention but also in the legal requirements for the description and claiming of the invention.

It is, therefore, advisable for the inventor to have the patent application prepared for him by a patent attorney who is trained in the art of drafting such applications. Although any telephone directory includes a listing of patent attorneys, it is preferable that the inventor select a patent attorney by recommendation, perhaps from his own attorney or from a colleague. Larger companies usually have their own patent attorneys and contact their engineers or other inventors on a regular basis to identify any new inventions.

To qualify for patent protection, the invention does not have to be reduced to practice if it follows sound and well-established engineering principles. Thus, if one is going to design a digital logic circuit to perform a function that has not been previously performed but which uses logic elements that are well known in the art, such as AND gates, OR gates, and the like, the circuit can be drawn and understood without having been actually reduced to practice. If an actual embodiment of the invention has not been reduced to practice, the filing of the patent application in the U.S. Patent and Trademark Office constitutes, under the law, a constructive reduction to practice for purposes of determining when the invention was reduced to practice by that particular inventor.

Once the attorney has drafted the specification and claims and has had rough drafts of the drawings prepared, he reviews those items with the inventor to ensure that the invention has been properly and adequately described as the inventor understands it. Once this is done, the application can be placed in final form and the drawings prepared in ink on bristol board. Again, the drawings must conform with statutory requirements [5] as to the type of paper on which the drawings are made, the type of ink used, the width of lines, the margins on pages, and the like, and so must be drafted by a competent draftsman.

The inventor must then sign an oath [6] or declaration [7] stating that his invention meets the requirements of the statute and also declaring that he is aware that if he makes false statements in the declaration or oath that it is punishable by fine or imprisonment or both under a certain section of the statute. Also, under the patent laws as amended, if he is a small inventor, (e.g., a company with less than 500 employees), he is entitled to a lower filing fee than a large entity. The application, including the declaration and statement of the small inventor status, is then sent to the U.S. Patent and Trademark Office along with the appropriate filing fee. The U.S. Patent and Trademark Office returns to the inventor or his designee (e.g., his patent attorney) a filing receipt that indicates the date of the filing of the patent application in its complete form.

How Much Does It Cost to Obtain a Patent? D

Obviously, the cost of obtaining a patent varies with the type of invention, the complexity of the art in which the invention is involved, the person drafting the application, and whether or not the inventor is a small entity under the patent laws.

Thus, in drafting the patent application, the two parts that require the expertise of a patent attorney are the specification and the claims. The specification may be very easy to draft for a simple invention such as a paper clip. On the other hand, the application may involve complex computer systems, complex methods of storing digital data, or complex chemical processes that require detailed technical explanation. Further, the length of the disclosures may range from three or four pages to several hundreds of pages. Since the attorney's charges for his services are usually based on time spent, the cost for drafting the applications can vary widely.

The claims are the important part of the patent application and must be drafted with great care. A considerable amount of time is usually required to draft claims properly, and again, the cost of the application rises.

As a general rule, a simple mechanical, electrical, or chemical case would cost approximately $1500 for attorneys' fees, one sheet of drawing would cost approximately $60, and filing fees of approximately $150 would be required for a small entity inventor. Thus, today's normal minimum total cost for a simple application would range between $1500 and $2000. Complex applications can cost between $5000 and $10,000, or more, depending upon their complexity and the nature of the invention.

Further, under the new patent laws, the number of claims submitted also affect the government fee portion of the cost of the filing of the application. All claims in excess of 20 claims are charged at the rate of $5 per claim for a small entity or $10 per claim for other than a small entity. In addition, claims are either independent or dependent. An independent claim stands by itself whereas a dependent claim relates to a prior claim. For independent claims, all those claims in excess of 3 are charged at the rate of $15 per claim for a small entity or $30 per claim for other than a small entity. The basic government filing fee for a small entity is $150, and for other than a small entity it is $300. Thus, the filing fees and costs of drawings vary depending on the complexity of the case, the number of sheets of drawing, the number of independent and dependent claims, and whether or not the inventor is a small entity or a large entity.

E To Litigate or Not to Litigate Patents?

It should be an axiom observed by individuals and companies alike that litigation is to be avoided if at all possible. It not only depletes the resources of the individual or company but it also wastes the time of the individuals involved with the litigation. Thus, it not only saps financial resources but also diverts the time and attention of individuals who are important to the business or company away from furthering the successful carrying on of their business.

Litigation also has some aspects not understood by the average person or businessman. In the first place, the plaintiff may not be able to drop the suit at a later time, even if he desires to do so, because the defendant may file a counterclaim which will prevent the plaintiff from escaping the litigation by dropping his own claim.

In the second place, the plaintiff may not be able to control the costs of the litigation. Once the lawsuit is instituted, the defendant may, and probably will, seek to have discovery through the taking of depositions, requests for production of documents, and the answering of interrogatories. Each of these methods of discovery is expensive in its own way. For instance, depositions require the presence of witnesses, attorneys for both parties, and a court reporter to make a record of those proceedings. Further, the depositions may take place at a remote location, necessitating expensive trips.

The need for documents can require hours of delving through records. Further, those documents will have to be reviewed by the attorneys for their possible substantive value and for determining whether or not they are properly discoverable materials.

Interrogatories are also burdensome because they require the attorney and the party receiving the interrogatories to review each question in detail and make a determination of how to answer the question and what documents would substantiate the answer. It is not enough to say that the party will refuse to respond to the discovery. If he does not, the court may order him to do so and could simply award judgment to the other side. Thus, once the litigation is started the parties are locked in if a counterclaim is filed, and only on a settlement basis can the litigation be terminated short of actual trial. In short, engaging in patent litigation is like grabbing an expense tiger by the tail—it's financial hell if you hold on, and many times you can't let go.

1 Cost of Litigation

If the litigation requires discovery and a trial and if the proceedings are of substantial commercial merit, each party can expect to expend a minimum of $200,000 in the litigation process through a decision at the district court level. Many cases have cost more than $750,000 at this stage. This does not include special proceedings such as appeals on specific questions or related legal matters, such as dealing with the re-examination of a patent during a litigation. If, for any reason, matters are raised that require rehearing or further discovery, the costs can double. Of course, these figures may be small and unrealistic for large companies such as IBM or Univac, because several patents may be included, and several millions of dollars and years of litigation may be spent by companies of this size.

How the Courts View Patents 2

Because patents are a statutory monopoly and because they involve legal matters involving technical arts, they are scrutinized critically by most judges.

Many judges do not understand the technical nature of patents, and most judges are inherently opposed to monopolies. Under such conditions, a wide divergence of results is obtained in patent litigation, depending on the judges who hear the cases. The percentage of litigated patents held valid varies from circuit to circuit, with the highest proportion of patents held valid being in the Fifth Circuit (where about 5 out of 10 litigated patents are held valid). However, this divergence should lessen or disappear with the creation of the Court of Appeals for the Federal Circuit, which took place on October 1, 1982. One of the purposes of this new court is to bring certainty and stability to patent law by having a single court of appeals for patent matters. Now all patent appeals will be brought before this new court, and the decisions will, hopefully, be more consistent.

Negotiated Settlements versus Litigation: Pros and Cons F

Negotiated settlements are always preferable to litigation and can usually be obtained by reasonable parties. There are those cases, the Hatfields and McCoys type of controversies in which parties cut off their noses to spite their faces. If one finds himself in that situation, he has only two choices—litigate or totally surrender to the other side. Negotiated settlements should provide a win/win situation whereby both parties are treated fairly to their mutual benefit. One-sided settlements normally create more problems than they solve; they inspire litigation because, in effect, they back one of the parties into a corner from which there is no escape except by capitulating or litigating. A fair settlement avoids the extremely complex, time consuming, and expensive litigation, and both parties benefit if the negotiations are approached reasonably.

G Other Economic Realities of Patents: Patent Program Effectiveness Measurement

Because patents are legally protected for a period of years, they provide the foundation of many new businesses and industries. The examples are numerous, but everyone is acquainted with the Xerox machine, the transistor, the integrated circuit, and the light bulb. Innovative patents are the life blood of industry, providing the incentive for investment.

An important measure of the effectiveness of a profit-oriented patent program is that it should provide a basis for exploitation under the company's patent licensing strategy and should serve one or more of the following purposes.

1 Assertive

(a) Support aggressive exploitation of company-developed technologies by preventing unauthorized exploitation of these technologies by competitors.

(b) Achieve recognition, for the company and the company's inventors, of leadership in the technologies of the company's businesses.

(c) Obtain royalty income through licensing of the company's innovations.

(d) Provide a royalty base for royalty income for use in technology transfer arrangements.

2 Defensive

(a) Protect the company's innovations through an aggressive and selective cost effective patent filing program, both domestically and outside the United States, to minimize opportunities for competitors to obtain patents covering the company's products and services.

(b) Provide a strong patent arsenal for negotiating license terms favorable to the company under adversely held patents, to remove patent obstacles in the company's path in exploiting technologies and inventions not developed by the company; avoid or at least minimize the company's obligation to pay royalties under its license agreements; and discourage patent infringement suits against the company.

COPYRIGHT PROTECTION II

What Is a Copyright? A

A copyright provides the copyright owner with the exclusive right to print, reprint, publish, copy, and vend the copyrighted work as well as translation and certain performance rights. The current copyright law, which went into effect on January 1, 1978, includes a general statement of what works of authorship are copyrightable, including four classes of works in which copyright may be claimed [8].

(1) Nondramatic literary works, including fiction, nonfiction, poetry, text books, reference works, directories, and catalogs.

(2) Works of the performing arts, including musical works and any accompanying words, dramatic works, and any accompanying music, pantomimes, choreography, motion pictures, and other audiovisual works.

(3) Works of the visual arts, including pictorial, graphic, and sculptural works such as two- and three-dimensional works, prints and art reproduction, maps, globes and charts, technical drawings, diagrams, and models.

(4) Sound recordings, including all published and unpublished sound recordings.

The act states that copyright protection extends to all original works of authorship fixed in any tangible medium of expression, now known or later developed, from which they can be perceived, reproduced, or otherwise communicated, either directly or with the aid of a machine or device. Purely useful objects are not now nor have they ever been subject to copyright protection except that works of art which form part of a useful object such as a statue used as a lamp base may obtain copyright protection. In general, a copyright protects original literary, musical, or other artistic work against unauthorized reproduction. It should be noted that copyright protection extends only to the particular expression, description, explanation, or illustration of the subject matter, and in no case does it extend to any idea, procedure, process, system, method of operation, concept, principle, or discovery revealed in the work regardless of the form in which it is described or explained.

For works already under statutory protection, the new law retains the present term of copyright of 28 years from first publication, renewable for a second period of protection of 47 years. Copyrights in their first term must still be renewed to receive the full new maximum term of 75 years.

For works created after January 1, 1978, the new law provides a term lasting for the author's life plus an additional 50 years after the author's death. For works made for hire (and certain other works), the new term will be 75 years from publication or 100 years from creation whichever is shorter.

B How Is Copyright Protection Obtained?

Copyright protection is obtained by placing a copyright notice on published copies of the work. Under the present law, registration is not a condition of copyright protection but is a prerequisite to an infringement suit. Subject to certain exceptions, the remedies of statutory damages and attorney's fees are not available for infringements occurring before registration. The notice of the copyright generally includes the symbol © or the word *copyright* or the abbreviation *COPR.*, the year of the publication, and the name of the copyright owner.

The procedure for registration requires that an application be filed in the Copyright Office accompanied by the requisite fee and copies of the copyrighted work. Special copyright registration forms are available from the Library of Congress for each of the seven works of authorship that can be protected by copyright.

C Scope of Copyright Protection and Commercial Practical Realities for VLSI

As stated previously, a copyright protects original literary, musical, or other artistic work against unauthorized copying or reproduction. Again, it should be noted that the copyright protection extends only to the particular expression, description, explanation or illustration of the subject matter. It does not extend to ideas, procedures, processes, systems, methods of operation, concepts, principles, or discoveries revealed in the work regardless of the form in which it is described or explained therein. Thus, according to Nimmer, a noted expert on the law of copyrights, mechanical devices that cannot qualify as pictorial, graphic, or sculptural works are not writings and, therefore, are not eligible for copyright protection. However, if the VLSI circuits form the object code of a computer program, then the situation may be different. In *Midway Mfg. Co. v. Strohon,* 26 PTCJ 165 at 166 (N.D. Ill.) June 1, 1983, the court stated that the current copyright legislation is intended to protect the object code as well as the source code. The court stated:

What legislative history there is of the 1980 amendments confirms that protection for object code was intended.... Whatever its merit as an original matter, the 1976 Copyright Revision Act has foreclosed the argument that the object code, meant to be read by machines rather than humans and incomprehensible to any but highly trained computer specialists, is not a proper subject for copyright protection. Since as we understand it, the object code is nothing other than a direct transformation of a computer program composed, at considerable expense of creative effort, in source code, the evident congressional policy determination that both forms of the computer program are protectable is an emminently sound one. To allow protection of the source code version of a program would by pyrrhic indeed if the object code version, the mechanical implementation of the same program, stored and marketed on disks or tapes, for example, could be freely reproduced without constituting an infringement.

Regarding the argument that silicon chips are themselves a form of computer circuitry or hardware and are, therefore, utilitarian objects for which copyright protection is not available, the court stated: "We see no basis for concluding in effect that object code stored on a silicon chip may be freely copied while the same code stored on a tape or disk may not be."

Further, in *Apple Computer Inc. v. Formula International Inc.*, 26 PTCJ 27, (U.S. D.C. C.D. CALIF.) April 25, 1983, the plaintiff, Apple Computer, accused the defendant, Formula, of infringing the copyrights covering five computer programs that are embodied in ROM and diskettes. The court granted the protection requested and stated that any doubt as to whether the Copyright Act protects all computer programs is removed by examining the legislative history of the 1980 amendment to the Copyright Act and particularly the report of the National Commission on New Technological Uses of Copyrighted Works (CONTU). The court found that CONTU recommended that *all* computer programs be included in copyright protection regardless of the method in which they were fixed or the function they performed. The court also stated that public policy also supports the ruling that all computer programs are subject to copyright protection.

Though these cases tend to point out that computer hardware representing object codes that store program information is protectable by copyright, it should be remembered that the copyright statute specifically includes computer programs. Whether this reasoning could be extended by analogy to VLSI topology is doubtful, at least at present. Since the courts are now holding that object codes are protectable by copyright even though they are embodied in a hardware configuration, it may be a logical extension of that doctrine to argue that the novel layout of VLSI topology is also copyrightable. No cases are known to the author that would sustain this extension. It is mentioned here simply as an exercise in conjecture that might be considered by one whose VLSI topology is being copied. At any rate, it is recommended that the owner of VLSI topology nevertheless place a copyright notice on the chip embodying the hardware technology and on the packages in which the chip is sold to preserve this avenue of protection for the future.

III TRADE SECRET PROTECTION

A What Is a Trade Secret?

A trade secret has been defined as "any formula, pattern, device, or compilation of information which is used in one's business and which gives him an opportunity to obtain an advantage over competitors who do not know or use it" [9]. The subject matter of a trade secret *must be secret*. Matters of public knowledge or of general knowledge in an industry cannot be appropriated by someone as a legally protectable trade secret. Matters that are completely disclosed by the goods that one markets cannot be a legally protectable trade secret [10].

B Scope of Trade Secret Protection

A trade secret protects its owner from the unauthorized conduct of another in learning, disclosing, and using the trade secret. The existence of a trade secret is a question of fact that must be determined in each particular case by the tryer of fact. Trade secrets could, therefore, be used as part of a license to use equipment involving the trade secrets but would not be used in a sale of the equipment, because the sale turns over to the buyer the trade secret in an unrestricted manner. It should be noted that if trade secrets are embodied in programs that are licensed, the licenses should not be called *leases* because the lease of a computer program may be construed as a publication that could destroy the trade secret form of protection.

C Advantages and Disadvantages of Trade Secret Protection

One disadvantage of trade secret law is that it is nonfederal and, therefore, varies widely from state to state. An advantage, of course, is that

52. How to Protect VLSI Intellectual Property

ideas can be protected as trade secrets under contract law and unfair competition law where patents and copyrights would not provide any protection. Another disadvantage of the trade secret law is that the contractual arrangement between two parties does not apply to an innocent third party. Therefore, unless the third party is somehow involved with the party bound to secrecy and induces that party to violate the secrecy agreement, then the party owning the trade secret may have no cause of action against the third party. Further, the third party may be an innocent user of the information given to them improperly by the party bound to secrecy. Again, in this context, the owner of the trade secret would have no cause of action against the innocent third party. Another disadvantage is that any agreement requiring secrecy may fail if the trade secret somehow becomes public through no wrongdoing of the contracting party.

Theories for trade secret protection include breach of confidence, property right, protection of discoveries, unfair competition, and implied contract [11].

The following forms of relief at law may be available to a plaintiff.

(1) Compensatory damages for breach of express or implied contract.
(2) Punitive damages.
(3) Damages for deceit or tortuous inducement to breach contract.
(4) Costs of the suit, which may include attorney's fees.

These remedies are at law. In addition, relief may be available in equity as follows.

(1) Preliminary and/or permanent injunction against use or disclosure of the trade secret.
(2) Enforcement of an express or constructive trust leading to an accounting.
(3) Specific performance of a contract not to compete.
(4) Specific performance of a contract not to reveal or use the trade secret.
(5) Reconveyance of the trade secret.
(6) An injunction to compel surrender of the embodiments of the trade secret.
(7) Action for restitution.
(8) Rescission of employment contract, if still in force.
(9) Refusal to pay employees compensation for services rendered.
(10) Costs.

There are however many defenses available in trade secret suits. The following is a list of some of these defenses.

(1) Trade secret could be independently ascertained from inspection or analysis of the marketed product or service [12].

(2) Secret was familiar to the industry (common knowledge) [13].
(3) Secret could be found in the prior art [14].
(4) Secret had been revealed by testimony in prior suits [15].
(5) Secret was merely an idea to which defendant had free access [16].
(6) Defendant learned secret prior to disclosure by alleged trade secret owner [17].
(7) Trade secret owner failed to safeguard the secret property [18].
(8) Defendant is a bona fide purchaser for value without notice of the secrecy [19].
(9) Latches [20].

There are many other defenses; those listed above are only typical and by way of example.

D Practical Do's and Don'ts To Establish and Maintain Enforceable Trade Secret Protection

Each company or the owner of a trade secret should provide a written general policy statement for all employees pointing out that the purpose of the policy is to state company policy regarding protection and preservation of company trade secrets and confidential and proprietary information (which includes inventions that may be considered patentable but on which patents have not yet been filed or issued).

The company policy in practice should be well-defined in clear and concise terms that clearly define *trade secret, confidential,* and *proprietary information*. It should be pointed out that these terms refer both to secret or proprietary information and inventions made by company employees and to sensitive financial information, methods of pricing, company business, research, new product plans, unpublished sales of the company, profits, pricing information, and methods, as well as any other unpublished financial or pricing information received or developed under contracts with third parties.

Each company employee has a clear obligation to protect company confidential information. Employees should know that confidential information may be disclosed to other employees of the company only on the strict "need to know" basis.

Any employee desiring to present papers at technical seminars or other meetings of such nature should be required to notify the company in writing and to secure the advance approval in writing of his or her cognizant superior prior to responding to a request for technical papers or other invitations.

The company should take the following steps to protect valuable trade secrets.

52. How to Protect VLSI Intellectual Property

(1) Enclose the plant with a patrolled gate with limited, guarded entries.
(2) Enclose security areas and post signs restricting entry.
(3) Shred and burn all confidential waste papers.
(4) Change door and desk locks periodically.
(5) Use code names or code numbers to identify special projects.
(6) Restrict circulation of confidential material.
(7) Number the copies of confidential materials and keep distribution lists.
(8) Destroy the copies of confidential information and materials when no longer necessary.
(9) Take additional precautions when planning a special project.
(10) Mark appropriate material *confidential*.
(11) Be cautious in filing foreign patent applications.
(12) Specifically instruct tour guides to avoid visiting unauthorized areas.
(13) Investigate employees who are in particularly sensitive positions.
(14) Inform future employers about trade secrets and agreements thereon when they check for references regarding terminated employees.
(15) Keep employees satisfied.
(16) Be alert to disgruntled or irresponsible employees.

In addition, employees should be required to do the following:

(1) Become educated about the importance of confidential information, with periodic reminders. This education should include entry and exit interviews.
(2) Upon initial employment have the employee execute agreements not to disclose trade secrets and, where appropriate, including covenants not to compete.
(3) Confidential information should not be discussed with visitors or employees of other companies.
(4) Confidential information should be disclosed only on a "need to know" basis.
(5) Question any unfamiliar person in restricted areas.
(6) Avoid discussing confidential matters over telephone or in public places (e.g., on airplanes).
(7) Submit all papers and publications for company approval screening.
(8) Keep confidential materials in locked places.
(9) Terminating employees should sign an agreement confirming that all company papers, reproductions, or descriptions thereof have been returned to the company.
(10) Wear visible identification tags.

Guards should be instructed to do the following:

(1) Be educated about measures to maintain security and its importance.
(2) Not permit themselves to be brow beaten by visitors.
(3) Patrol offices at night, and check that doors and desks are locked and that no papers are left lying about.

IV OTHER KEY ASPECTS IN THE PROTECTION PROGRAM

A Employee Relations Agreements: Exit Interviews

Each company supervisor should be held responsible for insuring that any terminating employee under his supervision is advised at the time of his termination during an exit interview of his continuing obligation to protect company confidential information after termination of employment. The supervisor should prepare a written record of such trade secret termination exit interviews, which should be signed by the terminating employee, and the interviewing supervisor should preserve a written record of what transpired at the exit interview.

B Visitor Agreements

Visitors should be required to sign visitor agreements acknowledging that information disclosed to them is proprietary and confidential and that the visitor is bound to maintain that confidentiality.

Visitors should also be required to do the following:

(1) Check in at the gate.
(2) Show personal identification.
(3) Wear visible identification tags.
(4) Leave unnecessary packages with guards.
(5) Be escorted at all times.
(6) Make appointments in advance.
(7) Not be taken to security areas.

C Vendor Agreements

Outside suppliers, contractors, and consultants should be required to execute specific written agreements concerning maintaining confidentiality of information given to them in requests for quotes, and the like.

52. How to Protect VLSI Intellectual Property

In addition, the owner of the confidential information should

(1) Avoid dealing with persons or companies who deal with competitors.
(2) Deal only with reputable firms.
(3) Do not give unnecessary information to outsiders.
(4) Restrict the movement of outside truckers, loaders, and vending machine service men.
(5) Advise suppliers that any plans given them contain secret or confidential information if such is the case.

D. Use Log Books

It should be imperative that all employees maintain engineering notebooks for recording ideas that are important to the company business. These notebooks should be the type that are bound and ruled, with each page numbered in sequence. The notebooks should include the name of the engineer or user of the notebook, and each page should be read, signed, and dated by another employee to indicate that on that date the information recorded on that page was explained to and understood by him, the one who is witnessing the page. Such notebooks are immensely valuable in determining who is the first inventor of a particular idea. This, of course, may include ideas or apparatus concerning VLSI circuits.

V. SUMMARY

Patent protection is the most advantageous manner of protecting patentable inventions including those involving VLSI circuits. It is, however, extremely expensive to litigate patents, and counsel should always attempt to negotiate a fair and reasonable settlement of disputes by the parties rather than engage in litigation.

Copyright protection applies to programs that may be associated with VLSI circuits but would probably be difficult to obtain on the VLSI topology itself.

Trade secret protection is a valuable and viable method of protecting VLSI circuits and information relating thereto, provided that the company takes adequate steps to establish and protect confidential information within the company so that it can be maintained as a trade secret.

Finally, by educating employees in their dealings with vendors and visitors and by reminding the employees of their obligations to the company during both entrance and exit interviews, valuable company confidential information can be reasonably protected.

Remember, protection of valuable VLSI technology requires constant and cost effective vigilance.

REFERENCES

1. 35 USC §154.
2. 35 USC §261.
3. 35 USC §111.
4. 35 USC §113.
5. 37 USC §84.
6. 37 USC §65.
7. 37 USC §68.
8. 37 USC §202.3.
9. Restatement, Torts (1939), Section 757, Comment (b).
10. Roger M. Millgram, "Trade Secrets", Vol. 1, §2.01 pp 208, (1982).
11. *Dupont v. Masland,* 244 US 100, 37 S.Ct. 575 (1917); *Glass v. Kottwitz,* 297 S.W. 573 (Court of Civil Appeals Texas, 1927); *Peabody v. Norfolk,* 98 Mass. 452, 96 AM. DEC. 664 (1868); *LaFrance v. Hart,* 8 Conn. Supp. 287 (1840); and *O & W THUM v. TLOCZYNSKI,* 114 Mich. 149, 72 N.W. 140 (1897).
12. *Carver V. Harr,* 132 N.J. eq. 207, 27 A.2nd 895 (1942).
13. *National Starch Products Inc. v. Polymer Industries, Inc.,* 273 app. div. 732, 79 NYS 2nd, 77 USPQ 644, (1948).
14. *Taber v. Hoffman,* 118 N.Y. 30, 23 N.E. 12 (1889).
15. *Palmeroy Inc. Co. v. Palmeroy,* 77 N.J. eq. 293, 78a 698 (1910).
16. *Heyman v. Ar. Winarick et al,* 140 USPQ 403 (CA 2nd 1963).
17. *Lee v. Sanburn,* 94 USPQ 153 (Cal. Superior Court 1952).
18. *Hamilton Mfg. Co. v. Tubbs Mfg. Co. et al,* 216 F. 401 (CC Mich. 1908).
19. *Chadwick v. Covell,* 151 Mass. 190, 23 N.E. 1068 (1890).
20. *Globe Ticket Co. v. International Ticket Co.,* 90 N.J. eq. 605, 104a 92 (1918).

Index

A

Aberrations, in electron optics, 359–360
Abstraction levels, in integrated circuit design automation, 40
Acoustic technology, radar applications, 719
ac parametric tests, 530
A/D conversion, for complex medical instrument, 770
Address decode faults, 542
Address multiplexing, dynamic RAM and, 159
Address sequencing faults, 542
Air, for circuit fabrication facility, 88–89
Algorithmic pattern generation, 542
Alignment marks, in direct writing, 354
Aluminum
　electromigration and, 441–443
　film deposition methods for, 446
　hillock formation and, 443–444
　reasons for choosing, 436
　silicon diffusion and junction spiking and, 436–441
　sputtering for alloy deposition, 498
　step coverage and, 444
Ammonia, as nitridating ambient, 402
Analog-to-digital converter
　for doppler flowmeter, 811–813
　radar applications, 719
Analog VLSI, radar applications, 714–719
AND-OR functions, junction-controlled logic gate and, 634–635
Anisotrophy, 488
　mechanisms for producing, 491–492
　parameters affecting, 491
Annealing, 458–459
　research in, 459–460
Anti-aliasing filter, for doppler flowmeter, 814
Aperture, numerical, in optical patterning, 330

Array(s)
　cell, 44
　for image processing
　　discrete Fourier transform, 795–796
　　LSI and VLSI processing systems, 797–799
　　spatial domain operations, 796–797
　minimizing types of, 672
　for pattern recognition
　　matrix operations, 789–791
　　statistical, 791–793
　　syntactic, 793–794
　switched-capacitor, 849
Array interfaces, understandability of, 672–673
Array yield, 674
Astigmatism, in electron optics, 359
Autodoping, 308
　in epitaxial growth, 292–293
Automobile applications, 851–852
　benefits of, 852–853
　body computers, 859–861
　cellular telephones, 862–863
　electronics market and, 853–854
　for engine controls, 855–859
　entertainment systems, 861–862
　government controls and fuel economy and, 852
　reliability of, 854–855

B

Backside metallizations, sputtering and, 499
Backup oscillator, for pacemaker, 752
Bake, input-output failures and, 554
Barrier layers
　interdiffusion of, 448–449
　lowering, 618
　practical, 449–452
　reason for using, 447

requirements for, 447–448
sputtering and, 499
tunneling through, 618–619
use in bipolar and MOS technologies, 447
Battery characteristics, for pacemaker, 748–749, 752–754
Battery life detection for pacemaker, 750–751
Beam writing, *see* Writing beam
Bias sputtering, 446
Binary searching, 540–541
Bipolar circuits
  digital gate, 113
    CML circuits, 116
    comparison logic circuits, 119–120
    ECL circuits, 115–116
    IIL (MTL) circuits, 116, 119
    STL circuits, 119
    TTL circuits, 113–115
  epitaxial films in, 287
  noise and, 608–610
  power-delay performance of MOS circuits and, 107
  use of barrier layers in, 447
Bipolar transistors
  junction (BJT), 109
  npn and *pnp,* 109–111
  Schottky barrier, 112
Bit growth, random access memory and, 151–152
Bit-slice processors, 833
BJT, *see* Bipolar transistors, junction
Block cryptographic method, 697–699
Block ping pong, 545
Blood circulation, pacemakers and, 738
Body computers, 859–861
Boron diffusion, blockage by nitride films, 411
Box, *see* Buried oxide
Breakdown, 617
  oxides and, 386–387
Breakdown voltage test, 539
Buffered FET logic circuits, 652–653
Buried channel formation, ion implantation and, 467–468
Buried layers, 287–289
  formation by ion implantation, 473–474, 480, 508–509
Buried oxide (Box) technique, 394–395
Burst mode, 146
Burst noise, 605
Bus scaling, 147–148
Butterfly pattern, 545

C

Cache, 145
  burst mode and, 146
  types of, 145–146
CAD, *see* Computer-aided design
CAE, *see* Computer-aided engineering
Capacitance–voltage (C–V) characteristics
  curve, oxide charges and, 389, 391
  of nitride films, 409–411
Capacitance–voltage (C–V) technique, 313–314
  pulsed, 314
Cardiac pacer systems, *see* Pacemaker
Carrier mobility, of epitaxial films, 298
CBC, *see* Cipher block chaining
CCD, *see* Charge coupled device
Cell array, 44
Cell faults, 542
Cell library, *see* Macro library
Cellular telephones, 862–863
CFB, *see* Cipher feedback
Channel length, drain current versus, 131–132
Characterization, to integrated circuit design, 46–47
Charge carrier mobility
  conductivity mobility, 181–184
  Hall mobility, 181
Charge coupled device (CCD), 582–583
  basic imager, 588–589
  frame transfer, 585
  interline transfer, 585
  line-by-line transfer, 587–588
  multiphase type, 591–592
  photosensitive, 583
  quadrilinear, 585
  radar applications, 715–716
  two phase, implanted barrier types, 589
  virtual phase type, 589–590
Charge injection device (CID), 588
Charge redistribution circuits, 124
Checkerboard pattern, 543
Chemical drainage, for circuit fabrication facility, 93
Chemical vapor deposition (CVD), 302–303, 306–307, 446
  plasma enhanced (PECVD), 499
    applications of, 501–502
    equipment configurations for, 499–500
    of nitrides, 404–405
    typical process, 500
Chip(s)
  clock, 148

Index

advantages over wait states, 148–149
  cost of, 134
  DSP, 834
  FSK, 833
Chip architectures, 582
  basic CCD area imagers and, 588–589
  charge coupled device and, 582–583
  line-scan imagers and, 583–585
  photoelements and, 582
  television imagers and, 585–588
Chronaxie, 740
CID, *see* Charge injection device
Cipher block chaining (CBC), 698–699
Cipher feedback (CFB), 699–701
Circuit(s), *see also* Bipolar circuits; Gallium arsenide circuit technology; Gate circuits
  charge redistribution, 124
  complexity of, density and, 5, 7
  ECL
    gate, 115–116
    noise and, 610
  integrated, low-volume, 354
  isolation of, by nitride films, 411–412
  logic
    buffered FET, 652–653
    comparison, 119–120
    direct-coupled, 653–654
    Schottky diode FET, 652–653
  protection of, for pacemaker, 751–752
Circuit characteristics, of pacemaker, 747–748
Circuit complexity, circuit density and, 5, 7
Circuit density, 1
  device and circuit complexity and, 3, 5, 7
  die size and, 2
  feature size and, 3
  parametric testing and, 517–518
  process complexity and, 7
Circuit design, 9–10, *see also* Integrated circuit design
  elements of, 14
    basic inverter, 16–21
    clocking schemes, 28
    distributed propagation delay, 23–24
    examples of building blocks, 24–25
    MOS transistor, 14–16
    NAND and NOR logic gates, 21
    other elements, 21–23
    regular design structures, 26–27
  future developments in, 35
  in hierarchical decomposition, 12

layout design and, 28–29
  automated layout and, 33–35
  design rules and, 29–30
  physical layout and, 30–32
methodology for, 10
  computer-aided design and, 13–14
  hierarchical decomposition and, 11–13
Circuit extraction, computer-aided design tools and, 62
Circuit fabrication facility
  clean air for, 88–89
  electrical power for, 91
  equipment for, 91–92
  gases for, 90–91
  management system and, 97
  materials for, 92
  personnel for, 92
    efficiency of, 96–97
    safety of, 95–96
  physical considerations for, 95
  protection for
    equipment fire protection, 96
    personnel safety, 95–96
    zones of cleanliness and, 96
  vacuum for
    house, 92
    housecleaning, 93
    process, 92–93
    process equipment vacuum pumps and, 93
  waste disposal for
    chemical drainage, 93
    floor drainage, 95
    fume exhaust, 93–94
    solvent exhaust, 95
  water for
    cooling, 90
    deionized, 89–90
Circuit modeling, parametric testing and, 517
Circuit simulation, 13–14
  computer-aided design tools and, 60
  foundry and, 71–73
Circuit timing, with CMOS technology, 122
Cleanliness zones, for circuit fabrication facility, 96
Clock chips, 148
  advantages over wait states, 148–149
Clocking schemes, inverter and, 28
Closed-loop controllers, for neural stimulators, 734–735
CML gate circuits, 116

Index

CMOS, *see* Complementary metal-oxide semiconductor
Code book, electronic, 697–698
Color technology, imagers and, 593
Column row surround disturb, 546
Communications, distributed, *see* Distributed communications
Comparison logic circuits, 119–120
Complementary metal-oxide semiconductor (CMOS), 121
  advantages of circuit design with, 122–124
  circuits, 99–100
    gate, 105–106
    noise and, 606–607
  discontinuities in, 135–138
  interconnects for, 133–135
  inverter and, 18–21, 104
  ion implantation and, 468
  optimization of processing and, 131–133
  for pacemakers, 743–744
  state-of-the-art process flow, 124–131
  transistors, 102–103
Computer-aided design (CAD)
  integrated systems for, 64
  tools of, 13–14
    automated placement and routing and, 61
    automatic test equipment, 63
    availability of, 63–64
    circuit extraction and, 62
    circuit simulators, 60
    design analysis and, 57, 59
    design capture and, 56–57
    design manipulation and, 62–63
    design processing and, 57
    interactive layout editors, 60–61
    layout verification and, 61–62
    logic simulators, 59–60
    netlists and, 59
    pattern generation and, 63
    schematic diagrams and, 58–59
    timing verification and, 60
    turnkey, 64
Computer-aided engineering (CAE) tools, 58–60
Computer-aided layout tools, 60–62
Computer-aided manufacturing, 62–63
Computerized tomography
  architecture of scanners for, 727
  types of scanners for, 726
  use of VLSI for, 728
Conduction, of nitride films, 407–409

Content addressable memory, 163
Contrast, in optical patterning, 329
Control(s), engine, 855–859
Controllers, peripheral, 149
Control logic, separating from data logic, 672
Cooling water, for circuit fabrication facility, 90
Copyrights
  definition of, 873–874
  how to obtain, 874
  scope of protection, 874–876
Corner frequency, 606
Correlation plots, 525–526
Coupling equations, for program and erase in EEPROM, 170–172
Crosstalk
  of imagers, 593
  noise and, 612
Cryptography, 696
  block method, 697–699
Current drain, neural stimulators and, 734
Current injection logic devices, 635–636
CVD, *see* Chemical vapor deposition
C-V technique, *see* Capacitance-voltage technique
Cycling, EEPROM and, 173–174

D

Damage gettering, ion implantation and, 476
Dark current, of imagers, 598–599
Dark signal, of imagers, 598–599
Data accumulation, for complex medical instrument, 771–772
Data base check, in hierarchical decomposition, 13
Data base formats, 71
Data base transfer, 70–71
Data encryption standard (DES)
  basic operation, 696–697
  block method, 697–699
  stream method, 699–702
Data logic, separating from control logic, 672
Data rates, as incentive for VLSI circuits in modems, 828
Data retention faults, 543
Data sensing faults, 542–543
DCL, *see* Direct-coupled logic
dc parametric tests, 530

Index

Decapsulation, input-output failures and, 554
Deception, as electronic countermeasure, 690–691
Decoders, static RAM and, 156–157
Dedicated-wafer processing, 73
Deep ultraviolet radiation (DUV), in optical patterning, 338
Defect(s), *see also* Error(s); Failure(s); Fault(s)
  in epitaxial films, 321–324
    measurement of, 324–325
    structural, 296
  nonfunctional, 561–563
  on masks, 353
Defect density, 387, 389, 396
Defibrillation protection, for pacemaker, 751
Deflection aberrations, in electron optics, 359–360
Deionized water, for circuit fabrication facility, 89–90
Delayering techniques
  input–output failures and, 556
  single node failures and, 557
Delay path analyzer, 14
Delid, input–output failures and, 554
Density, parametric testing and, 517–518
Depletion-mode devices, 102
Deposition chamber, 293
Deposition rate, 494
  in epitaxial growth, 290–292
DES, *see* Data encryption standard
Design analysis, computer-aided design tools and, 57, 59
Design capture, computer-aided design tools and, 56–57
Design errors
  locating and correcting, 673–674
  minimizing, 671–673
Design interface, foundry and, 71
Design manipulation, computer-aided design tools and, 62
Design processing, computer-aided design tools and, 57
Device complexity, circuit density and, 3, 5
Device development, parametric testing and, 517
Device schmooing, 540
Dielectrics
  isolation processes and, 573–574
  sputtering and, 499

Die size, circuit density and, 2
Differential sense amplifier, 25
Diffraction blur, in x-ray proximity printing, 367
Diffusion
  boron, blockage by nitride films, 411
  boundary versus bulk, in metallization layers, 448
  radiation enhanced, ion implantation and, 481–482
Diffusion length, silicon and, 187–188
Digital signal processor chips, speech processing and, 781
Digital VLSI, radar applications, 711–714
Direct-coupled logic (DCL) circuit, 653–654
Direct-coupled logic (DCL) gate, 636
Direct writing, on wafers, 354
Displacement, radiation-induced, 573
Distributed communications, 837–838
  future trends in, 849–850
  local area networks, 841–847
  signal processing and, 847–849
  VLSI technology and trends and, 838–841
Distributed propagation delay, inverter and, 23–24
Divergences, aluminum metallization and, 442
Documentation, 677
Dopant density, conversion between resistivity and, 178–181
Dopant profile
  of epitaxial films, 300–301
  measurement of, 312–314
Doping, 308
  in epitaxial growth, 292–293
Doping density, 391
Doppler flowmeter, 807
  implementation of system major blocks, 811–815
  in-vivo data, 819–820
  system description, 809
  system integration, 815–819
  ultrasound and, 807–809
Drain current, channel length versus, 131–132
Drift velocity, 442
  dependence on electric fields, 185–186
DSP chips, 834
Dual-port RAM, 161–162
Dynamic latch, 24
Dynamic memories, 124

E

E-beam, *see also* Electron beam lithography
  IX masks, 70
  pattern generator, 352
ECB, *see* Electronic code book
ECL circuits
  gate circuits, 115–116
  noise and, 610
Edge gradient, in optical patterning, 330
EEPROM, *see* Electrically erasable programmable read only memory
Efficiency, in x-ray lithography, 367–368
Electrical characterization
  of epitaxial films, 310–316
  testing and, 553
Electrical current, *see also* Voltage
  drain, neural stimulators and, 734
  operating, for pacemaker, 748
Electrically erasable programmable read only memory (EEPROM)
  in an array, 169
  coupling equations for program and erase and, 170–172
  FLOTEX, 168
  Fowler–Nordheim tunneling and, 169–170
  memory cell structure and, 167–169
  performance and reliability of
    read disturb and, 174–175
    reliability, 175
    retention and, 174
  programming characteristics of
    cycling and electron trapping, 173–174
    threshold voltage window as function of program voltage and time, 172–173
  scaling of, 175–176
  use of nitrides in, 413–414
Electrical power, for circuit fabrication facility, 91
Electrical properties
  of epitaxial film defects, 321, 323
  of epitaxial films, 297–301
Electrical retesting, input–output failures and, 555
Electrical stimulation, cardiac, 739–740
Electric fields
  dependence of drift velocity on, 185–186
  effects on C–V curves, 409–410
Electrocardiogram, 740–741
Electromigration, aluminum metallization and, 441–443
Electron backscattering, 355–356
Electron beam lithography, 343, 351
  direct writing and, 354
  electron optics and, 357–360
  mask making and, 352–353
  proximity effect in, 346–347
  raster scan and, 360–361
  resists and, 354–357
    multilevel, 347
    negative, 344–345
    positive, 345–346
  vector scan and, 362–364
Electron gun, in x-ray lithography, 369
Electronic code book (ECB), 697–698
Electronic warfare (EW)
  basic definitions, 681–682
  electronic counter–countermeasures and, 691–694
  electronic countermeasures and, 690–691
  functional divisions of, 682
Electronic warfare support measures (ESM), 682
  design considerations and, 687–689
  emitter fingerprinting and, 687
  emitter location and, 689
  probability of intercept emitters and, 687
  pulse sorting technology and, 683–685
  pulse testing and, 685–686
    memory and logic in, 686
  receiver characteristics and, 683
Electron optics
  aberrations and, 359–360
  electron beam generators and, 357–358
  light optics rules applicable to, 358
  spot sizes and, 358–359
Electron storage ring, in x-ray lithography, 369–370
Electron trapping, EEPROM and, 173–174
Electron wind effect, 441–442
Emission controls, 852
Emitter
  fingerprinting, electronic warfare support measures and, 687
  location of, electronic warfare support measures and, 689
Employee relations agreement, 880
Encryption applications, 695
  commercial device implementations and, 703–704
  cryptography and, 696

data encryption standard and, 696–702
future impact of, 704
public keys and, 702–703
End of battery life detection, for pacemaker, 750–751
Energy, minimum required per logic operation, 620–621
Engine controls, automotive, 855–859
Engineering, computer-aided, tools of, 58–60
Enhancement-mode devices, 100, 102
Entertainment systems, in automobiles, 861–862
Epitaxial films, 305–306
  defects in, 321–324
    measurement of, 324–325
    structural, 296
  development of, 286
  electrical characterization of, 310–316
  electrical properties of, 297–301
  physical and optical characterization, 316–321
  standards for, 296
  types of, 286
Epitaxial growth, 286–287, 306
  buried layers and, 287–289
  chemical processes in, 289–292
  doping and autodoping in, 292–293
  equipment for, 293
  fabrication steps, 287–288
  of gallium arsenide films, 301–303
  impurity transport during, 307–310
  lateral overgrowth, 510–511
  procedure for, 294–296, 306–307
Epitaxy
  chemical vapor, *see* Chemical vapor deposition
  definition of, 285–286
  liquid phase, 301–302
  molecular beam (MBE), 307
Equipment
  for chemical vapor deposition, 499–500
  for circuit fabrication facility, 91–92
  for epitaxial growth, 293
  for etching, 490–491
  for sputtering, 494–497
  testing
    computer-aided design tools and, 63
    parametric, 521–525
    of semiconductors, 531–538
Erase condition, EEPROM and, 170–171
Error(s), *see also* Defect(s); Failure(s);

Fault(s)
  locating and correcting, 623–624, 673–674
  for complex medical instrument, 773–774
  minimizing, 671–673
  on x-ray mask, 371–372
  soft, dynamic RAM and, 160
Error rate, as incentive for VLSI circuits in modems, 829
ESM, *see* Electronic warfare support measures
Etching, 488
  anisotrophy and, 488
    mechanisms for producing, 491–492
    parameters affecting, 491
  equipment for, 490–491
  etch profiles and, 488–490
  ion implantation and, 476–478
  selectivity and, 488
    mechanisms for producing, 491–492
    parameters affecting, 491
Evaporation deposition, 446
Evoked potentials, 731
  analysis of, 732
  VLSI versus hard-wired averagers for, 731–732
EW, *see* Electronic warfare
Exclusive OR gate, 25
Exit interviews, 880
Exposure systems, x-ray lithography and, 377–378

F

Fabrication, *see also* Circuit fabrication facility; Manufacturing
  in integrated circuit design automation, 38–39
Fabrication error detection, 673–674
Failure(s), *see also* Defect(s); Error(s); Fault(s)
  memory, 542–543
Failure analysis, 551–552
  initial nondestructive procedures in, 552–553
  input–output failures and, 553–556
  nonfunctional defects and, 561–563
  single node failures and, 556–561
Fault(s), *see also* Defect(s); Error(s); Failure(s)
  cell, 542

data retention, 543
data sensing, 542–543
Fault simulation, 43
Fault tolerance, 674
Feature size, circuit density and, 3
Feedback
 cipher, 699–701
 output, 701–703
Film formation
 chemical vapor deposition of nitrides and, 404–405
 deposition methods for aluminum, 446
 implantation and, 407
 thermal nitridation of silicon and, 402–403
 thermal nitridation of silicon dioxide and, 405–407
Filter bank estimation, speech processing and, 780–781
Filter design, electronic counter–countermeasures and, 692–693
FIPOS, see Full isolation by porous oxidized silicon
Fire protection, for circuit fabrication facility, 96
Flattening process, 43
Flicker noise, 605–606
Floating gate, in MOFSET, 168
Floor drainage, for circuit fabrication facility, 95
FM, see Frequency modulation detection
Formants, 777
Foundry, see Silicon foundry
Fourier transform infrared spectrometry, 320–321
Fowler–Nordheim tunneling, 169–170
Frequency modulation (FM) detection, 801–803
 system implementing example of, 807–820
 VLSI approach to, 803–807
Frequency-shift keying detection, 806–807
FSK modem chips, 833
Fuel economy, 852
Full isolation by porous oxidized silicon (FIPOS), 509–510
Fume exhaust, for circuit fabrication facility, 93–94
Functional simulation, 43
Functional testing, 39, 539
 on semiconductors, 530–531

G

Gallium arsenide circuit technology
 advanced, 655–656
 MESFET circuits, 651–654
 MESFET devices, 647–649
 performance and applications, 654
 planar fabrication process, 649–651
 properties of, 645–647
Gallium arsenide devices, radiation hardness of, 577–578
Gallium arsenide films, epitaxial growth of, 301–303
Galloping patterns, 545
Gamma (contrast), in optical patterning, 330–331
Gases, for circuit fabrication facility, 90–91
Gate array approach, to integrated circuit design, 45–46
Gate array layout systems, 34
Gate circuits
 bipolar digital, 113
  CML circuits, 116
  comparison logic circuits, 119–120
  ECL circuits, 115–116
  IIL (MTL) circuits, 116, 119
  STL circuits, 119
  TTL circuits, 113–115
 CMOS, 105–106
 NMOS, 104–105
Gate electrode resistance, reduction of, 134
Gate interconnect, sputtering and, 499
Gate length, threshold voltage versus, 132
Gate sidewall oxide, 135
Geometry effects, ion implantation and, 460–463
Grain boundary passivation, 512
Graphics, 848–849

H

Hearing aid, digital
 advantages over analog aids, 724
 input/output subsystems of, 725
 technologies and, 725–726
HEMTs, see High electron mobility transistors
Heteroepitaxy, 306
 silicon-on-insulator and, 505–506

Index

Heterojunction bipolar transistors, 656
Hierarchical decomposition, 11–13
High electron mobility transistors (HEMTs), 655
High-layout density, with CMOS technology, 123
Hillock formation, aluminum metallization and, 443–444
Hold register, 772
Homoepitaxy, 306
Horizontal scaling, in manufacturing process, 81–83
Hot electron injection and trapping, 397
Hot spot detection, input–output failures and, 555

I

IBM 5100 processor, 663
IBM PCjr, 664–665
IBM Personal Computer, 661, 663–664
IIL (MTL) gate circuits, 116, 119
Image processing, 785–786
　architectures and systems for, 795–799
　computer-vision levels and, 786
　image segmentation and description, 788
　image transforms and enhancement, 787
　system components, 786–787
Imagers, 581–582
　CCD area, 588–589
　chip architectures and, 582–589
　fabrication technologies for, 589–593
　line-scan, 583–585
　performance and design variables for, 593–601
　television, 585–588
Implantation, film formation and, 407
Impurities, electrically active, in epitaxial films, measurement of, 314, 316
Impurity atoms, transport during epitaxial growth, 307–310
Inductance, of superconducting circuits, 640–642
Infrared reflectance technique, for epitaxial films, 316, 319
Input impedance, pacemaker and, 749
Input noise, doppler flowmeter and, 815–818
Input–output failures, 553–554
　bake and, 554
　decapsulation and delid and, 554
　nondestructive tests and, 554–555
　second destructive testing and, 556
Input/output interface device, for complex medical instrument, 769
Input/output structure, of engine control module, 856–857
Input sensitivity test, for pacemaker, 755
Integrated circuit design
　automation of
　　building of macro library and, 46–47
　　design and, 37–38
　　fabrication and, 38–39
　　gate array approach, 45–46
　　goals of, 39–40
　　levels of abstraction and, 40
　　semicustom layout and, 47–48
　　semicustom methodologies for, 44–46, 48
　　standard cell approach, 44–45
　　trends in, 48–50
　　typical procedure for, 40–44
　computer tools for, 55–56
　　availability of, 63–64
　　design analysis and, 57
　　design capture and, 56–57
　　design processing and, 57
　　in engineering, 58–60
　　in layout, 60–62
　　in manufacturing, 62–63
Intel 4004, 658
Intel 8008, 658–659
Intel 80186, 662
Intel 80286, 662
Intel 8080, 659–660
Intel 8086, 661
Intel 8088, 661
Interactive delay calculator, 677
Intercept emitters, electronic warfare support measures and, low probability of, 687
Interconnection(s), 622–623
　for CMOS, 133–135
Interconnection delay, performance and, 668–670
Interference, 319
Interferometer, three junction, 635
Intermetallic compound formation, 448
Interval timer, for complex medical instrument, 769–770
Inversion layer, MOS, mobility in, 188–189

Inverter, 16–17
  clocking schemes and, 28
  CMOS, 18–21, 104
  design building blocks and, 24–25
  distributed propagation delay and, 23–24
  NAND and NOR logic gates and, 21
  NMOS, 17, 103–104
  other circuit elements and, 21–23
  regular design structures and, 26–27
Ion beam lithography, ion implantation and, 483–484
Ion implantation, 456
  anomalous distributions and annealing behavior and, 457–460
  beam writing and ion beam lithography and, 483–484
  buried insulator layer formation and, 480
  damage gettering and, 476
  doping applications in bipolar technology
    buried collector or subcollector implantation, 473–474
    implanted resistors, 474
    predisposition and, 472–473
    process simplification, 474–476
  doping applications in MOS
    buried channel formation, 467–468
    CMOS and, 468
    example, 471–472
    graded sources and drains, 471
    junction formation, 466
    punchthrough stopper, 468
    threshold voltage control, 466–467
  enhanced etching and, 476–478
  formation of buried insulating layers by, 508–509
  ion implantors and, 465–466
  local oxidation and, 478–480
  masking and, 463–465
  radiation enhanced diffusion and, 481–482
  range distribution and, 457
  silicidation of metal–silicon reaction and, 482–483
  silicon-on-insulator using oxygen implantation and, 480–481
  small geometry effects and, 460–463
Ionization, radiation-induced
  permanent, 569–572
  transient, 572–573
Isolation techniques, input–output failures and, 556

J

Josephson junction
  constituents of, 629
  control of, 630, 632
  damped pendulum analogy and, 630
  Josephson equations and, 630
  noise and, 642
  switching speed of, 638
    other limitations on, 639
    pair-breaking rate and, 638
    scattering rate of electrons and photons and, 639
  threshold characteristics of, 632
Junction-controlled logic gates, 632–635
Junction formation, ion implantation and, 466
Junction spiking, aluminum metallization and, 439–441

L

Labeled reference patterns, speech processing and, 781
LANs, see Local area networks
Large computers, 667–668
  design considerations and, 671–674
  design tools and staff and, 674–678
  performance of, 668–671
Latch, dynamic and static, 24
Layout design, 28–29
  automated layout and, 33–35, 38
  design rules and, 29–30
  in hierarchical decomposition, 12
  physical layout and, 30–32
Layout editors, computer-aided design tools and, 60–61
Layout tools, computer-aided, 60–62
Layout verification, computer-aided design tools and, 61–62
Lead, for pacemakers, 745–746
Leakage, oxides and, 385–386
Leakage test, 539
Lifetime, silicon and, 187–188
Linear phase differentiator, for doppler flowmeter, 814–815
Linear prediction coefficient analysis, speech processing and, 780–781
Linear searching, 540–541
Line-scan imagers, 583–585
Liquid phase epitaxy, 301–302

Index

Lithography, *see also* Electron beam lithography; X-ray lithography
  ion beam, ion implantation and, 483–484
Load register, 772
Local area networks (LANs), 841–842
  architectures, 842
  examples of, 846–847
  for intensive care unit, 730
    data transmission by, 730–731
    topology of, 730
    protocols and, 845–846
    security and protection and, 846
    transmission and, 842, 845
Local oxidation of silicon (LOCOS) process
  nitride films in, 411–412
  problems with, 412
LOCOS, *see* Local oxidation of silicon
Log books, 881
Logic, data and control, separating, 672
Logic circuits
  buffered FET, 652–653
  comparison, 119–120
  direct-coupled, 653–654
  Schottky diode FET, 652–653
Logic definition, in hierarchical decomposition, 12
Logic gates
  direct-coupled, 636
  junction-controlled, 632–635
  loop-controlled, 635–636
  NAND and NOR, inverter and, 21
  performance and, 670
Logic simulation, 13, 676
  computer-aided design tools and, 59–60
Logic specification, in integrated circuit design automation, 38
Logic testing, 539–540
  output relative timings and, 540–541
  speed binning and, 540
  synchronization at reset and, 540
Logic verification, 43
Loop-controlled logic gate, 635–636
Low resistance contacts, sputtering and, 499
Low-volume integrated circuits, 354

M

Macro library, 45
  building, 46–47

Macrosimulation, 13
Magnetron, 495
  techniques and, 446
Makeup air, for circuit fabrication facility, 88–89
Management system, for circuit fabrication facility, 97
Manufacturing
  computer-aided, 62–63
  facility for, *see* Circuit fabrication facility
Manufacturing process technology, 79–80
  control and, 84–85
  directions in
    horizontal scaling, 81–83
    vertical scaling, 83–84
Marching patterns, 544
Market, for automotive applications, 853–854
Mask(s)
  fabrication of, 352–353
  in hierarchical decomposition, 13
  x-ray, 371–372
Mask alignment, x-ray lithography and, 376–377
Masking, ion implantation and, 463–465
Maximum achievable velocity, 617
Maximum current test, 539
MBE, *see* Molecular beam epitaxy
Mean-frequency estimation, 804
  digital double-correlation, 804–807
Medical applications, 722
  complex medical instrument, 763
    A/D conversion and data accumulation, 770–772
    central data processor, 772–774
    sample handling and preparation, 765–770
    system architecture, 764–765
  computerized tomography, 726–728
  evoked potentials, 731–732
  hearing aid, 724–726
  local area network for an intensive care unit, 730–731
  neural stimulators, 733–735
  pacemakers, 722–724, 737–738
    cardiac cycle and implant and, 738–742
    design considerations for, 747–756
    future trends in, 762
    pacing modalities and, 756–761
    system description, 742–747

ultrasound imaging, 728–730
Memory(ies)
  allocation of, 143
  dynamic, 124
  management of, 142
    allocation and, 143
    protection and, 143–144
    virtual memory and, 144
  nonvolatile, use of nitrides in, 413–414
  of personal computers, 665
  protection of, 143–144
  random access, *see* Random access memory(ies)
  read only, 27, *see also* Electrically eraseable programmable read only memory
Memory cells, 632–635
  for EEPROM, 167–169
  of static RAM, 154–156
Memory testing, 541
  algorithmic pattern generation and, 542
  common test patterns, 543–546
  memory failure modes and, 542–543
  topological scrambling, 546
MESFET, *see* Metal semiconductor field effect transistor
Metallization
  backside, 499
  dual-layer, 133–134
Metal-oxide semiconductor field-effect transistor (MOSFET), 99
  CMOS transistors, 102–103
  floating gate in, 168
  NMOSFET, 100–102
  PMOSFET, 102
Metal-oxide semiconductor (MOS)
  circuit technology, 99–100
    CMOS gate circuit, 105–106
    CMOS inverter, 104
    CMOS transistors, 102–103
    MOSFET structures, 100–103
    NMOS gate circuit, 104–105
    NMOS inverter, 103–104
    NMOSFET structures, 100–102
    PMOSFET structures, 102
    power-delay performance of bipolar circuits compared with, 107
  inversion layer, mobility in, 188–189
  transistor, 14
    basic operation of, 15
    current characteristics of, 16
  use of barrier layers in, 447

Metal semiconductor field effect transistor (MESFET), gallium arsenide devices, 647–649
Microarchitectural design, in hierarchical decomposition, 12
Microcrack, step coverage and, 444
Microprocessors
  in automobiles, 857, 859
  evolution of, 658–662
Microscopic inspection, input–output failures and, 555
Microsimulation, 13
Minority carrier lifetime, of epitaxial films, 298, 300
MNOS, use of nitrides in, 413
Mobility
  carrier, of epitaxial films, 298
  of charge carriers
    conductivity mobility, 181–184
    Hall mobility, 181
  minority-carrier, 186–188
  in MOS inversion layer, 188–189
  temperature dependence of, 185
Mode codes, for pacemaker, 756–759
Modems, 823–825
  commercial devices, 833–834
  custom implementations for, 830–832
  future trends, 835
  high-speed, 825, 827
  incentives for VLSI use in, 827–829
  low-speed, 825
  medium-speed, 825
  VLSI and value-added features, 834–835
Modular approach, to parametric testing, 520–521
Modulation, in optical patterning, 329
Modulation transfer function, in optical patterning, 329
Moisture, film formation and, 402–403
Molecular beam epitaxy (MBE), 307
MOS, *see* Metal-oxide semiconductor
MOSFET, *see* Metal-oxide semiconductor field-effect transistor
MOS Technology 6502, 660
Motor control, for complex medical instrument, 767–768
Motorola 6800, 660
Motorola 68000, 661
Moving inversion, 545
MTF, of imagers, 596–598
MTL (IIL) gate circuits, 116, 119
Multiple address test, 543

Index

Multiple chip implementations, modems and, 831–832
Multiplexing, address, dynamic RAM and, 159
Multiprocessor systems, performance and, 670–671
Multiproduct-wafer processing, 73

N

Netlists, computer-aided design tools and, 59
Neural stimulators
  applications of, 733
  types of implants for, 733
  use of VLSI for, 733–735
Nitridating ambients, 402
Nitridation
  advantages and disadvantages of, 406–407
  thermal, of silicon dioxide, 405–407
Nitrides, 401
  applications of, 411–414
  electrical properties of, 407–411
  film formation, 402–407
Nitrogen
  as barrier layer, 452
  as nitridating ambient, 402
Nitrogen profiles, 406
NMOS, imager fabrication, 592
NMOS circuits, noise and, 607–608
NMOSFET, 100
  depletion-mode devices, 102
  enhancement-mode devices, 100, 102
NMOS gate circuit, 104–105
NMOS inverter, 17, 103–104
Noise, 603
  alpha-particle- and cosmic-ray-induced soft errors and, 612–613
  burst, 605
  cancellation of, 818–819
  circuits and
    CMOS, 606–607
    CTL bipolar, 608–609
    ECL, 610
    I^{2}L bipolar, 609–610
    NMOS with depletion mode load, 607–608
    NMOS with n-channel enhancement mode load, 608
  crosstalk and, 612
  doppler flowmeter and, 815–818
  flicker, 605–606
  in Josephson junction devices, 642
  shot, 604–605
  thermal, 604
  threshold voltages and, 611–612
Noise immunity, with CMOS technology, 122–123
Noise window, for pacemaker, 752
Nondestructive testing, 552
  electrical characterization and, 553
  input–output failures and, 554–555
  types of electrical failures and, 553
Nonelectrical tests, on semiconductors, 531
Nonfunctional defects, 561–563
Nonvolatile RAM
  self-contained battery backup, 164
  shadow, 164–165
Nyquist resolution, of imagers, 596–598

O

OFB, *see* Output feedback
Opens test, 538
Operating currents, for pacemaker, 748
Operating voltage, for pacemaker, 748
Optical microscopy, for detection of structural defects in epitaxial films, 296
Optical pattern generator, 352
Optical patterning, 328–331
  deep ultraviolet radiation and, 338
  negative resists and, 331–333
  positive resists and, 333–338
Optics, electron, 357–360
Oscillator, backup, for pacemaker, 752
Output amplitude test, for pacemaker, 756
Output driver, 25
Output driver test, 539
Output feedback (OFB), 701–702
Output impedance, pacemaker and, 749
Output relative timings, in semiconductor testing, 540–541
Oxidation, ion implantation and, 478–480
Oxide(s), 381–382
  breakdown and, 386–389
  field, 393–395
  gate, 396–397
  growth rates of, 383–384
  interlayer, 397–398
  leakage and, 385–386
  test structure for, 385
  thermal oxidation and, 382–384

wet oxygen ambient and, 384
Oxide charges, 389
  fixed oxide, 392
  interface-trapped, 392
  minimizing effects of, 392–393
  mobile, 391–392
  trapped, 392
  use of C–V curves and, 389, 391–392
Oxygen, film formation and, 402–403

P

Pacemaker, 737–738
  applied electrical stimulation and, 739–740
  architecture of, 722–723
  blood circulation and, 738
  capabilities of, 722
  CMOS technology and, 743–744
  CPU of, 723–724
  design considerations for
    battery characteristics, 752–754
    circuit characteristics, 747–748
    circuit protective measures, 751–752
    pacer testing, 754–756
  electrocardiogram and, 740–741
  future trends in, 762
  implantation of, 741
  method of operation of, 742–743
  natural, 738–739
  pacing modalities, 756
    pacer mode codes, 756–759
    pacer timing cycles, 759–761
  programmable pacing modes of, 724
  support instruments, 746–747
  system components, 744–746
Pacer systems analyzer, 746, 747
Pair-breaking rate, 638
Palladium, as barrier layer, 452
Parallelism, performance and, 670–671
Parametric testing, 515–516
  applications of, 518–519
  circuit modeling and, 517
  data analysis and, 525–526
  density and scaling and, 517–518
  improved analysis tools and, 519
  instrumentation, 521–525
  process and device development and, 517
  process control and, 516–517
  test structures, 520–521
  tighter control and, 519
  types of, 519–520
Passivation, of grain boundaries, 512
Patents
  cost of, 869
  definition of, 866
  effectiveness of, 872
  how to obtain, 867–869
  litigation and, 870–872
  reasons to obtain, 866–867
Pattern generation, computer-aided design tools and, 63
Pattern generation (PG) tapes, foundry interface and, 71
Pattern placement errors, on masks, 353
Pattern recognition, 785–786, 788–789
  architectures for, 789–794
  computer-vision levels, 786
  image segmentation and description, 788
  image transforms and enhancement, 787
  system components, 786–787
PECVD, *see* Plasma enhancement
Performance, scalability versus, 133
Performance limits, 615–616
  circuits and system limits, 622–627
  device, 618–622
  fundamental, 616–617
  materials, 617–618
Personal computers, 657
  evolution of, 662–665
  future of, 665–666
  microprocessor evolution and, 658–662
Personnel
  for circuit fabrication facility, 92
  clean air and, 89
  efficiency of, 96–97
  safety of, 95–96
  electronic counter–countermeasures and, 691
  importance to design, 678
PG, *see* Pattern generation
Photoconductive overlayer technology, 592–593
Photoelements, 582
Photoenhancement, for nitride deposition, 404–405
Ping pong, block, 545
Pipelining, 146–147
PLA, *see* Programmable logic array
Placement, 45, 47–48
  computer-aided design tools and, 61
Placers, 675–676
Plasma enhancement (PECVD), 487, 499

Index

applications of, 501–502
chemical vapor deposition and, 499–502
equipment configurations for, 499–500
etching and, 488–494
for nitride deposition, 404–405
sputtering and, 494–499
typical process in, 500
Plasma excitation, film formation and, 403
Plasma excitation frequency, 405
Plasma source, in x-ray lithography, 370–371
Platinum, as barrier layer, 452
PMOSFET, 102
Porosity, in films, 449
Post processing, 76–77
Power, active, reduction with CMOS technology, 122
Power-delay performance, of MOS and bipolar circuits, 107
Power-delay product, 620–621
Power dissipation, of IIL gate circuit, 119
Preamplifiers, imagers and, 600–601
Predisposition, ion implantation and, 472–473
Pricing incentives, for VLSI circuits in modems, 827–828
Process complexity
  circuit density and, 7
  ion implantation and, 474–476
Process control, parametric testing and, 516–517
Process development, parametric testing and, 517
Processing task, 38
Program condition, EEPROM and, 170–171
Program coupling ratio, EEPROM and, 171–172
Programmable logic array (PLA), 26–27
  layout systems, 34
Programming, for pacemakers, 746
Program voltage, threshold voltage window as function of, EEPROM and, 172–173
Protection, 846
  copyright and, 873–876
  employee relations agreements and, 880
  log books and, 881
  of memory, 143–144
  patents and, 866–872
  trade secret, 876–880
  vendor agreements and, 880–881
  visitor agreements and, 880
Proximity effect, 356
Proximity printing, x-ray, 366–367
Public keys, 702–703
Puddle develop, 336, 338
Pulsed capacitance–voltage technique, 314
Pulse noise, 605
Pulse sorting technology, electronic warfare support measures and, 683–685
Pulse testing, electronic warfare support measures and, 685–686
  memory and logic in, 686
Punchthrough, 618
  ion implantation and, 468

Q

Quiteron, 637–638

R

Radar system applications, 707–708
  functional requirements and radar overview, 708–711
  key microelectronic technologies, 711–719
Radiation effects
  in crystalline silicon, 567
  in devices and integrated circuits, 569
    permanent ionizing effects, 569–572
    permanent displacement effects, 573
    transient ionizing effects, 572–573
  in silicon dioxide–silicon interface, 568–569
Radiation hardening, 565–566
  of devices and integrated circuits, 573
    dielectric isolation processes, 573–574
    other techniques, 575
  of gallium arsenide integrated circuits, 577
    ionizing dose rate and, 578
    permanent displacement effects, 578
    single event upset and, 578
    total ionizing dose and, 577–578
  radiation requirements and, 566–567
  of silicon devices, 578–579
  single event upset and, 575–576
Radio frequency circuitry, radar applications, 715
Random access memory(ies), (RAM), 27
  basic organization of, 153

bipolar technologies and, 152
bit growth and, 151–152
comparison of technologies for, 153
definition of, 152
dynamic, 157–158
  address multiplexing and, 159
  available devices, 160
  outlook for, 160
  soft errors and, 160
specialty
  content addressable, 163
  dual-port, 161–162
  nonvolatile shadow, 164–165
  self-contained battery backup nonvolatile, 164
  video, 162
static, 154, 654
  decoders and, 156–157
  memory cell and, 154–156
voltage requirements of, 153
Raster scan, 360–361
Rate limit, for pacemaker, 752
Ratioless, 122–123
Rayleigh limit, 330
RC delays, 43
Reaction kinetics, for nitride growth, 403
Read condition, EEPROM and, 170
Read disturb, EEPROM and, 174–175
Reading data, static RAM and, 156
Read only memory(ies) (ROM), 27
  electrically erasable programmable, *see* EEPROM
Read–write (R/W) memory, *see* Random access memory(ies)
Receiver characteristics, electronic warfare support measures and, 683
Reference device testers, 532
Register transfer language, 13
Register transfer level (RTL), 40
Reliability
  of automotive applications, 854–855
  of EEPROM, 175
  as incentive for VLSI circuits in modems, 829
Repolarization period, 745–746
Resist(s)
  in electron beam patterning, 354
    available resists, 357
    limits on resolution and, 355–356
    multilevel, 347
    negative, 344–345
    positive, 345–346

solubility and, 354–355
thickness of, 356
multilevel
  bilevel, 341
  in electron beam patterning, 347
  reason for developing, 339–341
  trilevel, 343
negative, in optical patterning, 331–333
positive
  developers for, 335–336
  in optical patterning, 333–338
  puddle develop and, 336, 338
x-ray, 347–349, 372
  inorganic, 349
  materials for, 375–376
  negative, 374–375
  plasma developable, 349
  positive, 375
  x-ray absorption and, 373–374
Resistance
  gate electrode, 134
  source/drain, 134–135
Resistivity
  conversion between dopant density and, 178–181
  measurement of, 312
  temperature dependence of, 185
Resistors, ion implantation and, 474
Resolution, of resists, 355–356
Retention, EEPROM and, 174
Rheobase, 740
Ripple carry chain, 23–24
ROM, *see* Read only memory(ies)
Room materials, for circuit fabrication facility, 95
Routing, computer-aided design tools and, 61
RTL, *see* Register transfer level
Runout distortion, in x-ray proximity printing, 367
R/W memory, *see* Random access memory(ies)

S

Safety, *see also* Protection
  for circuit fabrication facility, 95–96
Sample(s)
  handling and preparation of, complex medical instrument and, 765–770
  testing of, 547–550
Scalability, performance versus, 133

Index

Scaling
  bus, 147–148
  parametric testing and, 517–518
Scan-in–scan-out, 673
Scanners, for computerized tomography, 726–728
Scattering rate, of electrons and photons, 639
Scatter plots, 525–526
Schematic diagrams, computer-aided design tools and, 58–59
Schmoo plot, 540
Schottky barrier transistor, 112
Schottky contacts, sputtering and, 499
Schottky diode FET logic circuits, 652–653
Searching, binary and linear, 540–541
Security, 846
Selectivity, 488
  mechanisms for producing, 491–492
  parameters affecting, 491
Semiconductor testing, 528
  with automatic test equipment, 531–538
  definition and scope of, 528
  describing control flow for, 535
  location in manufacturing process, 528–530
  logic testing, 539–541
  memory testing, 541–546
  quality assurance and sample testing, 547–550
  test descriptions, 538–539
  testing throughput, 546–547
  typical tests, 530–531
  writing program for, 534–535
Semicustom layout procedure, to integrated circuit design, 47–48
Sense amplifier, pacemaker and, 749–750
Sensitivity, in optical patterning, 331
SEU, see Single event upset
Shift register, 24
Shorts test, 538
Shot noise, 604–605
Signal processing, for communications, 847–849
Silane reactions, in epitaxial growth, 290
Silicidation, ion implantation and, 482–483
Silicides, 415–432
Silicon, 271
  bulk single crystal, 272
  crystalline, radiation effects on, 567
  electrical transport properties of conductivity mobility and, 181–184
  dependence of drift velocity on electric fields and, 185–186
  hall mobility and, 181
  minority-carrier mobility, lifetime, and diffusion length and, 186–188
  mobility in MOS inversion layer and, 188–189
  resistivity and dopant density and, 178–181
  temperature dependence of resistivity and mobility and, 185
  transport equation and, 178
  local oxidation of (LOCOS), 411–412
  material criteria, 191–193
    engineering properties, 194
    product characteristics, 193–194
    product trends, 194–196
    VLSI and ULSI product recommendation, 196
  polycrystalline, 271–272
  single crystal wafers, 272–273
  thermal nitridation of, 402–403
Silicon compilers, 34
Silicon devices, hardness trends of, 578–579
Silicon diffusion, aluminum metallization and, 436–439
Silicon dioxide–silicon interface, crystalline, 568–569
Silicon epitaxial films, see also Epitaxial films
Silicon foundry, 68–69
  interface of, 69–70
    circuit simulation and, 71–73
    data base transfer and, 70–71
    design interface, 71
  post processing and, 76–77
  processing and, 73
    mode selection and, 73–74
    technologies available and, 75–76
Silicon-on-insulator (SOI), 503–505
  epitaxial layer overgrowth and, 510–511
  full isolation by porous oxidized silicon, 509–510
  grain boundary passivation, 512
  heteroepitaxy and, 505–506
  by ion implantation, 508–509
  polysilicon thin film transistors, 511–512
  substrates, ion implantation and, 480–481
  by thin film recrystallization, 506–508
  three-dimensional integrated circuits

and, 512–514
Simulated signal, as electronic countermeasure, 691
Simulation, as computer-aided design tool, 13–14
Simulation models, foundry interface and, 72–73
Single event upset (SEU), 575–576, 578, *see also* Soft errors
Single node failures, 556
 analysis example, 557, 561
 delayering and, 557
 types of circuits and, 556–557
Slew rate, 739
Smearing, 831
Soft errors, alpha-particle- and cosmic-ray-induced, noise and, 612–613, *see also* Single event upset
SOI, *see* Silicon-on-insulator
Solid state diffusion, of impurity atoms during epitaxial growth, 308
Solubility, of resists, 354–355
Solvent exhaust, for circuit fabrication facility, 95
Source/drain diffusion resistance, reduction of, 134–135
Spectral responsivity, of imagers, 593
Speech
 analysis of, 779–780
 coding of, 779, 782–783
 production of, 777
  frequency response of vocal tract and, 777
  source-filter model of, 778
 recognition of, 665, 779
 synthesis of, 665, 779–780, 782–783
Speech processing, 775–776
 algorithms and hardware and, 780–781
 future of, 783–784
 main components of, 778–779
 speech analysis and synthesis and, 779–780
 speech coding and speech synthesis, 782–783
 speech production and, 777–778
 speech recognition, 781–782
Speed, of personal computers, 665
Speed binning, for logic testing, 540
Spreading resistance technique, 300
 probe and, 313
Sputtering, 446, 494
 bias, 446

equipment for, 494–497
typical applications of, 497–499
yield, 494
SQUID, *see* Superconducting quantum interference device
Stage motion, raster scan and, 361
Standard(s)
 for data encryption, 696–702
 for epitaxial films, 296
Standard cell approach, to integrated circuit design, 44–45
Standard cell layout system, 33–34
Standing wave effect, in optical patterning, 331
Static, 122
Static latch, 24
Step coverage, aluminum metallization and, 444
STL gate circuit, 119
Stored-response VLSI testers, 532
Straight data pattern, 543
Stream cipher method, 699–702
Substrate, for epitaxial growth, 305
Superconducting circuits
 digital devices, 632–638
 model of Josephson junction, 629–632
 switching speed of Josephson junctions and, 638–639
Superconducting quantum interference device (SQUID), junction-controlled logic gate and, 633–634
Surround disturb patterns, 545–546
Susceptor, 293
Swelling, of negative resists, 355
Switched-capacitor arrays, 849
Switched capacitor type circuits, 124
Symbolic layout tools, 33
Synchronization, for logic testing, 540
System timing, 147
 bus scaling and, 147–148
 clock chips and, 148
  advantages over wait states, 148–149
 wait states and, 148

T

Targets, in x-ray lithography, 369
Telemetry, for pacemakers, 746
Telephones, cellular, 862–863
Temperature
 pacemaker and, 748–749

Index

resistivity and mobility and, 185
Temperature acceleration, 396
Templates, speech processing and, 781–782
Testing, *see also* Parametric testing; Semiconductor testing
  breakdown voltage, 539
  as computer-aided design tool, 14
  electrical characterization and, 553
  functional, 39, 530–531, 539
  input–output failures and, 555
  leakage, 539
  logic, 539–541
  maximum current, 539
  memory, 541–546
  multiple address, 543
  nondestructive, 552–555
  opens, 538
  output driver, 539
  for oxides, 385
  for pacemaker, 754–756
  pulse, 685–686
  reference device, 532
  sample, 547–550
  shorts test, 538
  stored-response, 532
  throughput, 546–547
Testing equipment, computer-aided design tools and, 63
TFT, *see* Thin film transistors
Thermal noise, 604
Thermal oxidation, 382–384
Thickness measurements, of epitaxial films, 316
Thin film recrystallization, silicon-on-insulator by, 506–508
Thin film transistors (TFT), polysilicon, 511–512
Three-dimensional integrated circuits, 512–514
Threshold voltage(s), 391
  control of, ion implantation and, 466–467
  gate length versus, 132
  noise and, 611–612
Threshold voltage window, as function of program voltage and time, EEPROM and, 172–173
Throughput
  functional rate of, 623
  as incentive for VLSI circuits in modems, 829

testing, 546–547
Time
  threshold voltage window as function of, EEPROM and, 172–173
  transit, limiting, 617–618
Timing
  with CMOS technology, 122
  logic testing and, 540–541
  pacemaker and, 748, 759–761
  system, 147–149
  verification of, 43
    computer-aided design tools and, 60
  verifier, 675
Timing cycles, for pacemaker, 759–761
Titanium, as barrier layer, 449, 450–451, 452
Top–down design, 673
Top layer passivation, PECVD in, 501
Topological scrambling, 546
Trade secret protection
  advantages and disadvantages of, 876–878
  definition of, 876
  establishing and maintaining effectiveness of, 878–880
  scope of, 876
Transfer
  frame, 585
  interline, 585
  line-by-line, 587–588
Transistor(s)
  bipolar, 109
    npn and *pnp*, 109–111
    Schottky barrier, 112
  CMOS, 102–103
  heterojunction bipolar, 656
  high electron mobility, 655
  metal semiconductor field effect, 647–649
    GaAs circuits, 651–654
  MOS, 14–16
  thin film, polysilicon, 511–512
Transit time, limiting, 617–618
Transmitter, transtelephonic, 746–747
Transmitter implementation, modems and, 830–831
Transport equation, 178
Transtelephonic transmitter, 746–747
TTL circuits, 113–115
Tungsten, as barrier layer, 450–451, 452
Tunneling, 617
  Fowler–Nordheim, 169–170

Tunnel oxide thickness, EEPROM and, 171–172

U

Ultrasound
 doppler, 807–809
 imaging and
  architecture of, 729
  commercially available systems for, 728–729
  method of operation of, 728
  obtaining permanent records with, 730
Ultraviolet radiation, deep, in optical patterning, 338

V

Vacuum, for circuit fabrication facility, 92–93
Validation, in circuit design, 14
Vector scan, beam shapes and, 362–364
Velocity, maximum achievable, 617
Vendor agreements, 880–881
Vertical scaling, in manufacturing process, 83–84
Vertical yield maps, 525
Vibration control, for circuit fabrication facility, 95
Video, signal processing and, 848–849
Video RAM, 162
Virtual memory, 144
Visitor agreements, 880
Vocoder, 783
Voice, signal processing and, 847–848, *see also* Speech
Voltage
 flatband, 391
 operating, for pacemaker, 748
 RAM requirements for, 153
 threshold, 391
  control of, ion implantation and, 466–467
  as function of program voltage, 172–173
 gate length versus, 132
 noise and, 611–612
VTInet, 71

W

Wafer patterning, *see* Optical patterning
Wait states, 148
 advantages of clock chips over, 148–149
Walking patterns, 544–545
Warfare, *see* Electronic warfare
Waste disposal, for circuit fabrication facility, 93–95
Water, for circuit fabrication facility, 89–90
Wavelength, for x-ray lithography, 366
Weighing technique, for epitaxial films, 316
Wiring, of superconducting circuits, 640–642
Work flow, for circuit fabrication facility, 96–97
Workstations, 49
 CAE, 64
Writing beam
 ion implantation and, 483–484
 vector scan and, 362–364
Writing data, static RAM and, 155
Writing scheme, raster scan and, 360–361

X

X-ray lithography, 366
 applications of, 378–380
 exposure systems and, 377–378
 mask alignment and, 376–377
 masks and, 371–372
 proximity printing and, 366–367
 resists and, 372–376
 sources for, 367–371

Z

Zero bias barrier resistance, 300